TECHNICAL ENGINEERING AND DESIGN GUIDES
AS ADAPTED FROM THE
US ARMY CORPS OF ENGINEERS, No. 4

RETAINING AND FLOOD WALLS

Published by
ASCE Press
American Society of Civil Engineers
345 East 47th Street
New York, New York 10017-2398

ABSTRACT

The engineering manual Retaining and Flood Walls from the U.S. Army Corps of Engineers provides guidance for the safe design and economical construction of retaining walls and inland and coastal flood walls. Retaining walls are defined as any wall that restrains material to maintain a difference in elevation. Only those retaining walls subjected to hydraulic loadings such as flowing water, submergence, wave action, and/or severe climatic conditions are considered in this manual. Flood walls are defined as any wall having as its principal function the prevention of flooding of adjacent land. Besides discussing such issues as design considerations, forces, structural stability, and foundation analyses, the appendices present derivations of a variety of relevant equations.

Library of Congress Cataloging-in-Publication Data

Retaining and flood walls.
p.cm.—(Technical engineering and design guides as adapted from the US Army Corps of engineers; no. 4)
Includes bibliographical references and index.
ISBN 0-87262-968-6
1. Flood dams and reservoirs—Design and construction. 2. Retaining walls—Design and construction. I. American Society of Civil Engineers. II. United States. Army. Corps or Engineers. III. Series: Technical engineering and design guides as adapted from the U.S. Army Corps of Engineers; no. 4.
TC540.R48 1994 94-244
627'.4—dc20 CIP

The material presented in this publication has been prepared in accordance with generally recognized engineering principles and practices, and is for general information only. This information should not be used without first securing competent advice with respect to its suitability for any general or specific application.

The contents of this publication are not intended to be and should not be construed to be a standard of the American Society of Civil Engineers (ASCE) and are not intended for use as a reference in purchase specifications, contracts, regulations, statutes, or any other legal document.

No reference made in this publication to any specific method, product, process or service constitutes or implies an endorsement, recommendation, or warranty thereof by ASCE.

ASCE makes no representation or warranty of any kind, whether express or implied, concerning the accuracy, completeness, suitability or utility of any information, apparatus, product, or process discussed in this publication, and assumes no liability therefor.

Anyone utilizing this information assumes all liability arising from such use, including but not limited to infringement of any patent or patents.

Library of Congress Catalog Card No: 94-244
ISBN 0-87262-968-6
Manufactured in the United States of America.

TABLE OF CONTENTS

LIST OF FIGURES

DEPARTMENT OF THE ARMY
U.S. Army Corps of Engineers
WASHINGTON, D.C. 20314-1000

REPLY TO
ATTENTION OF:

3 SEP 1993

Mr. James McCarty
President, American Society
of Civil Engineers
345 East 47th Street
New York, New york 10017

Dear Mr. McCarty:

I am pleased to furnish the American Society of Civil Engineers (ASCE) a copy of the U. S. Army Corps of Engineers engineering manual, EM 1110-2-2502, Retaining and Flood Walls, and engineer technical letter, ETL 1110-2-322, Retaining Wall Design and Analysis. The Corps uses these manuals for design and analysis of walls for flood control and navigation projects. The manuals include requirements for stability, strength and serviceability.

I understand that ASCE plans to publish these manuals for public distribution. I believe this will benefit the civil engineering community by improving transfer of technology between the Corps and other engineering professionals.

Sincerely,

Arthur E. Williams
Lieutenant General, U. S. Army
Commanding

CHAPTER 1

INTRODUCTION

1-1. Purpose

A. GENERAL. This manual provides guidance for the safe design and economical construction of retaining and flood walls. This manual is intended primarily for retaining walls which will be subjected to hydraulic loadings such as flowing water, submergence, wave action, and spray, exposure to chemically contaminated atmosphere, and/or severe climatic conditions. For the design of retaining walls which will not be subjected to hydraulic loadings or severe environmental conditions as described above, TM 5-818-1 may be used for computing the loadings and evaluating the stability of the structure.

B. VARIATIONS FROM GUIDANCE. For the evaluation of existing retaining and flood walls which have been loaded up to or above the design loads and show no displacement problems or any other sign of weakening, consideration can be given to reducing the conservatism of the criteria contained in this manual. If variations from the guidance are necessary, justification for the variations should be submitted to HQUSACE to the attention of CECW-E for approval.

1-2. Applicability

This manual applies to all HQUSACE/OCE elements and to all field operating activities having responsibilities for the design of civil works projects.

1-3. References and Bibliography

References and computer program user guides cited in this manual are listed in Appendix A. Additional reference materials pertaining to the subject matter addressed in this manual are included in Appendix B, "Bibliography." Computer program abstract descriptions are shown in Appendix O.

1-4. Terms

Special terms used in this manual are explained in the Glossary.

1-5. Scope

A. TYPES OF WALLS. This manual presents design guidance for retaining walls and inland and coastal flood walls. A retaining wall is defined as any wall that restrains material to maintain a difference in elevation. A flood wall is defined as any wall having as its principal function the prevention of flooding of adjacent land. Not specifically covered in this manual are seawalls which are defined as structures separating land and water areas, primarily designed to prevent erosion and other damage due to wave action. They are frequently built at the edge of the water, but can be built inland to withstand periods of high water. Seawalls are generally characterized by a massive cross section and a seaward face shaped to dissipate wave energy. Coastal flood walls, however, are generally located landward of the normal high water line so that they are inundated only by hurricane or other surge tide and have the smooth-faced cantilever stems shown in this manual.

B. TYPES OF FOUNDATIONS. This manual describes procedures for the design of retaining and flood walls on shallow foundations, i.e., bearing directly on rock or soil. The substructure design of pile-founded walls is not included, but is covered in EM 1110-2-2906.

C. FLOOD WALL GUIDANCE. A flood wall is treated as a special case of a retaining wall. Unless specifically noted, the guidance herein applies to both retaining and flood walls.

D. GEOTECHNICAL AND STRUCTURAL ASPECTS. Both geotechnical and structural aspects of wall design are included. Coordination between geotechnical engineers, structural engineers, and geologists in the design of retaining and flood walls is essential.

CHAPTER 2

GENERAL DESIGN CONSIDERATIONS

Section I. Types of Retaining Walls

2-1. Common Types of Retaining Walls

The most common types of retaining walls are gravity concrete, cantilever T-type reinforced concrete, and cantilever and anchored sheet pile walls. Gravity and cantilever reinforced concrete walls are covered in this manual and illustrated in Figure 2-1. Alternate types of retaining walls, including mechanically stabilized backfill and precast modular gravity walls, are covered in Chapter 10. An example of one type of alternate retaining wall is shown in Figure 2-1. Counterfort and buttressed reinforced concrete walls are less commonly used and are not specifically discussed in this manual. Much of the conceptual information and the information in Chapters 3 and 9 is applicable to all types of walls.

2-2. Gravity Concrete Wall

A gravity wall (Figure 2-1) consists of mass concrete, generally without reinforcement. It is proportioned so that the resultant of the forces acting on any internal plane through the wall falls within, or close to, the kern of the section. A small tensile stress capacity is permissible for localized stresses due to extreme and temporary loading conditions.

2-3. Cantilever Reinforced Concrete Wall

A cantilever T-type reinforced concrete wall (Figure 2-1) consists of a concrete stem and base slab which form an inverted T. The structural members are fully reinforced to resist applied moments and shears. The base is made as narrow as practicable, but must be wide enough to ensure that the wall does not slide, overturn, settle excessively, or exceed the bearing capacity of the foundation. The bottom of the base should be below the zone subject to freezing and thawing or other seasonal volume changes. The T-type wall is usually the most economical type of conventional wall and is more widely used than any other type for common retaining wall heights.

2-4. Alternate Types of Retaining Walls

Retaining walls using mechanically stabilized backfill (Figure 2-1) and precast modular gravity walls can be substantially more economical to construct than conventional walls (Leary and Klinedinst 1984). However, a short life, serious consequences of failure, or high repair or replacement costs could offset a lower first cost. In addition, the design engineer must assure the overall adequacy of the design since the manufacturer of the wall may provide only that part of the design above the foundation. Chapter 10 covers mechanically stabilized backfill systems and precast modular gravity walls.

Section II. Types of Flood Walls

2-5. Common Types of Flood Walls

The most common types of flood walls are cantilever T-type and cantilever I-type walls. Examples of these walls are shown in Figure 2-2.

2-6. Cantilever T-Type Wall

Most flood walls are of the inverted T-type (Figure 2-2). These walls are discussed in detail in Chapter 7. The crossbar of the T serves as a base and the stem serves as the water barrier. When founded on earth, a vertical base key is sometimes used to increase resistance to horizontal movement. If the wall is founded on rock, a key is usually not provided. Where required, the wall can be supported on piles. A sheet pile cutoff can be included to control underseepage or provide scour protection for the foundation. T-type walls may be provided with a horizontal or sloped base. The advantages of sloped and horizontal bases are discussed in paragraph 7-5.

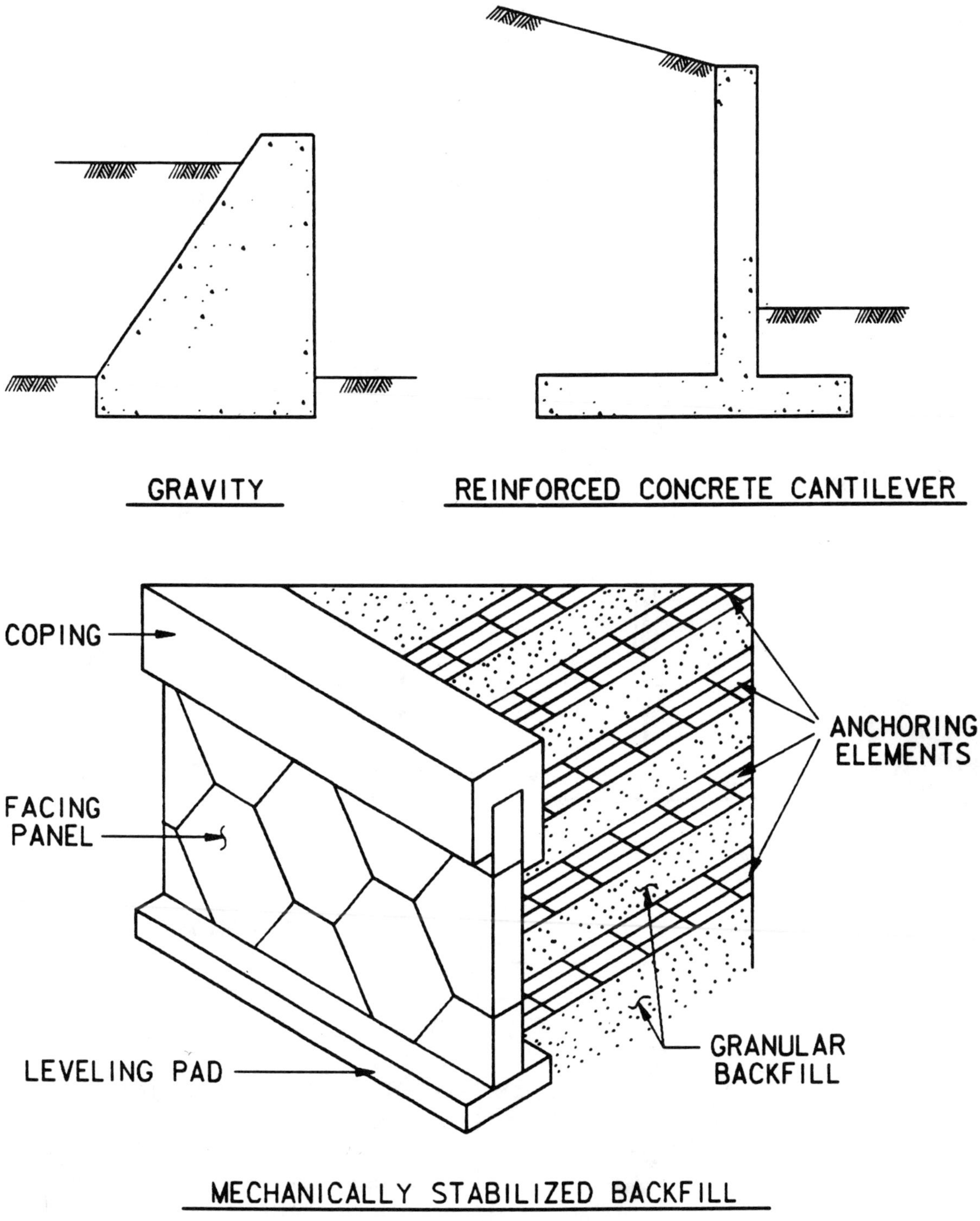

Fig 2-1. Types of retaining walls

2-7. Cantilever I-Type Wall

I-type flood walls consist of driven sheet piles capped by a concrete wall (Figure 2-2). I-walls are most often used in connection with levee and T-wall junctions or for protection in narrow restricted areas where the wall height is not over 8 to 10 feet, depending on soil properties and geometry. The design of these types of walls is beyond the scope of this manual.

2-8. Other Types of Flood Walls

A. BRACED SHEET PILE FLOOD WALL. This wall consists of a row of vertical prestressed concrete sheet piles, backed by batter piles connected to the sheet piles by a cast-in-place

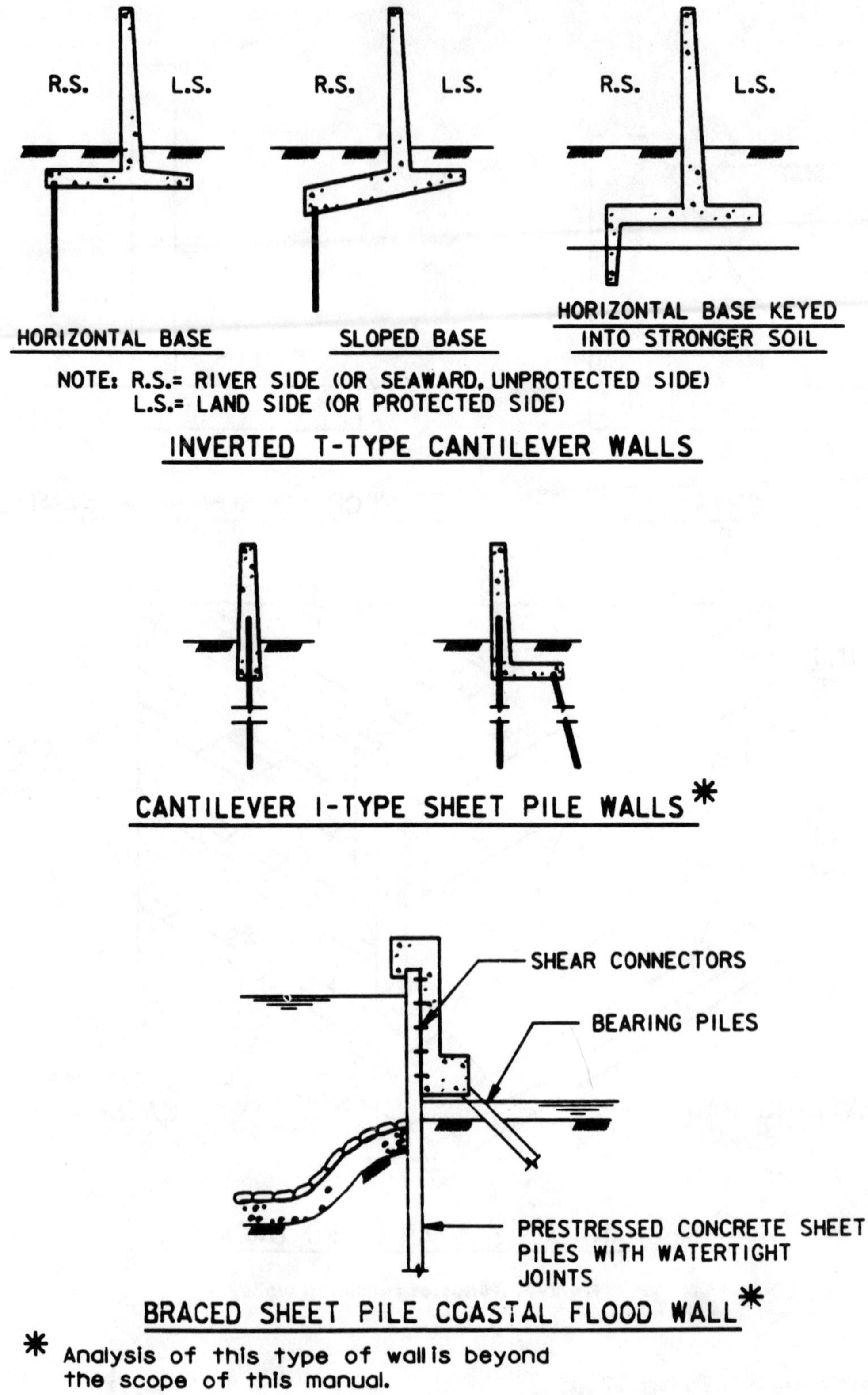

Fig 2-2. Types of flood walls

horizontal concrete beam with shear connectors as required to resist the vertical component of load in the batter pile (Figure 2-2). This type of wall has been used for coastal flood walls. It is ideal for wet areas because no excavation or dewatering is required to construct the wall. The disadvantage is that it is more indeterminate than other wall types. The design of this wall is beyond the scope of this manual.

B. LESS COMMONLY USED TYPES. There are various other types of walls that may be used for flood walls, such as buttress, counterfort, gravity, cellular, and cellular sheet pile, some of which are shown in Figure 2-3. These walls, except for the gravity wall, are beyond the scope of this manual.

Section III. Differences Between Retaining and Flood Walls

2-9. Purpose of Walls

A retaining wall is any wall that retains material to maintain a change in elevation whereas the principal function of a flood wall is to prevent flooding (inundation) of adjacent land. A floodwall is subject to water force on one side which is usually greater than any resisting earth force on the opposite side. A wall may be a retaining wall for one loading condition and a flood wall for another loading condition. The flood loading (surge tide, river flood, etc.) may be from the same or the opposite direction as the higher earth elevation.

2-10. Seepage and Leakage Control Requirements

All water-retaining structures may be subject to seepage through, under, and around them. Inadequate control of seepage may affect the stability of a flood wall regarding uplift or loss of support resulting from erosion. Properly controlled seepage, even if quantities of flow remain large, presents little or no hazard. Control of through-seepage is provided by water stops. Retaining walls rarely need seepage protection other than to relieve the hydrostatic load on the fill side of the wall. Water stops are used in retaining walls to prevent water passage from the backfill through the vertical joints. Seepage control and water stops are more fully discussed in paragraphs 3-23, 6-4e, 6-6, 7-4, and 7-13.

2-11. Wall Stability

Generally, it is more difficult to design stable flood walls than retaining walls. By their very nature, flood walls are usually built in a flood plain which may have poor foundation conditions. Uplift is always a critical item with flood walls but seldom a problem with retaining walls since the loads acting on a retaining wall are usually soil backfills. The water load on a flood wall can be more severe, especially when wave loadings are applicable. When the groundwater surface is near or above the wall footing, a common occurrence with flood walls, the allowable bearing capacity of the soil is reduced. The reduction of stability, due to the erosion of the earth cover over and beyond the base, must be considered.

2-12. Special Flood Wall Monoliths

Careful attention must be given to wall monoliths that have loading, support, or other conditions that vary along the length of the monolith. These monoliths, which may include closure structures, pipeline crossings, corner structures, etc., must be analyzed as complete three-dimensional entities instead of the usual two-dimensional unit slices.

2-13. Design Philosophy

Retaining walls are normally built as an appurtenance to other structures: dams, hydroelectric power houses, pump stations, etc. The consequences of failure of a retaining wall are often lower than for flood walls. Also, retaining walls are seldom more than a few hundred feet long; if they are designed conservatively, the added costs are of limited significance. Flood walls, on the other hand, are usually the primary feature of a local protection project. They must be designed for the most economical cross section per unit length of wall, because they often extend for great distances. Added to this need for an economical cross section is the requirement for safety. The consequences of failure for a flood wall are normally very great since it protects valuable property and human life. Thus, the design of retaining and flood walls is a complex process involving safety and economy factors, and design must be executed in a logical, conservative manner based on the function of the wall and the consequences of failure. Design documents should describe the decisions leading to the final degree of conservatism.

2-14. Stability Considerations

An adequate assessment of stability must include a rational assessment of loads and must account for the basic structural behavior, the mechanism of transmitting compressive and shearing

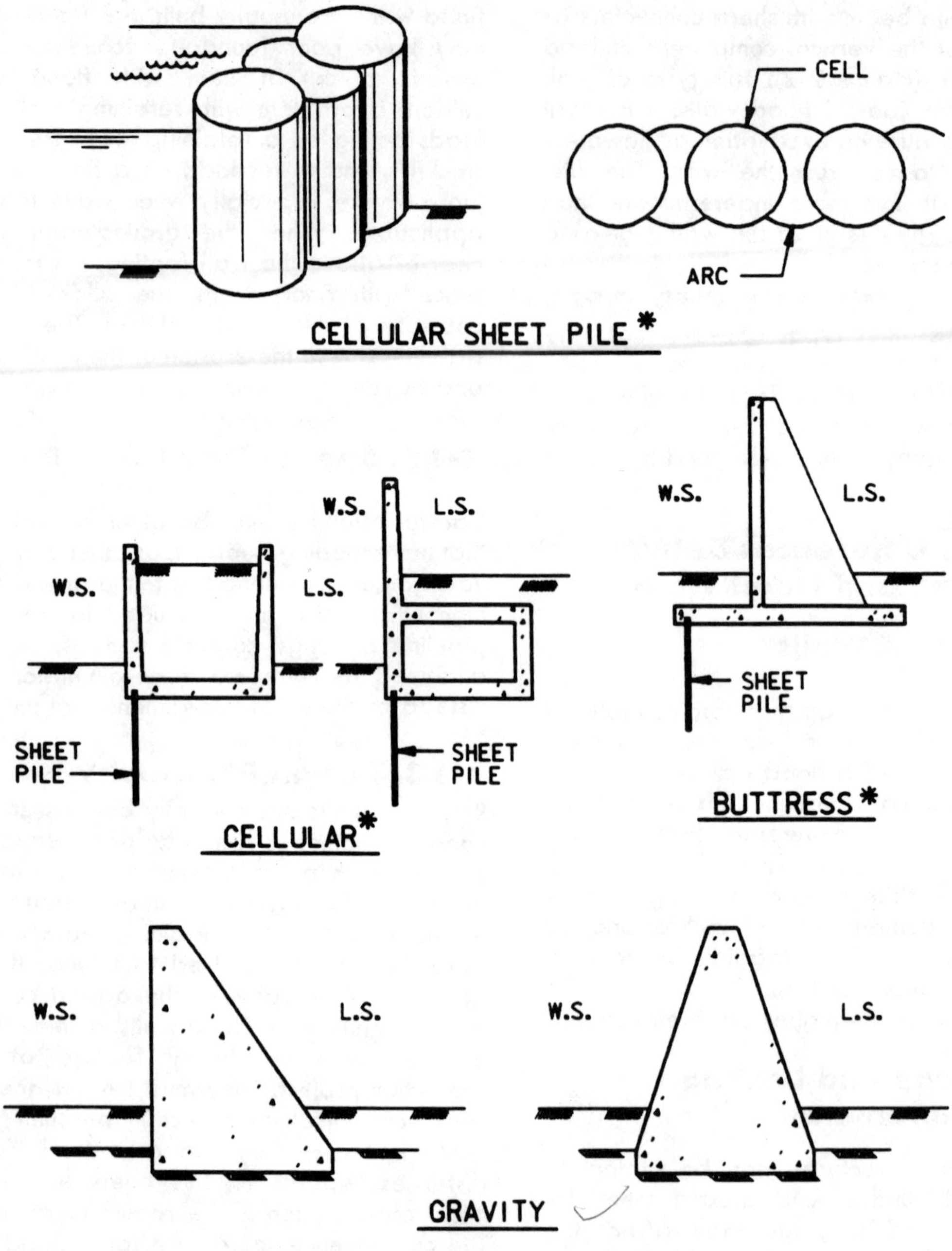

NOTES: W.S.= WATER SIDE (OR SEAWARD, UNPROTECTED SIDE)
L.S.= LAND SIDE (OR PROTECTED SIDE)

* Analysis of this type wall is beyond the scope of this manual.

Fig 2-3. Less commonly used flood wall types

loads to the foundation, the reaction of the foundation to such loads, and the secondary effects of the foundation behavior on the structure.

Section IV. Coordination Between Disciplines

2-15. Engineering Team

A fully coordinated team of geotechnical and structural engineers, and hydraulic engineers where applicable, should ensure that all pertinent engineering considerations are properly integrated into the overall design of a structure. Some of the critical aspects of design which require coordination are:

A. Preliminary estimates of geotechnical and hydraulic data, subsurface conditions, and types of structures which are suitable for the foundation.

B. Selection of design parameters, loading conditions, loading effects, potential failure mechanisms, and other related features of the analytical models.

C. Evaluation of the technical and economic feasibility of alternative types of structures.

D. Constructibility reviews in accordance with ER 1110-1-803.

E. Refinements of the preliminary structure configuration to reflect the results of detailed site explorations, material availability studies, laboratory testing, and numerical analysis.

F. Modification to the structure configuration during construction due to unexpected variations in the foundation conditions.

Section V. Geotechnical Investigations

2-16. Planning the Investigation

A. PURPOSE. The purpose of the geotechnical investigation for wall design is to identify the type and distribution of foundation materials, to identify sources and characteristics of backfill materials, and to determine material parameters for use in design analyses. Specifically, the information obtained will be used to select the foundation type and depth, design the foundation, estimate backfill pressures, locate the groundwater level, estimate settlements, and identify possible excavation problems. For flood walls, foundation underseepage conditions must also be assessed. Detailed information regarding subsurface exploration techniques may be found in EM 1110-1-1804 and EM 1110-2-1907.

B. REVIEW OF EXISTING INFORMATION. The first step in an investigational program is to review existing data so that the program can be tailored to confirm and extend the existing knowledge of soil and rock conditions. EM 1110-1-1804 provides a detailed listing of possible data sources; important sources include air photographs, geologic maps, surficial soil maps, and logs from previous borings. In the case of flood walls, study of old topographic maps can provide information on past riverbank or shore geometry and identify likely fill areas.

2-17. Foundation Exploration and Site Characterization

A. PRELIMINARY EXPLORATION. Where possible, exploration programs should be accomplished in phases, so that information obtained in each phase may be used advantageously in planning later phases. The results of each phase are used to "characterize" the site deposits for analysis and design by developing idealized material profiles and assigning material properties. For long, linear structures like flood walls, geophysical methods such as seismic and resistivity techniques often provide an ability to rapidly define general conditions during the preliminary phase at a modest cost. In alluvial floodplains, air photograph studies can often locate recent channel fillings or other potential problem areas. A moderate number of borings should be obtained at the same time to refine the site characterization and to "calibrate" geophysical findings. Borings should extend deep enough to sample any materials which may affect wall performance; a depth of twice the wall height below the ground surface can be considered a conservative "rule of thumb." For flood walls where underseepage is of concern, a sufficient number of the borings should extend deep enough to establish the thickness of any pervious strata.

B. DETAILED EXPLORATION. The purpose of this phase is the development of detailed material profiles and quantification of material parameters. The number of borings should typically be two to five times the number of preliminary borings. No exact spacing is recommended, as the boring layout should consider geologic conditions and the characteristics of the proposed structure. Based on the preliminary site characterization, borings should be situated to confirm the location of significant changes in foundation conditions as well as to confirm the continuity of apparently consistent foundation conditions. At this time, undisturbed samples should be obtained for laboratory testing and/or in situ tests should be performed.

C. ADDITIONAL EXPLORATION. In some cases, additional exploration phases may be useful to resolve questions arising during detailed design, and/or to provide more detailed information to bidders in the plans and specifications.

12-18. Testing of Foundation Materials

A. GENERAL. Procedures for testing soils are described in EM 1110-2-1906. Procedures for testing rock specimens are described in the *Rock Testing Handbook* (U. S. Army Engineer Waterways Experiment Station (WES) 1980). Much of the discussion on use of laboratory tests in EM 1110-1-1804 and EM 1110-2-1913 also applies to wall design. For wall design, classification and index tests (water content, Atterberg limits, grain size) should be performed on most or all samples and shear tests should be performed on selected representative undisturbed samples. Where settlement of fine-grained foundation materials is of concern, consolidation tests should also be performed. The strength parameters ϕ and c are not intrinsic material properties but rather are parameters that depend on the applied stresses, the degree of consolidation under those stresses, and the drainage conditions during shear. Consequently, their values must be based on laboratory tests that appropriately model these conditions as expected in the field.

B. COARSE-GRAINED MATERIALS. Coarse-grained materials such as clean sands and gravels are sufficiently pervious that excess pore pressures do not develop when stress conditions are changed. Their behavior can be modeled for static analyses (earth pressure, sliding, bearing) using parameters from consolidated-drained (S) tests. Failure envelopes plotted in terms of total or effective stresses are the same, and typically exhibit a zero c value and a ϕ value in the range of 25 to 40 degrees. Because of the difficulty of obtaining undisturbed samples of coarse-grained foundation materials, the ϕ value is usually inferred from in situ tests or is conservatively assumed based on material type. Where site-specific correlations are desired for important structures, laboratory tests may be performed on samples recompacted to simulate field density.

C. FINE-GRAINED MATERIALS.

1. When fine-grained materials such as silts and clays are subjected to stress changes, excess (positive or negative) pore pressures are induced because their low permeability precludes an instantaneous water content change. Undrained (Q or R) tests model such behavior. Shear strength envelopes for undrained tests plotted in terms of total stresses exhibit a nonzero c parameter. However, if plotted in terms of effective stresses, the c parameter is small (zero for all practical purposes) and the friction angle will be essentially equal to that from a drained test. Reasonable estimates of the drained friction angle ϕ' can often be made using correlations with the plasticity index (Figure 2-4).

2. At low stress levels, such as near the top of a wall, the undrained strength is greater than the drained strength due to the generation of negative pore pressures which can dissipate with time. Such negative pore pressures allow steep temporary cuts to be made in clay soils. Active earth pressures calculated using undrained parameters are minimum (sometimes negative) values that may be unconservative for design. They should be used, however, to calculate crack depths when checking the case of a water-filled crack.

3. At high stress levels, such as below the base of a high wall, the undrained strength is lower than the drained strength due to generation of positive pore pressures during shear. Consequently, bearing capacity and sliding analyses of walls on fine-grained foundations should be checked using both drained and undrained strengths.

4. Certain materials such as clay shales exhibit greatly reduced shear strength once shearing has initiated. For walls founded on such materials, sliding analyses should include a check using residual shear strengths.

2-19. In Situ Testing of Foundation Materials

A. ADVANTAGES. For designs involving coarse-grained foundation materials, undisturbed sampling is usually impractical and in situ testing is the only way to obtain an estimate of material properties other than pure assumption. Even where undisturbed samples can be obtained, the use of in situ methods to supplement conventional tests may provide several advantages: lower costs, testing of a greater volume of material, and testing at the in situ stress state. Although numerous types of in situ tests have been devised, those most currently applicable to wall design are the standard penetration test, the cone penetration test, and the pressuremeter test.

B. STANDARD PENETRATION TEST. The standard penetration test or SPT (ASTM D-1586) is routinely used to estimate the relative density and friction angle of sands using empirical correlations. To minimize effects of overburden stress, the penetration resistance, or N value, is

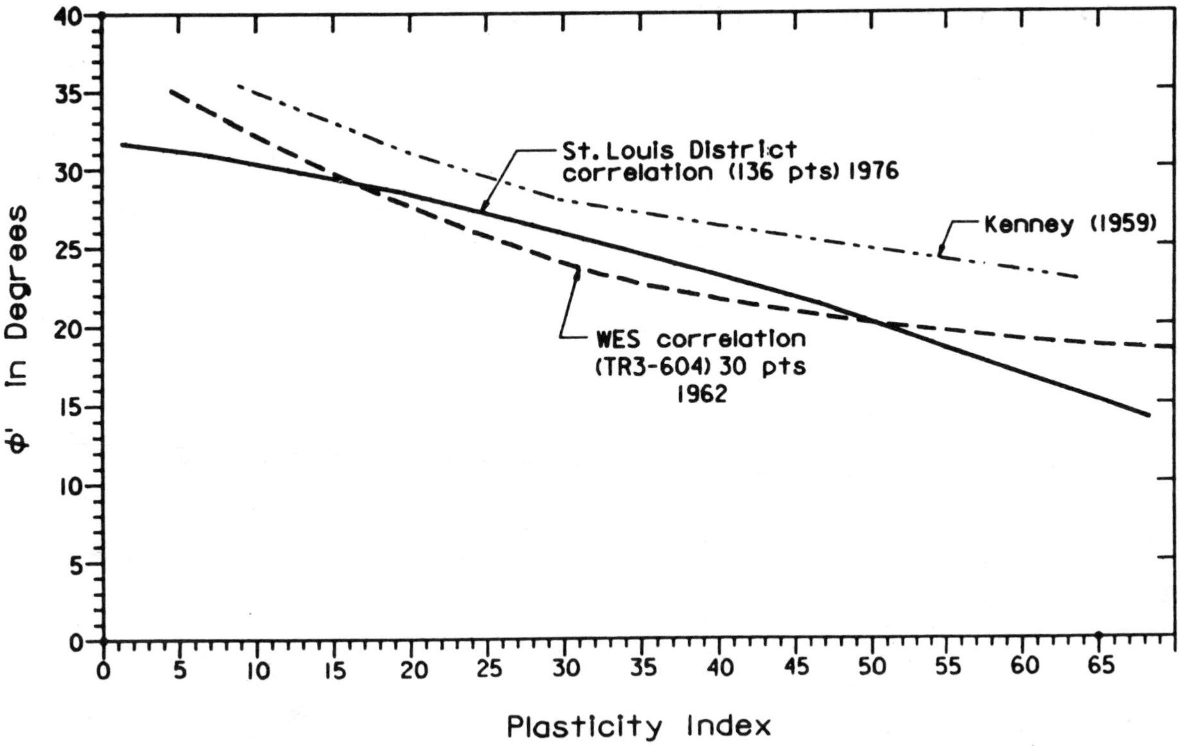

Fig 2-4. Drained friction angle versus plasticity index

usually corrected to an effective vertical overburden stress of 1 ton per square foot using an equation of the form:

$$N' = C_N N \qquad [2\text{-}1]$$

N' = corrected resistance
C_N = correction factor
N = measured resistance

Table 2-1 and Figure 2-5 summarize the most commonly proposed values for C_N. The drained friction angle ϕ' can be estimated from N' using Figure 2-6. The relative density of normally consolidated sands can be estimated from the correlation obtained by Marcuson and Bieganousky (1977):

$$D_r = 11.7 + 0.76[|222(N) + 1600 - 53(p'_{vo}) - 50(C_u)^2|]^{1/2} \qquad [2\text{-}2]$$

where

p'_{vo} = effective overburden pressure in pounds per square inch
C_u = coefficient of uniformity

Correlations have also been proposed between the SPT and the undrained strength of clays. However, these are generally unreliable and should be used only for very preliminary studies and for checking the reasonableness of SPT and lab data.

C. CONE PENETRATION TEST. The cone penetration test, or CPT (ASTM D 3441-79), is widely used in Europe and is gaining considerable acceptance in the United States. The interpretation of the test is described by Robertson and Campanella (1983). For coarse-grained soils, the cone resistance q_c has been empirically correlated with standard penetration resistance (N value). The ratio (q_c/N) is typically in the range of 2 to 6 and is related to median grain size (see Figure 2-7). The undrained strength of fine-grained soils may be estimated by using a modification of bearing capacity theory:

$$s_u = \frac{q_c - p_o}{N_k} \qquad [2\text{-}3]$$

where

p_o = the in situ total overburden pressure
N_k = empirical cone factor typically in the range of 10 to 20

The N_k value should be based on local experience and correlation to laboratory tests. Cone penetration tests also may be used to infer soil classification to supplement physical sampling. Figure 2-8 indicates probable soil type as a function of cone resistance and friction ratio. Cone penetration tests may produce erratic results in gravelly soils.

Table 2.1. SPT Correction to 1 tsf (2 ksf)

	Correction Factor C_N		
Effective Overburden Stress (kips/sq ft)	Seed, Arango, and Chan (1975) Seed	Peck and Bazaraa (1969) P&B	Peck, Hanson, and Thornburn (1974) PH&T
0.20	2.25	2.86	
0.40	1.87	2.22	1.54
0.60	1.65	1.82	1.40
0.80	1.50	1.54	1.31
1.00	1.38	1.33	1.23
1.20	1.28	1.18	1.17
1.40	1.19	1.05	1.12
1.60	1.12	0.99	1.08
1.80	1.06	0.96	1.04
2.00	1.00	0.94	1.00
2.20	0.95	0.92	0.97
2.40	0.90	0.90	0.94
2.60	0.86	0.88	0.91
2.80	0.82	0.86	0.89
3.00	0.78	0.84	0.87
3.20	0.74	0.82	0.84
3.40	0.71	0.81	0.82
3.60	0.68	0.79	0.81
3.80	0.65	0.78	0.79
4.00	0.62	0.76	0.77
4.20	0.60	0.75	0.75
4.40	0.57	0.73	0.74
4.60	0.55	0.72	0.72
4.80	0.52	0.71	0.71
5.00	0.50	0.70	0.70

D. PRESSUREMETER TEST. The pressuremeter test, or PMT, also originated in Europe. Its use and interpretation are discussed by Baguelin, Jezequel, and Shields (1978). Test results are normally used to directly calculate bearing capacity and settlements, but the test can be used to estimate strength parameters. The undrained strength of fine-grained materials is given by

$$s_u = \frac{p_l - p'_{ho}}{2K_b} \qquad [2\text{-}4]$$

where

p_l = limit pressure
p'_{ho} = effective at-rest horizontal pressure
K_b = a coefficient typically in the range of 2.5 to 3.5 for most clays

Again, correlation with laboratory tests and local experience is recommended.

2-20. Backfill Materials

Selection of backfill materials is discussed in Chapter 6. Every effort should be made to provide clean, free-draining backfill materials. Density and strength parameters should be determined from tests on laboratory-compacted samples over a range of densities consistent with expected specification requirements. Development of a local data base and correlations for the properties of locally obtained backfill materials may significantly reduce the need for testing. Figure 2-9 provides typical values of the friction angle for use in preliminary designs. The soil type codes are taken from the Unified Soil Classification System, shown in Technical Memorandum 3-357, prepared by the U. S. Army Engineer Waterways Experiment Station in 1960. The data for this figure were assembled from a wide variety of design references.

2-21. Design Strength Selection

As soils are heterogeneous (or random) materials, strength tests invariably exhibit scattered results. The guidance contained in EM 1110-2-1902 regarding the selection of design strengths at or below the thirty-third percentile of the test results is also applicable to walls. For small projects, conservative selection of design strengths near the lower bound of plausible values may be more cost-effective than performing additional tests. Where expected values of drained strengths (ϕ values) are estimated from correlations, tables, and/or experience, a design strength of 90 percent of the expected (most likely) value will usually be sufficiently conservative. In the case of rock foundations, the strength of intact rock, the strength and orientation of discontinuities, and the orientation of joints relative to the possible failure modes must all be considered in selecting design strengths.

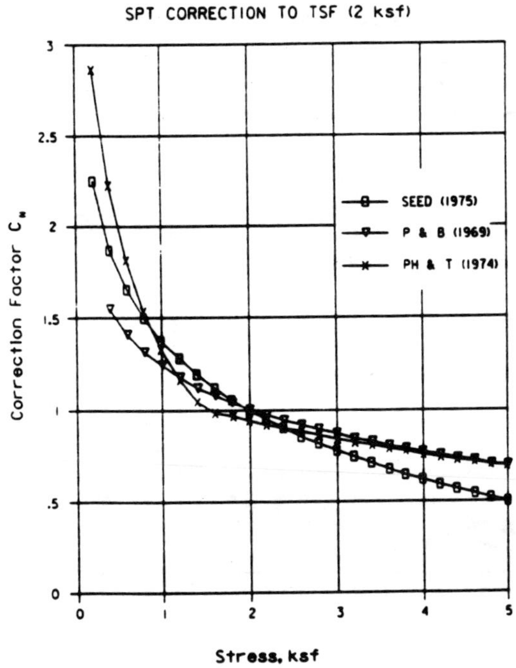

Fig 2-5. SPT correction to 1 tsf (2 ksf)

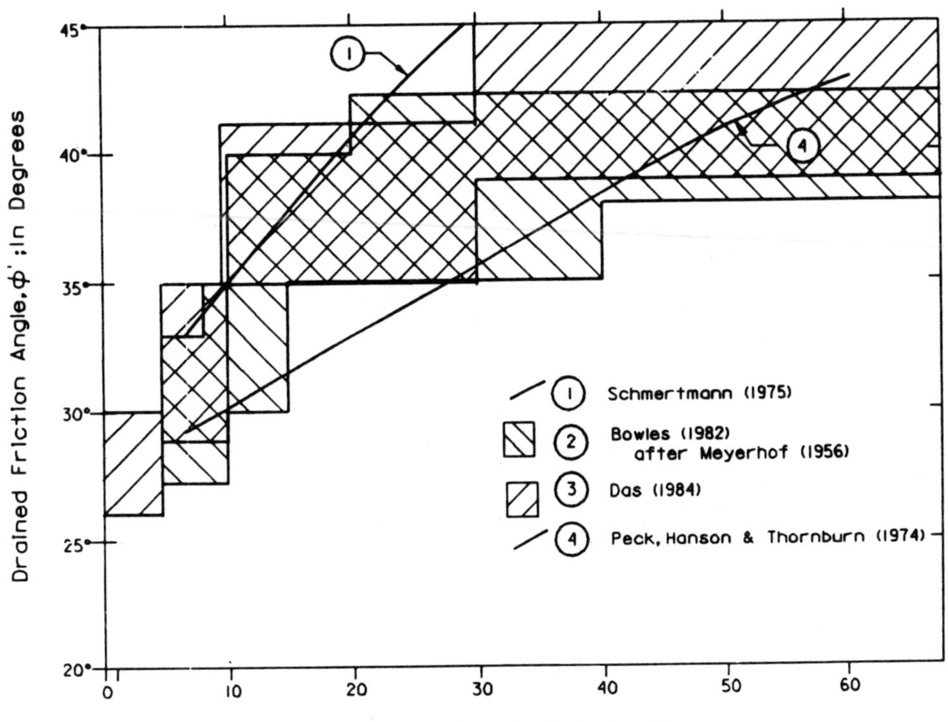

Fig 2-6. ϕ' versus N′ for granular materials

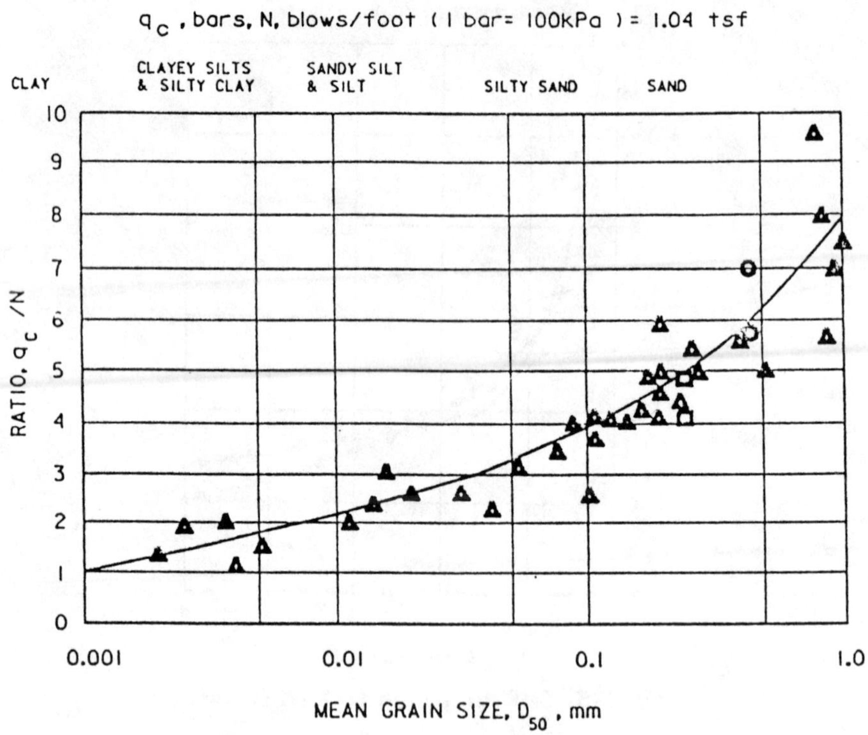

Fig 2-7. q_c/N versus D_{50} (after Robertson and Campanella 1983)

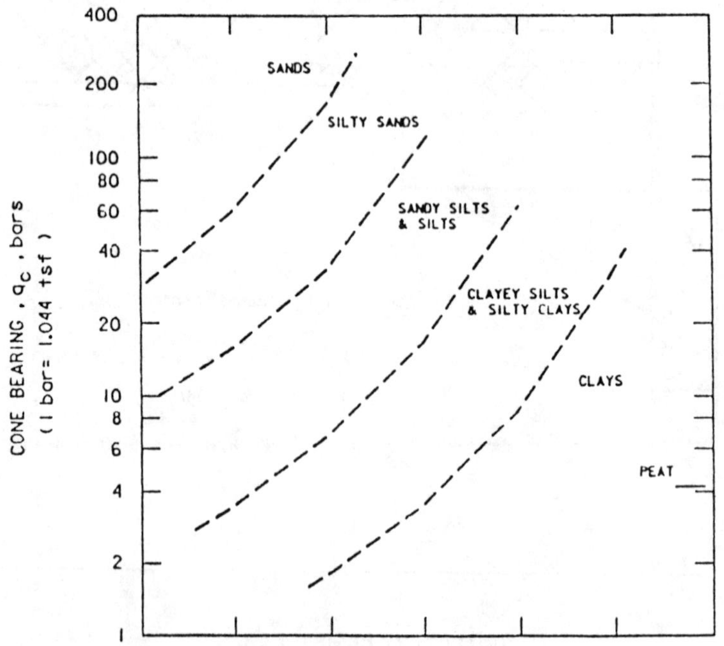

Fig 2-8. Soil classication from cone penetrometer (after Robertson and Campanella 1983)

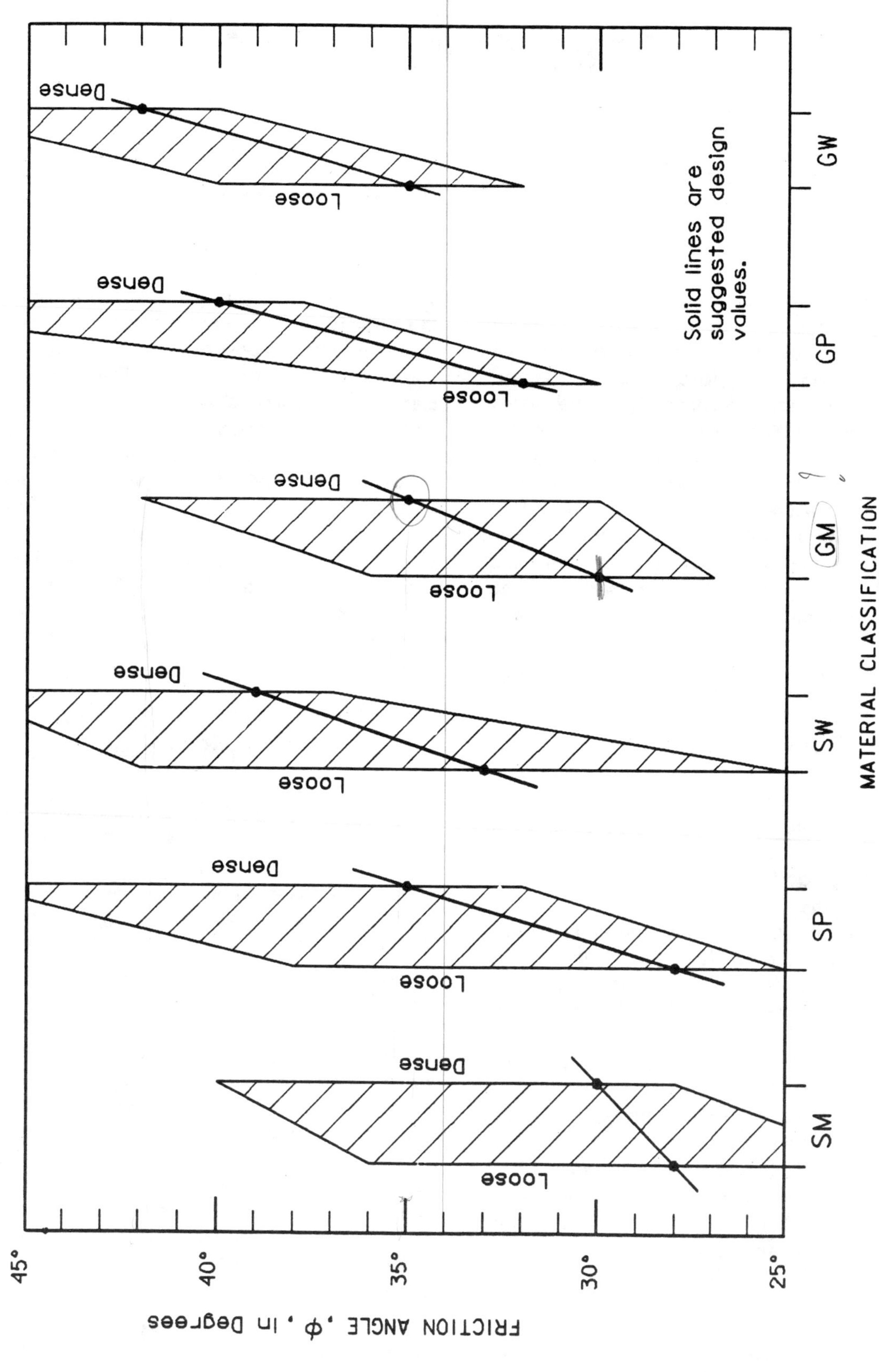

Fig 2-9. Friction angle of granular backfills

CHAPTER 3

FORCES ON WALLS

Section I. Introduction

3-1. General

Retaining walls and flood walls accommodate a difference in soil or water elevation over a typically short horizontal distance. On one side of the wall, the driving side, lateral forces exceed those on the opposite, resisting side; the force difference and resulting moment are accommodated by forces and pressures developed along the base. Lateral forces may be related to gravity, water seepage, waves, wind, and earthquakes. This chapter presents methods for calculating pressures and resulting forces on the driving and resisting sides of walls. These are necessary to calculate the magnitude and location of the base resultant force for overturning and bearing capacity analysis. They are also required for the design of the structural elements of the wall.

3-2. Limit-Equilibrium Analysis

The forces and pressures acting on a wall are in fact highly indeterminate. Static equilibrium equations are insufficient to obtain a solution for lateral forces; additional assumptions must be incorporated in the analysis. For nonlinear materials such as soils, this is commonly and conveniently done by assuming that a "limit" or failure state exists along some surface and that the shear force along the surface corresponds to the shear strength of the material. With these assumptions, equilibrium equations can be solved. Hence, this approach is commonly called "limit-equilibrium analysis." To assure that the assumed failure does not in fact occur, a factor (safety factor or strength mobilization factor) is applied to the material strength. It should be noted that this solution approach differs significantly from that commonly used for indeterminate structural analysis, where stress-strain properties and deformations are employed. This limit-equilibrium approach provides no direct information regarding deformations; it is implied that deformations are sufficient to induce the failure condition. Deformations are indirectly limited to tolerable values by judicious choice of a safety factor.

3-3. Relationship of Forces to Sliding Analysis

Forces calculated in accordance with this chapter are not always equal to those calculated in a sliding analysis (Chapter 4). The methods in this chapter are intended to produce reasonable and somewhat conservative estimates of actual forces operative on the wall. They can be used to perform a quick check on sliding stability as described in paragraph 4-15. The sliding analysis for general cases (paragraph 4-16) considers shear failure along the bases of a collection of interacting free bodies (or wedges) that include both the wall and surrounding soil. Sliding failure is prevented by applying a factor of safety on shear strength equally on all segments of the failure surface. The lateral forces calculated in the sliding analysis are a function of the sliding factor of safety.

Section II. Earth Pressures and Forces

3-4. Cohesionless Materials

A. ACTIVE EARTH PRESSURE. Cohesionless materials such as clean sand are the recommended backfill for retaining walls. Large-scale tests (e.g., Terzaghi 1934; Tschebatarioff 1949; Matsuo, Kenmochi, and Yagi 1978) with cohesionless ($c = 0$) backfills have shown that horizontal pressures are highly dependent on the magnitude and direction of wall movement. The minimum horizontal pressure condition, or active earth pressure, develops when a wall rotates about its base and away from the backfill an amount on the order of 0.001 to 0.003 radian (a top deflection of 0.001 to 0.003h , where h is the wall height). As the wall moves, horizontal stresses in the soil are reduced and vertical stresses due to backfill weight are carried by increasing shear stresses until shear failure is imminent (see Figure 3-1a).

B. PASSIVE EARTH PRESSURE. If a wall is moved toward the backfill, horizontal stresses increase and shear stresses reverse direction, first decreasing and then increasing to a maximum at failure (see Figure 3-1b). Because the horizontal

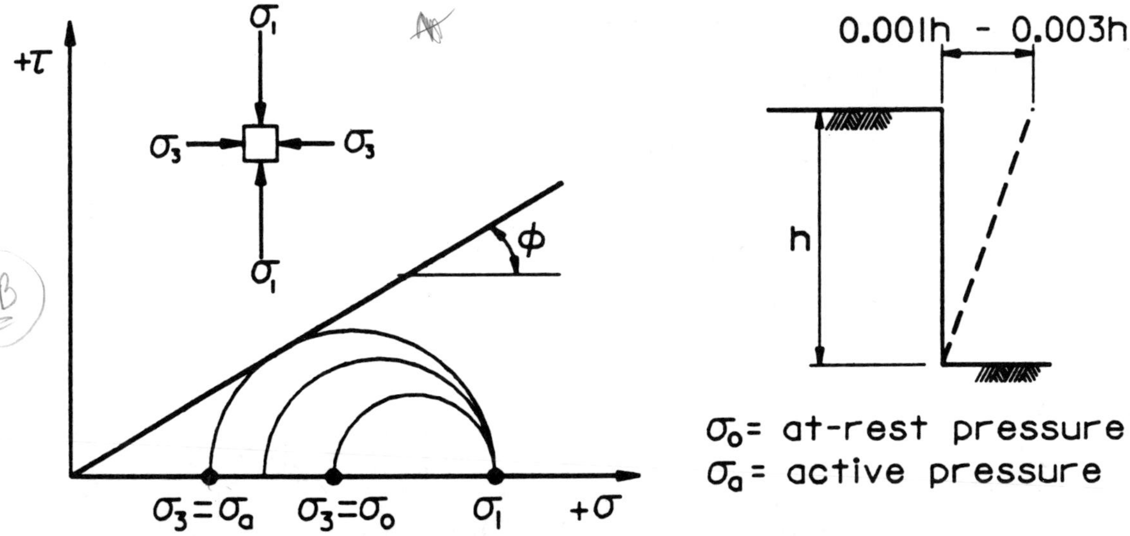

a. Development of active earth pressure

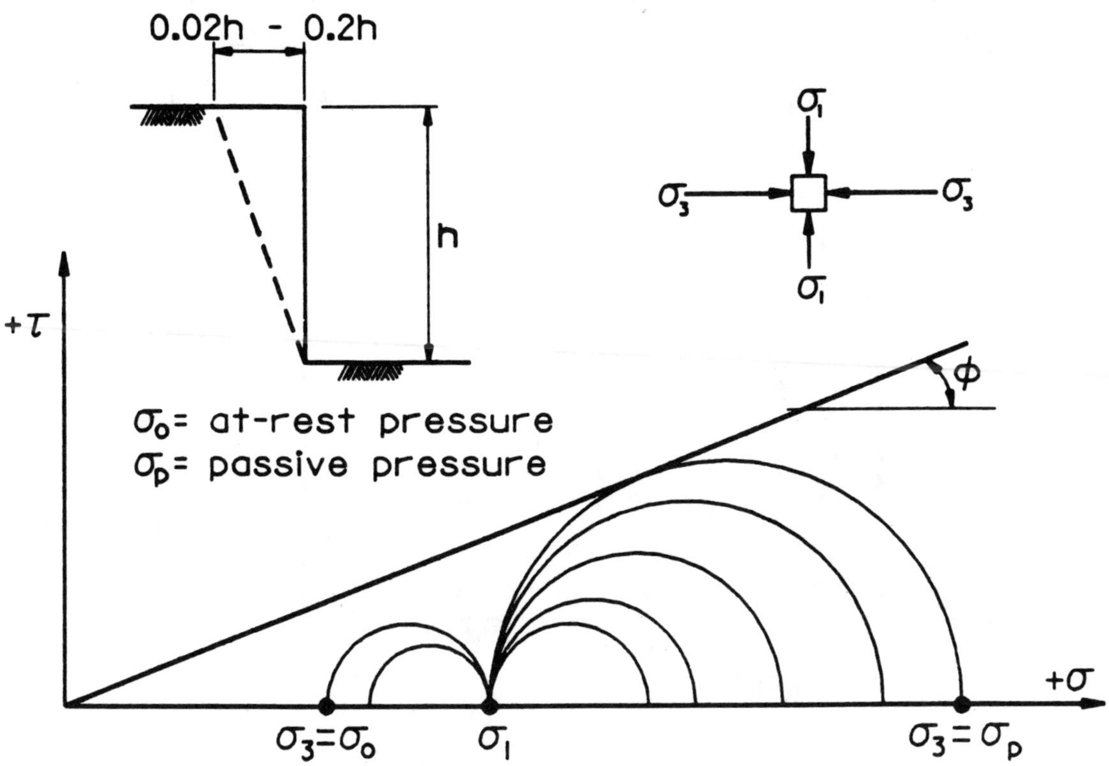

b. Development of passive earth pressure

Fig 3-1. Development of earth pressures for a cohesionless material

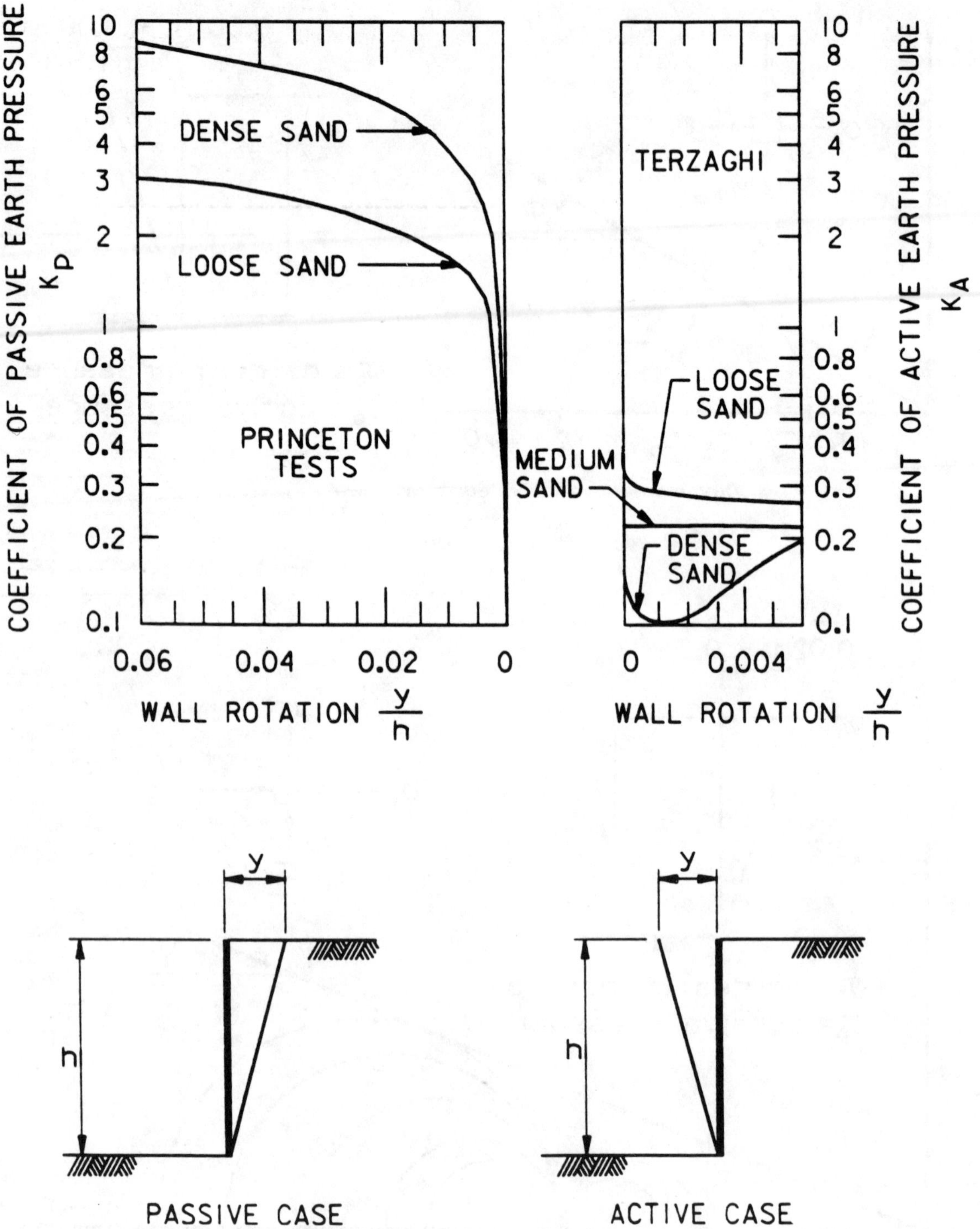

Fig 3-2. Relationship of earth pressures to wall movements (after Department of the Navy 1982b)

stress component along the shear planes is resisted by both shear stress and vertical stress components, higher horizontal stresses can be developed than for the active pressure case. Development of the maximum possible horizontal stress, or passive pressure, requires much larger wall rotations than for the active case, as much as 0.02 to 0.2 radian. It should be noted that the deformation required to mobilize one-half of the passive pressure is significantly smaller than that required for full mobilization.

C. AT-REST EARTH PRESSURE. If no wall movement occurs, the lateral pressure condition is termed the *at-rest pressure.*

D. LATERAL EARTH PRESSURE COEFFICIENT K. The ratio of the horizontal effective stress to the vertical effective stress in a cohesionless soil mass can be expressed by the earth pressure coefficient K. Typical relationships between the K value and wall movements are shown in Figure 3-2. The value of K can be obtained for active (K_A) and passive (K_P) conditions using limit-equilibrium methods. Empirical equations are available for the at-rest value (K_o) as described in paragraph 3-10.

E. CONDITIONS AFFECTING EARTH PRESSURE. For complicated backfill conditions, at-rest earth forces can be estimated using the general wedge method combined with factored soil strengths as described in paragraph 3-13. If the mode of wall movement is other than base rotation, the earth pressure and its distribution may differ considerably from any solutions herein and other analysis techniques are required (see paragraph 3-15g). Also, compaction of the backfill behind a wall can produce horizontal pressures in excess of at-rest pressures near the top of a wall as discussed in paragraph 3-17.

3-5. Cohesive Materials

A. STRENGTH PROPERTIES. So-called cohesive materials, typically fine-grained soils such as clay, exhibit shear strength under zero confining stress when loaded rapidly. The strength at zero confinement is expressed by the parameter c, or cohesion. Cohesive materials are usually saturated or nearly saturated because their small pore diameter attracts capillary water. When stress changes are imposed (such as by wall movement), the soil attempts to change volume. If low permeability prevents volume change from keeping pace with the external stress change, pressure changes are induced in the pore water. What appears to be stress-independent strength (cohesion) is, for the most part, the combined effects of frictional resistance between soil particles and induced pore pressure changes. Pore-water tension at low stresses permits vertical cuts in clay; however, such cuts eventually fail as negative pore pressures dissipate and water content increases. Horizontal pressures in cohesive materials are related to the soil's permeability and pore pressure response during shear in addition to wall movement. Therefore they are time dependent.

B. USE AS A BACKFILL MATERIAL. It is strongly recommended that cohesionless materials such as clean sands be used for wall backfill materials. Cohesionless materials have more predictable properties than cohesive materials, are less frost susceptible, and provide better drainage. However, there are certain instances (such as walls adjacent to impervious clay cutoffs in flood-control structures) where clay backfills may be unavoidable.

C. SHORT- AND LONG-TERM ANALYSIS. Solutions are included herein for earth pressures in terms of the general case involving both the c and ϕ parameters. Where cohesive backfills are used, two analyses (short-term and long-term) are usually required with different sets of strength parameters in order to model conditions that may arise during the life of the wall. Strength tests are further discussed in Chapter 2, Section V.

1. Short-Term Analyses. These analyses model conditions prevailing before pore pressure dissipation occurs, such as the end-of-construction condition. For these analyses, unconsolidated-undrained (Q) test parameters are appropriate. Often these tests yield a relatively high c value and a low or zero ϕ value. Calculations may indicate that the soil is in tension to significant depths and exerts zero pressure on the wall; thus, the short-term analysis alone will seldom govern wall design. However, the zone of theoretically negative soil pressure may correspond to cracking and should be assumed to crack as described in paragraphs 3-15f and 4-18. Water entering these cracks may exert significant horizontal pressure on a wall. Therefore, short-term stability analyses should include a check of the effect of water pressure in tension cracks.

2. Long-Term Analyses. These analyses model conditions prevailing after shear-induced excess pore pressures have dissipated. (Dissipation herein includes negative pore pressures increasing to zero.) For long-term analysis, consolidated-drained (S) test parameters are appropriate. These tests usually yield a relatively high ϕ value and a relatively low or zero c value.

D. OVERCONSOLIDATED "SWELLING" CLAY SOILS. For highly overconsolidated and/or "swelling" clay soils, lateral pressures may be developed in excess of those calculated using drained or undrained strength parameters. These pressures cannot generally be determined using limit-equilibrium techniques (see, e.g., Brooker and Ireland 1965). The use of such soils around retaining walls should be avoided.

3-6. Pressures in Soil-Water Systems

Soil grains are able to transmit shear stresses; water cannot. Consequently, effective pressures in soil may differ on horizontal and vertical planes but water pressures cannot. Effective soil pressures are therefore separated from water pressures in calculations. If the value of K is established, horizontal effective stresses may be calculated by multiplying the effective vertical stress at any point by the corresponding K value (see Figure 3-3). To obtain the total horizontal pressure, the effective horizontal pressure is added to the water pressure. Where more than one soil layer is present, vertical pressures increase continuously with depth but the horizontal pressure diagrams may be discontinuous as shown. Combining water pressures with effective earth pressures is further discussed in paragraphs 3-15 and 3-18.

3-7. Design Earth Pressures and Forces, Driving Side

A. USE OF AT-REST EARTH PRESSURE. The driving side of a retaining wall or flood wall is defined as that side on which soil and/or water exerts a horizontal force tending to cause instability. Designers have often assumed active earth pressure on the driving side because movements required to develop active pressures are small. However, several reasons exist to design walls for at-rest pressures. Because designs incorporate factors of safety, walls may be quite rigid and pressures may be greater than active. Hydraulic structures in particular are designed using conservative criteria that result in relatively stiff wall designs. Walls founded on rock or stiff soil foundations may not yield sufficiently to develop active earth pressures. Even for foundations capable of yielding, certain experiments with granular backfill (Matsuo, Kenmochi, and Yagi 1978) indicate that, following initial yield and development of active pressures, horizontal pressures may in time return to at-rest values. Another reference (Casagrande 1973) states that the gradual buildup of the backfill in compacted lifts produces greater-than-active pressures as do long-term effects from vibrations, water level fluctuations, and temperature changes.

B. ESTIMATION OF OPERATIVE PRESSURES. Design analyses require an estimate of the expected 'operative' (nonfailure) pressures on the wall for overturning and bearing capacity analyses and structural design. Therefore, walls should be designed to be safe against overturning and bearing failure for at-rest earth pressure conditions, and structural elements should be designed assuming at-rest earth pressures on the driving side. The lateral soil forces calculated using the multiple wedge sliding analysis described in Chapter 4 are in the at-rest pressure range when a safety factor of 1.5 is obtained.

C. COMPACTION AND SURCHARGE EFFECTS. Where significant compaction effort is specified for the backfill, design earth pressures should be increased beyond the at-rest values for depths above a "critical" depth as described in paragraph 3-17. Where surcharges are expected above the backfill (in stockpiles, rails, footings, etc.), the additional horizontal earth pressure due to the surcharge should be determined as discussed in paragraph 3-16 and superimposed on the at-rest pressure diagram. Examples of these effects are given in Appendix M.

3-8. Design Earth Pressures and Forces, Resisting Side

A. BACKGROUND. The resisting side of a wall is defined as that side where soil and/or water provide a lateral reaction tending to resist instability. The maximum earth force that can be developed is the passive earth force. However, for a wall in equilibrium, the actual resisting-side force will typically be smaller than the passive force as the forces on the driving side, base, and resisting side taken together must satisfy static equilibrium. The resistances to the driving-side force provided by the resisting-side force and the base shear force, respectively, are indeterminate. Allocation of the total resistance between these two forces is judgmental.

B. ESTIMATION OF PASSIVE RESISTANCE. A conservative and convenient design approach is to assume the resisting-side force is zero for overturning and bearing capacity analyses and for structural design. However, in some cases, such as walls with relatively deep foundations, it may be desirable to consider some lateral resistance for these analyses. To justifiably assume a non-zero resisting-side force, the material must not lose its resistance characteristics with any probable change in water content or environmental conditions

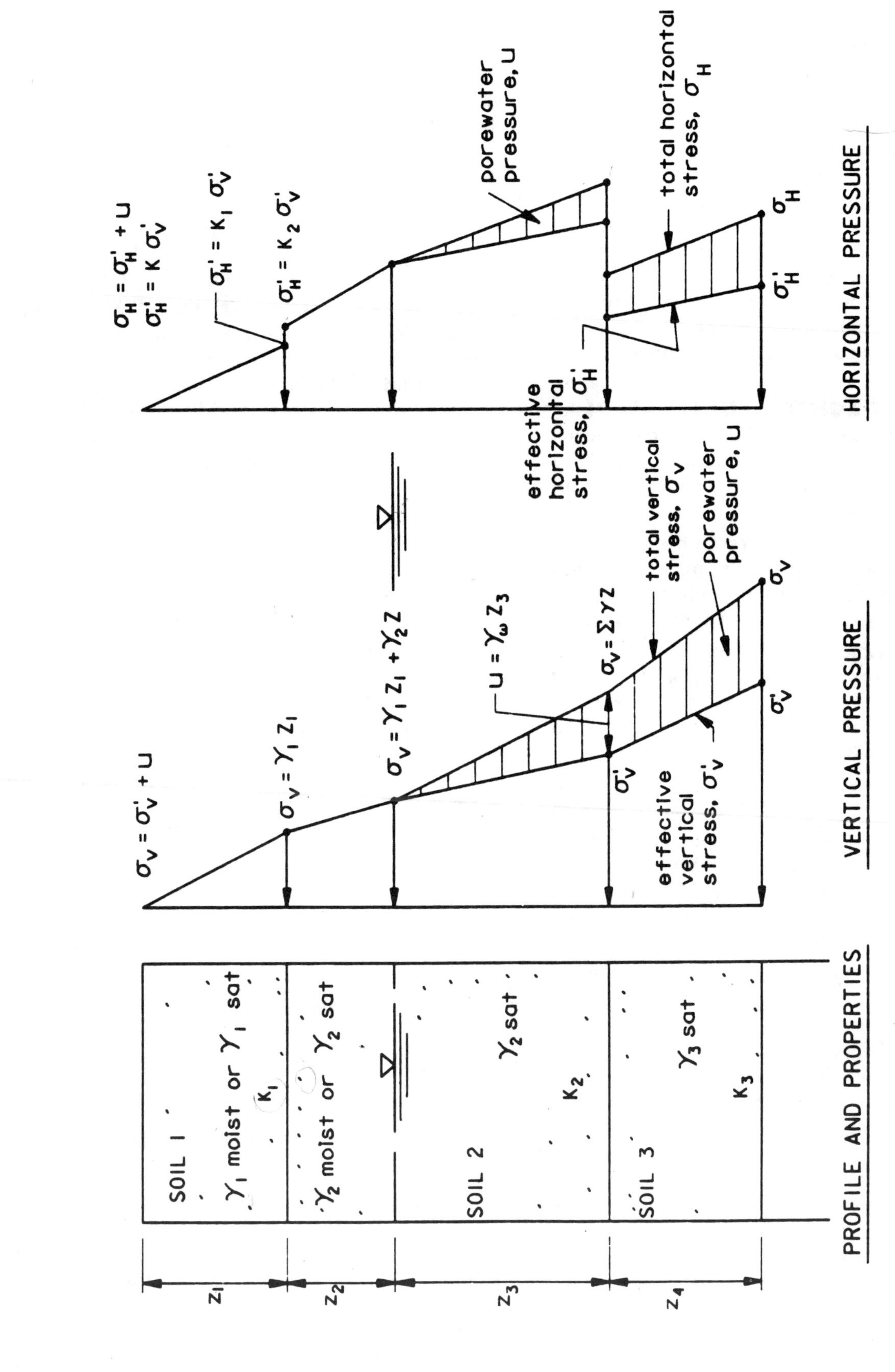

Fig 3-3. Vertical and horizontal pressures in a cohesionless soil mass

and must not be eroded or excavated during the life of the wall. If such assumptions can be justified, at-rest conditions may be conservatively assumed on the resisting side. Resisting-side pressures and forces generally should not be assumed to exceed at-rest conditions when the base resultant force and location and when structural components. However, if the driving-side earth force exceeds the sum of the resisting side at-rest earth force (if present) and the maximum available base shear force calculated using unfactored shear parameters, the additional required resistance should be assumed to be provided by additional resisting-side pressure. In no case should the resisting-side earth pressure exceed one-half the passive pressure calculated using unfactored shear strengths for overturning and bearing capacity analyses and structural design.

C. HORIZONTAL FORCE ALLOCATION. To summarize, the horizontal force allocation for overturning analysis, bearing capacity analysis, and design of structural components should be computed as follows:

1. Calculate the at-rest effective earth force on the driving side (paragraphs 3-10 through 3-13). Superimpose surcharge effects if present (paragraph 3-16). Add water pressures, if present.

2. Assume that the resisting-side earth force equals zero or calculate and apply the at-rest earth force on the resisting side of the wall, if justified (paragraphs 3-10 through 3-13). Add water forces if present.

3. Assume that the horizontal component of the base resultant is equal to the difference between the horizontal forces from (1) and (2).

4. If the maximum available base shear force is exceeded, assume that the remaining horizontal force is resisted by mobilizing a greater fraction of passive pressure so long as not more than one-half the available passive force is used. (This may occur where the resisting-side soil is strong relative to the driving-side and base soils.)

D. SLIDING STABILITY CHECK. Sliding stability should be checked using the single or multiple wedge methods found in paragraphs 4-15 and 4-16, respectively.

3-9. Design Earth Pressures and Forces on the Base

A. CALCULATION OF RESULTANT FORCE ON BASE. The resultant force on the base, its direction, and its location must be such that the wall is in static equilibrium for the "operative" loads (see Figure 3-4). In Figure 3-4a, the vertical component of the resultant is equal and opposite the summed weights of the "structural wedge," and the horizontal component is equal to the difference of the driving-side and resisting-side forces. Figure 3-4b illustrates a more complicated example including water and a sloping base with a key. The vertical and horizontal components of the base uplift force are calculated from base water pressures obtained from a seepage analysis. The remaining vertical and horizontal forces required for equilibrium are provided by components of the base shear force T and effective normal force N′. An overturning analysis as described in Chapter 4 must be performed in order to determine the effective normal force N′ and its location.

B. COMPUTATION OF BASE PRESSURES. The effective earth pressure on the base is assumed to vary linearly and N′ is applied at the centroid of the pressure diagram. When the resultant falls within the middle one-third of the base, the effective base pressures q′ are calculated by the following equation:

$$q' = \left(\frac{N'}{B}\right)\left(1 \pm \frac{6e}{B}\right) \qquad [3\text{-}1]$$

where

N' = effective normal force on base of structure

B = width of base of structure

e = eccentricity of N' from center of base

This is shown in Figure 3-5, a and b. If the resultant falls outside the middle one-third of the base i.e., e is greater than $B/6$, as shown in Figure 3-5c, the pressure distribution is triangular with a maximum pressure equal to

$$q'_{max} = \frac{4}{3}\left(\frac{N'}{B - 2e}\right) \qquad [3\text{-}2]$$

The base will be in compression over a distance b from the toe computed as

$$b = \tfrac{3}{2}(B - 2e) \qquad [3\text{-}3]$$

Refer to Appendix N for example computations.

3-10. At-Rest Earth Pressure Equations

A. HORIZONTAL BACKFILL. For the special case of a horizontal backfill surface and a normally consolidated backfill (no compaction or other prestress effects) the at-rest pressure coefficient K_o can be estimated from Jaky's (1944) equation

$$K_o = 1 - \sin\phi' \qquad [3\text{-}4]$$

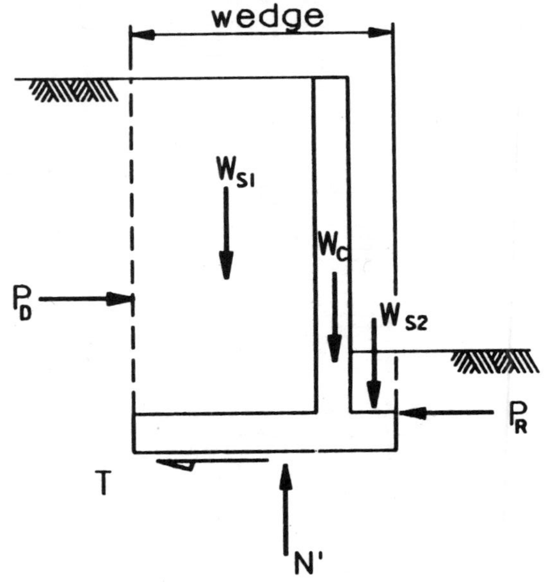

$N' = \Sigma W$

$T = P_D - P_R$

a. Simple retaining wall

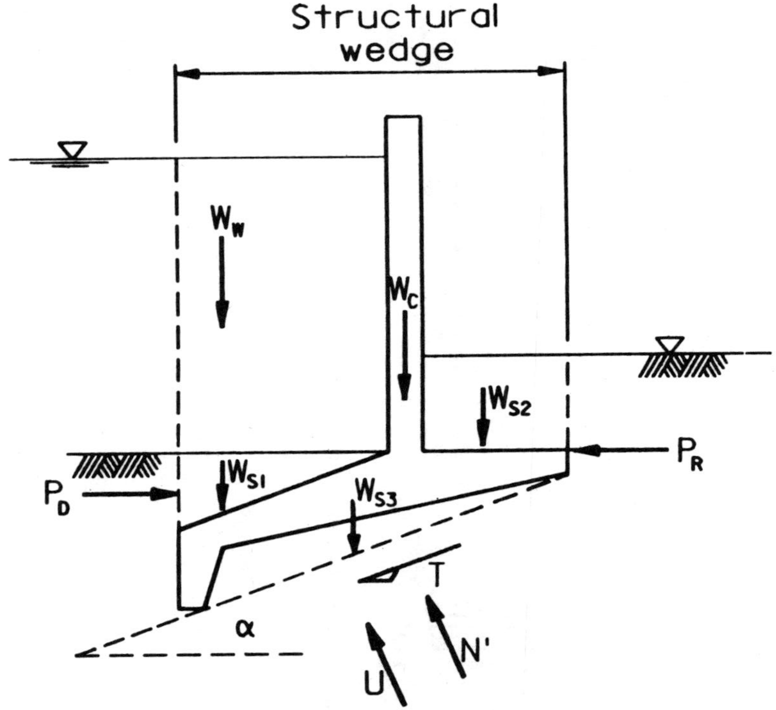

$T = \Sigma H\cos\alpha - \Sigma V\sin\alpha$ $N' = \Sigma V\cos\alpha + \Sigma H\sin\alpha$

ΣH and ΣV are summation of all forces not including T and N'

b. Flood wall with sloping base and key

Fig 3-4. Forces on base of wall

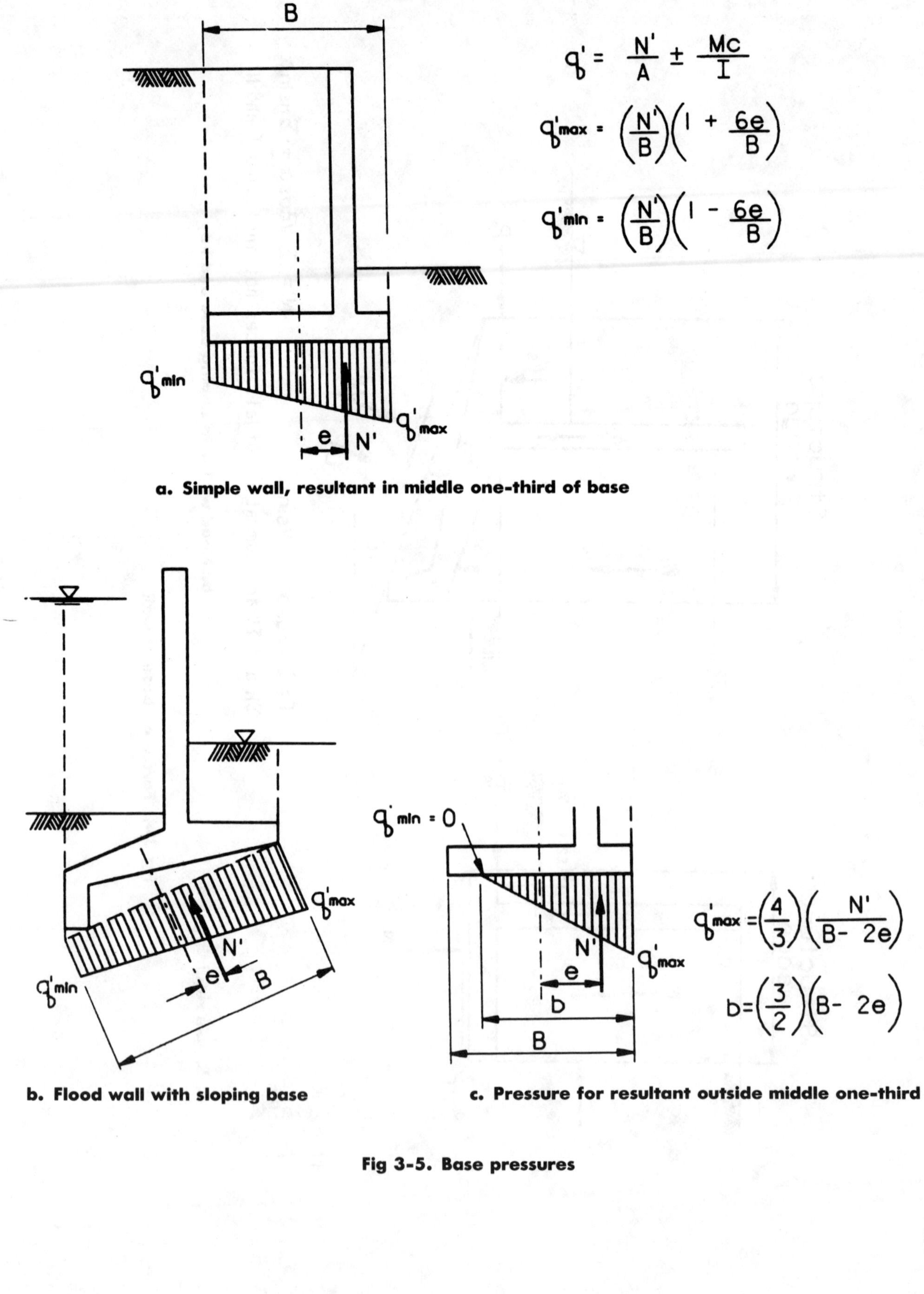

a. Simple wall, resultant in middle one-third of base

b. Flood wall with sloping base

c. Pressure for resultant outside middle one-third

Fig 3-5. Base pressures

and the lateral earth pressure computed by

$$p_o = \gamma' K_o z$$

where

ϕ' = drained internal friction angle

γ' = effective unit weight (moist or saturated above water table, submerged or buoyant below water table)

z = depth below surface of backfill along a vertical plane

B. SLOPING BACKFILLS. For normally consolidated sloping backfills, results of experiments to measure K_o are quite variable. The following equation proposed by the Danish Code (Danish Geotechnical Institute 1978) is recommended:

$$K_{o\beta} = K_o(1 + \sin\beta) \qquad [3\text{-}5]$$

Substituting Equation 3-4 in 3-5 gives

$$K_{o\beta} = (1 - \sin\phi')(1 + \sin\beta) \qquad [3\text{-}6]$$

and the lateral earth pressure computed

$$p_o = \gamma' K_{o\beta} z$$

where β is the slope angle from the horizontal. Beta is positive for a soil layer that slopes upward and away from the structure. Values for K_o and $K_{o\beta}$ are given in Appendix E.

C. GENERAL CONDITIONS. For walls with irregular backfill surfaces, non-homogeneous backfills, surcharge loadings, and/or other complicating conditions, empirical relationships for the at-rest pressure are not generally available. For routine designs, an approximate solution for the horizontal earth force may be obtained using Coulomb's active force equation or the general wedge method with values of c and tan ϕ multiplied by a strength mobilization factor (defined in paragraph 3-11). Because this is an empirical approach, results will differ slightly from calculations using Equations 3-4 through 3-6 where companion solutions can be obtained. Appendix E includes a comparison of K_o values so obtained for both horizontal and sloping backfills. Figure 3-6 shows a comparison of Jaky's equation with Coulomb's equation for a horizontal backfill.

D. RESISTING SIDE. Jaky's equation and the Danish Code equation may be used to compute at-rest pressures for the resisting side for horizontal and sloping soil surfaces, respectively. Example computations are shown in example 7 of Appendix M and in Appendix N.

3-11. Strength Mobilization Factor

A. DEFINITION. The strength mobilization factor (SMF) is defined as the ratio of the assumed mobilized or developed shear stress τ along an assumed slip surface to the maximum shear strength τ_f of the soil material at failure. If an appropriate SMF value is assumed and applied to c and tan ϕ, it allows calculation of greater-than-active earth pressures using Coulomb's active force equation (paragraph 3-12) or the general wedge equation (paragraph 3-13). Alternatively, the safety against sliding may be assessed by calculating the average SMF along an assumed sliding surface from an equilibrium analysis and comparing it to a recommended maximum value. These concepts are illustrated in Figure 3-7. In equation form, the strength mobilization factor may be expressed as:

$$\text{SMF} = \frac{\tau}{\tau_f} \qquad [3\text{-}7]$$

B. DEVELOPED SHEAR STRESS. According to the Mohr-Coulomb failure criterion (Figure 3-8) the shear strength on the failure plane is defined as

$$\tau_f = \sigma'_n \tan\phi + c \qquad [3\text{-}8]$$

where

σ'_n = effective normal stress

ϕ, c = shear strength parameters of soil (where ϕ and c in the above equation are drained strengths ($\phi = \phi'$, $c = c'$) for long-term analysis and undrained ($\phi = 0$, $c = S_u$) for short-term analysis of cohesive materials)

The failure plane is inclined $45 + \phi/2$ degrees from the plane of the major principal stress. For limit-equilibrium analyses to be valid, the assumed slip surface must be inclined at this angle relative to the principal stresses. In the Coulomb and general wedge methods, a plane slip surface is assumed. Discontinuities in the backfill surface, surcharges, and wall friction all cause variation in the principal stress directions and induce curvature in the slip surface. Assuming that the plane slip surface approximation is valid and is properly oriented relative to the principal stresses, the shear stress on it is

$$\tau = \text{SMF}(c) + \sigma'_n\,\text{SMF}(\tan\phi) \qquad [3\text{-}9]$$

Thus, the shear stress on a presumed slip surface is taken to be a function of the shear strength parameters, the effective normal stress, and the strength mobilization factor.

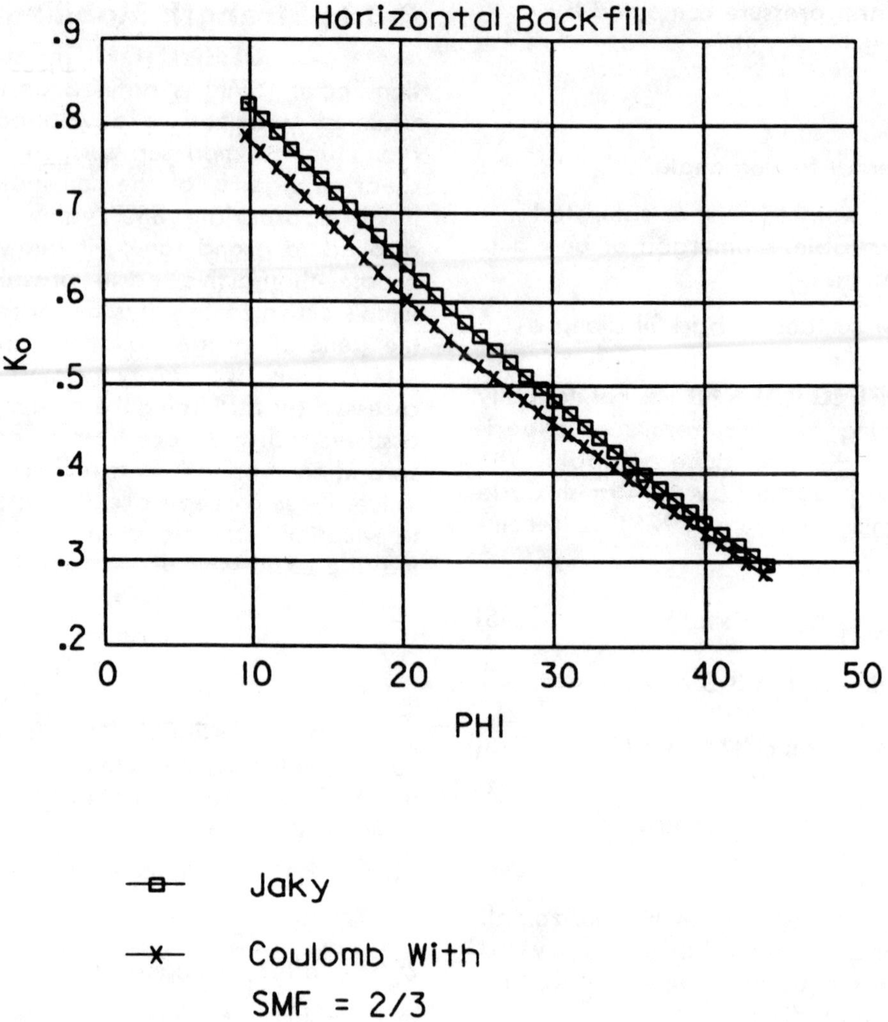

Figure 3-6. At-rest earth pressure coefficients

C. DEVELOPED SHEAR STRENGTH PARAMETERS. Multiplying the shear strength parameters (c and tan ϕ) by the appropriate SMF reduces them to the 'developed' values (c_d and tan ϕ_d) assumed to be operative in equilibrium conditions. The developed shear strength parameters, the actual shear strength parameters, and the SMF are related as follows:

$$SMF = \frac{\tan \phi_d}{\tan \phi} = \frac{c_d}{c} \qquad [3\text{-}10]$$

To estimate at-rest pressures for design using Coulomb's active earth pressure equation or the general wedge equation, the SMF should be taken as 2/3 (0.667). K_o values so obtained are compared with Jaky's equation in Figure 3-6. The Coulomb equation with an SMF of 2/3 is compared to the Danish Code and Jaky equations in Appendix E. It should be noted that as the ratio, tan β/tan ϕ, exceeds 0.56, the lateral earth force computed by the Coulomb or general wedge equations using an SMF = 2/3 will be increasingly larger than that given by computing the earth force using a K_o given by the Danish Code equation, for those conditions where the Danish Code equation applies. Therefore, computing at-rest earth loadings using the Coulomb or general wedge equations for a sloping backfill when tan β/tan ϕ exceeds 0.56 will be conservative (see Appendix E).

3-12. Earth Force Calculation, Coulomb's Equations

A. GENERAL

1. Coulomb's equations solve for active and passive earth forces by analyzing the equilibrium of a wedge-shaped soil mass. The mass is assumed to be a rigid body sliding along a plane slip surface. Design (at-rest) earth pressures and forces may be estimated using developed shear strength parameters (Equation 3-10) corresponding to an SMF of 2/3 in the Coulomb active earth force equation. The Coulomb equations have the advantage of providing a direct solution where the following conditions hold:

a. There is only one soil material (material properties are constant). There can be more than one soil layer if all the soil layers are horizontal.

b. The backfill surface is planar (it may be inclined).

c. The backfill is completely above or completely below the water table, unless the top surface is horizontal, in which case the water table may be anywhere within the backfill.

d. Any surcharge is uniform and covers the entire surface of the driving wedge.

e. The backfill is cohesionless, unless the top surface is horizontal, in which case the backfill may be either cohesionless or cohesive.

2. Although Coulomb's equation solves only for forces, it is commonly expressed as the product of a constant horizontal pressure coefficient K and the area under a vertical effective stress diagram. Assuming the concept of a constant K is valid, horizontal earth pressures can be calculated as the product of K times the effective vertical stress. The variation of the Coulomb solution from a more rigorous log-spiral solution is generally less than 10 percent, as shown in Figure 3-9.

B. DRIVING-SIDE EARTH FORCE

1. The total active force P_A on a unit length of wall backfilled with a cohesionless material (c = 0) is given by:

$$P_A = \tfrac{1}{2}\gamma' \frac{1}{\sin\theta \cos\delta} K_A h^2 \qquad [3\text{-}11]$$

and acts at an angle δ from a line normal to the wall. In the above equation (refer to Figure 3-10):

γ' = effective unit weight (moist or unsaturated unit weight if above the water table, submerged or buoyant unit weight if below the water table)

θ = angle of the wall face from horizontal (90 degrees for walls with a vertical back face or structural wedge)

δ = angle of wall friction

K_A = active earth pressure coefficient

h = height of fill against gravity wall or height of fill at a vertical plane on which the force is being computed

where

$$K_A = \frac{\sin^2(\theta+\phi)\cos\delta}{\sin\theta\,\sin(\theta-\delta)\left[1+\sqrt{\dfrac{\sin(\phi+\delta)\sin(\phi-\beta)}{\sin(\theta-\delta)\sin(\theta+\beta)}}\right]^2} \qquad [3\text{-}12]$$

Examples 1 and 2 in Appendix M and the examples in Appendix N demonstrate the use of Equation 3-12.

2. When wall friction is neglected ($\delta = 0$), Equation 3-12 reduces to:

$$K_A = \frac{\sin^2(\theta+\phi)}{\sin^2\theta\left[1+\sqrt{\dfrac{\sin\phi\,\sin(\phi-\beta)}{\sin\theta\,\sin(\theta+\beta)}}\right]^2} \qquad [3\text{-}13]$$

3. For the case of no wall friction ($\delta = 0$) and a vertical wall ($\theta = 90$ degrees)

$$K_A = \frac{\cos^2\phi}{\left[1+\sqrt{\dfrac{\sin\phi\,\sin(\phi-\beta)}{\cos\beta}}\right]^2} \qquad [3\text{-}14]$$

4. For the special case of no wall friction, horizontal backfill surface, and a vertical wall, Coulomb's equation for K_A reduces to:

$$K_A = \frac{1-\sin\phi}{1+\sin\phi} = \tan^2\left(45^\circ - \frac{\phi}{2}\right) \qquad [3\text{-}15]$$

which is identical to Rankine's equation for this special case.

5. As stated in paragraph 3-11c and demonstrated in Figure 3-6 and Appendix E, a developed ϕ angle computed by Equation 3-10 using an SMF of 2/3 can be used in Coulomb's equation to compute an earth pressure coefficient close to that given by the Jaky or Danish Code equations.

6. For the horizontal component of the earth force acting on a vertical plane, with no wall friction, the term $(1/\sin\theta\cos\delta)$ in Coulomb's equation is equal to unity. Thus, Equation 3-11 reduces to

$$P_{AH} = (\tfrac{1}{2})K_A\gamma' h^2 \qquad [3\text{-}16]$$

7. If total stress or undrained strength parameters are used and there is a cohesion term

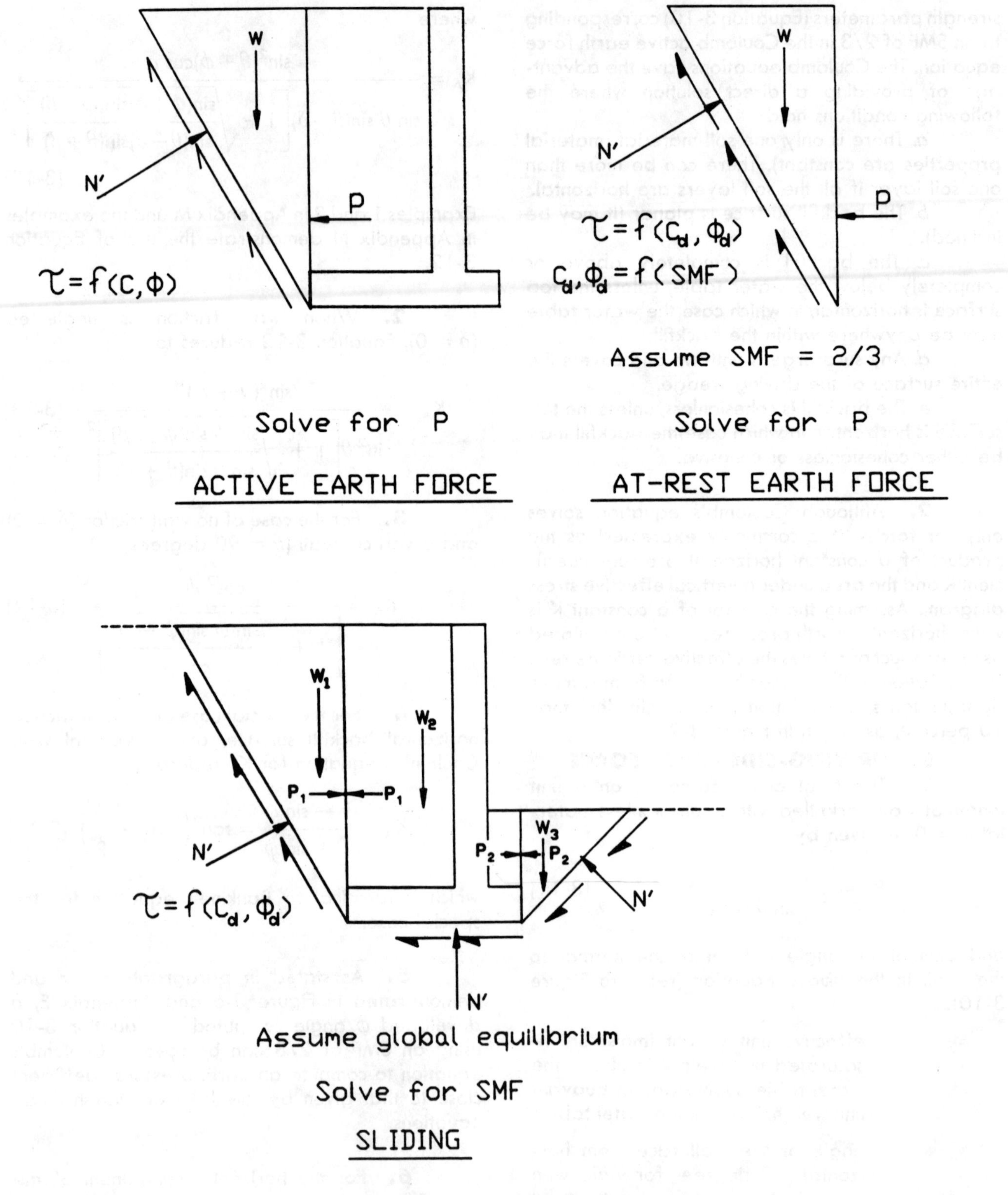

Fig 3-7. Application of the strength mobilization factor

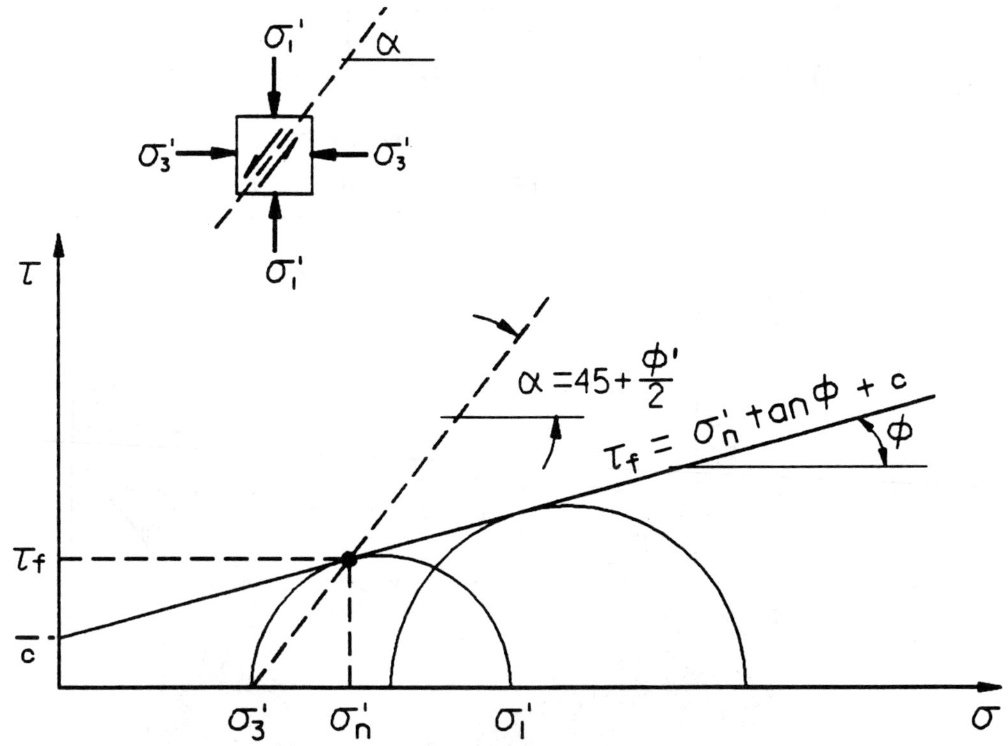

Fig 3-8. Mohr-Coulomb failure criterion

c it has the effect of reducing the active earth force P_{AH}:

$$P_{AH} = (\tfrac{1}{2})K_A\gamma'h^2 - 2c\sqrt{K_A}\,h + \frac{2c^2}{\gamma'} \quad [3\text{-}17]$$

For a backfill with a horizontal surface, K_c given in Appendix H, paragraph H-2c, equals $\sqrt{K_A}$. The second term is the reduction in the active force due to the effect of cohesion on the slip plane and the third term accounts for the shortened length of slip plane due to the effect of a tension crack. If the third term is neglected, and K is assumed constant with depth, the active pressure can be obtained as the derivative of P_{AH} with respect to the depth from the top of the wall z:

$$P_{AH} = K_A\gamma'z - 2c\sqrt{K_A} \quad [3\text{-}18]$$

Refer to examples 5 and 8 of Appendix M for examples involving cohesion.

8. Estimation of at-rest pressures using the SMF concept with Coulomb's equation may give unreliable results for medium to highly plastic cohesive materials. If these materials cannot be avoided in the area of the driving side wedge, the at-rest pressure should be taken as the overburden pressure times an empirical K value, such as from Massarsch's (1979) or Brooker and Ireland's (1965) correlation of K with the plasticity index. Because of the number of uncertainties about the behavior of cohesive materials, a degree of conservatism should be exercised in the selection of the K values. Also, the effects of short- and long-term conditions (paragraph 3-5c) and compaction (paragraph 3-17) should be included in estimating the at-rest pressure.

C. RESISTING-SIDE EARTH FORCE

1. The Coulomb and general wedge equations assume a plane slip surface. However, wall friction effects cause the actual slip surface at failure to be curved. For active pressure calculations, the magnitude of error introduced by the plane surface assumption is not significant, as shown in Figure 3-9 (Driscoll 1979). Coulomb's passive force equation, however, is grossly unconservative where wall friction is present as shown in Figure 3-11 (Driscoll 1979). However, where δ is less than about one-third ϕ, the error is small. If wall friction is neglected, Coulomb's equation is therefore acceptable. The Coulomb passive pressure coefficient for the case of no wall friction ($\delta = 0$) and a vertical wall ($\theta = 90$ degrees) is:

$$K_P = \frac{\cos^2\phi}{\left[1 - \sqrt{\dfrac{\sin\phi\,\sin(\phi+\beta)}{\cos\beta}}\right]^2} \quad [3\text{-}19]$$

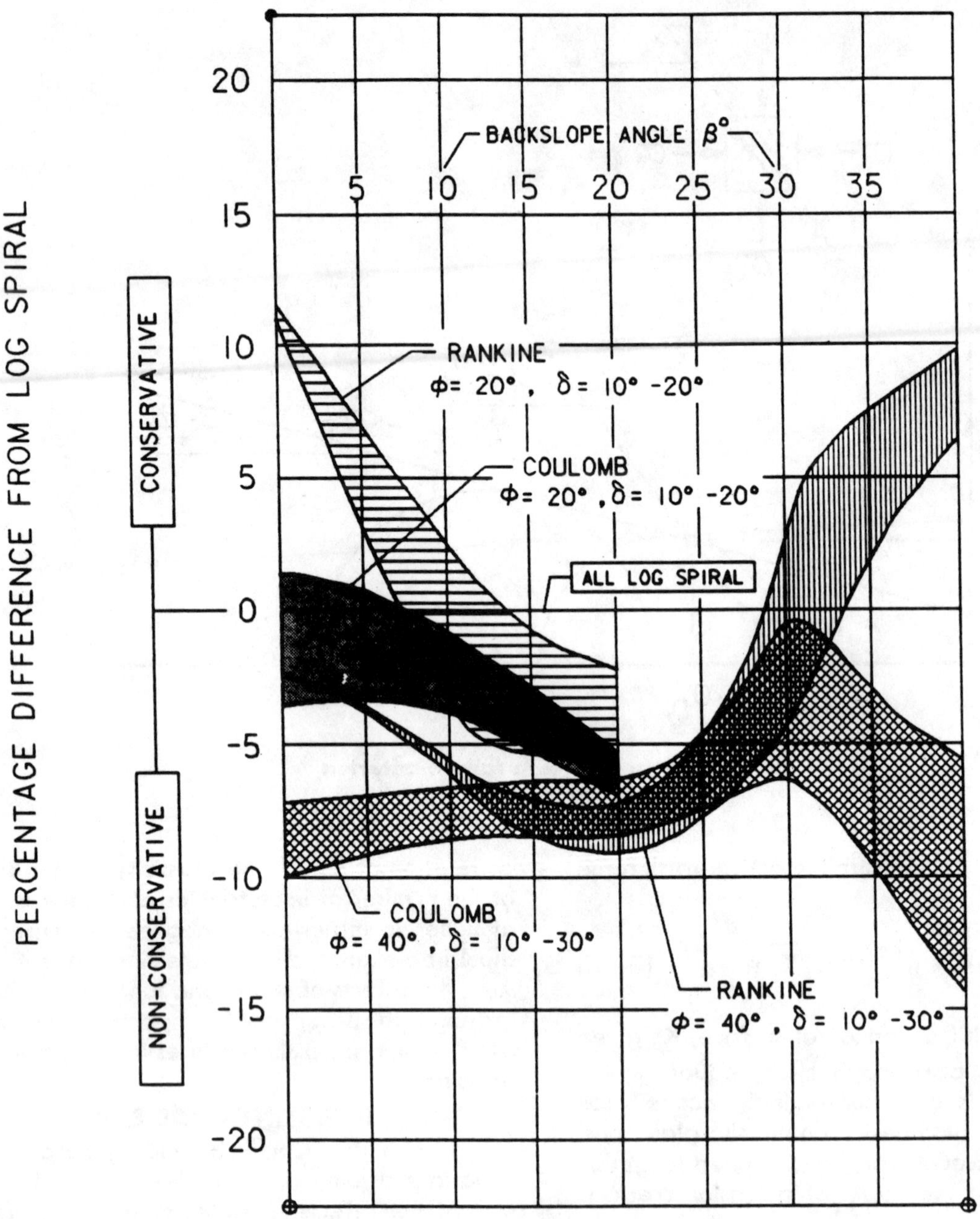

NOTE: Log spiral calculations based upon Caquot and Kerisel coefficients. Range of values for Rankine is from influence of wall friction on log spiral.

Fig 3-9. Comparison of active earth pressures (after Driscoll 1979)

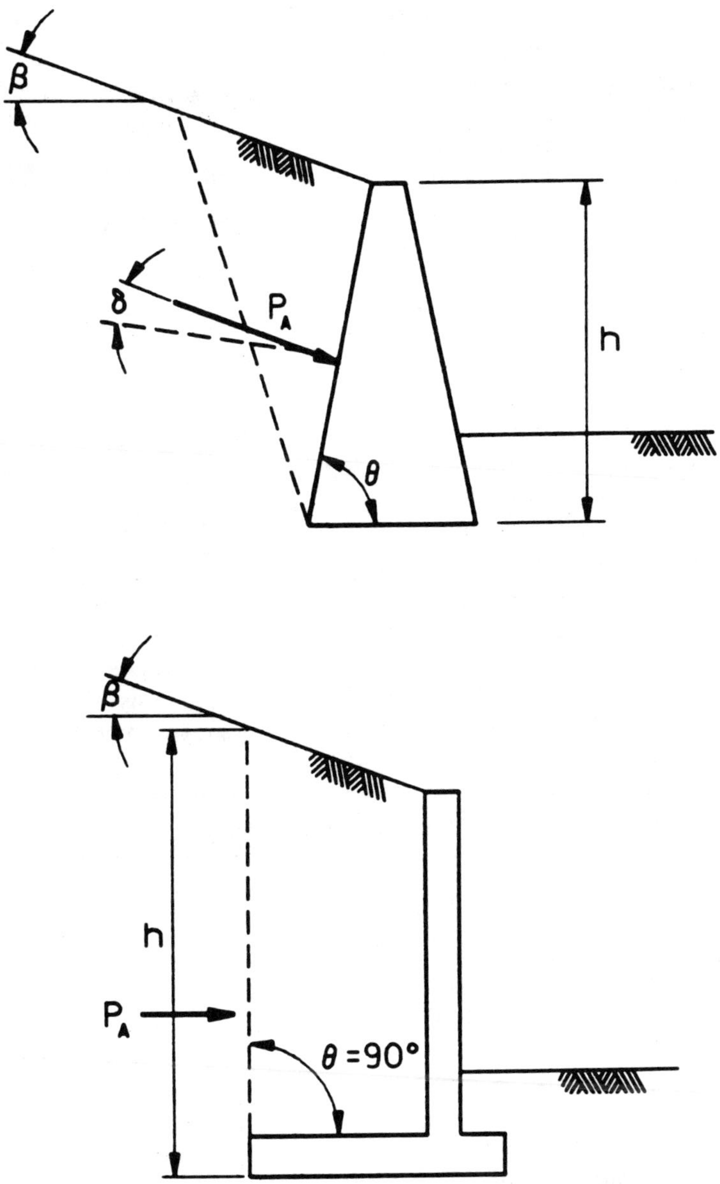

Fig 3-10. Variables used in Coulomb's equation

For a horizontal backfill ($\beta = 0$), this reduces to

$$K_P = \frac{1 + \sin\phi}{1 - \sin\phi} = \tan^2\left(45^\circ + \frac{\phi}{2}\right) = \frac{1}{K_A} \quad [3\text{-}20]$$

2. If total stress or undrained strength parameters are used and there is a cohesion term c, it has the effect of increasing the passive earth forceP_{PH}:

$$P_{PH} = \tfrac{1}{2}K_P\gamma' h^2 + 2c\sqrt{K_P}h \quad [3\text{-}21]$$

By differentiating P_{PH} with respect to the depth from the top of the resisting wedge z, the passive pressure may be obtained as:

$$P_{PH} = K_P\gamma' z + 2\sqrt{K_P}c \quad [3\text{-}22]$$

3-13. Earth Force Calculation, General Wedge Method

A. GENERAL. The general wedge method refers to a limit equilibrium analysis of a set of assumed rigid bodies (soil and/or structural elements) termed *wedges.* The horizontal earth force on the driving or resisting side of a retaining structure may be estimated by such an analysis employing properly chosen strength parameters.

Where the special conditions listed in paragraph 3-12a(1) apply, the weight of the sliding mass and orientation of the critical sliding plane are unique functions of the backfill geometry and soil properties, and Coulomb's equations provide direct solutions for the driving and resisting earth forces. Where one or more of the variables in Coulomb's equation cannot be accommodated as a single value (such as the case with multiple soils where not all of the soil layers are horizontal, location of the water table, irregular backfills, or where nonuniform surcharges are present), the critical inclination of the sliding surface and, in turn, the gravity forces (weight plus surcharges) on the sliding mass must be solved in order to calculate the horizontal earth force. In these cases, this requires a trial-and-error solution using the general wedge equation.

B. USE IN PRACTICE. When used with unfactored soil strength parameters, the general wedge equations yield the active and passive earth forces. When c and $\tan \phi$ are factored by an SMF value of 2/3, solution of the driving-side wedge provides an estimate of the at-rest earth forces (see paragraph 3-12). An SMF of 2/3 is not used to compute the resisting wedge force for the overturning, bearing, and structural design of the wall since a larger resisting force than is acceptable would be computed. See paragraph 3-8 for the procedure recommended to determine the resisting force for overturning and bearing capacity analyses and structural design of the wall.

C. DRIVING-SIDE EARTH FORCE, GENERAL WEDGE METHOD

1. Wedge Geometry and Forces. The geometry of a typical driving-side wedge and its free-body diagram are shown in Figure 3-12. The angle of wall friction and the shear force between vertical wedge boundaries are assumed to be zero. The inclination of the slip surface α is that which maximizes the earth force. Calculation of α is discussed in paragraphs 3-13c(2) and 3-13c(4). If force equilibrium is satisfied, the forces on the wedge form a closed-force polygon as shown in Figure 3-13. The equation for the effective horizontal earth force P_{EE} exerted by a driving-side wedge on a wall or an adjacent wedge is given by the general wedge equation as:

$$P_{EE} = \frac{(W + V)(1 - \tan \phi_d \cot \alpha)\tan \alpha}{1 + \tan \phi_d \tan \alpha} + \frac{U \tan \phi_d - c_d L}{\cos \alpha(1 + \tan \phi_d \tan \alpha)} + H_L - H_R - P_W \quad [3\text{-}23]$$

where

P_{EE} = effective horizontal earth force contributed by wedge or wedge segment

W = total wedge weight, including water

V = any vertical force applied to wedge

α = angle between slip plane and horizontal

U = uplift or buoyancy force acting on and normal to wedge slip plane

L = length along the slip plane of the wedge

H_L = any external horizontal force applied to the wedge from the left, acting to the right

H_R = any external horizontal force applied to the wedge from the right, acting to the left

P_W = internal water force acting on the side of the wedge free body (P_W is equal to the net difference of the water force for wedge segments with water on two vertical sides as shown in Figures 3-12 and 3-13.)

The developed strength parameters $\tan \phi_d$ and c_d are as defined in paragraph 3-11. Equation 3-23 is derived for failure occurring from left to right. All values are positive in the directions indicated in Figure 3-12. Refer to Appendix M for examples using Equation 3-23.

2. Critical Value of Slip-Plane Angle

A. The critical value of α for a driving-side wedge with a horizontal top surface and a uniform surcharge or no surcharge is:

$$\alpha = 45^\circ + \frac{\phi_d}{2} \quad [3\text{-}24]$$

B. For the special case of a backfill with a planar (flat or inclined) top surface and a strip surcharge V, the following equation can be used to compute the critical α value:

$$\alpha = \tan^{-1}\left(\frac{c_1 + \sqrt{c_1^2 + 4c_2}}{2}\right) \quad [3\text{-}25]$$

The above equation for α assumes that the backfill is completely above or completely below the water table, but can be used when the water table is anywhere within the backfill with sufficient accuracy for design. The surcharge V can have any arbitrary shape but must be contained entirely within the driving wedge. The equations for c_1 and c_2 are:

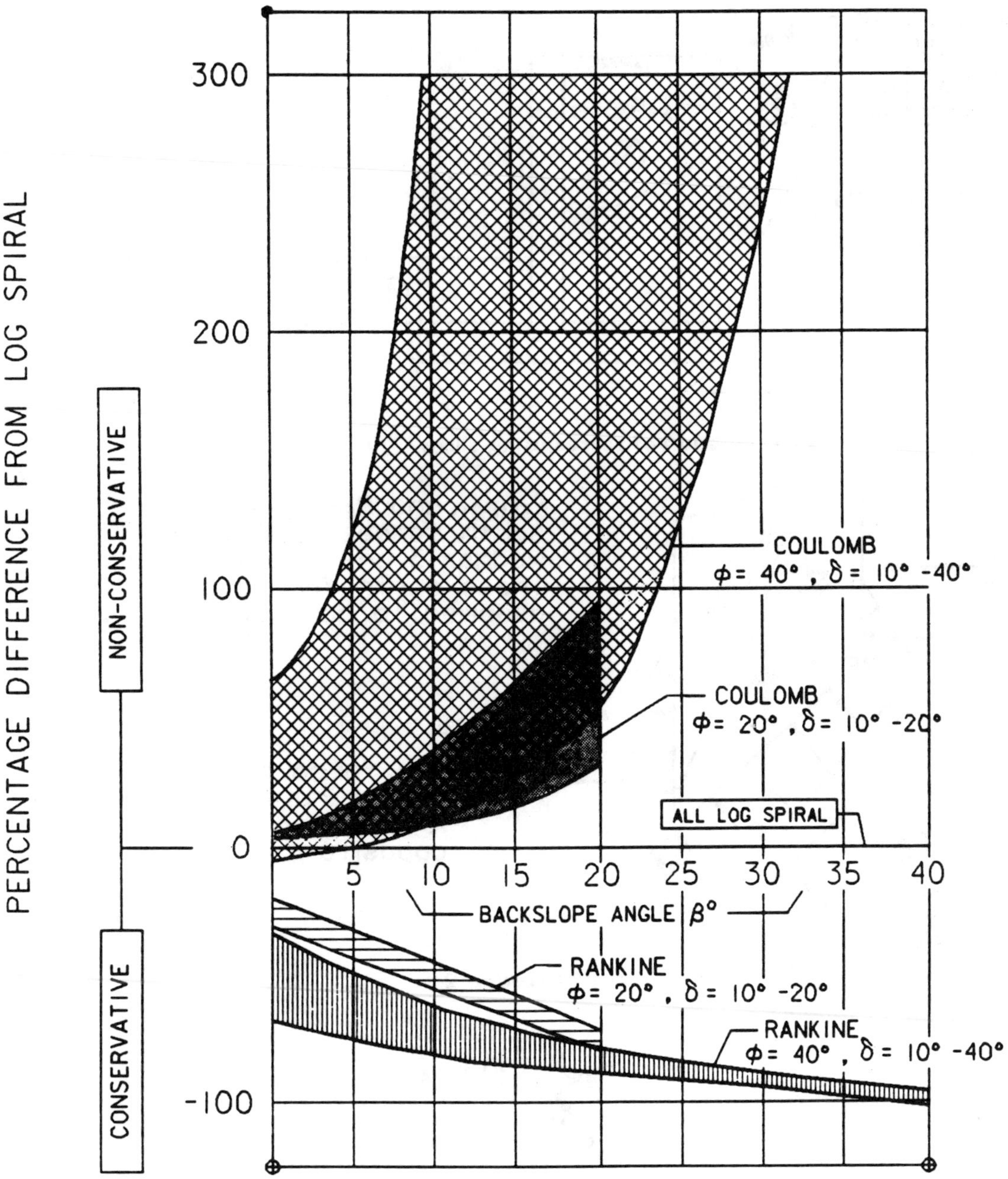

NOTE: Log spiral calculations based upon Caquot and Kerisel coefficients.

Fig 3-11. Comparison of passive earth pressures (after Driscoll 1979)

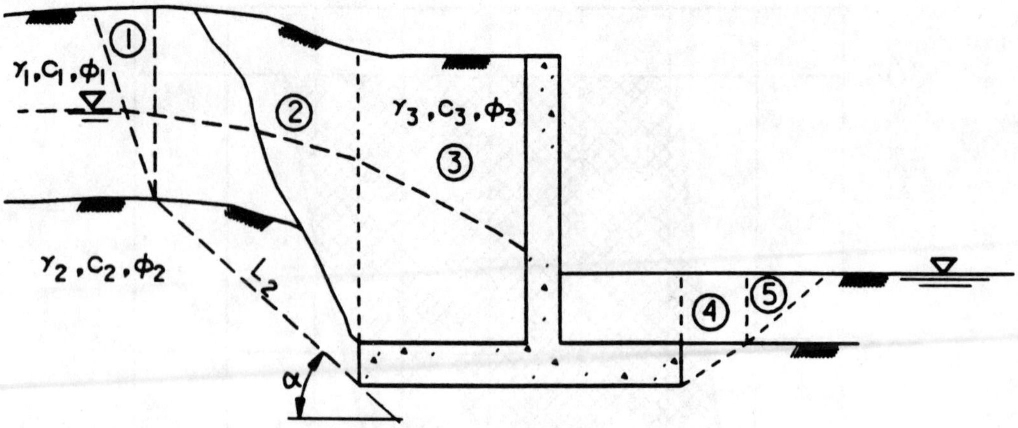

a. Estimation of forces using slices

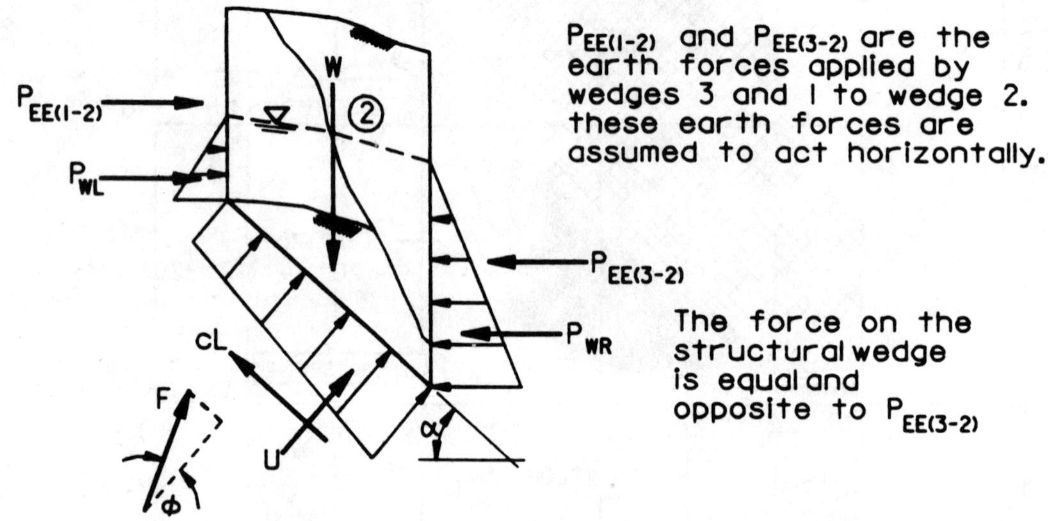

b. Free body of wedge 2

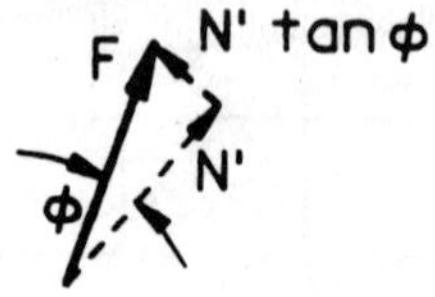

c. Derivation of force F

Fig 3-12. Wedge method on driving side

i. For a cohesionless backfill without a strip surcharge:

$$c_1 = 2 \tan \phi_d \quad [3\text{-}26]$$

$$c_2 = 1 - \tan \phi_d \tan \beta - \left(\frac{\tan \beta}{\tan \phi_d}\right) \quad [3\text{-}27]$$

ii. For a cohesive or a cohesionless backfill with a strip surcharge:

$$c_1 = \frac{2 \tan^2 \phi_d + \dfrac{4c_d(\tan \phi_d + \tan \beta)}{\gamma(h + d_c)} - \dfrac{4V \tan \beta(1 + \tan^2 \phi_d)}{\gamma(h^2 - d_c^2)}}{A} \quad [3\text{-}28]$$

$$c_2 = \frac{\tan \phi_d(1 - \tan \phi_d \tan \beta) - \tan \beta + \dfrac{2c_d(1 - \tan \phi_d \tan \beta)}{\gamma(h + d_c)} + \dfrac{2V \tan^2 \beta(1 + \tan^2 \phi_d)}{\gamma(h^2 - d_c^2)}}{A} \quad [3\text{-}29]$$

where

$$A = \tan \phi_d + \frac{2c_d(1 - \tan \phi_d \tan \beta)}{\gamma(h + d_c)} - \frac{2V(1 + \tan^2 \phi_d)}{\gamma(h^2 - d_c^2)} \quad [3\text{-}30]$$

These equations when applied to a cohesive backfill are subject to the limitations described in paragraph 3-12b(8). The derivation of these equations is shown in Appendix G. Examples using these equations are shown in Appendix M.

c. For irregular backfills, obtaining the critical inclination of the driving-side slip surface may require a trial-and-error solution. As a first approximation, the backfill surface may be bounded by two inclined lines originating from the top of the wall and the value of α may be calculated by using an "average" β value between the two bounding lines (Figure 3-14), or by introducing a surcharge as shown in Example 9 of Appendix M.

3. Limitations of Critical Slip Plane Equations. The equations for c_1 and c_2 are valid except when the strip surcharge V is too large or when the slope of the top surface is too great. The maximum value for the strip surcharge is determined by setting the denominator of the equation for c_1 or c_2 equal to zero and solving for V. This value is:

$$V_{max} = \frac{\gamma(h^2 - d_c^2)\tan \phi_d + 2c_d(h - d_c)(1 - \tan \phi_d \tan \beta)}{2(1 + \tan^2 \phi_d)} \quad [3\text{-}31]$$

When $V \geq V_{max}$ the value of α is set by the location of the strip surcharge as shown in Figure 3-15, and given by the equation

$$|\alpha| = \tan^{-1} \frac{h - d_c + (S)\tan \beta}{S} \quad [3\text{-}32]$$

Even when $V < V_{max}$, a check should be made to be certain that the entire strip surcharge lies on the top surface of the wedge as defined by the calculated value of α. Also, when $c_1^2 + 4c_2 < 0$, α is indeterminate. This is an indication that the slope of the top surface is too great to be sustained by the developed strength parameters $\tan \phi_d$ and c_d. See example 8 in Appendix M for a solution to this problem.

4. Layered Soils. The wedge equations imply a single set of strength parameters along the wedge base. For layered soils, the wedge must be divided into wedge segments, each with its base in a single soil. The wedge base inclinations α are theoretically different in every soil (Figure 3-16a); calculation of an optimum solution (maximized earth force) for the set of α values is tedious and cumbersome. Three approximate methods may be used:

a. The critical inclination in each layer may be calculated according to Equation 3-25 using the developed shear strength parameters for the soil along the wedge base and using the slope angle β at the top of each wedge segment (see Figure 3-16a).

b. The wedge segment bases may be assumed to have a constant inclination α through all materials and the critical value (corresponding to the maximum driving side force) may be calculated by trial using Equation 3-23 (see Figure 3-16b).

c. Alternatively, the critical slip-plane angle may be calculated (for each layer below the top layer) by using the procedure presented in paragraph G-7 of Appendix G (see Example 6 in Appendix M).

If the surfaces of all layers are horizontal, the critical slip plane may be determined using Equation 3-24.

5. Surcharges. The wedge method incorporates surcharge effects into the resultant earth force if the surcharge force is added to the wedge

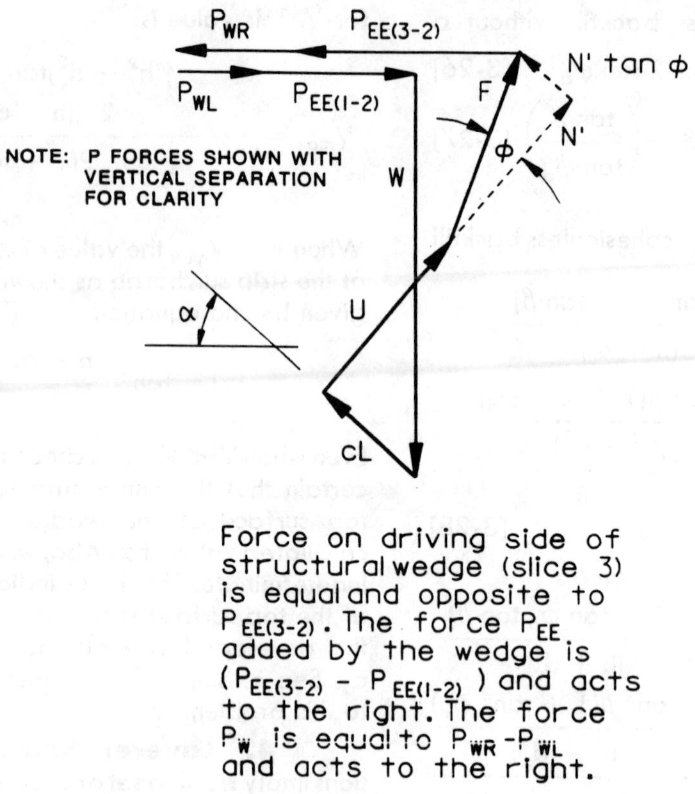

Fig 3-13. Force polygon for wedge method on driving side

weight. However, it is preferable to calculate horizontal pressures due to surcharges separately for the following reasons:

a. The presence of a nonuniform surcharge alters the principal stress directions, increasing the curvature of the slip surface, and increasing the error associated with assuming a plane surface.

b. Stresses induced by surcharges are distributed throughout a soil mass in a manner that may considerably alter the point of application and distribution of earth pressure as further described in paragraph 3-16. Limit-equilibrium techniques and the earth pressure coefficient concept do not accurately predict such distributions.

c. The additional pressures developed on the wall depend on the amount of wall movement and may be twice as great for nonyielding walls as for yielding walls.

The intent of this manual is to consider walls to be relatively rigid and to design for at-rest conditions. Therefore, pressures and forces due to non-uniform surcharges should be calculated in accordance with paragraph 3-16, adding the results to the pressures and/or forces obtained from Coulomb's equation or the general wedge equation. For the sliding analysis, surcharge effects may be included directly in the wedge method weight calculations because the sliding analysis considers only force equilibrium; thus, the point of application of the forces does not matter. Examples 4 and 10 of Appendix M demonstrate the calculation of horizontal pressures involving surcharges.

6. Pressure Coefficients

a. Structural engineers are familiar with the use of Coulomb's equations (paragraph 3-12) for the determination of earth pressure coefficients and the use of these coefficients in determining pressures and forces acting on retaining walls. These equations suffer from several limitations as discussed in paragraph 3-12a(1). The general wedge equation (Equation 3-23) is not subject to any of the limitations of Coulomb's equations and may be used to solve for the lateral earth force on a wedge due to complicated geometry and surface loading. If lateral earth pressure coefficients are derived from the general wedge equation, these coefficients may be used in a rather simple manner to solve complex earth pressure problems.

b. Earth pressures can be calculated from general wedge method solutions by assuming that pressures vary in a piecewise linear fashion and that the slopes of the pressure diagrams are the product of densities and pressure coefficients (K).

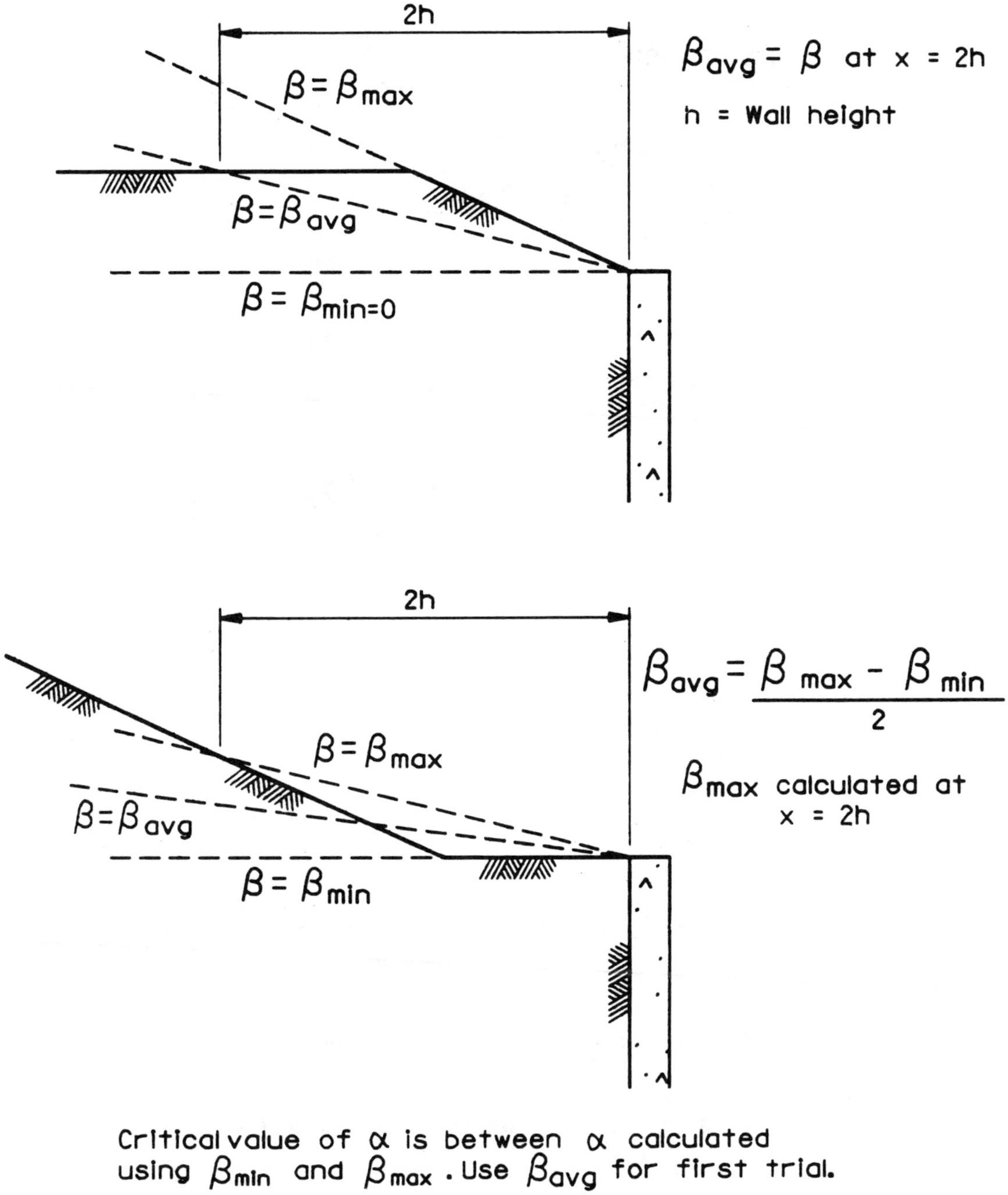

Fig 3-14. Wedge analysis for irregular backfill

The slopes may be considered the density of an "equivalent fluid" loading the wall. These pressure coefficients are dependent on the problem geometry and are derived in Appendix H. It should be noted that pressure coefficients (K values) below the water table may differ from those above the water table in the same material as shown in Appendix H. One example where the K value is different above and below the water table is the case of a sloping backfill. Examples using pressure coefficients are shown in Appendix M.

D. RESISTING-SIDE EARTH FORCE, GENERAL WEDGE METHOD

1. Wedge Geometry and Forces.

The geometry of a typical resisting-side wedge and its free-body diagram are shown in Figure 3-17. The angle of wall friction and the shear force between vertical wedge boundaries are assumed to be zero. The inclination of the slip surface α is that which minimizes the earth force. Calculation of α is discussed in paragraph 3-13d(2). If force equilibrium is satisfied, the forces on the wedge

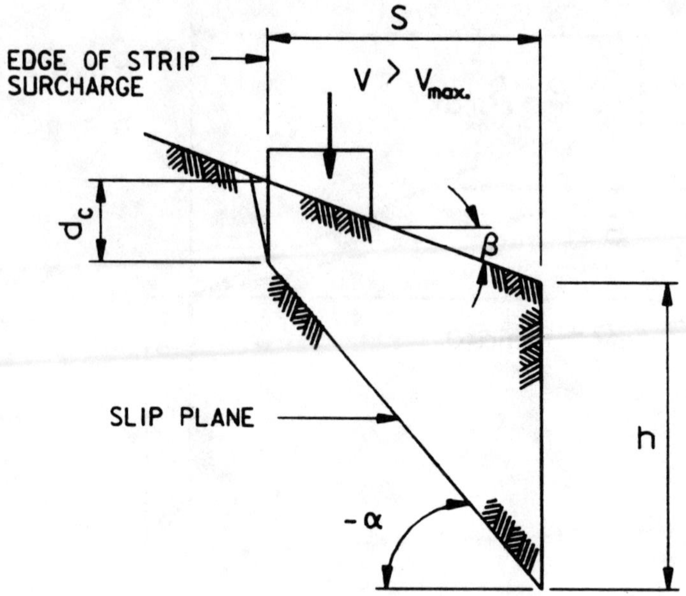

$$S = \frac{h - d_c}{\tan\alpha - \tan\beta}$$

$$\tan\alpha = \frac{h - d_c}{S} + \tan\beta = \frac{h - d_c + S\tan\beta}{S}$$

$$|\alpha| = \tan^{-1} = \left(\frac{h - d_c + S\tan\beta}{S}\right)$$

Fig 3-15. Surcharge effect on critical slip plane

form a closed force polygon as shown in Figure 3-17. The equation for the horizontal effective earth force P_{EE} exerted by a resisting-side wedge on a wall or an adjacent wedge is:

$$P_{EE} = \frac{(W + V)(1 + \tan\phi_d \cot\alpha)\tan\alpha}{1 - \tan\phi_d \tan\alpha} - \frac{U\tan\phi_d - c_d L}{\cos\alpha(1 - \tan\phi_d \tan\alpha)} - H_L + H_R - P_W \qquad [3\text{-}33]$$

where the terms are the same as for the driving-side wedge equation (Equation 3-23). Equation 3-33 is derived for failure occurring from left to right. All values are positive in the directions indicated in Figure 3-17.

2. Critical Value of Slip-Plane Angle

a. For a resisting-side wedge with a horizontal top surface, α can be computed as follows:

$$\alpha = 45^\circ - \frac{\phi_d}{2} \qquad [3\text{-}34]$$

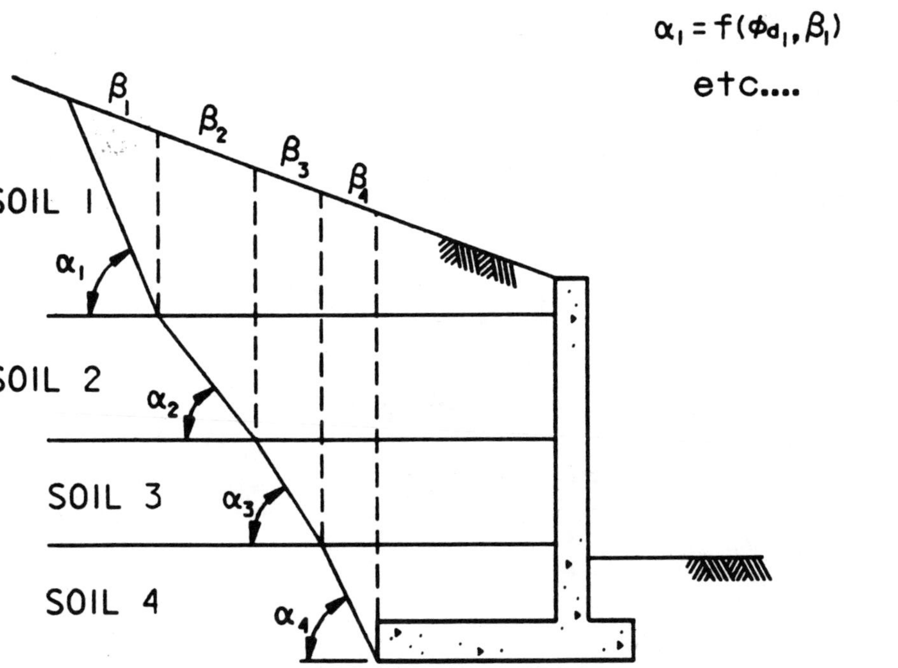

a. Wedge 1

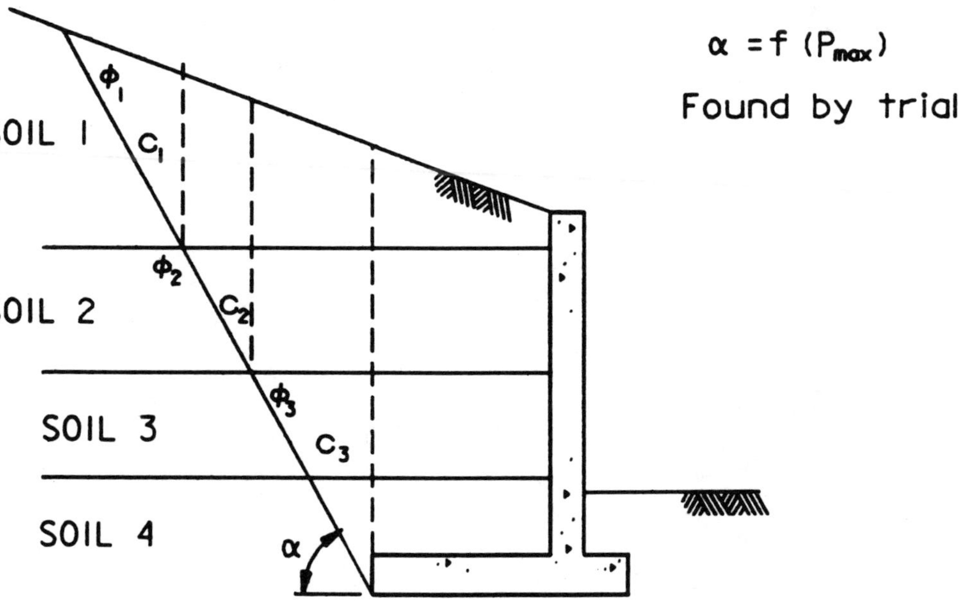

b. Wedge 2

Fig 3-16. Wedge analysis in layered soil

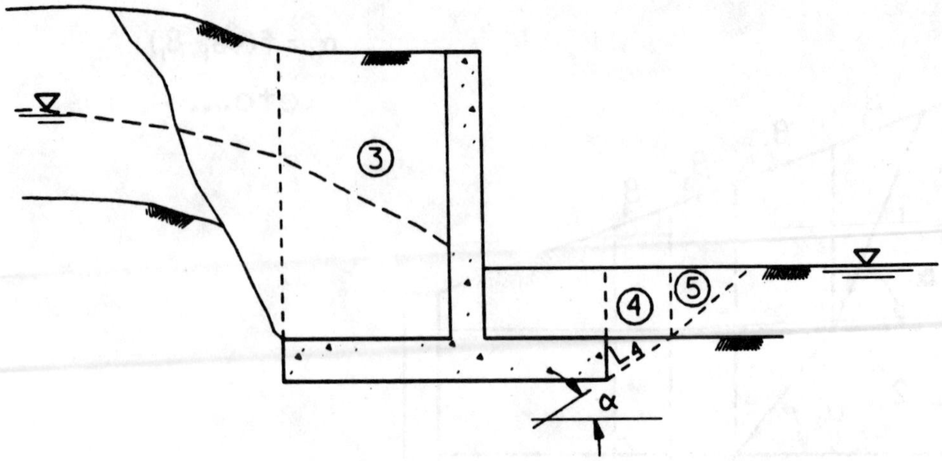

a. Estimation of forces using slices

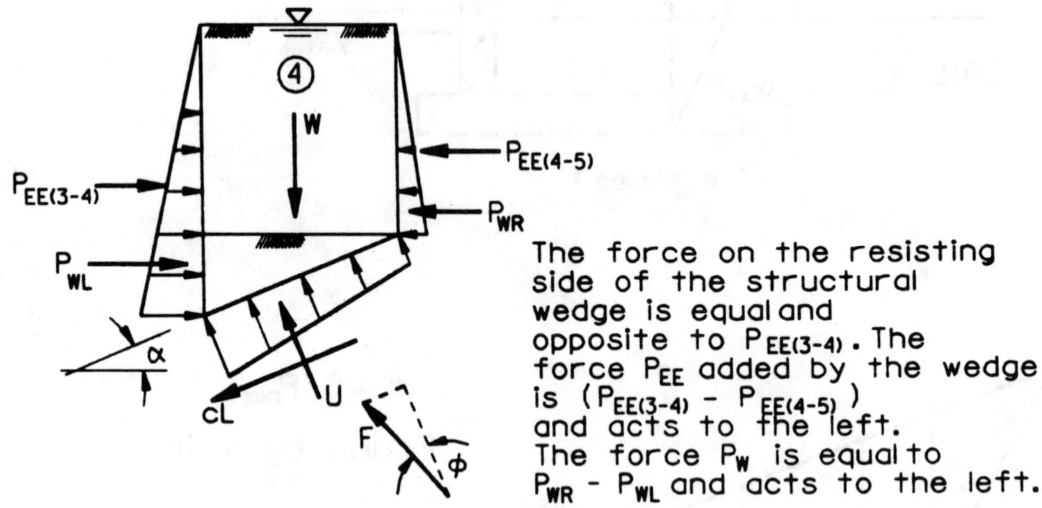

b. Free-body diagram of slice 4

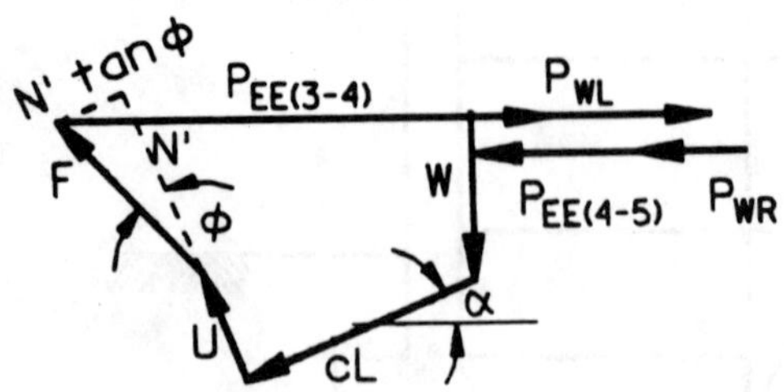

c. Force polygon for slice 4

Fig 3-17. Wedge method for resisting-side wedge

b. The critical angle α for a resisting-side wedge with a planar (flat or inclined) top surface, with no surcharge or with a strip surcharge V , is given by the equation:

$$\alpha = \tan^{-1}\left[\frac{-c_1 + \sqrt{c_1^2 + 4c_2}}{2}\right] \quad [3\text{-}35]$$

For a resisting-side wedge, the equations for c_1 and c_2 are

$$c_1 = \frac{2\tan^2\phi_d + \dfrac{4c_d(\tan\phi_d - \tan\beta)}{\gamma h} - \dfrac{4V\tan\beta(1+\tan^2\phi_d)}{\gamma h^2}}{A} \quad [3\text{-}36]$$

$$c_2 = \frac{\tan\phi_d(1 + \tan\phi_d\tan\beta) + \tan\beta + \dfrac{2c_d(1+\tan\phi_d\tan\beta)}{\gamma h} - \dfrac{2V\tan^2\beta(1+\tan^2\phi_d)}{\gamma h^2}}{A} \quad [3\text{-}37]$$

where

$$A = \tan\phi_d + \frac{2c_d(1+\tan\phi_d\tan\beta)}{\gamma h} + \frac{2V(1+\tan^2\phi_d)}{\gamma h^2} \quad [3\text{-}38]$$

3. Surcharges. The comments regarding surcharges in paragraph 3-13c(5) relative to analysis of driving-side wedges also apply in general to resisting-side wedges. However, surcharges on resisting-side wedges tend to enhance stability and therefore it is conservative to neglect them in analysis. If resisting-side surcharges are not neglected, it must be assured that the surcharge loading will be in place for the condition analyzed.

4. Pressure Coefficients. Earth pressures for the resisting side may be calculated as equivalent fluid pressures in a manner similar to that for the driving side. See paragraph 3-13c(6) and Appendix H for further discussion.

3-14. Earth Pressure Calculations Including Wall Friction

A. DRIVING SIDE. Friction between the backfill and wall, or on a plane within the backfill, of up to one-half of the internal friction angle (unfactored) of the backfill material may be used in the design.

B. RESISTING SIDE. When wall friction is included in the analysis, assuming the slip surface to be a log-spiral or other curved surface provides lower and more reasonable values for the passive force and passive pressure coefficient K_P (see Figure 3-11). Although the angle of wall friction should generally be taken as zero, it may be assumed greater than zero where movement and settlement of the wall are expected and permissible. Figure 3-18 provides earth pressure coefficients for horizontal backfills based on the work of Caquot and Kerisel (1948) and Shields and Tolunay (1973).

3-15. Distribution of Horizontal Earth Pressure

A. SUPERPOSITION OF PRESSURES. The distribution of total horizontal pressure on the driving or resisting side is obtained by superposing the distributions due to horizontal effective earth pressure, water, and surcharges. Where compaction efforts are specified, horizontal earth pressures should be calculated in accordance with paragraph 3-17.

B. SOILS COMPLETELY ABOVE OR COMPLETELY BELOW THE WATER TABLE. The effective earth pressure may be assumed to have a triangular distribution when all of the following conditions hold:

1. The wall will not move or it will rotate about the base.

2. The water table is at or below the base of the wall or at or above the top of the wall (submerged soil).

3. Water conditions are hydrostatic (no seepage).

4. There is only one soil material.

5. There is no cohesion ($c = 0$).

6. The backfill surface is plane (it may be inclined).

The distribution is given by:

$$p'_{hz} = K\gamma' z \quad [3\text{-}39]$$

where

$K = K_o$ on the driving side. K for the resisting side could vary between K_P and K_o or could be taken as zero.

γ' = the effective unit weight (total, saturated or moist unit weight if above the water table, buoyant or submerged unit weight if below the water table)

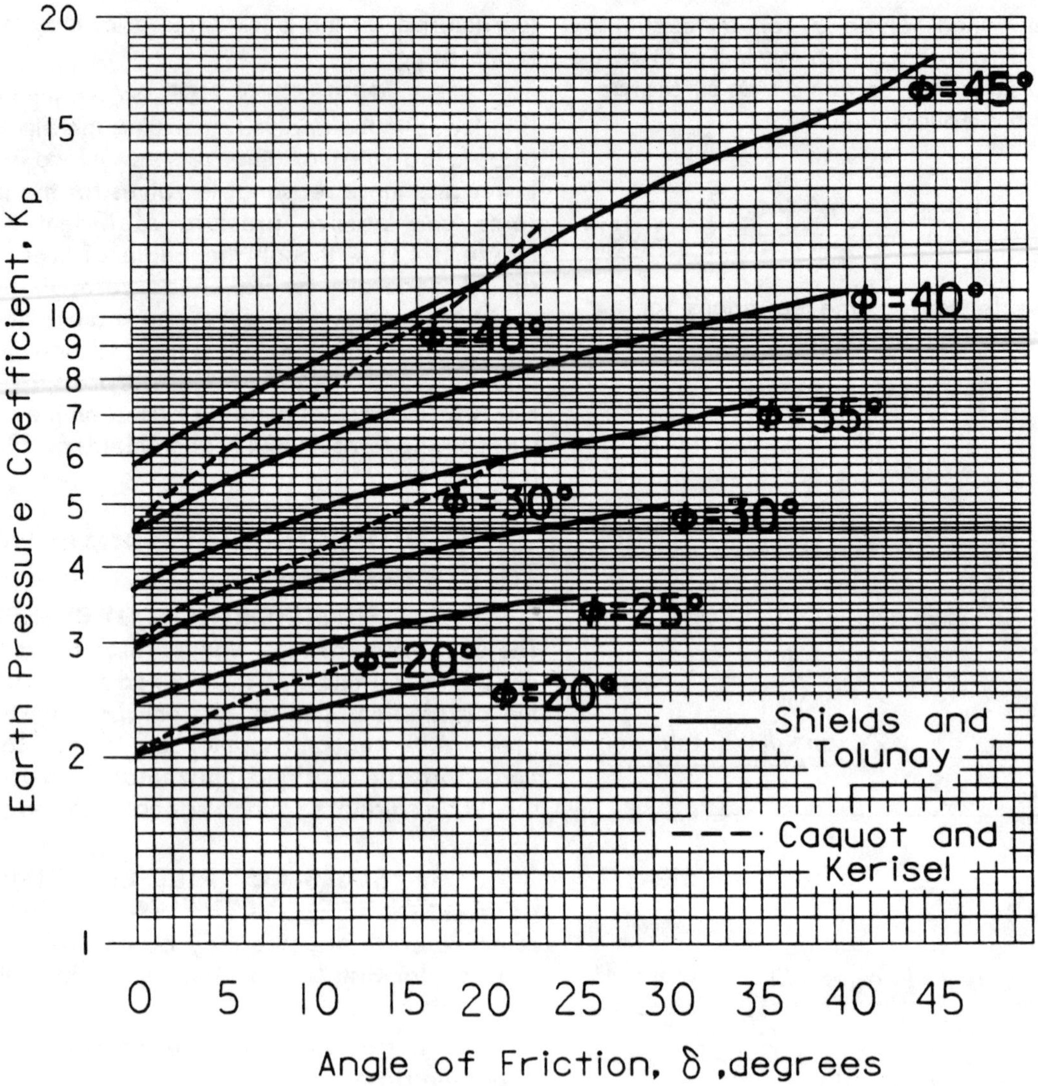

Fig 3-18. Passive earth pressure coefficients

z = vertical distance measured down from the backfill surface

See Figure 3-19 for an example.

C. PARTLY SUBMERGED SOILS. Where the water table occurs between the top and the base of the wall, and only one soil is present, the top portion of the pressure diagram is a triangle given by Equation 3-39 and the bottom is a trapezoid given by

$$p'_{hz} = K[\gamma z_w + \gamma'(z - z_w)] \qquad [3\text{-}40]$$

where

z_w = depth to water table
$\gamma' = (\gamma - \gamma_w)$ below the water table

An example is shown in Figure 3-20. See Appendix H where water table and backfill surface are not parallel.

D. LAYERED SOILS. Where layered soils are present, the pressure diagram is a triangle underlain by a series of trapezoids given by

$$p'_{hz} = K_i(p'_{vi} + \gamma'_i z_i) \qquad [3\text{-}41]$$

where

K_i = horizontal earth pressure coefficient for the ith layer

p'_{vi} = vertical effective earth pressure at the top of the ith layer

γ'_i = effective unit weight of the ith layer

z_i = vertical distance measured down from the top of the ith layer

An example is shown in Figure 3-21.

E. IRREGULAR BACKFILLS. Where the backfill is irregular, the pressure diagram may be

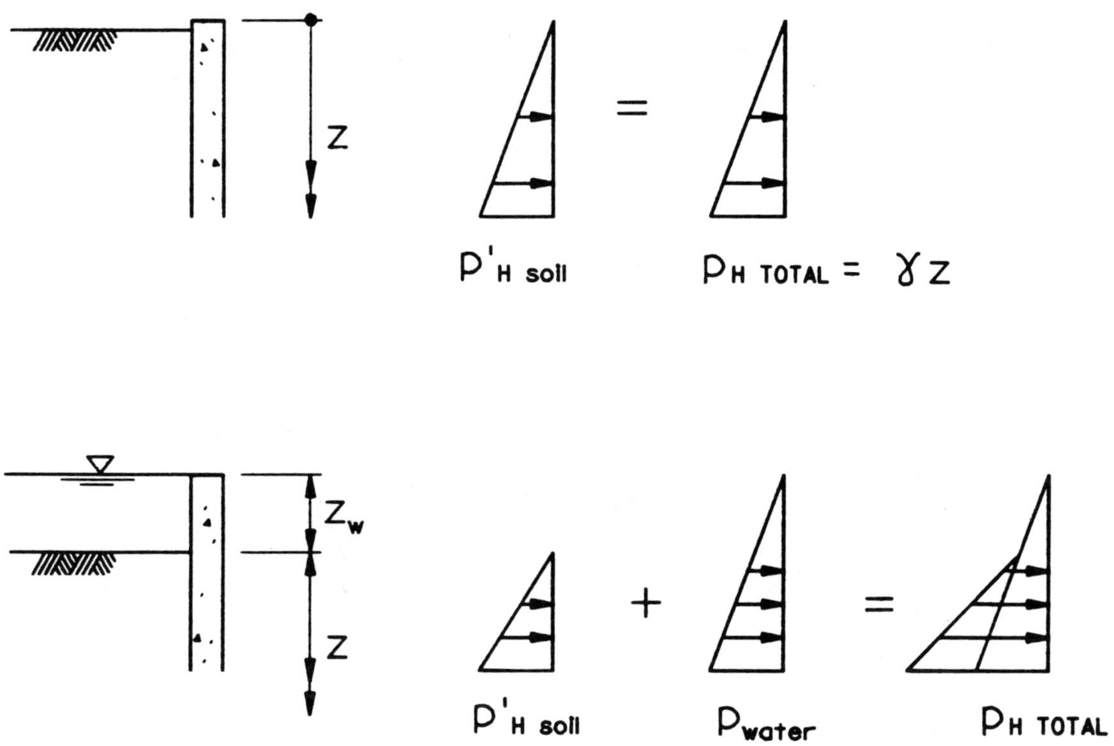

$$P_v = P_v + P_{water}$$
$$= \gamma_w z_w + \gamma z$$
$$P'_v = P_v - P_{water}$$
$$= \gamma_w z_w + \gamma z - \gamma_w (z_w + z)$$
$$= (\gamma - \gamma_w) z$$
$$= \gamma' z$$
$$P'_H = K \gamma' z$$

Fig 3-19. Lateral pressures, one soil completely above water table or completely below water table

estimated by performing successive wedge analyses at incremental depths from the top of the wall and applying the force difference from successive analyses over the corresponding vertical area increment (Agostinelli et al. 1981). Since this procedure is approximate, increasing the number of calculation points does not necessarily increase accuracy. An example of this procedure is shown in Figure 3-22. The pressure diagram may also be estimated by the use of pressure coefficients (see paragraphs 3-13c[6]) as shown in Examples 7, 8, and 9 of Appendix M.

F. COHESION EFFECTS

1. Where the backfill is horizontal and where cohesion is present, its theoretical effect is to reduce the driving side earth pressure by $2c_d\sqrt{K_A}$ for the entire depth of the soil layer (see Equation 3-18). This infers tension in the soil to a "crack depth" d_c where

$$d_c = \frac{2c_d}{\gamma'\sqrt{K_A}} \qquad [3\text{-}42]$$

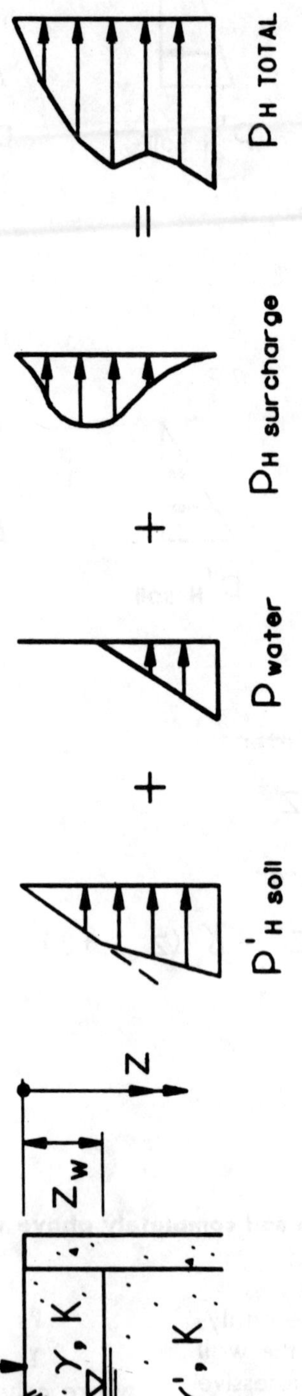

Fig 3-20. Lateral pressures, one soil, water, finite surcharge

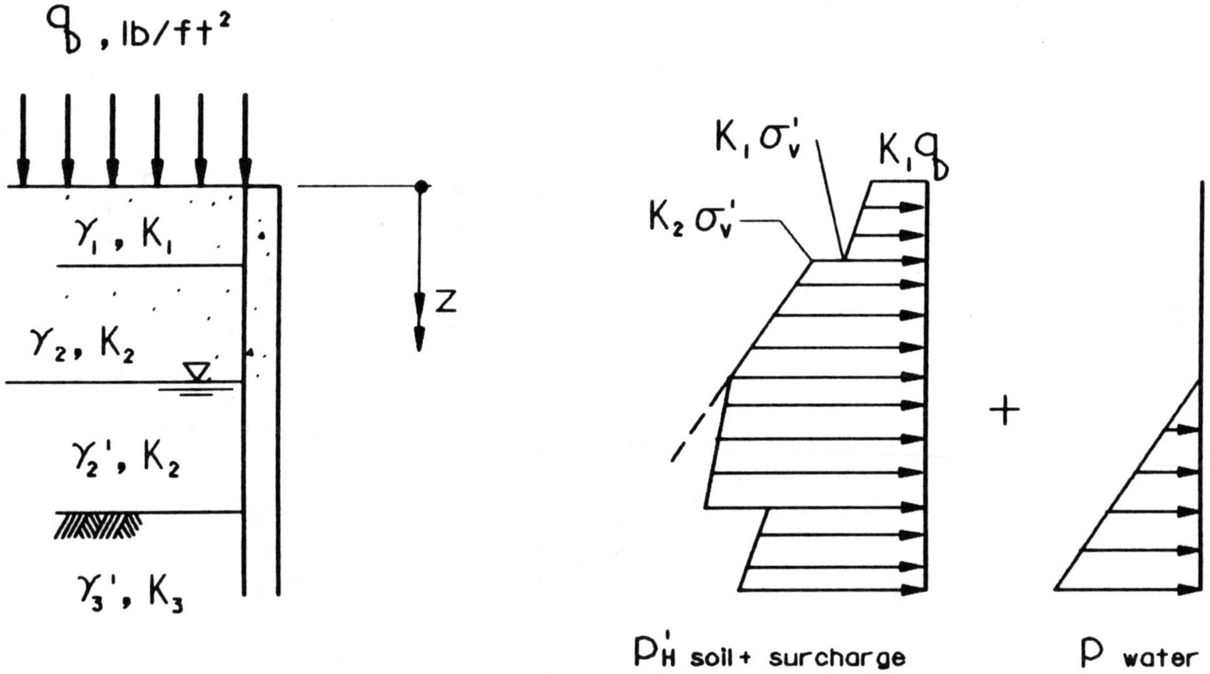

Fig 3-21. Lateral pressures, three soils, water

Consequently there is zero load on the wall in this region. For sloping backfills, see Appendices H and I. Where cohesion is present, a water-filled tension crack should be considered in the inferred tension zone. The maximum crack depth using the unfactored c value should also be checked. Where the horizontal earth force is calculated from a pressure diagram that includes negative pressure, the force reduction due to the inferred negative pressure zone should be taken as zero. The pressure on the driving side should be computed using Equation 3-18 by setting the 2nd term equal to 0 and K_A equal to K_o and using the pressure distribution as shown in Figure 3-23.

2. For the resisting side, passive pressure theory indicates no tension crack will form and the pressure would be calculated using Equation 3-22. The pressure distribution for a cohesive soil on the resisting side of a structure is shown in Figure 3-24. However, for operating conditions without movement, a tension crack may form due to moisture loss reducing or eliminating the resisting side pressure. See paragraph 3-8 for resisting pressure to be used for design.

3. Refer to guidance on use of cohesive materials in paragraph 3-5.

G. WALL MOVEMENT EFFECTS

1. Where the expected mode of wall movement is translation and/or rotation about a point other than the base (such as for braced walls) the value of K varies with depth and the horizontal earth pressure distribution will be parabolic rather than triangular. Solution methods for such conditions are less reliable than those for rotation about the base. Available methods include Rendulic's procedure (Winterkorn and Fang 1975), Dubrova's procedure (Harr 1977), and a procedure given by Wu (1966).

2. Where the expected mode of wall movement is translation and/or rotation about a point other than the base, the force may be assumed the same as that obtained for rotation about the base, but the point of application should be taken at 45 percent of the wall height above the base.

3-16. Surcharge Effects

A. UNIFORM SURCHARGES. Where uniform surcharges (q) are present, the vertical effective stress increases by the amount of the surcharge and the horizontal earth pressure diagram is a trapezoid given by

$$p'_{hz} = K(q + \gamma' z) \qquad [3\text{-}43]$$

An example is shown in Figure 3-21.

B. FINITE SURCHARGES

1. Pressure Increase Due to Finite Surcharges. The distribution of the horizontal pressure increase due to finite surcharges should be calculated using experimentally modified elastic

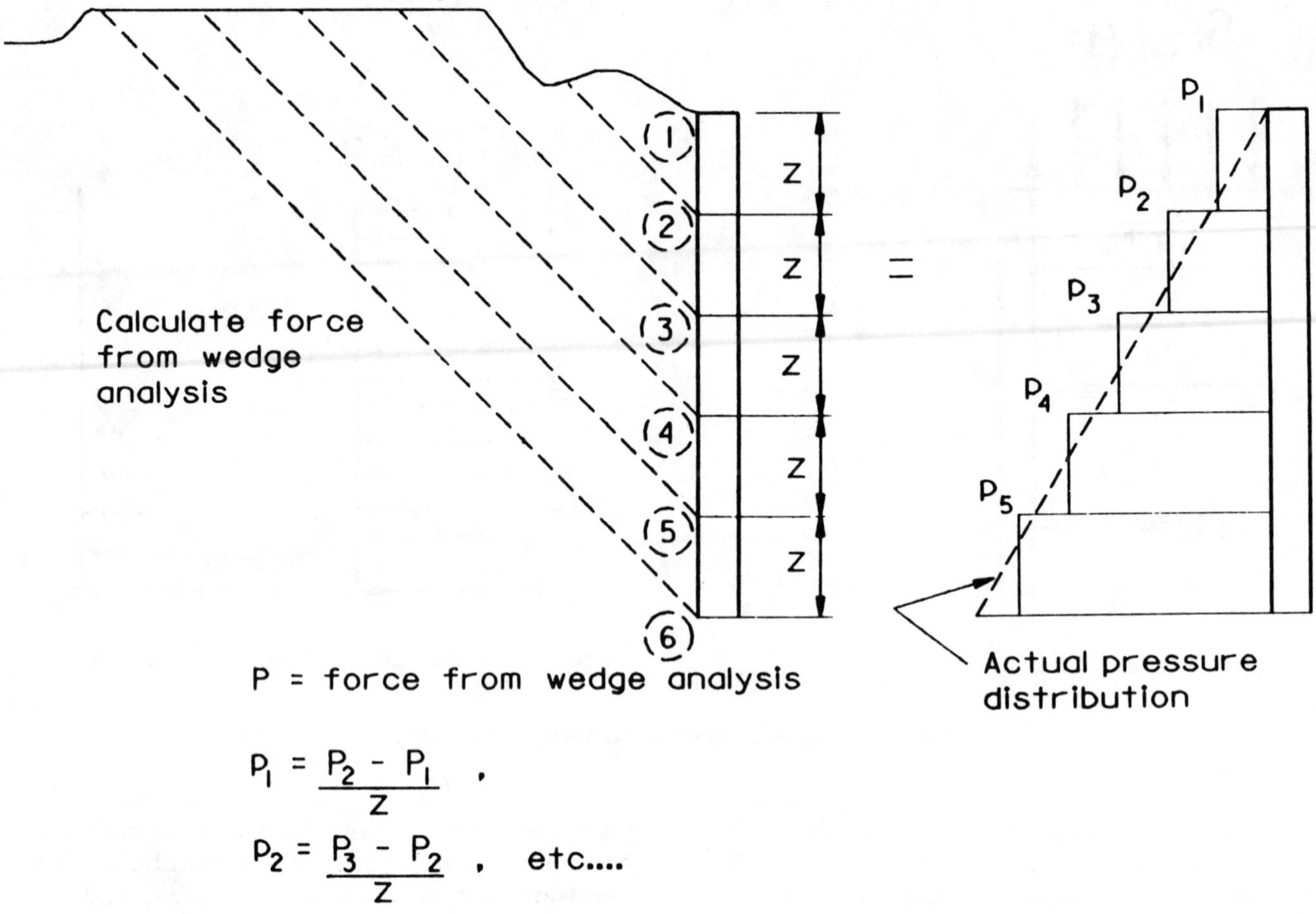

Fig 3-22. Pressure distribution due to irregular backfill

theory where expected (or allowable) strains due to the surcharge are small. Pressures due to point and line loads can be calculated using Figures 3-25 and 3-26, respectively. The resulting pressures are about twice as great as would be obtained from either unadjusted elastic solutions or limit-equilibrium solutions. This difference is due to wall rigidity not considered in elastic or limit-equilibrium methods. Pressures due to strip loads can be calculated using Figure 3-27. Pressures due to strip surcharge loads of more general shapes can be calculated by applying the principle of superposition to these solutions; a pressure-intensity curve of any shape can be modeled to any desired degree of accuracy as the sum of point, line, or strip loads. Example computations involving surcharges are shown in examples 4 and 10 of Appendix M.

2. Force Due to Finite Surcharges. Point, line, or nonuniform (finite) surcharge loads are supported by distribution or "diffusion" of stresses within the backfill material. These result in a curved pressure diagram; the point of application for the horizontal force resultant due to point or line loads is given in Figure 3-28. Where surcharge pressure distributions of a general shape have been modeled by superposition of these basic solutions, the point of application is found by dividing the total moment due to the surcharge resultants by the sum of the surcharge resultants. Where surcharge loadings are included in a wedge-method analysis, the difference in resultant force due to the surcharge (ΔP_H) should be applied at a different point on the wall from the resultant due to backfill weight. An approximate method for locating the line of action for a line load (Terzaghi 1943) is shown in Figure 3-29. An example using this approximate method is shown in example 4 of Appendix M.

3-17. Earth Pressures Due to Compaction

The use of heavy rollers for compaction adjacent to walls can induce high residual pressures against the wall. Although a reasonable degree of compaction is necessary to provide adequate shear

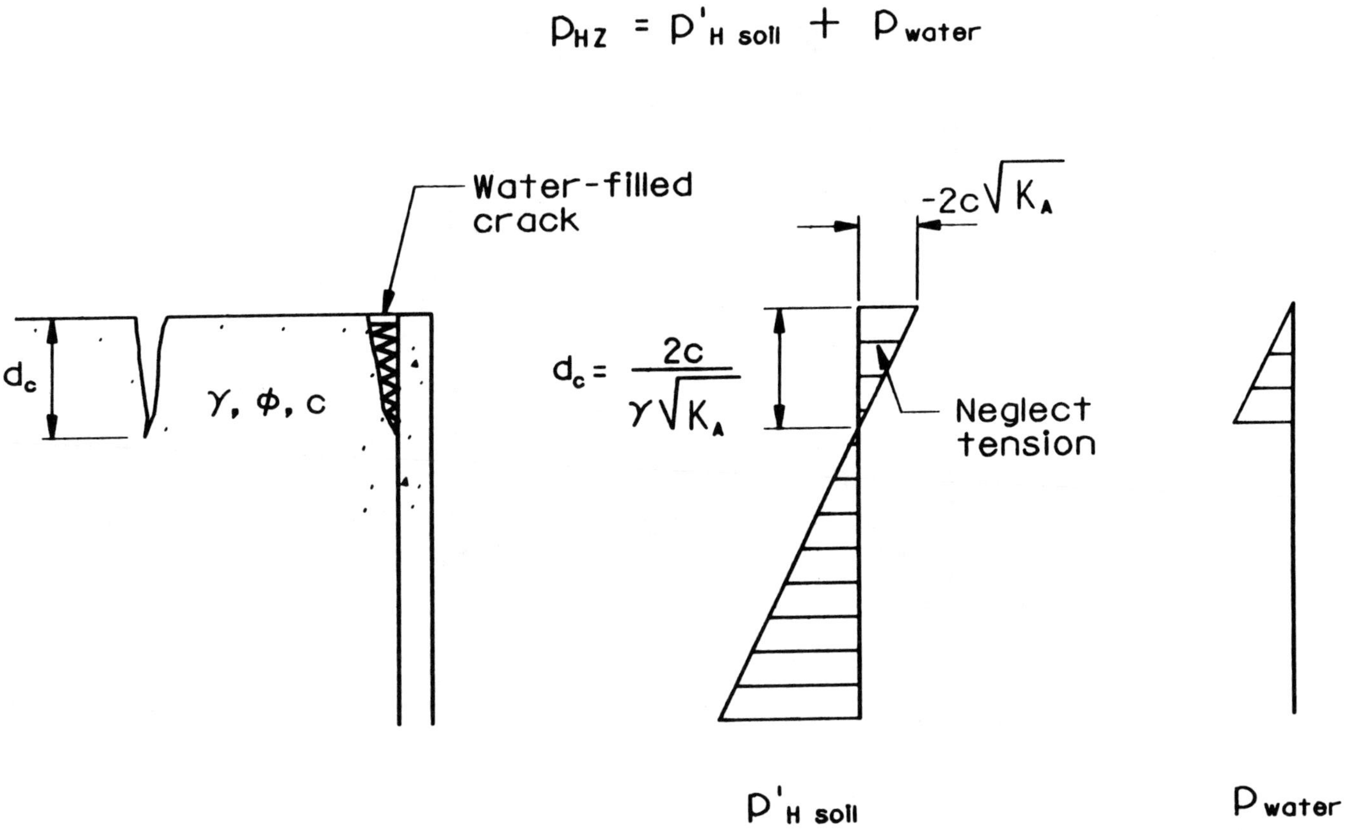

Fig 3-23. Lateral pressure distribution; active case, soil with cohesion

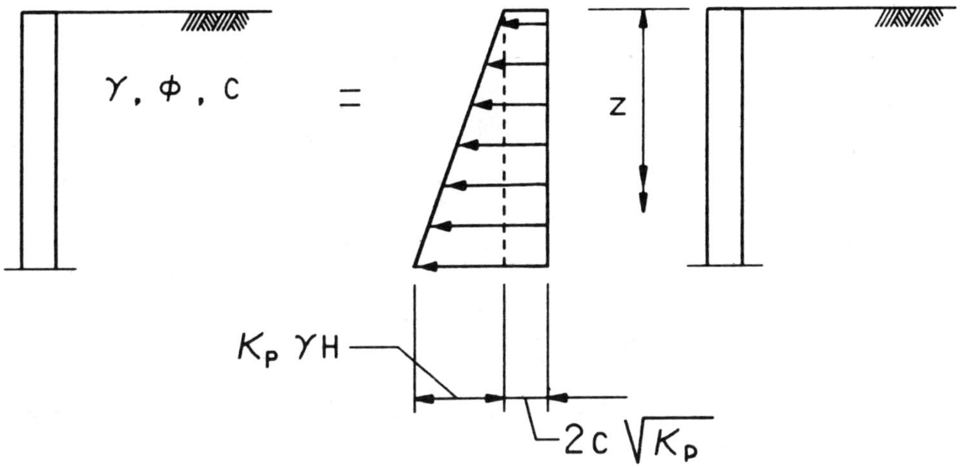

$$p_{PH} = K_p\,\gamma' z + 2c\sqrt{K_p}$$

Fig 3-24. Lateral pressure distribution; passive case, soil with cohesion

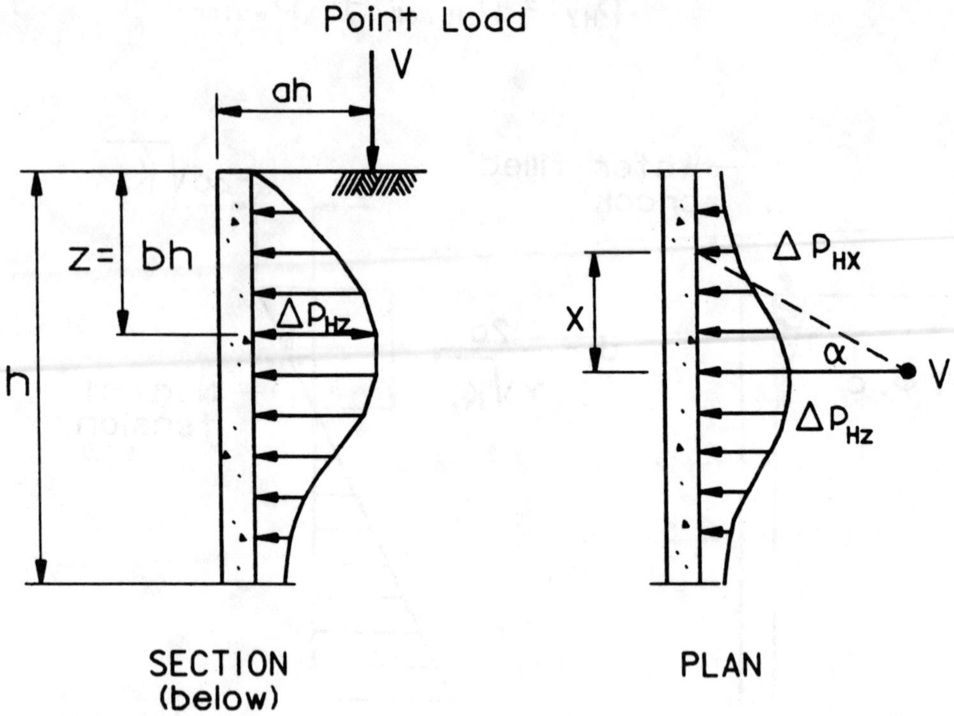

Increase in horizontal pressure, ΔP_{Hz}, on a section through point load, V

$$\Delta P_{Hz} = \left(\frac{V}{h^2}\right)\left(\frac{a^2 b^2}{(a^2 + b^2)^3}\right) \quad \text{for } a > 0.4$$

$$\Delta P_{Hz} = \left(\frac{0.28V}{h^2}\right)\left(\frac{b^2}{(0.16 + b^2)^3}\right) \quad \text{for } a \leq 0.4$$

Increase in horizontal pressure, ΔP_{HX}, at distance X from plane of load, V

$$\alpha = \tan^{-1}\left(\frac{X}{ah}\right)$$

$$\Delta P_{HX} = \Delta P_{Hz} \cos^2(1.1\alpha)$$

Fig 3-25. Increase in pressure due to point load (after Spangler and Mickle 1956)

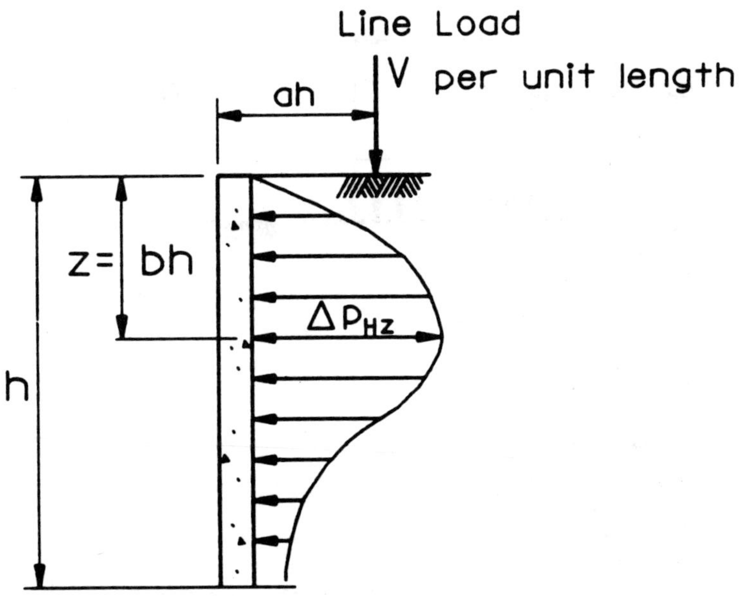

Increase in horizontal pressure, ΔP_{Hz}, at depth bH due to line load, V/ft at distance aH from wall.

$$\Delta P_{Hz} = \left(\frac{4V}{\pi h}\right)\left(\frac{a^2 b}{(a^2 + b^2)^2}\right) \quad \text{for } a > 0.4$$

$$\Delta P_{Hz} = \left(\frac{V}{h}\right)\left(\frac{0.203\, b}{(0.16 + b^2)^2}\right) \quad \text{for } a \leq 0.4$$

Fig 3-26. Increase in pressure due to line load (after Spangler and Mickle 1956)

strength and minimize settlement, excess backfill compaction should be avoided. Ingold (1979a,b) proposed a procedure for estimating lateral pressures due to compaction that has been modified herein (Appendix J) to account for walls designed for at-rest conditions. The roller is assumed to exert a line load of P lb/ft obtained from the roller weight and drum dimensions; double this value is recommended for vibratory rollers. The design pressure diagram (Figure 3-30) is composed of three linear segments:

A Starting at the top of the wall, the pressure increases linearly to a value of p'_{hm} at a depth z_{cr}. In this region, the horizontal stress is increased during compaction due to the roller pressure but then the horizontal stress is reduced by passive failure when the roller is removed.

B. The horizontal pressure is constant with depth from z_{cr} to z_2 and is compaction induced.

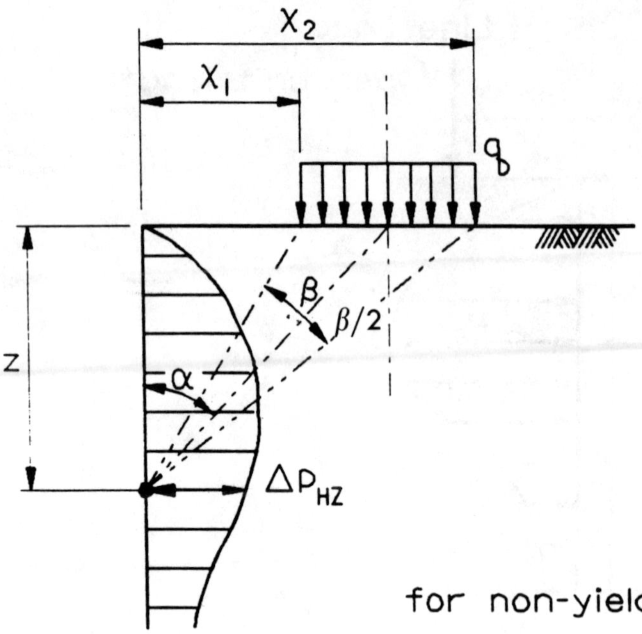

$$q = \frac{V}{(X_2 - X_1)}$$

for non-yielding walls:

$$\Delta P_{HZ} = \frac{2q}{\pi}\left(\beta - \sin\beta \ \cos 2\alpha\right)$$

for yielding walls,
(walls at failure):

$$\Delta P_{HZ} = \frac{q}{\pi}\left(\beta - \sin\beta \ \cos 2\alpha\right)$$

β in radians

$$\alpha = \frac{\tan^{-1}\left(\frac{\chi_2}{z}\right)}{2} + \frac{\tan^{-1}\left(\frac{\chi_1}{z}\right)}{2}$$

$$\alpha = \tan^{-1}\left(\frac{\chi}{z}\right) + \beta/2$$

Fig 3-27. Increase in pressure due to strip load

C. At depth z_2, the compaction-induced pressure equals the horizontal pressure due to soil weight (at-rest pressure). The pressure increases linearly below this depth according to the equations in paragraph 3-15.

Compaction-induced pressures need only be considered for structural design. For overturning, bearing, and sliding analyses, any wall movement due to compaction-induced pressures would be accompanied by a reduction in the pressure. As shown by the calculations in Appendix J, horizontal pressures due to compaction may exceed the at-rest pressure in only the upper few feet unless roller loads are particularly high. The effects of compaction are shown in Example 1 of Appendix M.

Section III. Water Pressures

3-18. Pressure Calculations

In all cases, water pressures at a point may be calculated by multiplying the pressure head at the point by the unit weight of water (62.4 lb/cu ft). As water has no shear strength, water pressures are equal in all directions (K = 1.0). The pressure

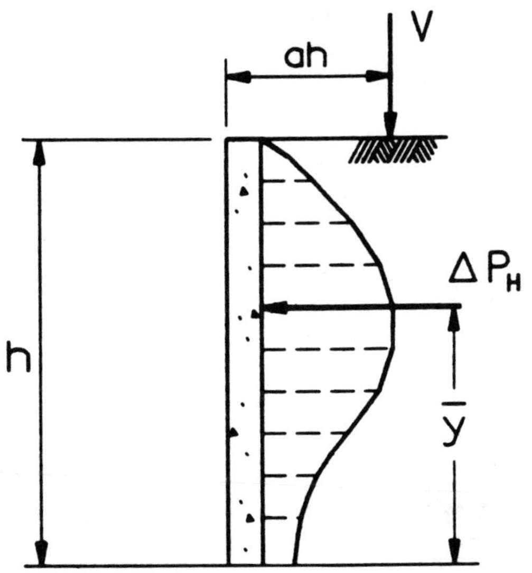

Point Load, V

a	ΔP_H	$\bar{y}$
0.4	0.78 (V/h)	.59 h
0.5	0.60 (V/h)	.54 h
0.6	0.46 (V/h)	.48 h

Line Load, V/ft

for $a \leq 0.4$ $\quad \Delta P_H = 0.55\ V$

for $a > 0.4$ $\quad \Delta P_H = \dfrac{0.64\ V}{(a^2 + 1)}$

a	$\bar{y}$
0.4	0.60 h
0.5	0.56 h
0.6	0.52 h
0.7	0.48 h

Fig 3-28. Resultant forces, point and line loads (after Spangler and Mickle 1956)

head is equal to the total head minus the elevation head. The pressure head at a point is the height water would rise in a piezometer placed at the point. The elevation head is the height of the point itself above an arbitrary datum. Water pressures must be added to effective earth pressures to obtain total pressures.

A. Static Pressures. For static water (no seepage) above or below the ground surface, the total head is constant and the pressure head at any point is the difference in elevation between the water surface and the point.

B. Water Pressures with Earth Pressure Equations. Where Coulomb or at-rest

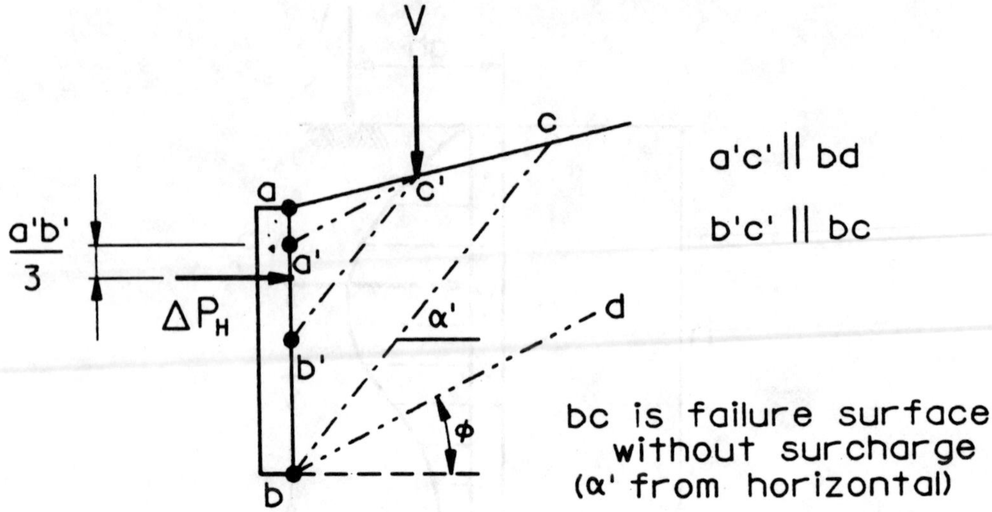

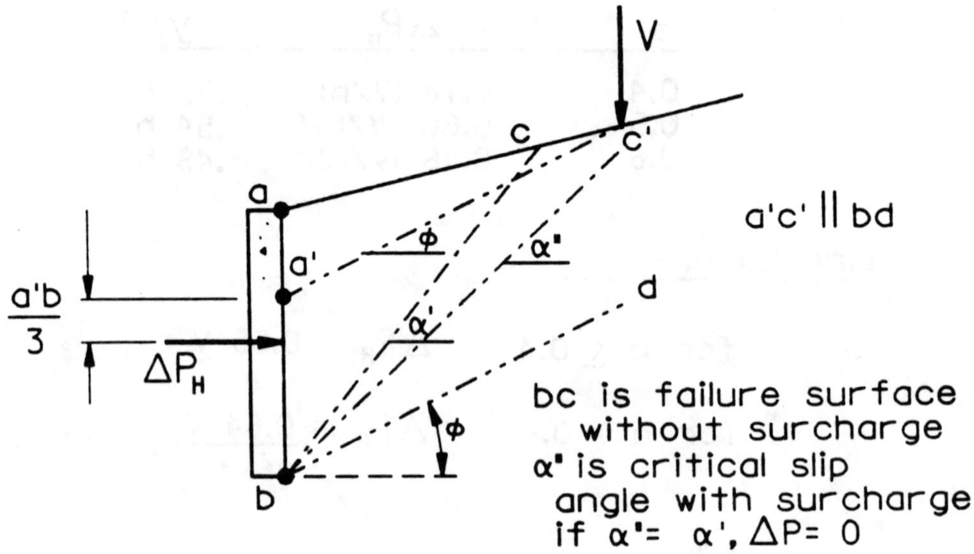

$$\Delta P_H = P(\gamma, V) - P(\gamma)$$

Fig 3-29. Approximate line of action for line loads (Terzaghi 1943)

equations are used to calculate the driving-side earth pressures for a totally submerged soil mass, the buoyant soil weight ($\gamma' = \gamma_{sat} - \gamma_w$) is used in the earth pressure equations and the calculated effective earth pressures are added to the calculated water pressures.

C. Water Forces with Wedge Analysis. The wedge method (Equations 3-23 and 3-33) uses total densities, uplift forces, and horizontal water forces on the vertical sides of the wedge. Consequently, it gives the effective earth force, and water forces must be added to obtain total forces (see Example 3 in Appendix M).

D. WATER PRESSURES WHERE SEEPAGE IS PRESENT. Where seepage occurs, the pressure head at points of interest must be obtained from a seepage analysis. Such an analysis must consider the types of foundation and backfill materials, their possible range of horizontal and vertical permeabilities, and the effectiveness of drains. Techniques of seepage analysis are discussed in EM 1110-2-1901, Casagrande (1937),

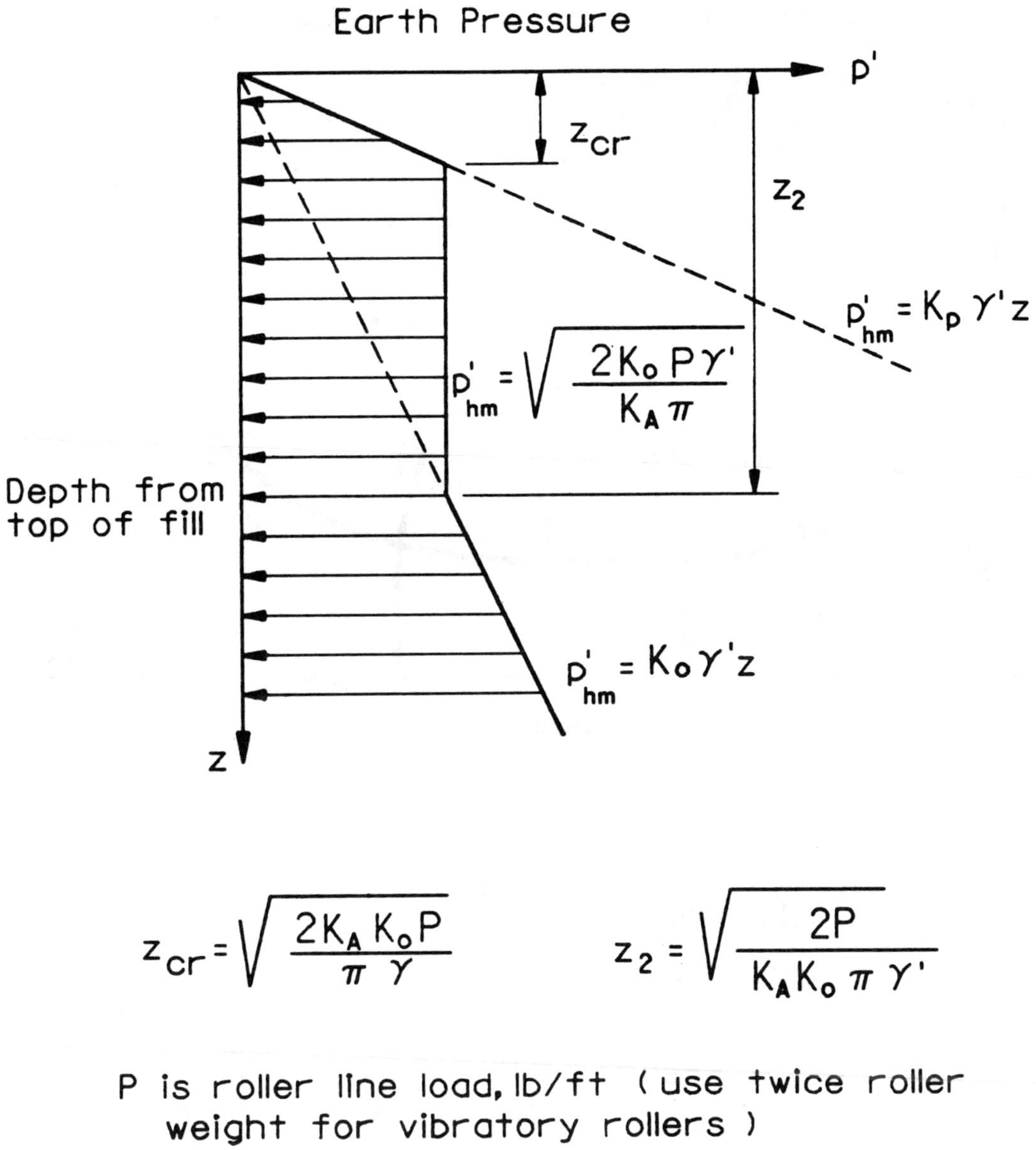

Fig 3-30. Design pressure envelope for nonyielding walls with compaction effects

Cedergren (1967), Harr (1962), and other references. Techniques applicable to wall design include flow nets and numerical methods such as the finite element method and the method of fragments. An example of pressure calculations using a flow net is shown in Figure 3-31. Where soil conditions adjacent to and below a wall can be assumed homogeneous (or can be mathematically transformed into equivalent homogeneous conditions) simplified methods such as the line-of-creep method may be used. Simplified methods are advantageous for preliminary studies to size wall elements or compare alternate wall designs; however, designers should ensure that the final design incorporates water pressures based on appropriate consideration of actual soil conditions.

3-19. Seepage Analysis by Line-of-Creep Method

Where soil conditions can be assumed homogeneous, the line-of-creep (or line-of-seepage) method provides a reasonable approximate method for estimating uplift pressures that is particularly useful for preliminary or comparative designs. The line of creep may underestimate uplift pressures on the base and thus be unconservative. Therefore, final design should be based on a more rigorous analysis. The method is illustrated in Figure 3-32. The total heads at the ends of the base (points B and C) are estimated by assuming that the total head varies linearly along the shortest possible seepage path (A'BCD'). Once the total head at B and C is

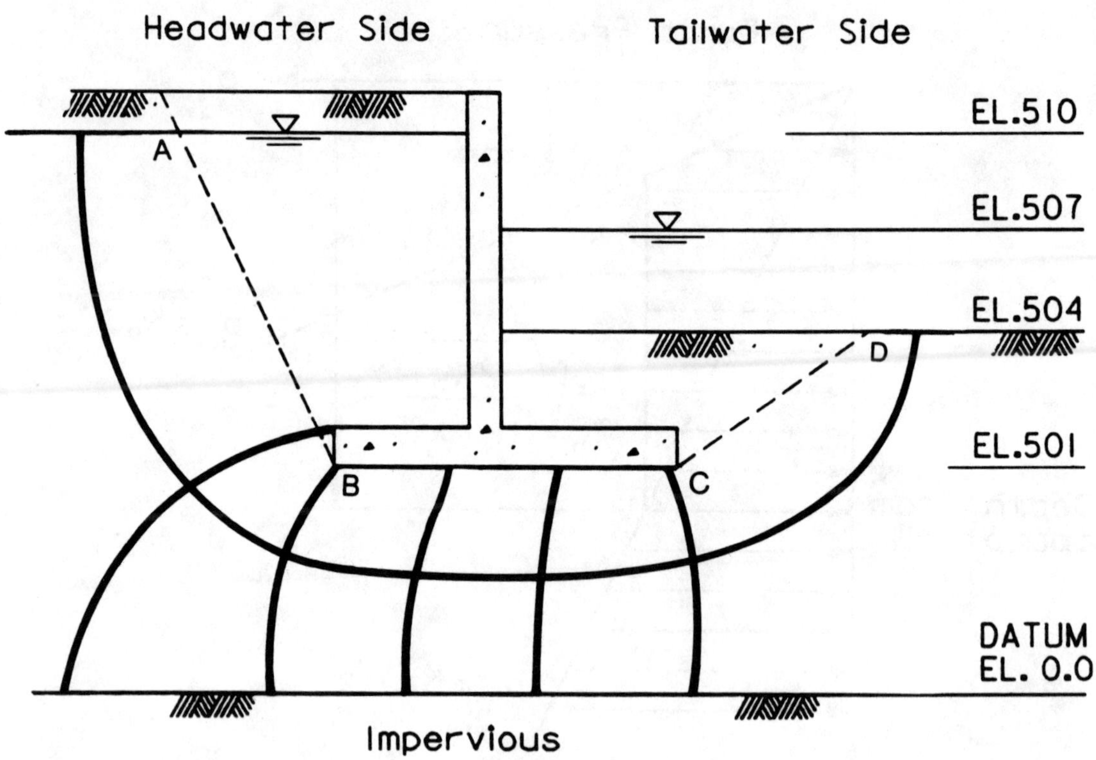

Pt. A Total Head: 510
Elevation Head: 510
Pressure Head: 0
Pressure: 0

Pt. B Total Head: $510 - (2/6)(510 - 507) = 509$
$507 + (4/6)(3) = 509$

Elevation Head: 501
Pressure Head: $509 - 501 = 8'$

Pressure: $(8)(62.4) = 499.2\ \text{lb/ft}^2$

Fig 3-31. Water pressures from flow net

known, the uplift pressures U_B and U_C are calculated by subtracting the elevation head from the total head at each point and multiplying the resulting pressure head by the unit weight of water. The total uplift diagram along the failure surface is completed in a similar manner. Where a key is present (Figure 3-33), point B is at the bottom of the key and line segment BC is drawn diagonally. Examples using the line-of-creep method are contained in Appendix N.

3-20. Seepage Analysis by Method of Fragments

Another approximate method applicable to homogeneous soil conditions is the method of fragments. It is more accurate than the line-of-creep method. The soil is divided into a number of regions or fragments for which exact solutions of the seepage conditions exist. The head loss through each fragment is calculated by mathematically combining the

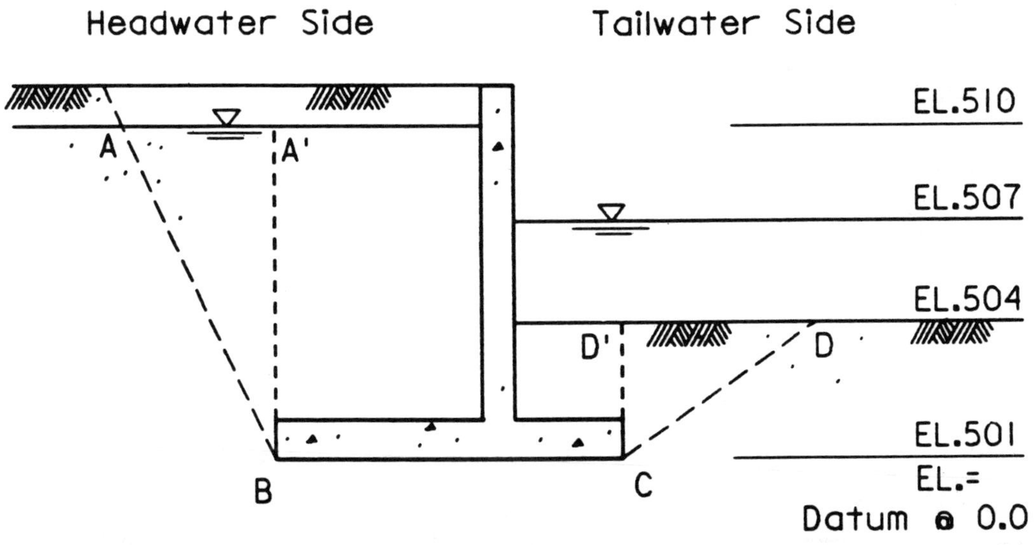

Length of Shortest Seepage Path: A'BCD'

(A') Total Head at A': 510 ft.
Elevation Head at A': 510 ft.
Pressure Head at A': 0 ft.

(B) Total Head at B: $510 - \left[\frac{(A'B)}{(A'BCD')}\right](510 - 507)$

Elevation Head at B: 501
Pressure Head at B = Total Head - Elevation Head

(C) Total Head at C: $510 - \left[\frac{(A'BC)}{(A'BCD')}\right](510 - 507)$

Elevation Head at C: 501

etc.

Water Pressure: $P_w = \gamma_w \cdot (\text{Pressure Head})$

Fig. 3-32. Water pressure by line-of-creep method

assemblage of fragments. The method assumes that fragment boundaries are equipotential lines (contours of equal total head) and provides an exact solution where this assumption is true (I-walls and single sheet piles). Details of the method and instructions for the computer program CFRAG (Appendix O) are presented by Pace, et al. (1984). Further background on the method is presented by Harr (1962, 1977). Keyed bases should be modeled by treating the key as a sheet pile and the soil below the base as a type IV. Accuracy is improved by minimizing the number of fragments, not by maximizing them to incorporate minor changes in geometry. Where water pressure on the base of a wall is calculated by the method of fragments, the water pressure along the driving-

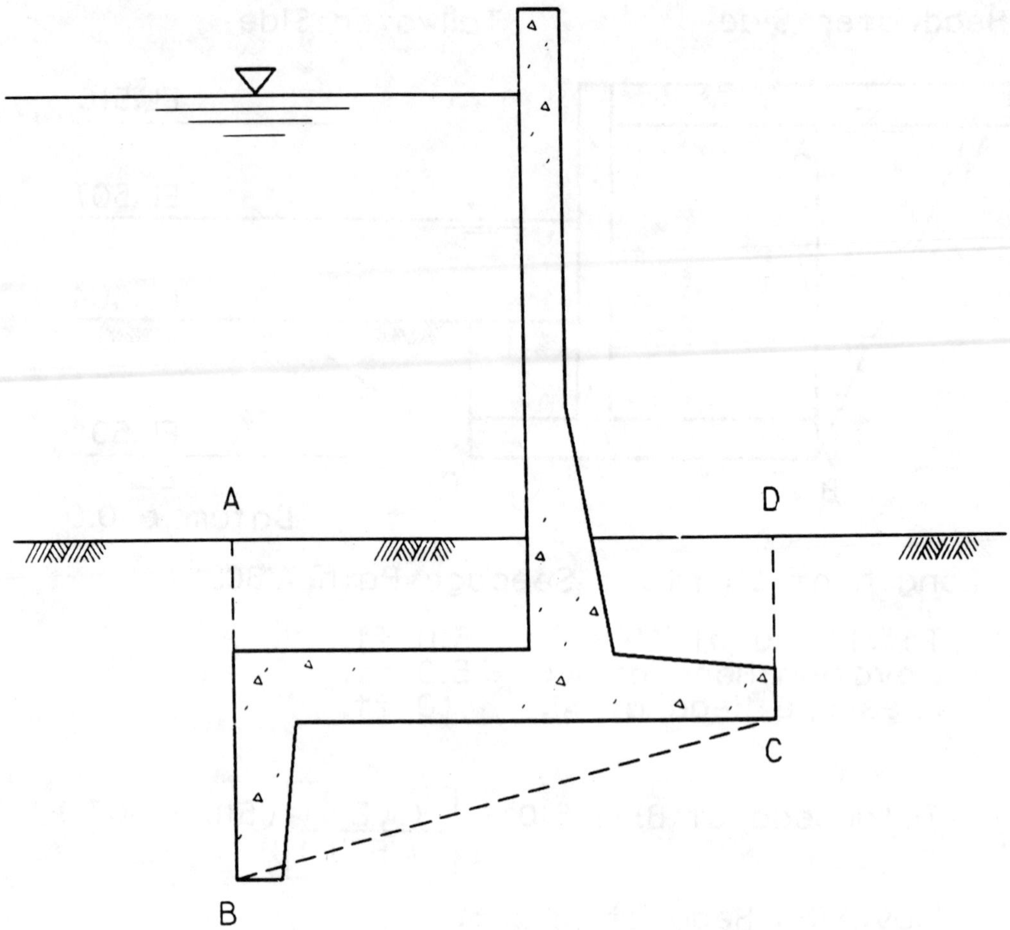

Fig 3-33. Seepage path for line of creep

side and resisting-side wedge bases may be taken to vary linearly as described in the preceding paragraph.

3-21. Seepage Analysis by the Finite Element Method

The finite element method provides a powerful tool to solve confined or unconfined seepage problems involving multiple soils with isotropic or anisotropic permeabilities. It is particularly useful for evaluating the effect of drains and analyzing walls with complicated foundation and backfill geometry. The WES computer program for the finite element method is described by Tracy (1983). Pre- and postprocessors for the program are also available (Tracy 1977a, 1977b).

3-22. Uplift Calculations for Rock Foundations

Seepage beneath flood walls founded on competent rock typically occurs in joints and fractures, not uniformly through pores as assumed for soils. Consequently, the assumptions of isotropy and homogeneity and the use of two-dimensional analysis models commonly employed for soil foundations will generally be invalid. Total head, uplift pressure, and seepage quantities may be highly dependent on the type, size, orientation, and continuity of joints and fractures in the rock and the type and degree of treatment afforded the rock foundation during construction. Since any joints or fractures in the rock can be detrimental to underseepage control, the joints and fractures should be cleaned out and filled with grout before the concrete is placed, as discussed in paragraph 7-4g. For walls on a rock foundation, the total seepage path can be assumed to be the length of the base which is in compression. An example of a wall on a rock foundation is shown in Example 2 of Appendix N.

3-23. Effect of Drains

Water pressures for design analyses should consider both working drains and blocked drainage conditions. Achieving an adequate factor of safety for an analysis considering blocked drainage is usually not good justification for omitting drains. Preferred practice is to provide drains; lower factors of safety than specified herein may be justified where blocked drainage assumptions are combined with rare and/or conservative loading assumptions. All such deviations from recommended safety factors should be supported by an assessment of expected drain reliability, and a justification that the factor of safety is reasonable in light of the analyzed conditions. Drains are discussed further in paragraphs 6-6 and 7-4.

3-24. Surge and Wave Loads

A. GENERAL CRITERIA. Wave and water level predictions for the analysis of walls should be determined with the criteria presented in the *Shore Protection Manual* (U. S. Army Engineer Waterways Experiment Station 1984). Design forces acting on the wall should be determined for the water levels and waves predicted for the most severe fetch and the effects of shoaling, refraction, and diffraction. A distinction is made between the action of nonbreaking, breaking, and broken waves, where the methods recommended for calculation of wave forces are for vertical walls. Wave forces on other types of walls (i.e., sloping, stepped, curved, etc.) are not sufficiently understood to recommend general analytical design criteria. In any event, a coastal engineer should be involved in establishing wave forces for the design of important structures.

B. WAVE HEIGHTS. Wave heights for design are obtained from the statistical distribution of all waves in a wave train, and are defined as follows:

H_s = average of the highest one-third of all waves
$H_1 = 1.67H_s$ = average of highest 1 percent of all waves
H_b = height of wave which breaks in water depth d_b

C. NONBREAKING WAVE CONDITION. When the depth of water is such that waves do not break, a nonbreaking condition exists. This occurs when the water depth at the wall is greater than approximately 1.5 times the maximum wave height. The H_1 wave shall be used for the nonbreaking condition. Design nonbreaking wave pressures shall be computed using the Miche-Rudgren Method, as described in Chapter 7 of the *Shore Protection Manual* (U. S. Army Engineer Waterways Experiment Station 1984). Whenever the maximum stillwater level results in a nonbreaking condition, lower stillwater levels should be investigated for the possibility that shallow water may produce breaking wave forces which are larger than the nonbreaking forces.

D. BREAKING WAVE CONDITION. The breaking condition occurs when the steepness of the wave and the bottom slope on the front of the wall have certain relationships to each other. It is commonly assumed that a structure positioned in a water depth d_s will be subject to the breaking wave condition if $d_s \leq 1.3$ H where H is the design wave height. Study of the breaking process indicates that this assumption is not always valid. The height of the breaking wave and its breaking point are difficult to determine, but breaker height can be equal to the water depth of the structure, depending on bottom slope and wave period. Detailed determination of breaker heights and distances for a sloping-approach grade in front of the wall are given in the *Shore Protection Manual* (U. S. Army Engineer Waterways Experiment Station 1984). Special consideration must be given to a situation where the fetch shoals abruptly (as with a bulkhead wall submerged by a surge tide) near the wall, but at a distance more than an approximate 0.7 wavelength away from the wall, and then maintains a constant water depth from that point to the wall. In this case waves larger than the water depth can be expected to have broken at the abrupt shoaling point, leaving smaller, higher frequency waves to reach the wall. Design breaking wave pressure should be determined by the Minikin method presented in Chapter 7 of the *Shore Protection Manual* (U. S. Army Engineer Waterways Experiment Station 1984). Breaking wave impact pressures occur at the instant the vertical force of the wave hits the wall and only when a plunging wave entraps a cushion of air against the wall. Because of this dependence on curve geometry, high impact pressures are infrequent against prototype structures; however, they must be recognized and considered in design. Also, since the high impact pressures caused by breaking waves are of high frequency, their importance in design against sliding and overturning may be questionable relative to longer lasting lower dynamic forces. An example involving a breaking wave condition is shown in Example 7 of Appendix N.

E. BROKEN WAVE CONDITION. Broken waves are those that break before reaching the wall but near enough to have retained some of the forward momentum of breaking. The design breaker height in this case (H_b) is the highest wave that will be broken in the break zone. Design wave forces for the height H_b should be determined by the method presented in Chapter 7 of the *Shore Protection Manual* (U. S. Army Engineer Waterways Experiment Station 1984).

F. SEEPAGE PRESSURES. Seepage pressures are based on the elevation of the surge stillwater level (paragraph 4-5).

Section IV. Supplemental Forces

3-25. Wind Load

Wind loads should be considered for retaining and flood walls during construction, prior to placing backfill. Wind loads can act any time in the life of a flood wall. In locations subjected to hurricanes, a wind load of 50 lb/sq ft can be used conservatively for walls 20 feet or less in height for winds up to 100 miles per hour (mph). In locations not subjected to hurricanes, 30 lb/sq ft can be used conservatively for the same height of wall and wind velocity conditions. For more severe conditions, the wind loads should be computed in accordance with ANSI A58.1 using a coefficient C_f equal to 1.2.

3-26. Earthquake Forces

A. GENERAL. For retaining walls which are able to yield laterally during an earthquake, the calculation of increased earth pressures induced by earthquakes can be approximated by the Mononobe-Okabe pseudo-static approach outlined below. In addition, the inertial forces of the wall, plus that portion of the adjacent earth and/or water which is assumed to act with the wall, should be included.

B. MONONOBE–OKABE ANALYSIS. This analysis is an extension of the Coulomb sliding-wedge theory taking into account horizontal and vertical inertial forces acting on the soil. The analysis is described in detail by Seed and Whitman (1970) and Whitman and Liao (1985).

1. Assumptions. The following assumptions are made by the Mononobe-Okabe analysis:

a. The wall is free to yield sufficiently to enable full soil strength or active pressure conditions to be mobilized.

b. The backfill is completely above or completely below the water table, unless the top surface is horizontal, in which case the backfill can be partially saturated.

c. The backfill is cohesionless.

d. The top surface is planar (not irregular or broken).

e. Any surcharge is uniform and covers the entire surface of the soil wedge.

f. Liquefaction is not a problem.

2. Equations. Equilibrium considerations of the soil wedge on the driving and resisting sides lead to the following Mononobe-Okabe equations for computing the active and passive forces exerted by the soil on the wall when the soil mass is at the point of failure (total shear resistance mobilized) along the slip plane of the Mononobe-Okabe wedge shown in Figure 3-34:

For driving (active) wedges (Figure 3-34a)

$$P_{AE} = \tfrac{1}{2} K_{AE} \gamma (1 - k_v) h^2 \qquad [3\text{-}44]$$

$$K_{AE} = \frac{\cos^2(\phi - \Psi - \theta)}{\cos\Psi \cos^2\theta \cos(\Psi - \theta + \delta)\left[1 - \sqrt{\dfrac{\sin(\phi + \delta)\sin(\phi - \Psi + \beta)}{\cos(\beta - \theta)\cos(\Psi - \theta + \delta)}}\right]^2} \qquad [3\text{-}45]$$

For resisting (passive) wedges (Figure 3-34b)

$$P_{PE} = \tfrac{1}{2} K_{PE} \gamma (1 - k_v) h^2 \qquad [3\text{-}46]$$

$$K_{PE} = \frac{\cos^2(\phi - \Psi + \theta)}{\cos\Psi \cos^2\theta \cos(\Psi - \theta + \delta)\left[1 - \sqrt{\dfrac{\sin(\phi + \delta)\sin(\phi - \Psi + \beta)}{\cos(\beta - \theta)\cos(\Psi - \theta + \delta)}}\right]^2} \qquad [3\text{-}47]$$

P_{AE} and P_{PE} are the combined static and dynamic forces due to the driving and resisting wedges, respectively. The equations are subject to the same limitations that are applicable to Coulomb's equations. Definitions of terms are as follows:

γ = unit weight of soil

k_v = vertical acceleration in g's

h = height of wall

ϕ = internal friction angle of soil

$\Psi = \tan^{-1}\left(\dfrac{k_h}{1 - k_v}\right)$ = seismic inertia angle

k_h = horizontal acceleration in g's

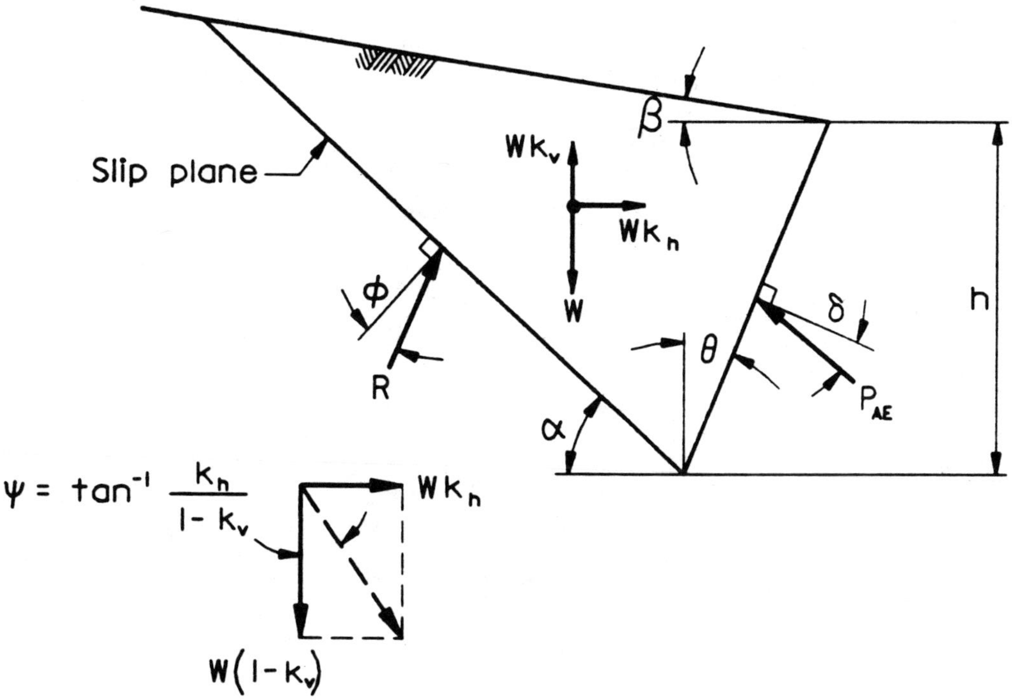

a. Mononobe-Okage (active) wedge

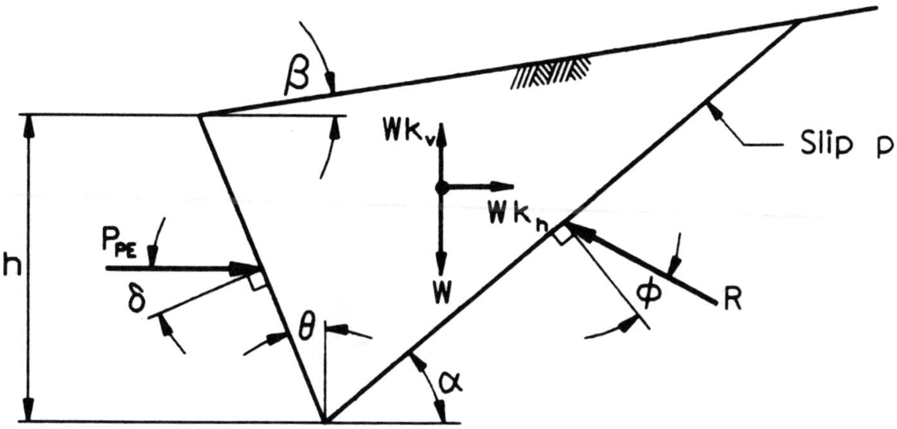

b. Passive wedge

Fig 3-34. Driving and resisting seismic wedges, no saturation

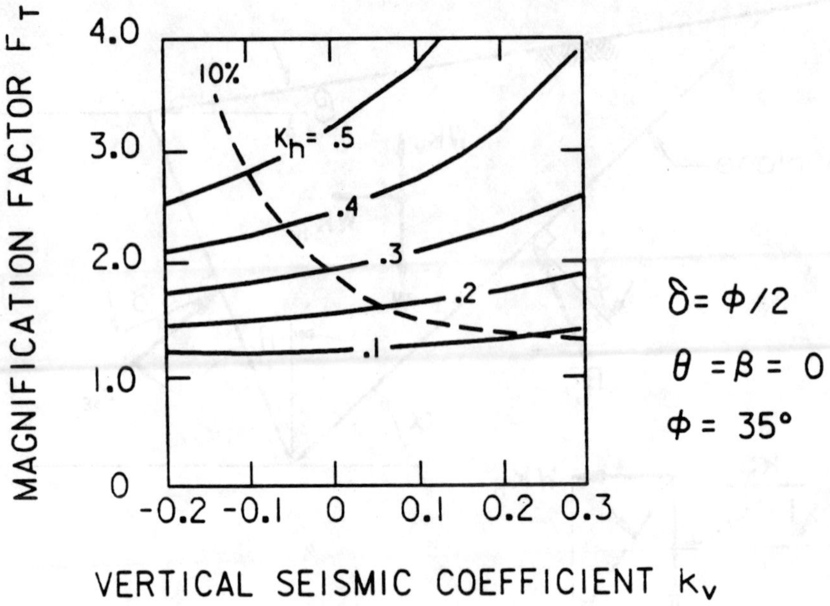

a. Influence of vertical seismic coefficient on magnification factor

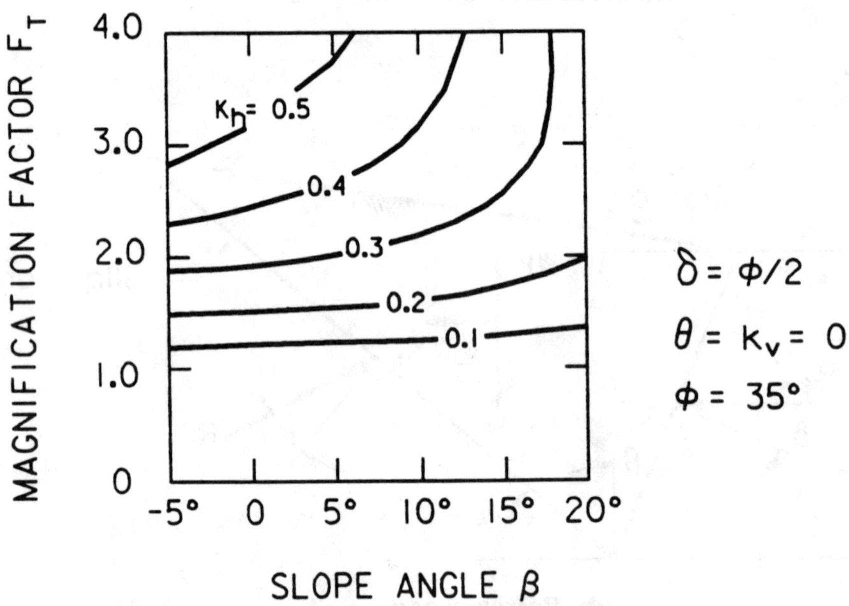

b. Influence of backfill slope angle on magnification factor

Fig 3-35. Influence of k_v and β on magnification factor (after Applied Technology Council 1981)

θ = inclination of wall with respect to vertical (this definition of θ is different from θ in Coulomb's equations)

δ = wall friction angle

β = inclination of soil surface (upward slopes away from the wall are positive)

3. Simplifying Conditions. For the usual case where k_v, δ, and θ are taken to be zero, the equations reduce to

$$K_{AE} = \frac{\cos^2(\phi - \Psi)}{\cos^2\Psi\left[1 + \sqrt{\dfrac{\sin\phi\,\sin(\phi - \Psi - \beta)}{\cos\beta\,\cos\Psi}}\right]^2} \qquad [3\text{-}48]$$

$$K_{PE} = \frac{\cos^2(\phi - \Psi)}{\cos^2\Psi\left[1 - \sqrt{\dfrac{\sin\phi\,\sin(\phi - \Psi + \beta)}{\cos\beta\,\cos\Psi}}\right]^2} \qquad [3\text{-}49]$$

where

$$\Psi = \tan^{-1}(k_h)$$

and

$$P_{AE} = \tfrac{1}{2}K_{AE}\gamma h^2$$
$$P_{PE} = \tfrac{1}{2}K_{PE}\gamma h^2$$

For the case when the water table is above the backfill, P_{AE} and P_{PE} must be divided into static and dynamic components for computing the lateral forces. Buoyant soil weight is used for computing the static component below the water table, with the hydrostatic force added, and saturated soil weight is used for computing the dynamic component (see paragraph 3-26c[3]).

4. Observations. General observations from using Mononobe-Okabe analysis are as follows:

a. As the seismic inertia angle Ψ increases, the values of K_{AE} and K_{PE} approach each other and, for a vertical backfill face ($\theta = 0$), become equal when $\Psi = \theta$.

b. The locations of P_{AE} and P_{PE} are not given by the Mononobe-Okabe analysis. Seed and Whitman (1970) suggest that the dynamic component ΔP_{AE} be placed at the upper one-third point, ΔP_{AE} being the difference between P_{AE} and the total active force from Coulomb's active wedge without the earthquake. The general wedge earthquake analysis described in paragraph 3-26c places the dynamic component ΔP_{AE} at the upper one-third point also, but computes ΔP_{AE} as being the difference between P_{AE} and the total active force from the Mononobe-Okabe wedge. The latter method for computing ΔP_{AE}, which uses the same wedge for computing the static and dynamic components of P_{AE}, is preferred.

c. Another limitation of the Mononobe-Okabe equation is that the contents of the radical in the equation must be positive for a real solution to be possible, and for this it is necessary that $\phi \geq \Psi + \beta$ for the driving wedges and $\phi \geq \Psi - \beta$ for the resisting wedges. This condition could also be thought of as specifying a limit to the horizontal acceleration coefficient that could be sustained by any structure in a given soil. The limiting condition for the driving wedge is

$$k_h \leq (1 - k_v)\tan(\phi - \beta) \qquad [3\text{-}50]$$

and for the resisting wedge

$$k_h \leq (1 - k_v)\tan(\phi - \beta) \qquad [3\text{-}51]$$

d. Figure 3-35a (Applied Technology Council 1981) shows the effect on the magnification factor F_T (equal to K_{AE}/K_A) of changes in the vertical acceleration coefficient k_v. Positive values of k_v have a significant effect for values of k_h greater than 0.2. The effect is greater than 10 percent above and to the right of the dashed line. For values of k_h of 0.2 or less, k_v can be neglected for all practical purposes.

e. K_{AE} and F_T are also sensitive to variations in backfill slope, particularly for higher values of horizontal acceleration. This effect is shown in Figure 3-35b.

C. GENERAL WEDGE EARTHQUAKE ANALYSIS. When the Coulomb wedge assumptions cannot be met, the following wedge analysis can be used. The equations for the dynamic force given below for various conditions are simply the horizontal acceleration coefficient multiplied by the weight of the wedge defined by the critical slip-plane angle. See Example 11 of Appendix M for sample calculations.

1. Assumptions. The equations for determining the critical slip-plane angle for driving and resisting wedges subjected to a horizontal acceleration are developed with the following assumptions:

a. The shear on the vertical face of the wedge is zero.

b. The shear strength along the potential slip planes in the soil has not been mobilized to any extent (i.e., for static loading prior to an earthquake).

2. Equations for Cohesionless, Dry Backfill Above the Water Table. Driving and resisting forces for cohesionless, dry, sloping, planar-surfaced backfill below the water table where k_v, δ, and $\theta = 0$ can be computed as follows:

a. Static Components. The static components for a driving and resisting wedge are:

$$P_A = \tfrac{1}{2}K_A\gamma h^2 \quad [3\text{-}52]$$

$$P_P = \tfrac{1}{2}K_P\gamma h^2 \quad [3\text{-}53]$$

where

$$K_A = \left(\frac{1 - \tan\phi\cot\alpha}{1 + \tan\phi\tan\alpha}\right)\left(\frac{\tan\alpha}{\tan\alpha - \tan\beta}\right) \quad [3\text{-}54]$$

$$K_P = \left(\frac{1 + \tan\phi\cot\alpha}{1 - \tan\phi\tan\alpha}\right)\left(\frac{\tan\alpha}{\tan\alpha - \tan\beta}\right) \quad [3\text{-}55]$$

as derived in paragraphs H-2 and H-3, Appendix H.

For an active wedge

$$\alpha = \tan^{-1}\left(\frac{c_1 + \sqrt{c_1^2 + 4c_2}}{2}\right) \quad [3\text{-}56]$$

$$c_1 = \frac{2(\tan\phi - k_h)}{1 + k_h\tan\phi} \quad [3\text{-}57]$$

$$c_2 = \frac{\tan\phi\,(1 - \tan\phi\tan\beta) - (\tan\beta + k_h)}{\tan\phi\,(1 + k_h\tan\phi)} \quad [3\text{-}58]$$

For a passive wedge

$$\alpha = \tan^{-1}\left(\frac{-c_1 + \sqrt{c_1^2 + 4c_2}}{2}\right) \quad [3\text{-}59]$$

$$c_1 = \frac{2(\tan\phi - k_h)}{1 + k_h\tan\phi} \quad [3\text{-}60]$$

$$c_2 = \frac{\tan\phi\,(1 + \tan\phi\tan\beta) + (\tan\beta - k_h)}{\tan\phi\,(1 + k_h\tan\phi)} \quad [3\text{-}61]$$

If $k_v > 0$, replace γ with $(1 - k_v)\gamma$.

b. Dynamic Components. The dynamic component for each wedge is:

$$\Delta P_{AE} = \Delta P_{PE} = k_h\left[\frac{\gamma h^2}{2(\tan\alpha - \tan\beta)}\right] \quad [3\text{-}62]$$

c. Total Driving Force. The total driving force is:

$$P_{AE} = P_A + \Delta P_{AE} \quad [3\text{-}63]$$

which is equal to:

$$P_{AE} = \tfrac{1}{2}K_{AE}\gamma h^2 \quad [3\text{-}64]$$

from the Mononobe-Okabe analysis.

The line of action for P_{AE} may be found as:

$$Y_{AE} = \frac{P_A\left(\frac{h}{3}\right) + \Delta P_{AE}\left(\frac{2h}{3}\right)}{P_{AE}} \quad [3\text{-}65]$$

It should be noted that for large values of k_h, which cause α to be small, P_A can be negative causing the line of action of P_{AE} to lie above the upper one-third point.

d. Total Resisting Force. The total resisting force is:

$$P_{PE} = P_P - \Delta P_{PE} \quad [3\text{-}66]$$

which is equal to:

$$P_{PE} = \tfrac{1}{2}K_{PE}\gamma h^2 \quad [3\text{-}67]$$

from the Mononobe-Okabe analysis.

The line of action for P_{PE} may be found as:

$$Y_{PE} = \frac{P_P\left(\frac{h}{3}\right) - \Delta P_{PE}\left(\frac{2h}{3}\right)}{P_{PE}} \quad [3\text{-}68]$$

3. Equations for Cohesionless Backfill with Water Table. Driving and resisting forces for cohesionless, sloping, planar-surfaced backfill with water table where k_v, δ, and $\theta = 0$ can be computed as follows:

a. Driving Force. The static components for a driving wedge are (see Figures 3-36a and 3-37a):

$$P_A = P_{A1} + P_{A2} = \tfrac{1}{2}K_A\gamma(h - h_s)^2 + \tfrac{1}{2}h_s[2K_A\gamma(h - h_s) + K_b\gamma_b h_s] \quad [3\text{-}69]$$

$$P_{ws} = \tfrac{1}{2}\gamma_w h_s^2 \quad [3\text{-}70]$$

and the dynamic components are (see Figures 3-36a and 3-37a):

$$\Delta P_{AE} = \Delta P_{AE1} + \Delta P_{AE2} = k_h\left[\frac{\gamma h^2}{2(\tan\alpha - \tan\beta)}\right] + k_h\left[\frac{(\gamma_s - \gamma)h_s^2}{2\tan\alpha}\right] \quad [3\text{-}71]$$

giving a total force

$$P_{AE} = P_A + P_{ws} + \Delta P_{AE} \qquad [3\text{-}72]$$

where

γ_s = saturated unit weight of fill

γ = moist unit weight of fill

γ_b = buoyant unit weight of fill

γ_w = unit weight of water

$$K_A = \left(\frac{1 - \tan\phi \cot\alpha}{1 + \tan\phi \tan\alpha}\right)\left(\frac{\tan\alpha}{\tan\alpha - \tan\beta}\right)$$

$$K_b = \left(\frac{1 - \tan\phi \cot\alpha}{1 + \tan\phi \tan\alpha}\right) \times \left[1 + \left(\frac{\tan\alpha}{\tan\alpha - \tan\beta} - 1\right)\frac{\gamma}{\gamma_b}\right]$$

and α is defined in Equation 3-56.

b. Resisting Force. The static components for the resisting wedge are (see Figures 3-36b and 3-37b):

$$\begin{aligned} P_P &= P_{P1} + P_{P2} \\ &= \tfrac{1}{2}K_P\gamma(h - h_s)^2 \\ &\quad + \tfrac{1}{2}h_s[2K_P\gamma(h - h_s) + K_b\gamma_b h_s] \qquad [3\text{-}73] \end{aligned}$$

$$P_{ws} = \tfrac{1}{2}\gamma_w h_s^2 \qquad [3\text{-}74]$$

and the dynamic components are

$$\begin{aligned} \Delta P_{PE} &= \Delta P_{PE1} + \Delta P_{PE2} \\ &= k_h\left[\frac{\gamma h^2}{2(\tan\alpha - \tan\beta)}\right] \\ &\quad + k_h\left[\frac{(\gamma_s - \gamma)h_s^2}{2\tan\alpha}\right] \qquad [3\text{-}75] \end{aligned}$$

giving a total force of

$$P_{AE} = P_P + P_{ws} - \Delta P_{PE} \qquad [3\text{-}76]$$

where γ, γ_b, γ_s, and γ_w are defined in paragraph 3-26c(3)(a), and

$$K_P = \left(\frac{1 + \tan\phi \cot\alpha}{1 - \tan\phi \tan\alpha}\right)\left(\frac{\tan\alpha}{\tan\alpha - \tan\beta}\right) \qquad [3\text{-}77]$$

$$K_b = \left(\frac{1 + \tan\phi \cot\alpha}{1 - \tan\phi \tan\alpha}\right) \times \left[1 + \frac{\tan\alpha}{\tan\alpha - \tan\beta} - 1\,\frac{\gamma}{\gamma_b}\right] \qquad [3\text{-}78]$$

and the equations for α are given in Equation 3-59.

4. Equations for Cohesive Backfill with Water Table. Driving and resisting forces for a cohesive, sloping, planar-surfaced backfill with water table where k_v, δ, and $\theta = 0$ can be computed as follows:

a. Driving Force. The static components for the driving wedge are (see Figure 3-38a):

$$\begin{aligned} P_A &= P_{A1} + P_{A2} \\ &= \tfrac{1}{2}K_A\gamma[(h - d_c) - h_s]^2 \\ &\quad + \tfrac{1}{2}h_s[2K_A\gamma(h - d_c - h_s) + K_b\gamma_b h_s] \qquad [3\text{-}79] \end{aligned}$$

$$P_{ws} = \tfrac{1}{2}\gamma_w h_s^2 \qquad [3\text{-}80]$$

and the dynamic components are (see Figure 3-38a)

$$\begin{aligned} \Delta P_{AE} &= \Delta P_{AE1} + \Delta P_{AE2} \\ &= k_h\left[\frac{\gamma(h^2 - d_c^2)}{2(\tan\alpha - \tan\beta)}\right] \\ &\quad + k_h\left[\frac{(\gamma_s - \gamma)^2 h_s^2}{2\tan\alpha}\right] \qquad [3\text{-}81] \end{aligned}$$

giving a total force of

$$P_{AE} = P_A + P_{ws} + \Delta P_{AE} \qquad [3\text{-}82]$$

where

γ = moist unit weight of fill

γ_b = buoyant unit weight of fill

γ_s = saturated unit weight of fill

γ_w = unit weight of water

$$K_A = \left(\frac{1 - \tan\phi \cot\alpha}{1 + \tan\phi \tan\alpha}\right)\left(\frac{\tan\alpha}{\tan\alpha - \tan\beta}\right) \qquad [3\text{-}83]$$

$$K_b = \left(\frac{1 - \tan\phi \cot\alpha}{1 + \tan\phi \tan\alpha}\right) \times \left[1 + \left(\frac{\tan\alpha}{\tan\alpha - \tan\beta} - 1\right)\frac{\gamma}{\gamma_b}\right] \qquad [3\text{-}84]$$

$$\alpha = \tan^{-1}\left(\frac{c_1 + \sqrt{c_1^2 + 4c_2}}{2}\right) \qquad [3\text{-}85]$$

$$c_1 = \frac{2\tan\phi\,(\tan\phi - k_h) + \dfrac{4c(\tan\phi + \tan\beta)}{\gamma(h + d_c)}}{A} \qquad [3\text{-}86]$$

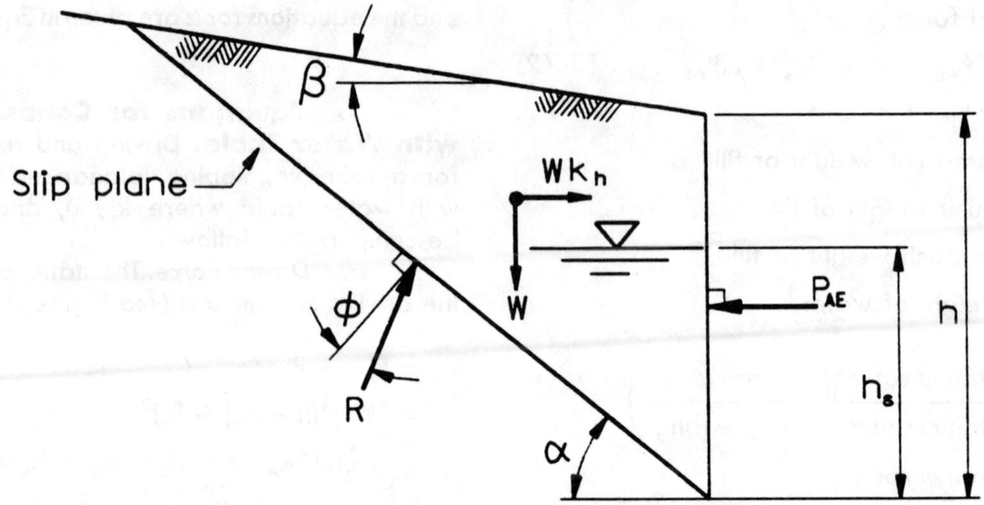

a. Driving wedge

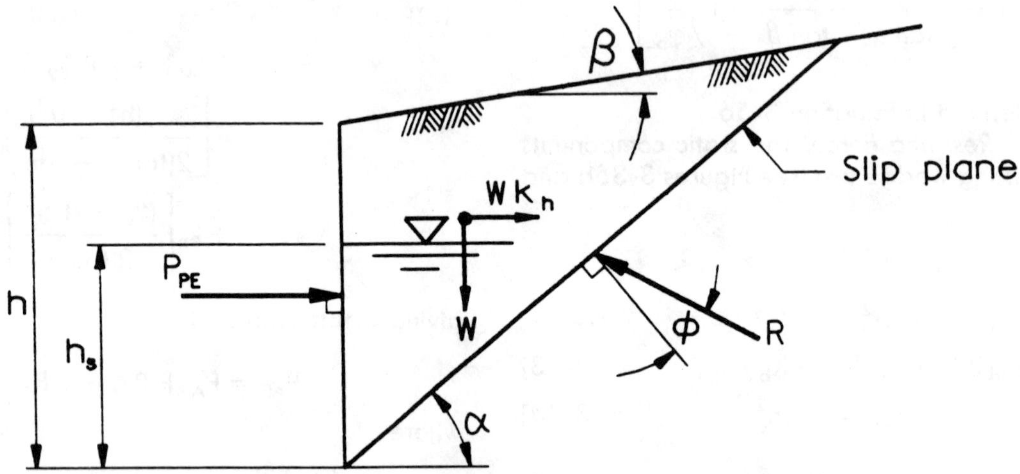

b. Resisting wedge

Fig 3-36. Seismic wedges, water table within wedge

$$c_2 = \frac{\tan\phi\,(1-\tan\phi\tan\beta) - (\tan\beta + k_h) + \dfrac{2c(1-\tan\phi\tan\beta)}{\gamma(h+d_c)}}{A} \qquad [3\text{-}87]$$

$$A = (1 + k_h\tan\phi)\tan\phi + \frac{2c(1-\tan\phi\tan\beta)}{\gamma(h+d_c)} \qquad [3\text{-}88]$$

$$d_c = \frac{c/\gamma}{\cos\alpha\,(\sin\alpha - \tan\phi\cos\alpha)} \qquad [3\text{-}89]$$

b. Resisting Force. The static components for the resisting wedge are (Figure 3-38b):

$$\begin{aligned} P_P &= P_{P1} + P_{P2} \\ &= \tfrac{1}{2}K_P\gamma(h-h_s)^2 \\ &\quad + \tfrac{1}{2}h_s[2K_P\gamma(h-h_s) + K_b\gamma_b h_s] \\ &\quad + 2K_c ch \end{aligned} \qquad [3\text{-}90]$$

$$P_{ws} = \tfrac{1}{2}\gamma_w h_s^2 \qquad [3\text{-}91]$$

and the dynamic components are (see Figure 3-38b):

$$\begin{aligned} \Delta P_{PE} &= \Delta P_{PE1} + \Delta P_{PE2} \\ &= k_h\left[\frac{\gamma h^2}{2(\tan\alpha - \tan\beta)}\right] \\ &\quad + k_h\left[\frac{(\gamma_s - \gamma)h_s^2}{2\tan\alpha}\right] \end{aligned} \qquad [3\text{-}92]$$

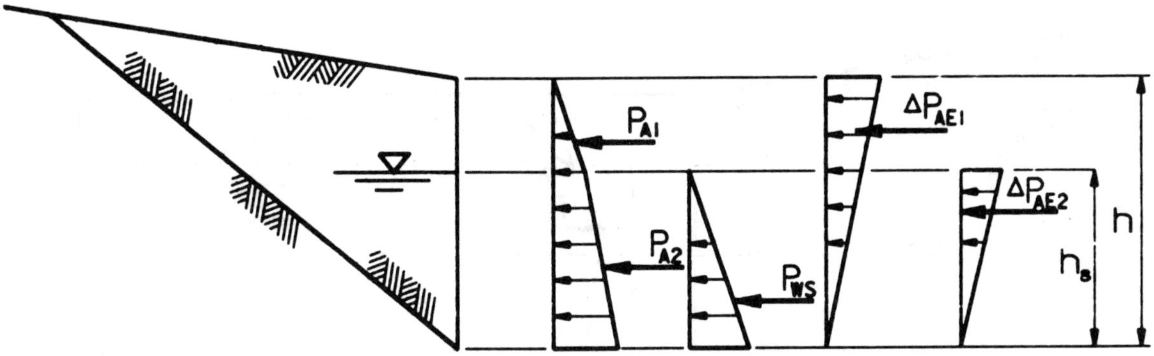

a. Driving wedge

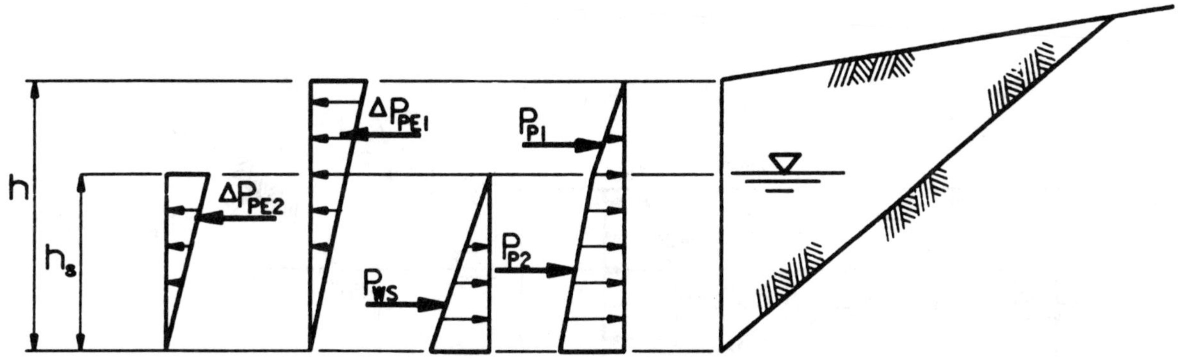

b. Resisting wedge

Fig 3-37. Static and dynamic pressure diagrams, water table within wedge

giving a total force of:

$$P_{PE} = P_P + P_{ws} + \Delta P_{PE} \qquad [3\text{-}93]$$

where γ, γ_b, γ_s, and γ_w are defined in paragraph 3-26c(4)(a), and

$$K_P = \left(\frac{1 + \tan\phi \cot\alpha}{1 - \tan\phi \tan\alpha}\right)\left(\frac{\tan\alpha}{\tan\alpha - \tan\beta}\right) \qquad [3\text{-}94]$$

$$K_b = \left(\frac{1 + \tan\phi \cot\alpha}{1 - \tan\phi \tan\alpha}\right) \times \left[1 + \left(\frac{\tan\alpha}{\tan\alpha - \tan\beta} - 1\right)\frac{\gamma}{\gamma_b}\right] \qquad [3\text{-}95]$$

$$\alpha = \tan^{-1}\left(\frac{-c_1 + \sqrt{c_1^2 + 4c_2}}{2}\right) \qquad [3\text{-}96]$$

$$c_1 = \frac{2\tan\phi(\tan\phi - k_h) + \dfrac{4c(\tan\phi - \tan\beta)}{\gamma h}}{A} \qquad [3\text{-}97]$$

$$c_2 = \frac{\tan\phi(1 + \tan\phi\tan\beta) + (\tan\beta - k_h) + \dfrac{2c(1 + \tan\phi\tan\beta)}{\gamma h}}{A} \qquad [3\text{-}98]$$

$$A = (1 + k_h\tan\phi)\tan\phi + \frac{2c(1 + \tan\phi\tan\beta)}{\gamma h} \qquad [3\text{-}99]$$

$$K_c = \frac{1}{2\sin\alpha\cos\alpha(1 - \tan\phi\cos\alpha)} \cdot \frac{\tan\alpha}{\tan\alpha - \tan\beta} \qquad [3\text{-}100]$$

D. INERTIA FORCE OF WALL. The inertia force of the wall, including that portion of the backfill above the heel or toe of the wall and any water within the backfill which is not included as part of the Coulomb wedge, is computed by multiplying the selected acceleration coefficient by

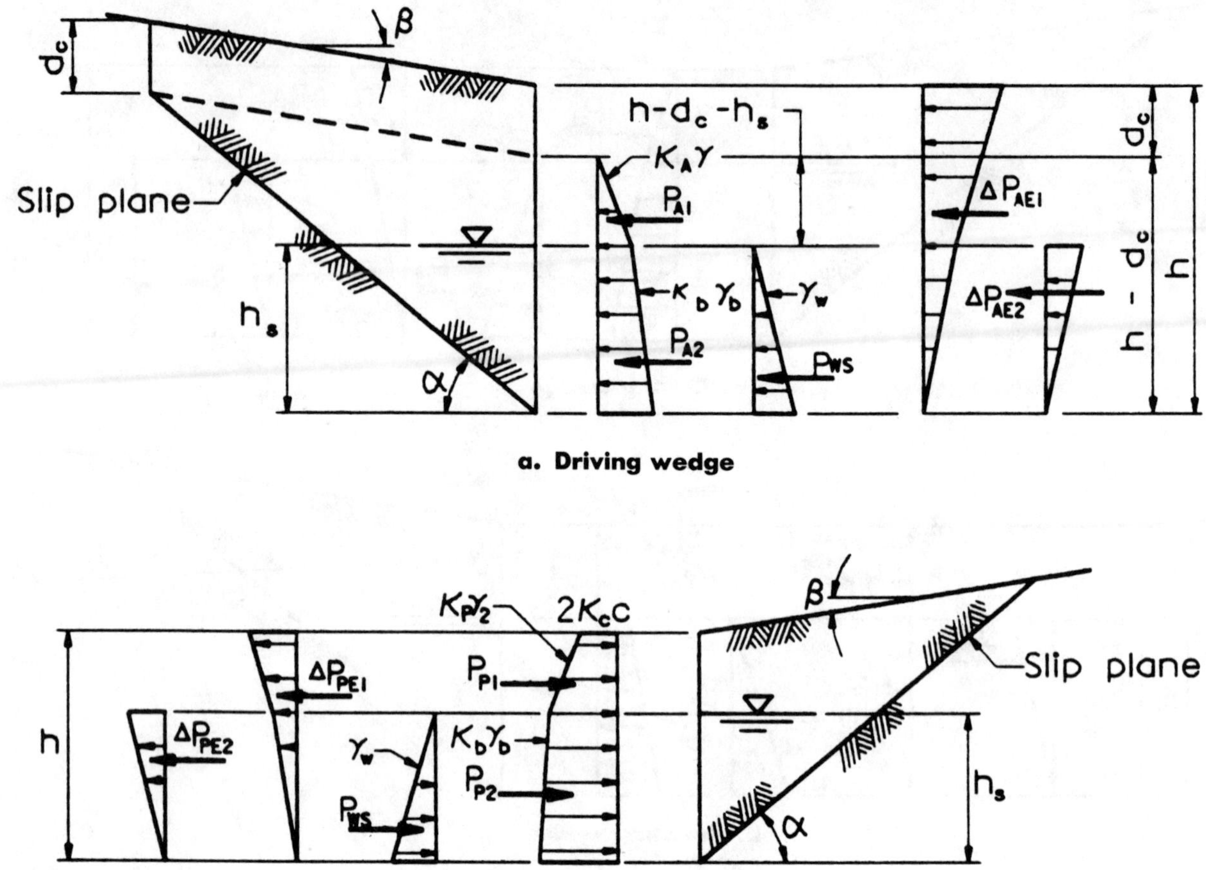

a. Driving wedge

b. Resisting wedge

Fig 3-38. Static and dynamic pressure diagrams, cohesive fill, water table within wedge

the weight of the wall and backfill. This force is obtained by multiplying the mass by acceleration as follows:

$$F = ma = ma\left(\frac{g}{g}\right) = \frac{a}{g}w = k_h W \quad [3\text{-}101]$$

E. HYDRODYNAMIC FORCE DUE TO WATER ABOVE GROUND LEVEL. Water standing above ground can have its static pressure, acting against a wall, increased or decreased due to seismic action. Figure 3-39 shows the pressures and forces due to earthquakes for freestanding water. The dynamic force is given by Westergaard's (1933) equation as:

$$P_E = (\tfrac{2}{3})C_E k_h h^2 \quad [3\text{-}102]$$

where C_E is a factor depending upon the depth of water, h, in feet, and the earthquake period of vibration, T, in seconds. Westergaard's approximate equation for C_E in kip-second-foot units is

$$C_E = \frac{0.051}{\sqrt{1 - 0.72\,(h/1000T)^2}} \quad [3\text{-}103]$$

Normally, for retaining and flood walls, C_E can be taken as 0.051. The pressure distribution is parabolic, and the pressure at any point y below the top surface is:

$$p_E = C_E k_h \sqrt{hy} \quad [3\text{-}104]$$

The line of action of force P_E is 0.4h above the ground surface.

F. SELECTION OF ACCELERATION COEFFICIENTS

1. Minimum Acceleration Coefficients. Minimum horizontal acceleration coefficient

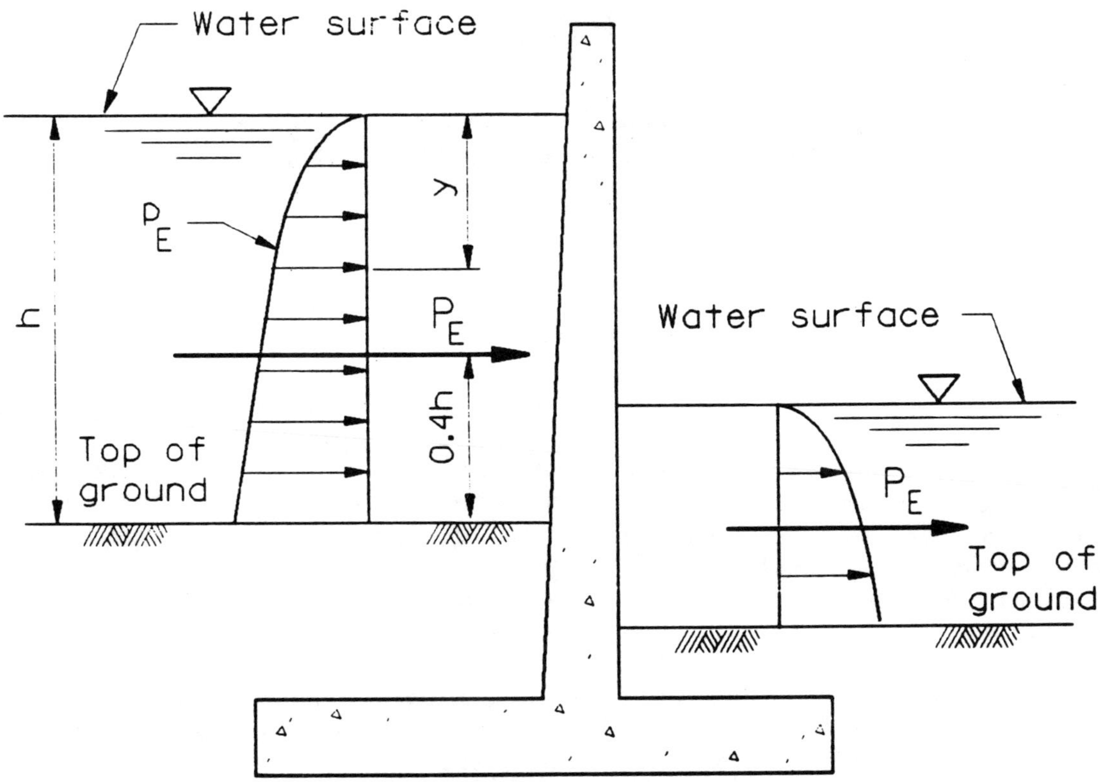

Fig 3-39. Hydrodynamic forces for freestanding water

values for the United States and its territories are listed in ER 1110-2-1806. In the absence of more accurate data, these values can be used as a guide for determining the acceleration coefficient to be used in the calculation of lateral earthquake forces on retaining and flood walls. As discussed in paragraph 3-26b(4)(d) where the horizontal ground acceleration is 0.2 g or less, the vertical ground acceleration can be neglected for all practical purposes. When the vertical acceleration coefficient is included in the analysis, it is normally taken as two-thirds of the horizontal acceleration coefficient.

2. Acceleration Coefficients Greater than 0.2. When the design acceleration coefficient exceeds 0.2, the Mononobe-Okabe analysis may require the size of the wall to be excessively great. To provide a more economical structure, design for a small tolerable lateral displacement rather than no lateral displacement may be preferable (Applied Technology Council 1981). A method for computing the magnitude of relative wall displacement during a given earthquake is described by Whitman and Liao (1985).

3. Acceleration Coefficients for Walls Forming Part of a Dam. For retaining walls forming part of a dam, where failure of the wall would jeopardize the safety of the dam, the selection of the acceleration coefficients for the design of the wall should be consistent with those used for the stability analyses and concrete design of the dam, where required (ER 1110-2-1806).

CHAPTER 4

STRUCTURE STABILITY

4-1. Scope

This chapter presents information for stability analysis of retaining walls and inland and coastal flood walls. The methods of analysis to determine overturning and sliding stability and to assess bearing capacity are discussed. The forces as determined in Chapter 3 are used to assess overturning stability and bearing capacity. In certain cases as described in this chapter, the same forces computed for overturning may be used to check sliding stability. In other cases, sliding stability should be computed by the multiple wedge iterative method or by an adjustment of the driving and resisting wedge forces based on the factor of safety required, both of which are discussed in this chapter. Loading conditions for the various types of walls and the acceptable criteria for each loading condition are given for each of the stability analyses.

Section I. Loading Conditions

4-2. Representative Loading Conditions

The following loading conditions are generally representative of conditions affecting retaining walls and inland and coastal flood walls. The loading cases for a specific wall should be chosen, as applicable, from the lists below. Loading conditions which are not listed below should be analyzed, where applicable. Note that some walls may require consideration of loadings from both lists, as discussed in paragraph 2-9.

4-3. Retaining Walls

A. CASE R1, USUAL LOADING. The backfill is in place to the final elevation; surcharge loading, if present, is applied (stability should be checked with and without the surcharge); the backfill is dry, moist, or partially saturated as the case may be; any existing lateral and uplift pressures due to water are applied. This case also includes the usual loads possible during construction which are not considered short-duration loads.

B. CASE R2, UNUSUAL LOADING. This case is the same as Case R1 except the water table level in the backfill rises, for a short duration, or another type of loading of short duration is applied; e.g., high wind loads (paragraph 3-25), equipment surcharges during construction, etc.

C. CASE R3, EARTHQUAKE LOADING. This is the same as Case R1 with the addition of earthquake-induced lateral and vertical loads, if applicable; the uplift is the same as for Case R1.

4-4. Inland Flood Walls

A. CASE I1, DESIGN FLOOD LOADING. The backfill is in place to the final elevation; the water level is at the design flood level (top of wall less freeboard) on the unprotected side; uplift is acting.

B. CASE I2, WATER TO TOP OF WALL. This is the same as Case I1 except the water level is at the top of the unprotected side of the wall.

C. CASE I3, EARTHQUAKE LOADING. The backfill is in place to the final elevation; the water is at the usual level during the non-flood stage; uplift, if applicable, is acting; earthquake-induced lateral and vertical loads, if applicable, are acting. (Note: This case is necessary only if the wall has a significant loading during the non-flood stage.)

D. CASE I4, CONSRUCTION SHORT DURATION LOADING. The flood wall is in place with the loads added which are possible during the construction period, but are of short duration such as from strong winds (paragraph 3-25) and construction equipment surcharges.

4-5. Coastal Flood Walls

A. CASE C1, SURGE STILLWATER LOADING. The backfill is in place to the final elevation; the water is at the surge stillwater level on the unprotected side; wave forces are excluded; uplift is acting.

B. CASE C2A, NONBREAKING WAVE LOADING. This is the same as Case C1 with a nonbreaking wave loading added, if applicable; uplift is the same as for Case C1.

C. CASE C2B, BREAKING WAVE LOADING. This is the same as Case C1 with a breaking wave loading added, if applicable; uplift is the same as for Case C1.

D. C2C, BROKEN WAVE LOADING. This is the same as Case C1 with a broken wave loading added, if applicable; uplift is the same as for Case C1.

E. CASE C3, EARTHQUAKE LOADING. The backfill is in place to the final elevation; water is at the usual (nonstorm) level; uplift, if applicable, is acting; earthquake-induced lateral and vertical loads, if applicable, are acting. (Note: If the wall has no significant load during the usual [nonstorm] stage, no earthquake case is necessary.)

F. CASE C4, CONSTRUCTION SHORT-DURATION LOADING. The flood wall is in place with the loads added which are possible during the construction period but are of short duration, such as from strong winds and construction equipment surcharges.

G. CASE C5, WIND LOADING. The backfill is in place to the final elevation; water is at the usual [non-storm] level on the unprotected side; a wind load of 50 lb/sq ft on the protected side of the wall is applied (paragraph 3-25).

Section II. Stability Considerations

4-6. General Requirements

Figure 4-1 illustrates the potential failure modes for which stability must be analyzed. The basic requirements for the stability of a retaining or flood wall for all loading conditions are discussed below.

A. The wall should be safe against sliding at its base, through any soil layer or rock seam below the base.

B. The wall should be safe against overturning at its base, and, in the case of gravity walls, at any horizontal plane within the wall.

C. The wall should be safe against bearing failure and excessive differential settlement in the foundation.

4-7. Stability Criteria

The stability criteria for retaining walls and inland and coastal flood walls are listed, by loading case, in Tables 4-1 through 4-3.

Section III. Overturning Stability

4-8. Resultant Location

A. GENERAL COMPUTATIONS. To assess the overturning stability of a wall, such as the one with a horizontal base shown in Figure 4-2 (see Examples 1, 2, 3, 5, 6, and 7 of Appendix N), all operative forces must be applied to a free body of the structural wedge wall/soil system. Methods to calculate the lateral and uplift forces are discussed in Chapter 3. The moments of these forces are summed about point O as shown in Figure 4-2 and the distance x_R is calculated as:

$$x_R = \frac{\text{summation of moments about point O}}{\sum V} \qquad [4\text{-}1]$$

where

$\sum V$ = resultant base force required for vertical equilibrium

A ratio defined as the resultant ratio is computed as follows:

$$\text{Resultant ratio} = \frac{x_R}{\text{horizontal width of base}} \qquad [4\text{-}2]$$

Equations 4-1 and 4-2 are valid for a wall with a horizontal base with or without a key and for a wall with a sloped base and a key. If a wall has only a sloped base (no key), as shown in Figure 4-3 (see Example 4 of Appendix N), x_R is calculated as:

$$x_R = \frac{\text{summation of moments about point O}}{\text{effective normal base force, } N'} \qquad [4\text{-}3]$$

The resultant ratio is defined as

$$\text{Resultant ratio} = \frac{x_R}{\text{sloped base width}} \qquad [4\text{-}4]$$

The resultant ratio is related to the percent of the base in compression as shown in Figure 4-4. The percent of the base of the structure which is in compression should be checked for compliance with the overturning stability criteria discussed in paragraph 4-9.

B. WALLS WITH KEYS

1. Performing an overturning stability analysis on a wall with a key requires determining the resisting forces acting along the key and along the base. Since these forces are indeterminate and

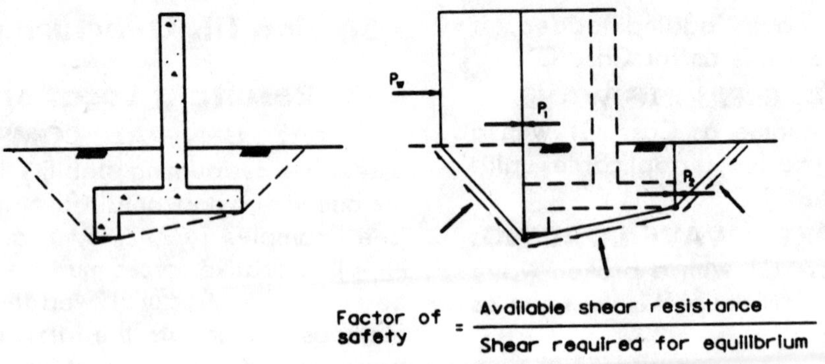

a. Sliding

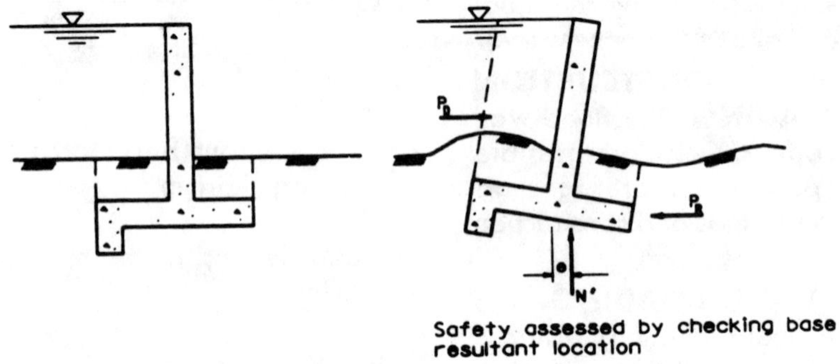

b. Overturning

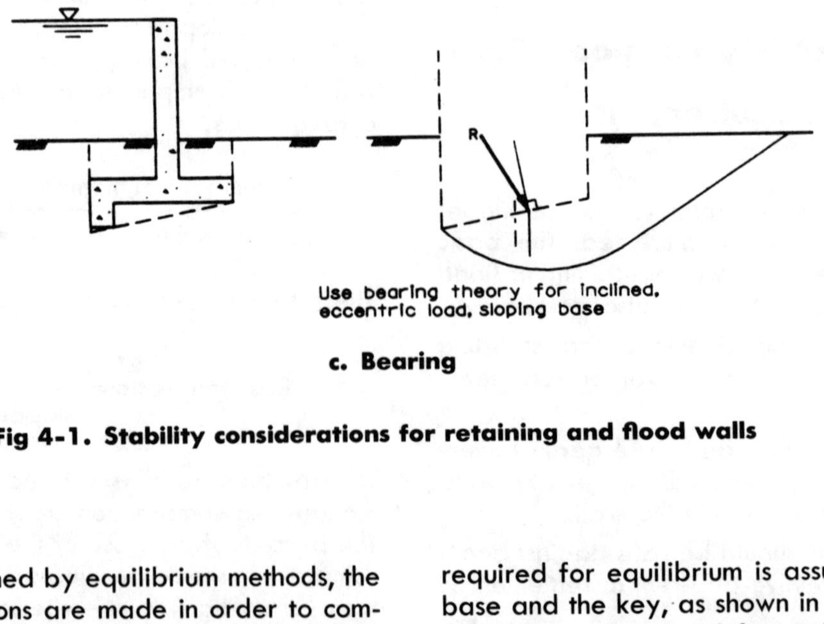

c. Bearing

Fig 4-1. Stability considerations for retaining and flood walls

cannot be determined by equilibrium methods, the following assumptions are made in order to compute the overturning stability. For a wall with a horizontal base and a key, the shearing resistance of the base is assumed to be zero and the horizontal resisting force acting on the key is that required for equilibrium, as shown in Figure 4-5. For a wall with a sloping base and a key, the horizontal force required for equilibrium is assumed to act on the base and the key, as shown in Figure 4-6. In both cases the resisting soil force down to the bottom of the toe may be computed using at-rest earth pressure if the material on the resisting side will not lose its resistance characteristics with any probable change in water content or environmental conditions and will not be eroded or excavated

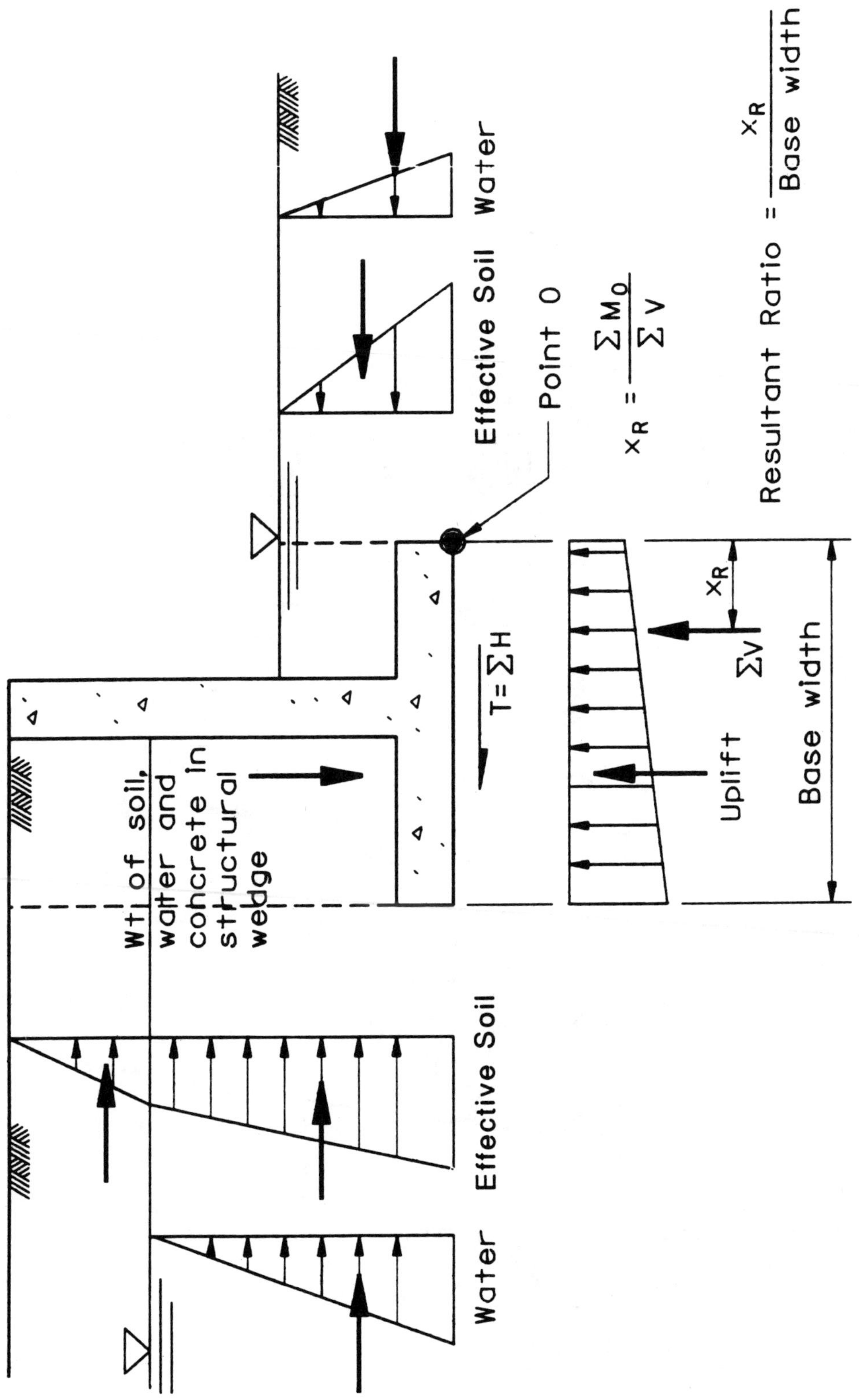

Fig 4-2. Forces for overturning analysis for wall with horizontal base

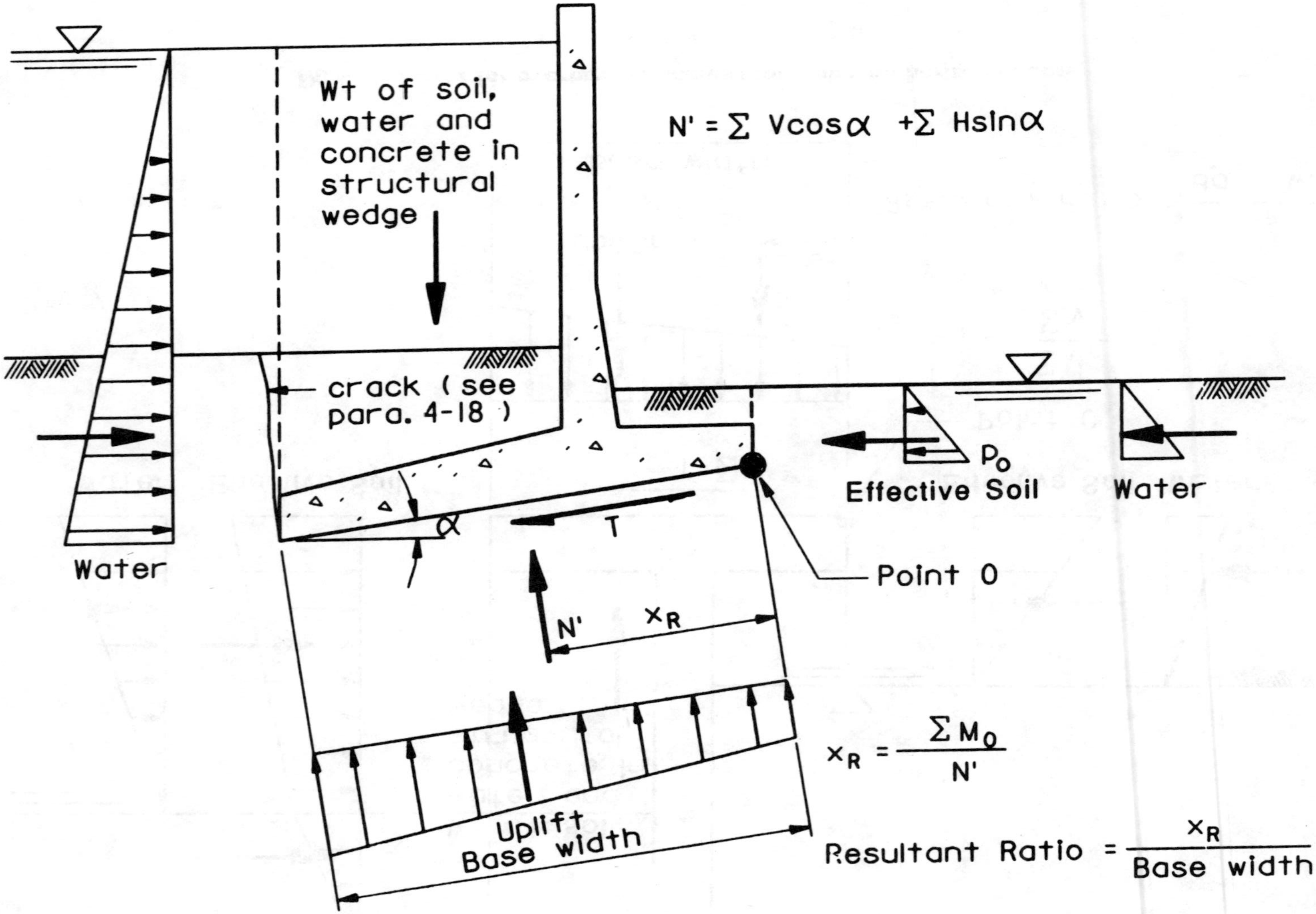

Fig 4-3. Forces for overturning analysis for wall with a sloping base

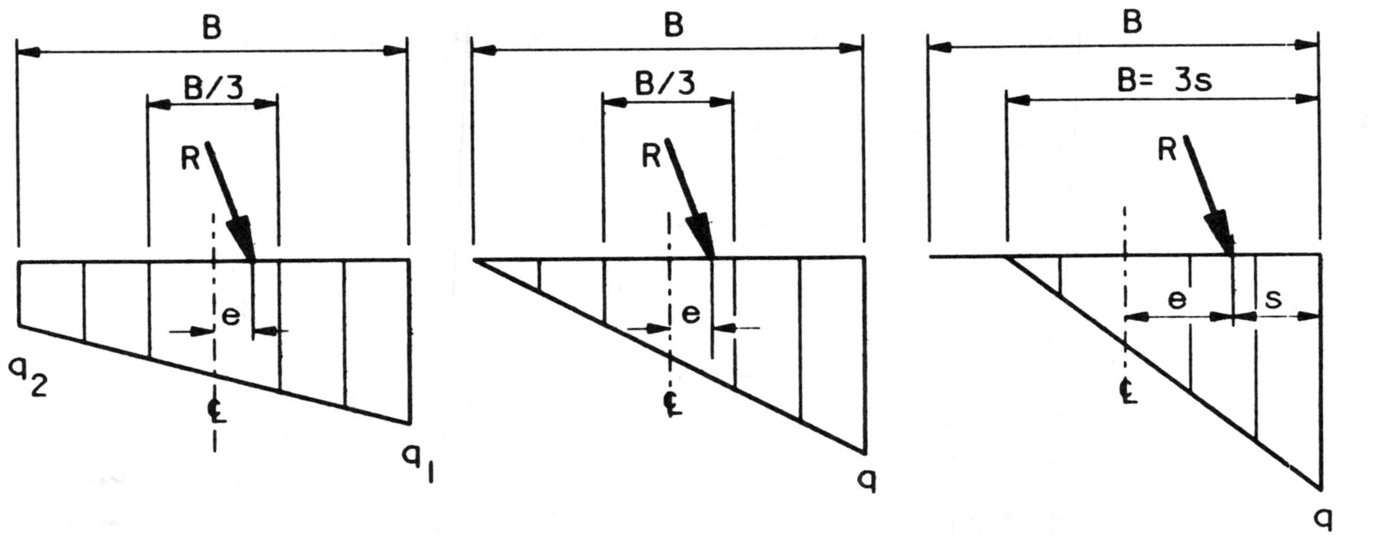

100% base in compression. Resultant within central third of base.

100% base in compression. Resultant at 1/3 point.

Less than 100% base in compression. Resultant outside central third.

BASE AREA IN COMPRESSION	RESULTANT RATIO $= \dfrac{x_R}{\text{Base width}}$
100%	0.500 to 0.333
75%	0.25
60%	0.20
50%	0.1667
40%	0.1333

Fig 4-4. Relationship between base width in compression and resultant location

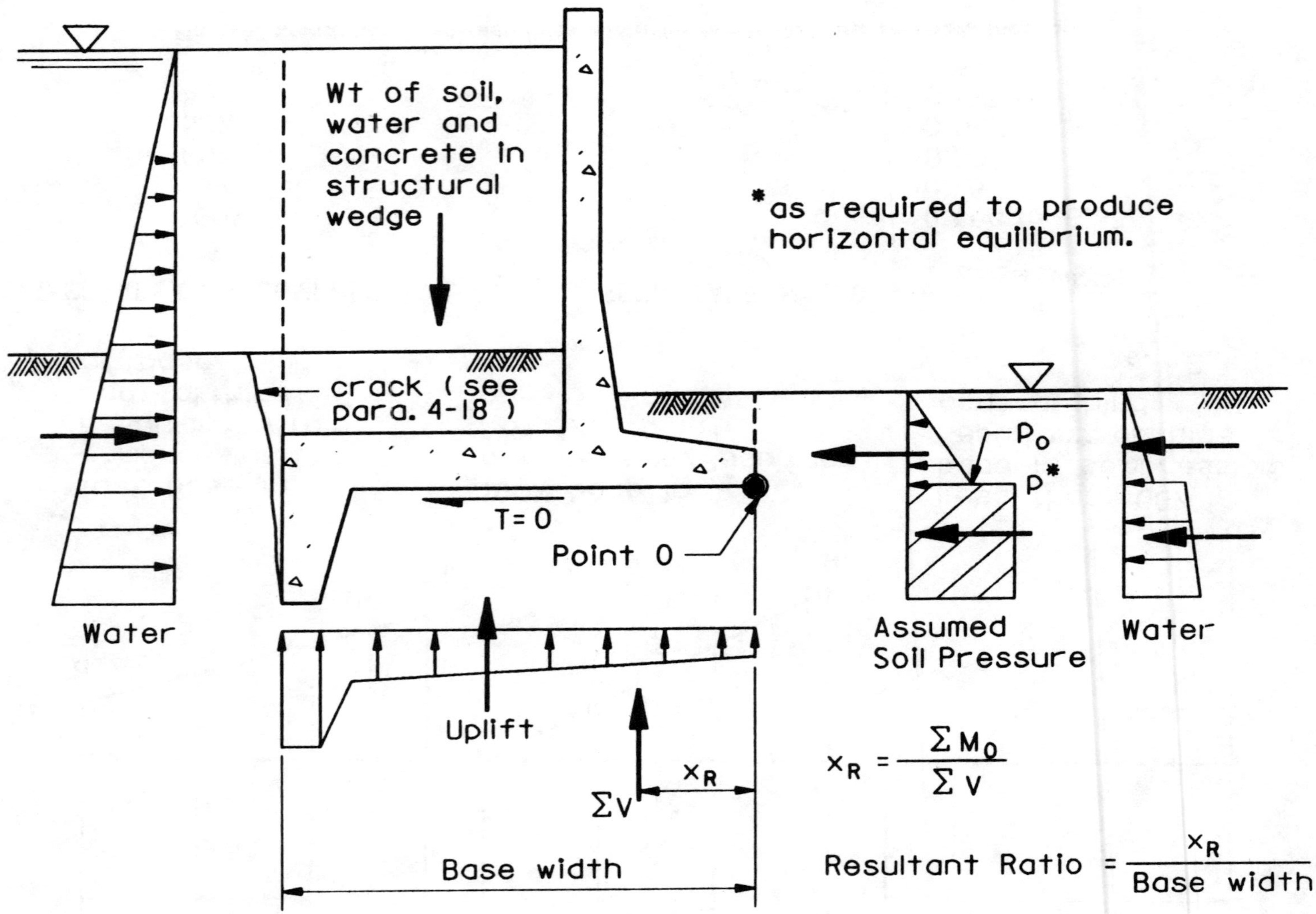

Fig 4-5. Forces for overturning analysis for wall with horizontal base and key

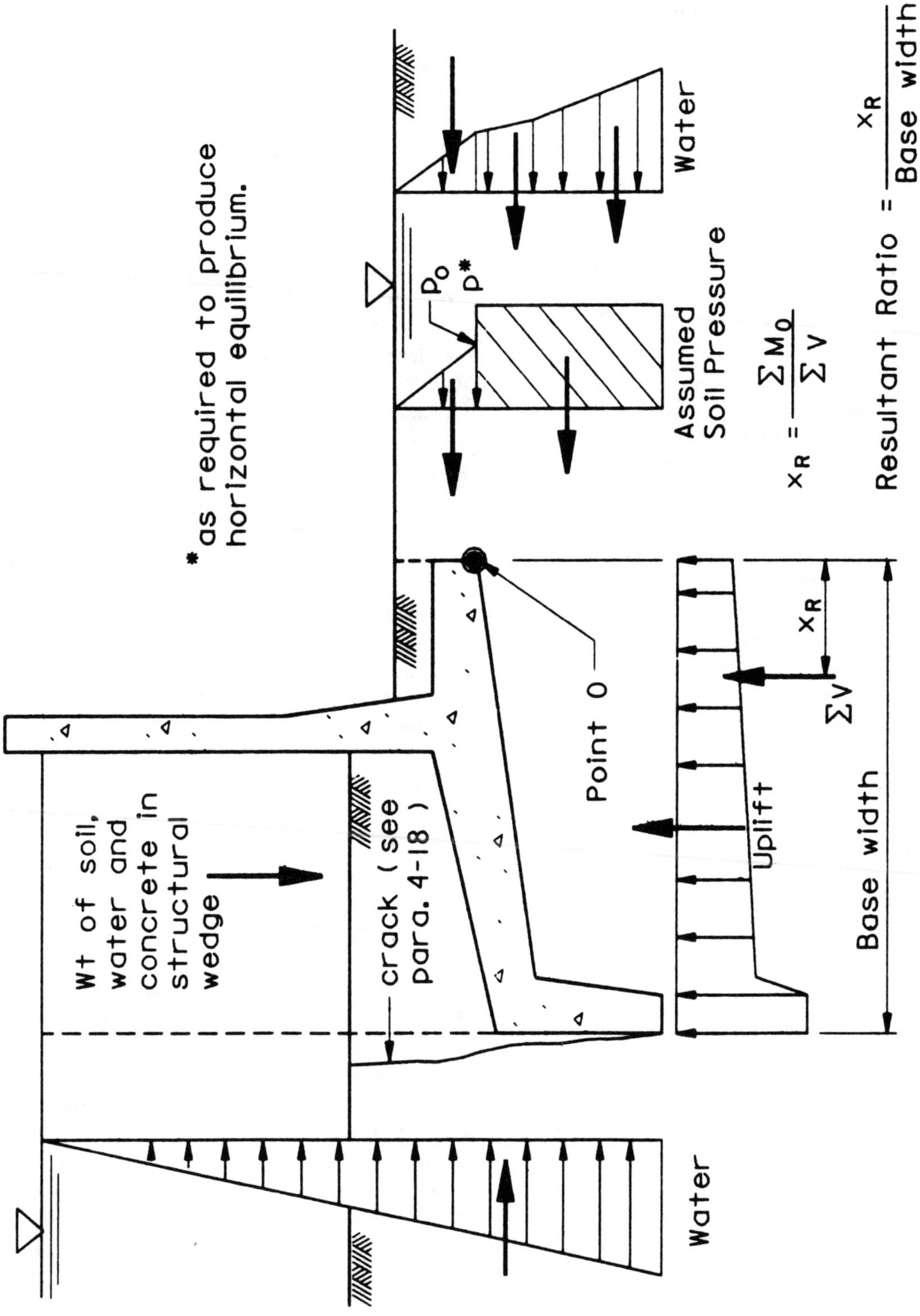

Fig 4-6. Forces for overturning analysis for wall with sloping base and key

during the life of the wall. See Examples 3 and 6 of Appendix N for stability analyses of walls with keys.

2. Prior to performing an overturning analysis, the depth of the key and width of the base should be determined from a sliding stability analysis.

C. SLOPING BACKFILLS. For an upward-sloping backfill, an additional shear force can be taken advantage of in the overturning analysis. The calculation of this shear force is shown in Figure 4-7. The magnitude of this shear force is just large enough to cause the horizontal forces acting on the stem to be equal to the part of the horizontal wedge force that lies above the heel of the wall. This will cause the force used for the structural design of the stem to be equal to the force used in the stability analyses. This force will also cause the summation of moments about the stem-toe-heel joint to equal zero for the structural design. The derivation of this shear force is given in Appendix K. A wall with a sloping backfill is shown in Example 1 of Appendix N.

D. UPLIFT FOR WALLS WITH KEYS. For walls with keys, the soil may be assumed to remain in contact with the key and head loss to occur around the perimeter of the key and along the base even if the overturning analysis shows less than 100 percent of the base in compression.

4-9. Overturning Stability Criteria

The overturning stability requirements in Tables 4-1 through 4-3 are given as minimum percent base areas in compression. Figure 4-4 illustrates the relationship between the percent of the base area in compression and the resultant location.

Section IV. Structure Sliding Stability

4-10. Overview of Sliding Stability Analysis

A. PURPOSE. The purpose of a sliding stability analysis is to assess the safety of a structure against a potential failure due to excessive horizontal deformations. The potential for a sliding failure may be assessed by comparing the applied shear forces to the available resisting shear forces along an assumed failure surface. A sliding failure is imminent when the ratio of the applied shear forces to the available resisting shear forces is equal to 1.

B. ANALYSIS MODEL

1. The shape of the failure surface may be irregular depending on the homogeneity of the backfill and foundation material. The failure surface may be composed of any combination of plane and curved surfaces. However, for simplicity all failure surfaces are assumed to be planes which form the bases of wedges as shown in Figure 4-8.

2. Except for very simple cases, most sliding stability problems encountered in engineering practice are statically indeterminate. To reduce a problem to a statically determinate one, the problem is simplified by dividing the system into a number of rigid body wedges, arbitrarily assuming the direction of the moment equilibrium forces which act between the wedges, and neglecting any frictional forces between adjacent wedges.

3. Figure 4-8 also illustrates how the failure surface would be divided into wedges. The base of a wedge is formed from either a section of the failure surface that lies in a single soil material or along the base of the structure. The interface between any two adjacent wedges is assumed to be a vertical plane which extends from the intersection of the corners of the two adjacent wedges upward to the top soil surface. The base of a wedge, the vertical interface on each side of the wedge, and the top soil surface between the vertical interfaces define the boundaries of an individual wedge.

4. In the sliding analysis, the retaining or flood wall and the surrounding soil are assumed to act as a system of wedges as shown in Figure 4-8. The soil-structure system is divided into one or more driving wedges, one structural wedge, and one or more resisting wedges.

5. Depending on the geologic conditions of the foundation material, the the location of the total failure surface or parts of the failure surface may be predetermined. The inclination of some of the failure planes or the starting elevation of the failure planes adjacent to the structure may be known due to natural constraints at the site. Conditions which warrant the predetermination of parts of the failure surface include bedding planes or cracks in a rock foundation.

C. ANALYSIS PROCEDURE OF THE SOIL-STRUCTURE SYSTEM. An iterative procedure can be used to find the critical failure surface. For an assumed factor of safety, the inclination of the base of each wedge is varied to produce a maximum driving force for a driving wedge or a minimum resisting force for a resisting wedge. The assumed factor of safety affects the critical inclina-

Table 4-1. Retaining wall stability criteria

Case No.	Loading Condition	Sliding Factor of Safety, FS	Shear Strength Test Required: Soil Foundation	Shear Strength Test Required: Rock Foundation[3]	Overturning Criteria Minimum Base Area in Compression: Soil Foundation	Overturning Criteria Minimum Base Area in Compression: Rock Foundation	Minimum Bearing Capacity Safety Factor
R1	Usual	1.5	(Q &/or S)[1,2]	Direct shear	100%[4]	75%[4]	3.0
R2	Unusual	1.33	(Q &/or S)[1,2]	Direct shear	75%[4]	50%[4]	2.0
R3	Earthquake	1.1	(Q)	Direct shear	Resultant within base	Resultant within base	>1.0

Notes:

1. For soil foundations which are not free draining (permeability $< 10 \times 10^{-4}$ cm/sec), analyze for both Q and S strengths and design for the worst condition. For free-draining soil foundations (permeability $> 10 \times 10^{-4}$ cm/sec), analyze for S strengths only.
2. For construction loadings in Cases R1 and R2, use Q strengths when excess porewater pressure in the soil foundation is anticipated and use S strengths when it is not anticipated.
3. The sliding analysis of a wall on rock should be based on the frictional resistance (tan ϕ) of concrete on rock or rock on rock. The values should be obtained from direct shear tests of precut samples of concrete on rock and rock on rock, or direct shear tests of natural rock joints or bedding planes.
4. Less base area in compression than the minimum shown may be acceptable provided adequate safety against unacceptable differential settlement and bearing failure is obtained.

Table 4-2. Inland flood wall stability criteria

Case No.	Loading Condition	Sliding Factor of Safety, FS	Shear Strength Test Required: Soil Foundation	Shear Strength Test Required: Rock Foundation[3]	Overturning Criteria Minimum Base Area in Compression: Soil Foundation	Overturning Criteria Minimum Base Area in Compression: Rock Foundation	Minimum Bearing Capacity Safety Factor
I1	Design flood	1.5	(Q &/or S)[1]	Direct shear	100%[4]	75%[4]	3.0
I2	Water to top of wall	1.33	(Q &/or S)[1]	Direct shear	75%[4]	50%[4]	2.0
I3	Earthquake	1.1	(Q)	Direct shear	Resultant within base	Resultant within base	>1.0
I4	Construction	1.33	(Q &/or S)[2]	Direct shear	75%[4]	50%[4]	2.0

Notes:

1. For soil foundations which are not free draining (permeability $< 10 \times 10^{-4}$ cm/sec), analyze for both Q and S strengths and design for the worst condition. For free-draining soil foundations (permeability $> 10 \times 10^{-4}$ cm/sec), analyze for S strengths only.
2. For construction loadings cases, use Q strengths when excess porewater pressure in the soil foundation is anticipated and S strengths when it is not anticipated.
3. The sliding analysis of a wall on rock should be based on the frictional resistance (tan ϕ) of concrete on rock or rock on rock. The values should be obtained from direct shear tests of precut samples of concrete on rock and rock on rock, or direct shear tests of natural rock joints or bedding planes.
4. Less base area in compression than the minimum shown may be acceptable provided adequate safety against unacceptable differential settlement and bearing failure is obtained.

Table 4-3. Coastal flood wall stability criteria

Case No.	Loading Condition	Sliding Factor of Safety, FS	Shear Strength Test Required: Soil Foundation	Shear Strength Test Required: Rock Foundation[3]	Overturning Criteria Minimum Base Area in Compression: Soil Foundation	Overturning Criteria Minimum Base Area in Compression: Rock Foundation	Minimum Bearing Capacity Safety Factor
C1	Surge stillwater	1.5	(Q &/or S)[1]	Direct shear	100%[5]	75%[5]	3.0
C2	Wave						
C2a	Nonbreaking	1.33	(Q &/or S)[1]	Direct shear	75%[5]	50%[5]	2.0
C2b	Breaking	1.25	(Q)[4]	Direct shear	60%[5]	40%[5]	1.5
C2c	Broken	1.33	(Q)[4]	Direct shear	75%[5]	50%[5]	2.0
C3	Earthquake	1.1	(Q)	Direct shear	Resultant within base	Resultant within base	>1.0
C4	Construction	1.33	(Q &/or S)[2]	Direct shear	75%[5]	50%[5]	2.0
C5	Wind	1.33	(Q &/or S)[2]	Direct shear	75%[5]	50%[5]	2.0

Notes:

1. For soil foundations which are not free draining (permeability $< 10 \times 10^{-4}$ cm/sec), analyze for both Q and S strengths and design for the worst condition. For free-draining soil foundations (permeability $> 10 \times 10^{-4}$ cm/sec), analyze for S strengths only.
2. For construction loadings cases, use Q strengths when excess porewater pressure in the soil foundation is anticipated and S strengths when it is not anticipated.
3. The sliding analysis of a wall on rock should be based on the frictional resistance (tan ϕ) of concrete on rock or rock on rock. The values should be obtained from direct shear tests of precut samples of concrete on rock and rock on rock, or direct shear tests of natural rock joints or bedding planes.
4. Less base area in compression than the minimum shown may be acceptable provided adequate safety against unacceptable differential settlement and bearing failure is obtained.

tion of the base of each wedge. The factor of safety is varied until a failure surface is produced that satisfies equilibrium. The failure surface which results from this procedure will be the one with the lowest factor of safety. Several base inclinations of the structural wedge, such as those shown in Figure 4-8, should be evaluated to determine the failure surface which has the lowest factor of safety.

4-11. Sliding Factor of Safety

A. GENERAL. Limit equilibrium analysis is used to assess the stability against sliding. A factor of safety (FS) is applied to the factors which affect the sliding stability and are known with the least degree of certainty. These factors are the material strength properties. An FS is applied to the material strength properties in a manner that places the forces acting on the structure and soil wedges into equilibrium. Since the in situ strength parameters of rock and soil are never known exactly, one role of the FS is to compensate for the uncertainty that exists in assigning single values to such important parameters. In other words, the FS compensates for the difference between what may be the real shear strength and the shear strength assumed for the analysis.

B. DEFINITION

1. A state of limiting equilibrium is said to exist when the applied shear stresses are equal to the maximum shear strength along a potential failure surface. Therefore, a structure is stable against sliding along a potential failure surface when the applied shear stress is less than the available shear strength along that surface. The ratio of the shear strength to the applied shear stress along a potential failure surface is defined as the FS, as shown in Equation 4-5.

$$FS = \frac{\tau_f}{\tau} = \frac{\sigma'(\tan\phi) + c}{\tau} \qquad [4\text{-}5]$$

M Tan φ

where

τ_f = maximum shear strength according to the Mohr-Coulomb failure criterion

τ = applied shear stress

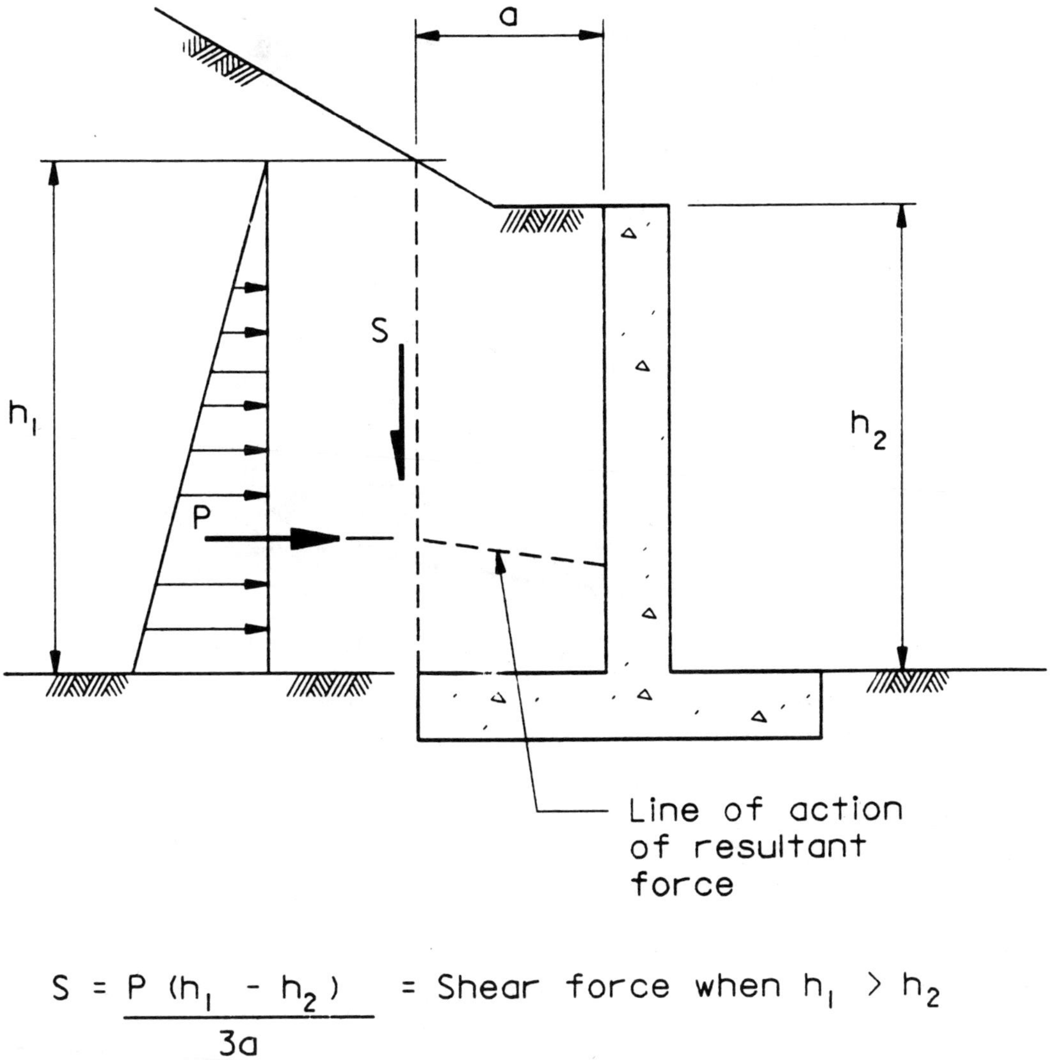

Fig 4-7. Shear force for upward-sloping backfill

2. The sliding FS can also be defined as the ratio of the shear force (T_F) that would cause failure along the slip plane to the corresponding shear force (T) along the slip plane at service conditions (see Figure 4-9):

$$FS = \frac{T_F}{T} = \frac{N' \tan \phi + cL}{T} \quad [4\text{-}6]$$

where L is the length of base in compression for a 1-foot strip of wall. For $c = 0$,

$$FS = \frac{N' \tan \phi}{N' \tan \phi_d} = \frac{\tan \phi}{\tan \phi_d} \quad [4\text{-}7]$$

where $\tan \phi_d$, c_d is that portion of the shear strength considered to be mobilized or developed along the slip plane(s).

4-12. Assumptions and Simplifications

A. SLIP SURFACE. The slip surface can be a combination of planes and curved surfaces, but for simplicity, all slip surfaces are assumed to be planes. These planes form the bases of the wedges. It should be noted that for the analysis to be realistic, the assumed slip planes have to be kinematically possible. In rock, the slip planes may be predetermined by discontinuities in the foundation. If alternate planes are possible, all must be considered to find the most critical.

B. TWO-DIMENSIONAL ANALYSIS. The sliding equilibrium method presented is a two-dimensional analysis. This method should be extended to a three-dimensional analysis if unique three-dimensional geometric features and loads

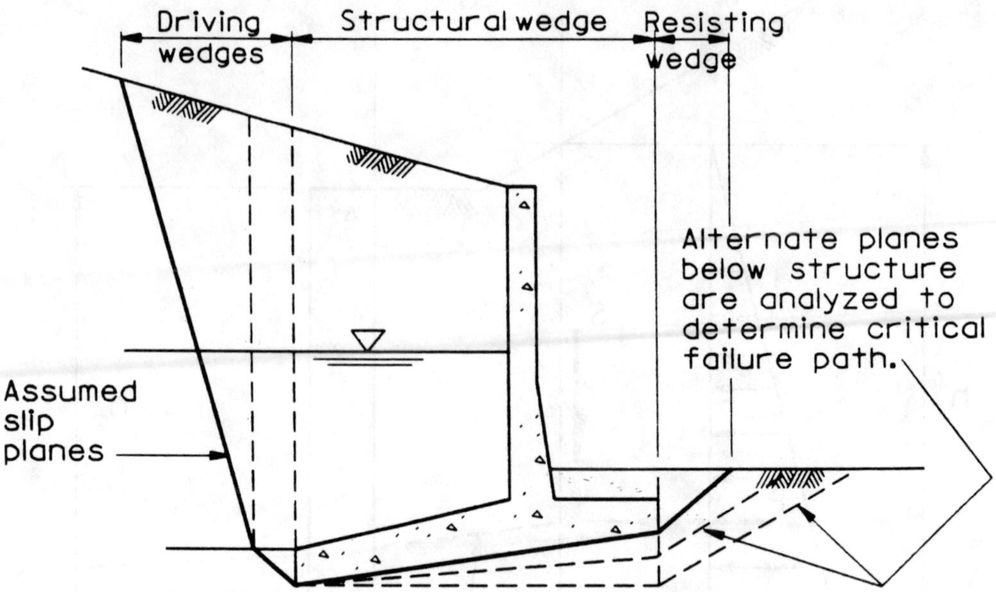

a. Failure plane for wall without a key

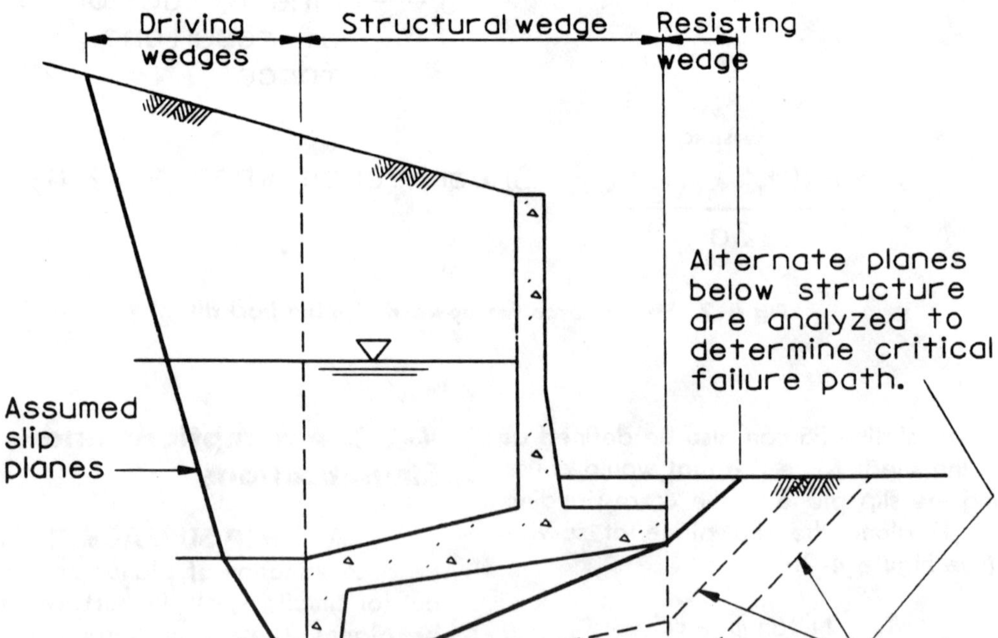

b. Failure plane for wall with a key

Fig 4-8. Typical soil-structure system with an assumed failure surface

critically affect the sliding stability of a specific structure.

C. FORCE EQUILIBRIUM ONLY. Only force equilibrium is satisfied. Moment equilibrium is not considered. The shearing force acting parallel to the interface of any two wedges is assumed to be negligible. Therefore, the portion of the slip surface at the bottom of each wedge is loaded only by the forces directly above or below it. There is no interaction of vertical effects between the wedges. The resulting wedge forces are assumed horizontal.

D. DISPLACEMENTS. Considerations regarding displacements are excluded from the limit equilibrium approach. The relative rigidity of different foundation materials supporting the structure and the concrete structure itself may influence the results of the sliding stability analysis. Such complex structure-foundation systems may require a more intensive sliding investigation than a limit equilibrium approach. The effects of strain compatibility along the assumed slip surface may be approximated in the limit equilibrium approach by selecting the shear strength parameters from in situ or laboratory tests consistent with the failure strain selected for the stiffest material.

E. RELATIONSHIP BETWEEN SHEARING AND NORMAL FORCES. A linear relationship is assumed between the resisting shearing force and the normal force acting on the slip plane beneath each wedge. This relationship is determined by the Mohr-Coulomb failure criterion.

F. STRUCTURAL WEDGE. The general wedge equation is based on the assumption that shearing forces do not act on the vertical wedge boundaries. Hence, there can only be one structural wedge since concrete structures transmit significant shearing forces across vertical internal planes. Discontinuities in the slip path beneath the structural wedge should be modeled by assuming an average slip plane along the base of the structural wedge.

G. INTERFACE OF OTHER WEDGES WITH STRUCTURAL WEDGE. The interface between the group of driving wedges and the structural wedge is assumed to be a vertical plane located at the heel of the structural wedge and extending to the base of the structural wedge. The interface between the group of resisting wedges and the structural wedge is assumed to be a vertical plane located at the toe of the structural wedge and extending to the base of the structural wedge.

4-13. General Wedge Equation

A. SIGN CONVENTION

1. The geometry and sign convention of a typical ith wedge and adjacent wedges are shown in Figure 4-10. The equations for the sliding stability of a general wedge system are derived using a right-hand coordinate system. The origin of each wedge is located at the lower left corner of the wedge. The x-axis is horizontal and the y-axis is vertical.

2. Axes which are tangent (t) and normal (n) to a failure plane are inclined at an angle (α) to the $+x$- and $+y$-axes. A negative angle is formed from a clockwise rotation of the axes. A positive angle is formed from a counterclockwise rotation of the axes.

B. DERIVATION

1. By writing equilibrium equations normal and parallel to the slip plane for a typical wedge as shown in Figure 4-9, solving for N_i' and T_i, and substituting the expressions for N_i' and T_i into Equation 4-6 for the factor of safety of the ith wedge, the following equation results. (Refer to Appendix L for a detailed derivation.)

$$FS = \{[(W_L + V_i) \cos \alpha_i + (H_{Li} - H_{Ri}) \sin \alpha_i = + (P_{i-1} - P_i) \sin \alpha_i - U_i] \tan \phi_i + c_i L_i\} / [(H_{Li} - H_{Ri}) \cos \alpha_i + (P_{i-1} - P_i) \cos \alpha_i - (W_i + V_i) \sin \alpha_i] \quad [4\text{-}8]$$

Solving for $(P_{i-1} - P_i)$ gives the general wedge equation

$$(P_{i-1} - P_i) = [(W_i + V_i)(\tan \phi_{di} \cos \alpha_i + \sin \alpha_i) - U_i \tan \phi_{di} + (H_{Li} - H_{Ri})(\tan \phi_{di} \sin \alpha_i - \cos \alpha_i) + c_{di} L_i] / (\cos \alpha_i - \tan \phi_{di} \sin \alpha_i) \quad [4\text{-}9]$$

where

i = number of wedge being analyzed

$(P_{i-1} - P_i)$ = summation of applied forces acting horizontally on the ith wedge. (A negative value for this term indicates that the applied forces acting on the ith wedge exceed the forces resisting sliding along the base of the wedge. A positive value for this term indicates that the applied forces acting on the ith wedge are less than the forces resisting sliding along the base of the wedge.)

W_i = total weight of water, soil, rock, or concrete in the ith wedge

V_i = any vertical force applied above the top of ith wedge

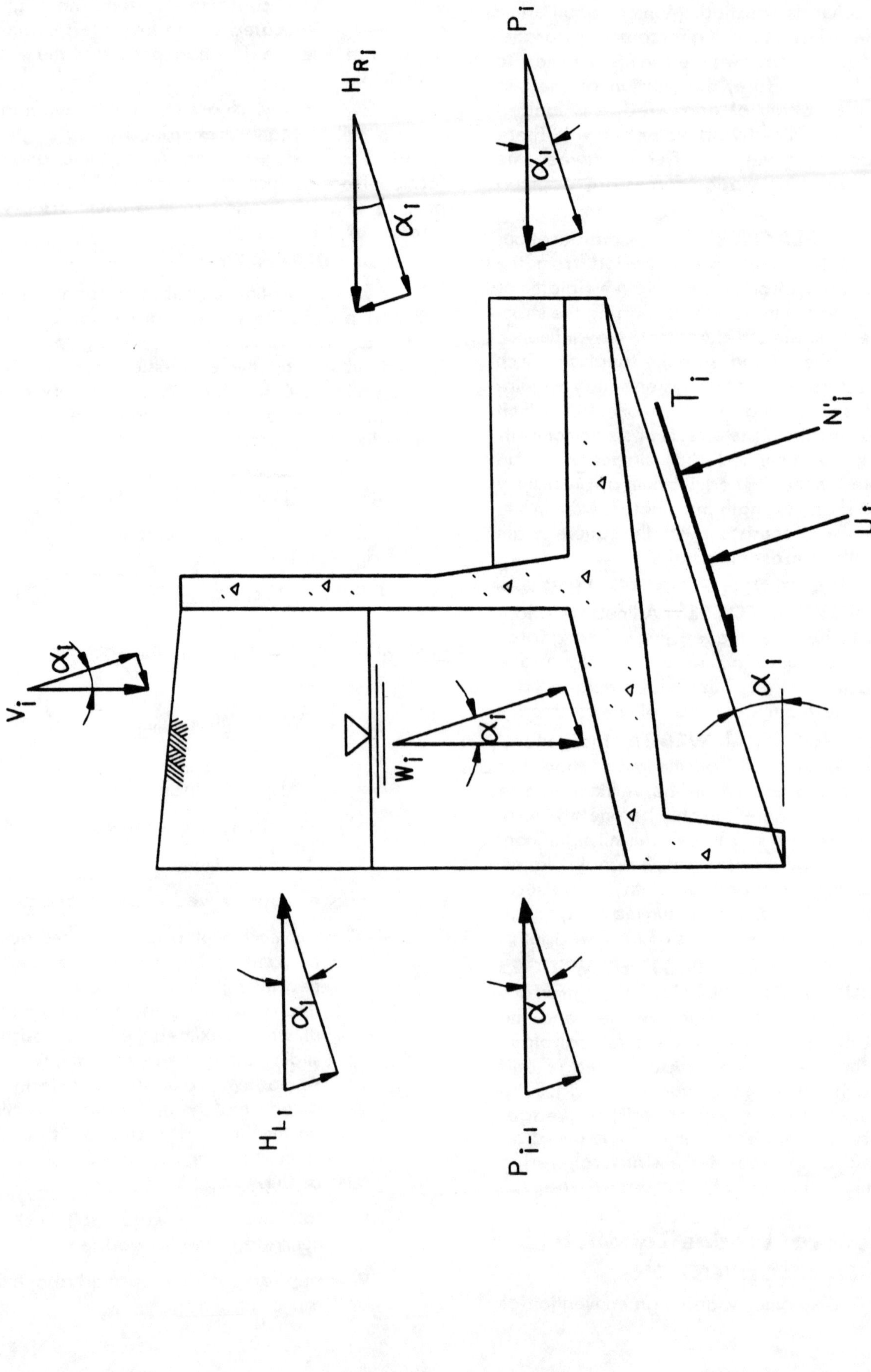

Fig 4-9. Free-body diagram of ith wedge

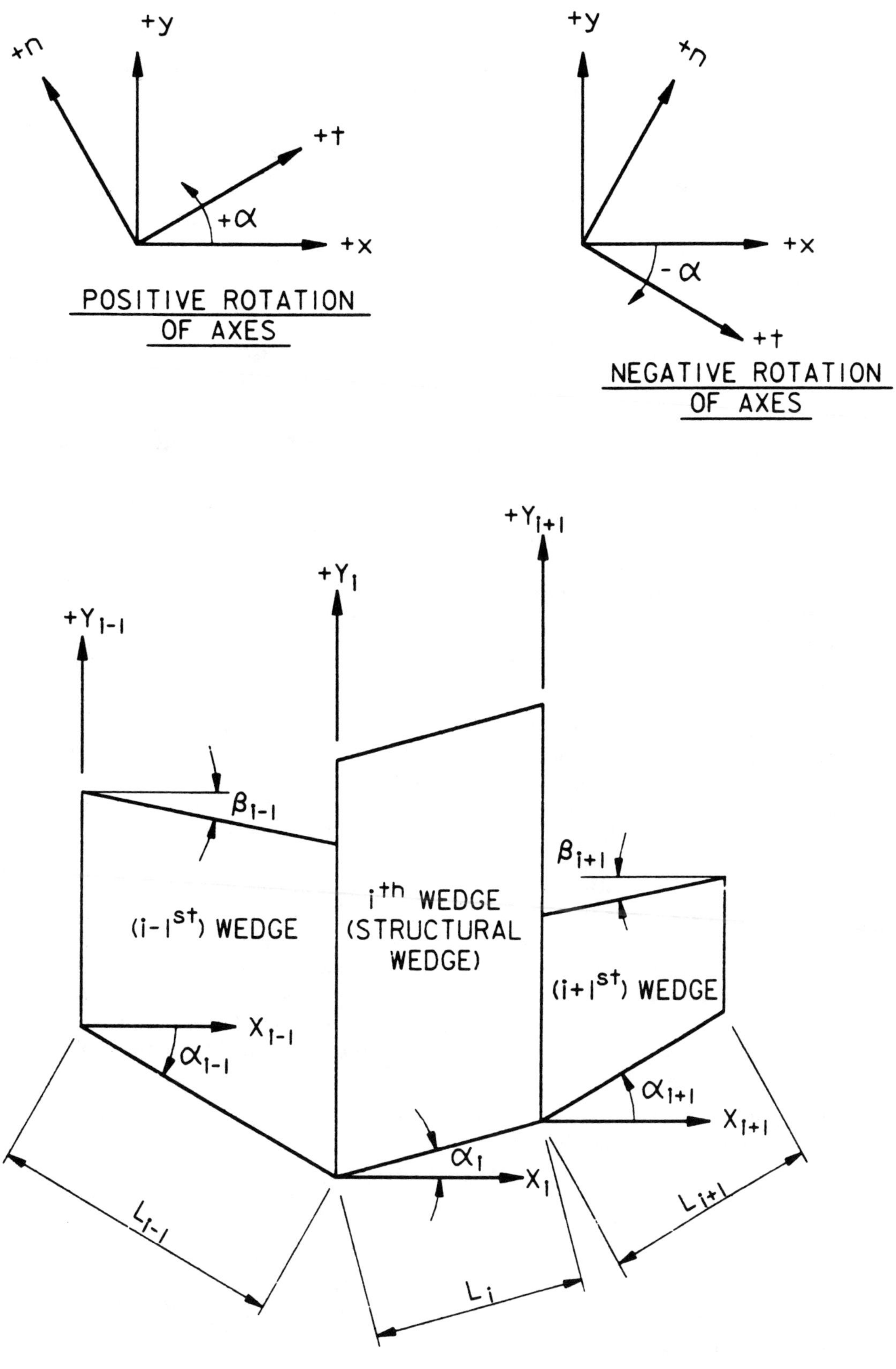

Fig 4-10. Geometry of typical ith wedge and adjacent wedges

$\tan \phi_{di} = \tan \phi_i / FS$

α_i = angle between slip plane of the ith wedge and the horizontal (positive is counterclockwise)

U_i = uplift force exerted along slip plane of the ith wedge

H_{Li} = any horizontal force applied above the top or below the bottom of the left side adjacent wedge

H_{Ri} = any horizontal force applied above the top or below the bottom of the right side adjacent wedge

$c_{di} = c/FS$

L_i = Length along the slip plane of the ith wedge

2. This equation is used to compute the sum of the applied forces acting horizontally on each wedge for an assumed FS. The same FS is used for each wedge. The system of wedges is in equilibrium if the horizontal forces calculated from Equation 4-9, for all wedges, sum to zero.

4-14. Slip-Plane Angle

A. DEFINITION OF CRITICAL SLIP-PLANE ANGLE. The slip-plane angle α varies with the value of the FS. For a driving wedge, the critical α would be the angle that produces a maximum driving force as calculated using Equation 4-9. For a resisting wedge, the critical α would be the angle that produces a minimum resisting force as calculated using Equation 4-9. Since the determination of α is a trial-and-error procedure, for an initial trial the slip-plane angle α for a driving wedge can be approximated by

$$\alpha = 45^\circ + \frac{\phi_d}{2} \qquad [4\text{-}10]$$

where $\phi_d = \tan^{-1}(\tan \phi / FS)$. For a resisting wedge, the slip-plane angle can be approximated by

$$\alpha = 45^\circ - \frac{\phi_d}{2} \qquad [4\text{-}11]$$

B. COMPUTATION OF CRITICAL SLIP-PLANE ANGLE. The above equations for the slip-plane angle are the exact solutions for wedges with a horizontal top surface with or without a uniform surcharge. Other methods to calculate the critical slip angle, for conditions other than a horizontal top surface with or without a uniform surcharge, may be found in paragraph 3-13.

4-15. Single Wedge Analysis

A. INTRODUCTION. A quick check of the sliding stability of a structure can be obtained by performing a single wedge analysis of the structural wedge using the same loadings computed from an overturning analysis if the minimum required sliding FS is no greater than 1.5. If a minimum sliding FS greater than 1.5 is used, driving forces would be larger than the forces calculated from the overturning analysis, which uses an SMF (paragraph 3-11) of two-thirds. In this case, the single wedge equation might incorrectly indicate the structure to satisfy criteria for the larger FS (see paragraph 4-15b[5] for removing this restraint). Example calculations are shown in Appendix N.

B. PROCEDURE FOR SINGLE WEDGE ANALYSIS

1. Compute the sliding resistance required for equilibrium parallel to the assumed sliding plane beneath the structural wedge. Use the forces computed from the overturning analysis for the same loading case being analyzed for sliding. The sliding resistance required for equilibrium is calculated as shown in Figure 4-11.

2. Compute the total sliding resistance available along the assumed sliding plane beneath the structural wedge using the unfactored shear strength parameters and divide the total sliding resistance by the minimum factor of safety required for the case being analyzed.

3. If the sliding resistance needed, as computed in step (1), is equal to or less than the available sliding resistance divided by the minimum sliding factor of safety as computed in step (2), a multiple wedge analysis is not required. A multiple wedge analysis would give a sliding FS equal to or greater than the minimum required. This check on the sliding stability can be expressed by:

$$T \le \frac{N' \tan \phi + cL}{FS} \qquad [4\text{-}12]$$

where

T = resultant of sliding resistance parallel to the assumed sliding plane required for equilibrium

N' = resultant of forces normal to the assumed sliding plane

$\tan \phi$ and c = unfactored shear strength parameters of the foundation material through which the sliding plane passes

L = length of sliding plane beneath the structure

FS = minimum sliding factor of safety required

If the assumed sliding plane is horizontal, T would equal the resultant of the horizontal forces and N′ would equal the resultant of the vertical forces. See Example 1 in Appendix N.

4. If Equation 4-12 is not satisfied, perform a multiple wedge analysis to determine the actual sliding factor of safety (see the following paragraph).

5. The necessity for a multiple wedge solution may be eliminated if the driving and resisting wedge forces are calculated using the minimum FS required. If Equation 4-12 is not satisfied for the FS required, a multiple wedge solution will show the same results. If Equation 4-12 is satisfied, the system has an FS equal to or greater than the minimum FS required.

4-16. Multiple Wedge Analysis

A. PROCEDURE

1. Divide the assumed sliding mass into a number of wedges, including a single structural wedge, based on the configuration and discontinuities of the backfill, wall proportions, and discontinuities of the foundation.

2. Estimate the FS for the first trial.

3. Compute the critical sliding angles for each wedge. For a driving wedge, the critical angle is the angle that produces a maximum driving force. For a resisting wedge, the critical angle is the angle that produces a minimum resisting force.

4. Compute the uplift pressures, if any, along the slip plane. The effects of seepage should be included.

5. Compute the weight of the wedges, including any water and surcharges.

6. Compute the summation of the lateral forces for each wedge using the general wedge equation. In certain cases where the loadings or wedge geometries are complicated, the critical angles of the wedges may not be easily calculated. The general wedge equation may be used to iterate and find the critical angle of a wedge by varying the angle of the wedge to find a minimum resisting or maximum driving force.

7. Sum the lateral forces for all the wedges.

8. If the sum of the lateral forces is negative, decrease the FS and recompute the sum of the lateral forces. By decreasing the FS, a greater percentage of the shearing strength along the slip planes is mobilized. If the sum of the lateral forces is positive, increase the FS and recompute the sum of the lateral forces. By increasing the FS, a smaller percentage of the shearing strength is mobilized.

9. Continue this trial-and-error process until the sum of the lateral forces is approximately zero for the FS used. This will determine the FS that causes the sliding mass to be in horizontal equilibrium.

10. If the FS is less than the minimum required, redesign by widening or sloping the base or by providing a key.

B. COMPUTER PROGRAM. The computer program CSLIDE (Appendix O) can assist in performing a multiple wedge sliding analysis.

4-17. Sliding Stability Criteria

The sliding stability criteria are given in terms of a minimum factor of safety for the various loading conditions as shown in Tables 4-1 through 4-3. Guidance on deep-seated sliding is given in Chapter 5.

4-18. Design Considerations

A. EFFECTS OF CRACKS IN FOUNDATION. Sliding analyses should consider the effects of cracks on the active side of the structural wedge in the foundation material due to differential settlement, shrinkage, or joints in the rock mass. The depth of cracking in cohesive foundation material with a plane ground surface can be estimated with the following equations.

$$d_c = \frac{2c_d}{\gamma'\sqrt{K_A}} = \frac{2c_d}{\gamma'}\tan\left(45^\circ + \frac{\phi_d}{2}\right) \quad [4\text{-}13]$$

where

$c_d = c/FS$

$\phi_d = \tan^{-1}(\tan\phi/FS)$

γ', K_A (see Equation 3-11)

For sloping backfills see Appendix I. The value d_c in a cohesive foundation and the depth of cracking in massive strong rock foundations should be assumed to extend to the base of the structural wedge. The depth of cracking in a level clay blanket should be computed using Equation 4-13. Full hydrostatic pressure should be assumed to act at the bottom of the crack. The hydraulic gradient across the base of the structural wedge should reflect the presence of a crack at the heel of the structural wedge. Examples showing the calculation of d_c are found in Appendix N in Examples 3, 4, 5, 6, and 7.

B. PASSIVE RESISTANCE. When passive resistance is used, special considerations must be made. Rock or soil that may be subjected to high velocity water scouring should not be used unless amply protected. Also, the compressive strength of rock layers must be sufficient to develop the wedge resistance. In some cases, wedge resistance should not be assumed without resorting to special treatment, such as rock anchors.

Section V. Bearing Capacity Analysis

4-19. General Computations

The bearing capacity is checked for the same loading conditions as determined by the overturning analysis for each case analyzed. The bearing capacity should be checked along the same plane assumed in the sliding analysis. A normal (N′) and tangent (T) force are calculated for the structural wedge along the assumed bearing plane. These forces are shown in Figure 4-11. T and N′ are used in combination to check the bearing capacity. The bearing capacity analysis discussed in Chapter 5 and in the CBEAR user's guide (Mosher and Pace 1982) (see Appendix O) considers both the normal and tangent components of the resultant force at the base of the structure. The factor of safety against a bearing failure can be computed by dividing the normal component of the ultimate bearing capacity by the effective normal force applied to the structural wedge as shown below:

$$FS = \frac{Q}{N'} \qquad [4\text{-}14]$$

where

Q = normal component of the ultimate bearing capacity

N' = effective normal force applied to the structural wedge

The value computed from the general bearing capacity equation in Chapter 5 is the bearing capacity normal to the base of the structure. The computer program CBEAR (Appendix O) can assist in performing a bearing capacity analysis. Example calculations are shown in Appendix N.

4-20. Inadequate Bearing Capacity

If the factor of safety against bearing failure is insufficient, consideration should be given to increasing the width of the base, lowering the base of the wall, or founding the wall on piles.

4-21. Bearing Capacity Criteria

The criteria for bearing capacity are given in terms of a factor of safety as defined in paragraph 4-19 and shown in Tables 4-1 through 4-3.

Section VI. Summary of Design Procedures

4-22. Design Procedures

Figure 4-12 presents a summary of the design procedures discussed in this chapter.

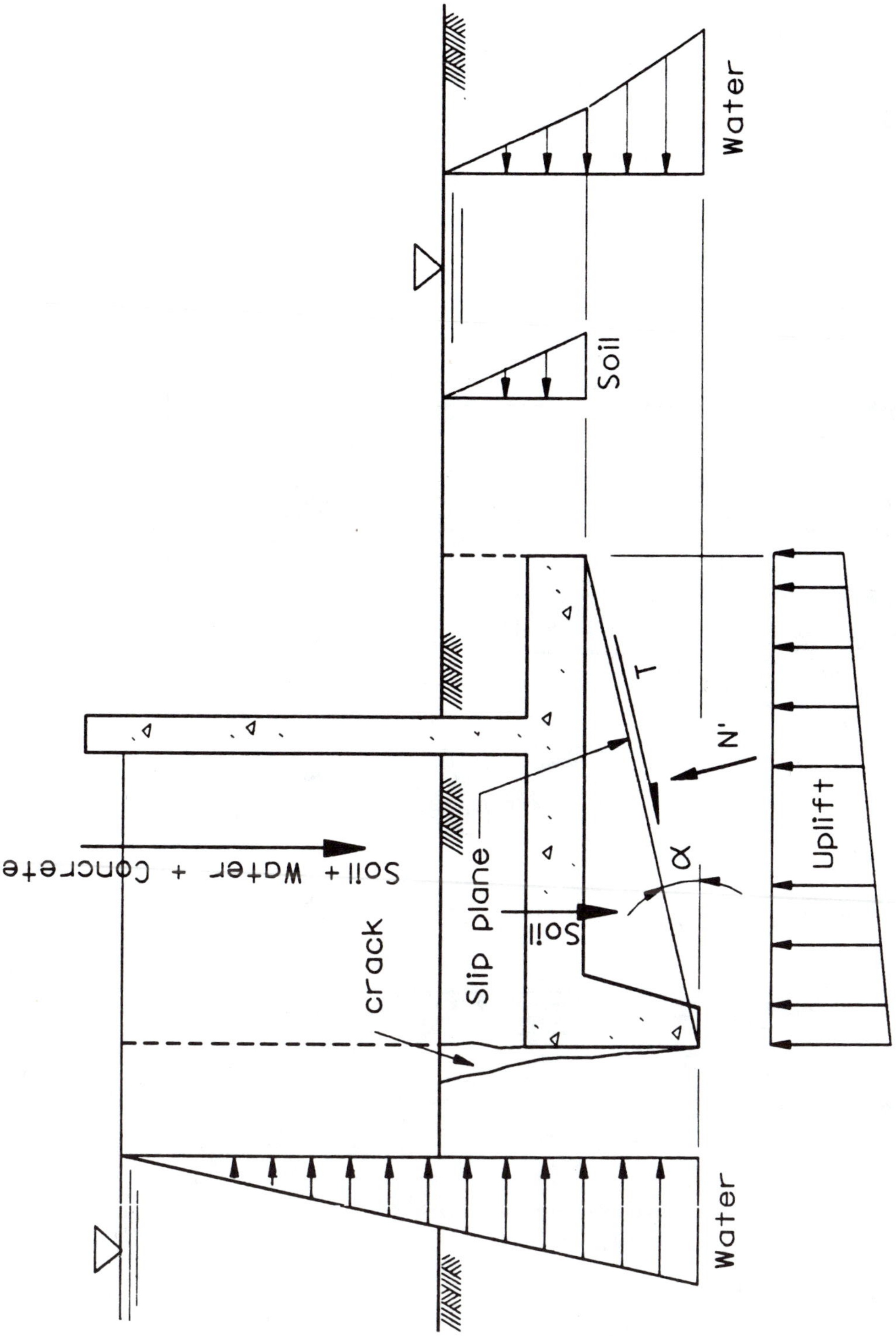

$$N' = \sum V\cos\alpha + \sum H\sin\alpha \quad , \quad T = \sum H\cos\alpha - \sum V\sin\alpha$$

Fig 4-11. Single wedge analysis of sliding stability

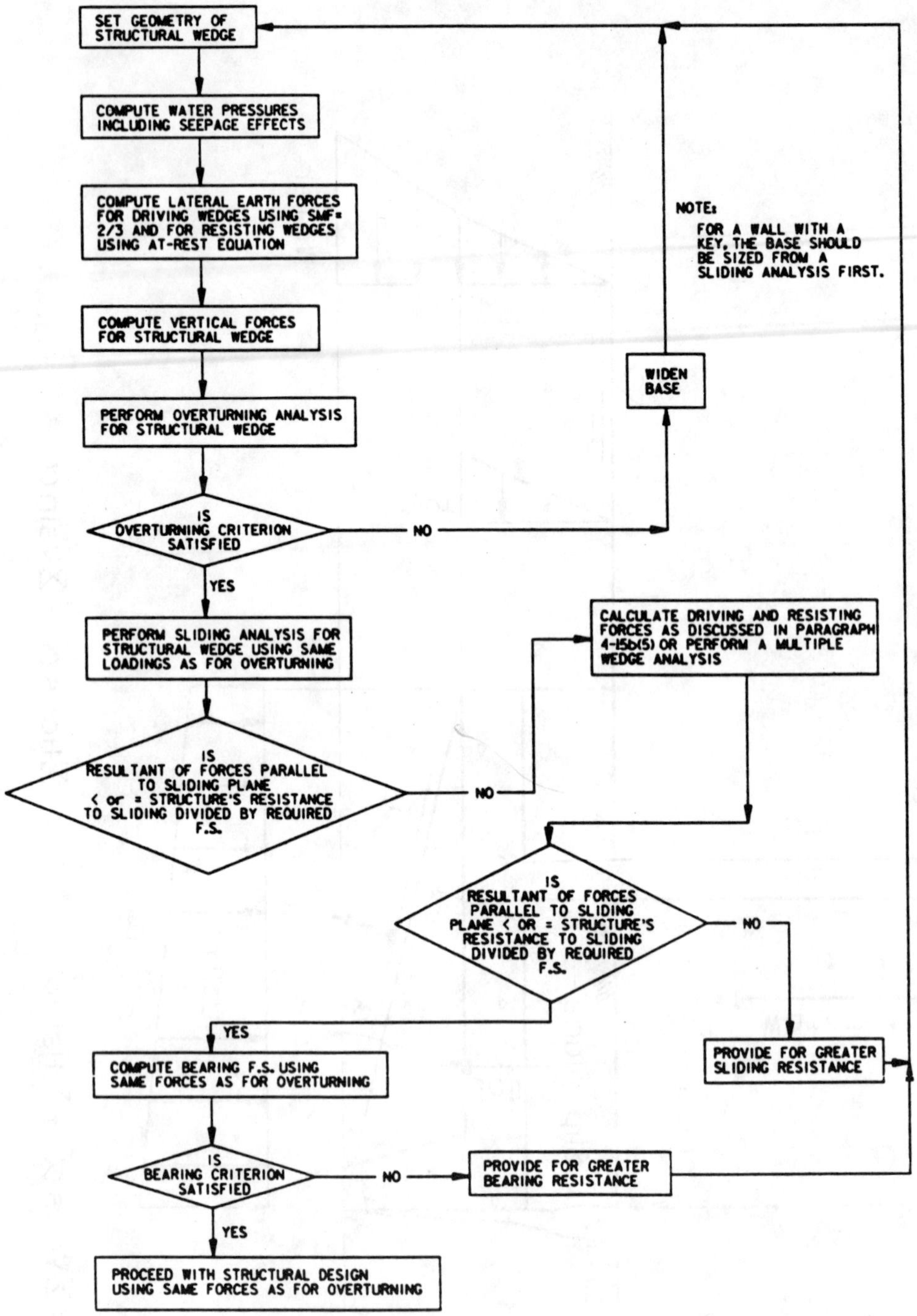

Fig 4-12. Wall design flowchart

CHAPTER 5

FOUNDATION ANALYSES

Section I. Bearing Capacity of Wall Foundations

5-1. Analysis Principles and Methods

A. EM 1110-2-1903. A discussion of the principles and methods involved in analyzing bearing capacity is contained in EM 1110-2-1903. The manual concludes that Terzaghi's general bearing capacity equation, $q = CN_c + wzN_q + WbN_w$, is preferred. However, the manual does not address modifying the general equation for effects of embedment, inclined loads, sloping bases, passive-type wedges with sloping surfaces, overburden pressure, and eccentric loads (moment-induced stresses), all of which are needed for computing the bearing capacity of retaining and flood walls. The computer program CBEAR (Appendix O) can assist in these computations.

B. MODE OF FAILURE. The mode of failure depends on the relative compressibility of the soil, loading conditions, and geometric considerations (Vesic 1975). This manual is restricted to general shear failure of shallow strip foundations, i.e., those whose widths are greater than their embedment. A general shear failure normally exists for dense sand and stiff clay. However, for loose sand and soft clay, which may occur more frequently for flood walls constructed in a flood plain, the bearing capacity should be computed based upon local shear conditions (Vesic 1975).

C. FACTOR OF SAFETY. The FS is calculated as follows;

$$FS = \frac{Q}{N'} \qquad [5\text{-}1]$$

where

N' = effective normal force applied to the base of the structure
Q = normal component to the base of the structure of the ultimate bearing capacity

The minimum acceptable bearing capacity factors for retaining walls and inland and coastal flood walls are listed by loading case in Tables 4-1 through 4-3. For each loading case, the same loadings as determined by the overturning analysis should be used. Options to consider in the event of inadequate bearing capacity have been presented in paragraph 4-20.

5-2. General Bearing Capacity Equation

The general bearing capacity equation for a strip footing is

$$Q = \bar{B}\left[(\xi_{cd}\,\xi_{ci}\,\xi_{ct}\,\xi_{cg}\,cN_c) + (\xi_{qd}\,\xi_{qi}\,\xi_{qt}\,\xi_{qg}\,q_0 N_q) + \frac{(\xi_{\gamma d}\,\xi_{\gamma i}\,\xi_{\gamma t}\,\xi_{\gamma g}\,\bar{B}\gamma N_\gamma)}{2}\right] \qquad [5\text{-}2]$$

where

Q = normal component of the ultimate bearing capacity of the foundation
$\bar{B}$ = effective width of the base ($B - 2e$, as shown in Figure 5-1)
B = width of the geometric base (as shown in Figure 5-1)
e = cohesion parameter of the foundation
ξ = factors as explained in paragraphs 5-4 through 5-8
N_c, N_q, N_γ = bearing capacity factors for a strip load
q_0 = effective overburden pressure on the plane passing through the base of the footing
γ = effective unit weight of the foundation material, $\gamma_{buoyant}$ below water table, γ_{moist} above

Figure 5-1 illustrates the meanings of all of the terms required to use the information given in paragraphs 5-3 through 5-8. The general bearing capacity equation is taken from the CBEAR user's guide (Mosher and Pace 1982) (see also Appendix O). The appropriate soil foundation shear strength for retaining walls and inland and coastal flood walls is listed, by loading case, in Tables 4-1 through 4-3.

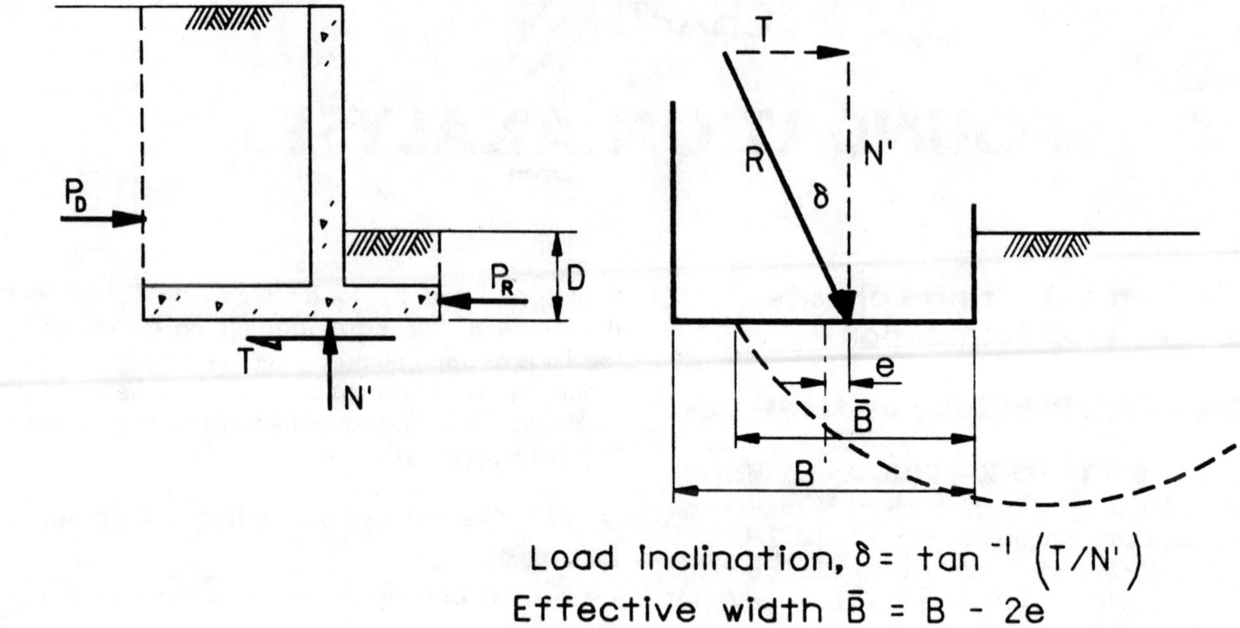

a. Horizontal base

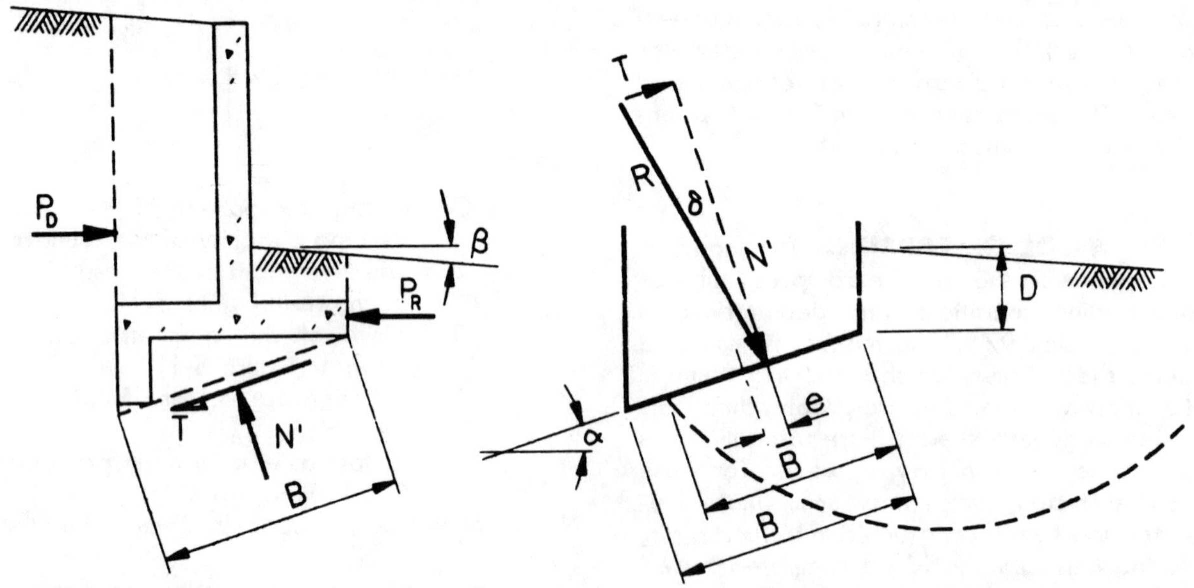

b. Keyed base, sloping ground

Fig 5-1. Terms used in bearing capacity equation

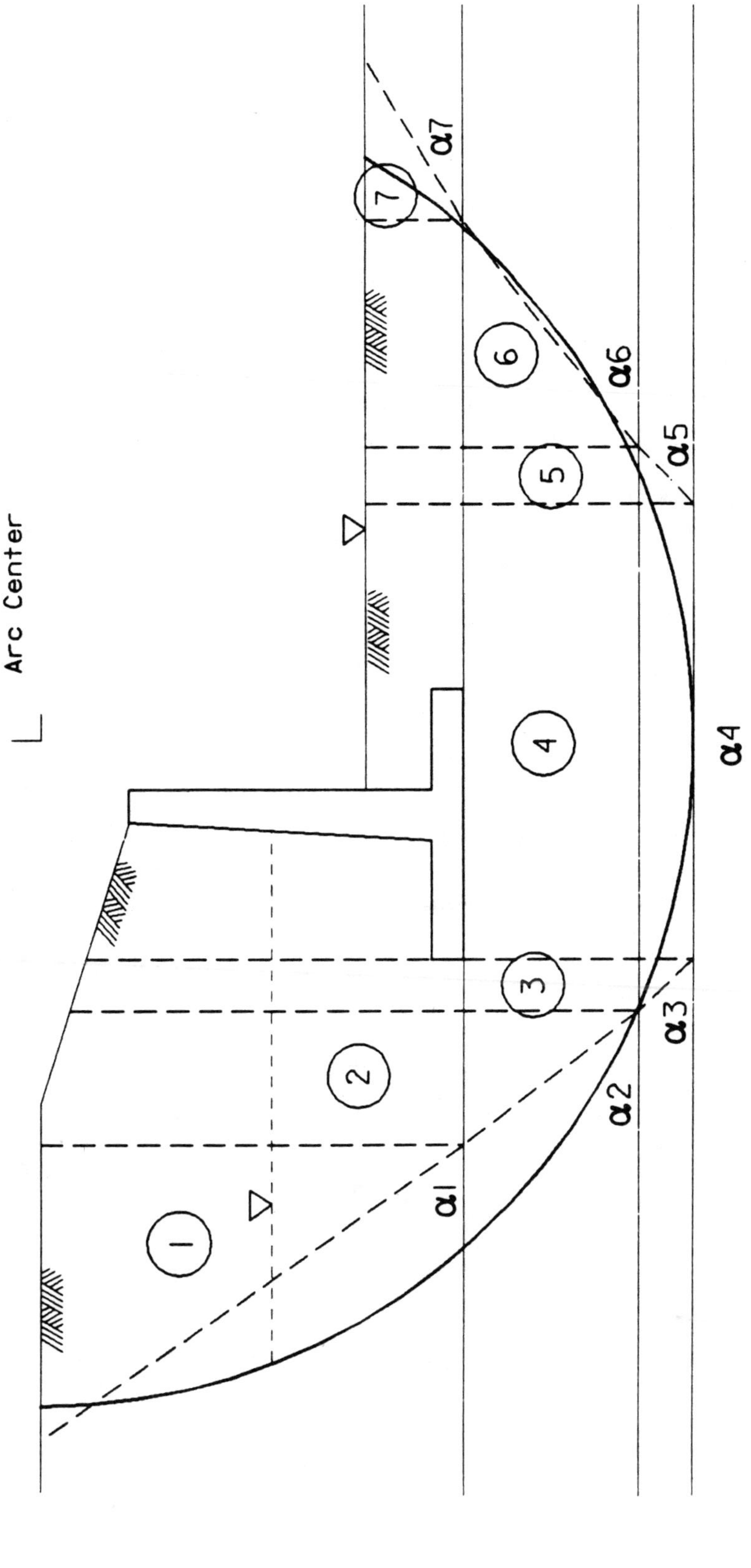

Fig 5-2. Deep-seated sliding analysis with vertical face of driving wedge at heel

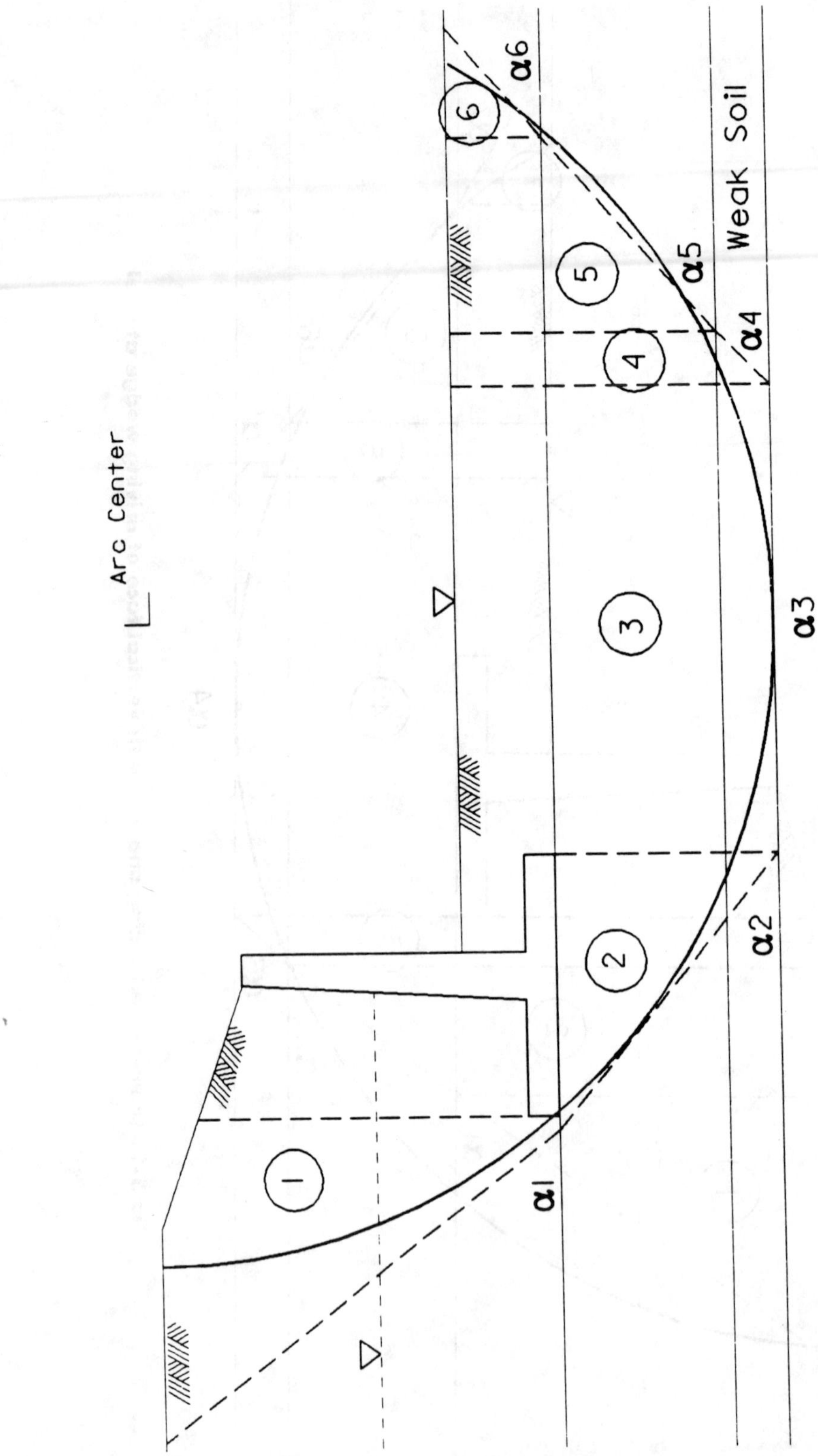

Fig 5-3. Deep-seated sliding analysis with vertical face of driving wedge at toe

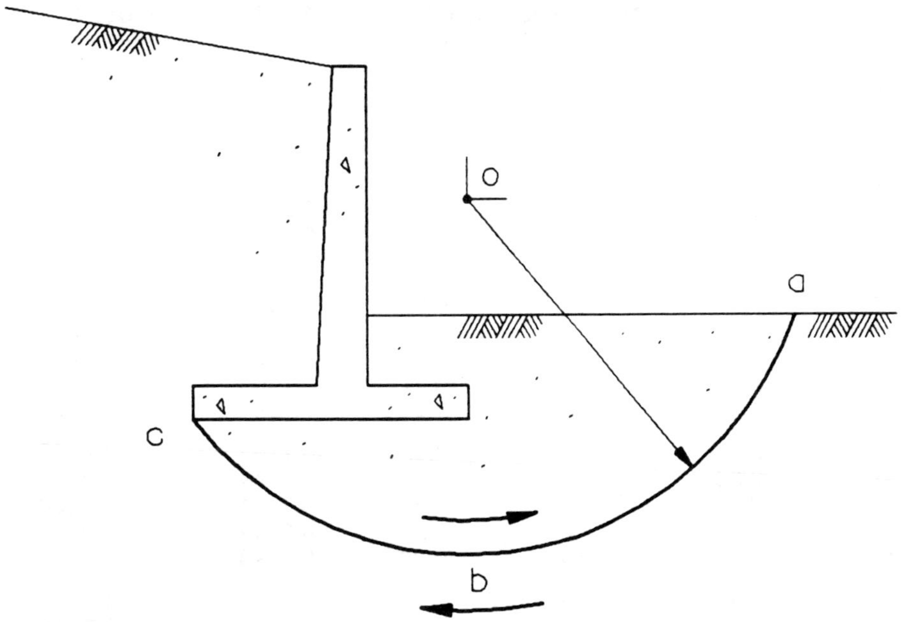

Fig 5-4. Shallow shear surface

5-3. Bearing Capacity Factors

Bearing capacity factors for a horizontal strip footing under vertical loading are:

$$N_q = [e^{(\pi \tan \phi)}]\tan^2\left(45^\circ + \frac{\phi}{2}\right) \qquad [5\text{-}3a]$$

$$N_c = (N_q - 1)\cot \phi \qquad \text{(when } \phi > 0) \qquad [5\text{-}3b]$$

or

$$N_c = 5.14 \qquad \text{(when } \phi = 0) \qquad [5\text{-}3c]$$

$$N_\gamma = (N_q - 1)\tan(1.4\phi) \qquad [5\text{-}3d]$$

Bearing capacity factor values for ϕ, ranging between 0 and 50 degrees, are given in Table 5-1.

5-4. Embedment Factors

Embedment factors take into consideration the shearing resistance along the foundation slip plane that exists in the soil above the base of the footing, on the toe side of a wall. These factors can be computed as

$$\xi_{cd} = 1 + 0.2\left(\frac{D}{\bar{B}}\right)\tan\left(45^\circ + \frac{\phi}{2}\right) \qquad [5\text{-}4a]$$

$$\xi_{qd} = \xi_{\gamma d} = 1 \qquad \text{(when } \phi = 0^\circ) \qquad [5\text{-}4b]$$

or

$$\xi_{qd}\xi_{\gamma d} = 1 + 0.1\left(\frac{D}{\bar{B}}\right)\tan\left(45^\circ + \frac{\xi}{2}\right) \qquad \text{(when } \phi > 10^\circ) \qquad [5\text{-}4c]$$

When ϕ lies between 0 and 10 degrees, a linear interpolation can be made for $\xi_{\gamma d}$ between 1 for $\phi = 0^\circ$, and $1 + 0.1(D/\bar{B})\tan(45^\circ + \phi/2)$ for $\phi = 10^\circ$. Embedment factors account for the shear strength above the base of the footing. Their use may be unconservative if the shear strength does not exist.

5-5. Inclination Factors

Inclination factors account for the effect of load inclination for concentrically loaded foundations. They are computed as follows:

$$\xi_{qi} = \xi_{ci} = \left(1 - \frac{\delta^\circ}{90^\circ}\right)^2 \qquad [5\text{-}5a]$$

$$\xi_{\gamma i} = \left(1 - \frac{\delta^\circ}{\phi}\right)^2 \qquad [5\text{-}5b]$$

Where δ is the angle that the line of action of the load makes with a line drawn normal to the base. if $\delta > \phi$, $\xi_{\gamma i}$ should be set equal to zero.

Table 5-1. Bearing capacity factors (CBEAR user's guide)*

ϕ	N_c	N_q	N_γ	$\tan \phi$	$\tan^2\left(45^\circ + \frac{\phi}{2}\right)$
0	5.14	1.00	0.00	0.0000	1.0000
1	5.38	1.09	0.00	0.0175	1.0355
2	5.63	1.20	0.01	0.0349	1.0723
3	5.90	1.31	0.02	0.0524	1.1105
4	6.19	1.43	0.04	0.0699	1.1500
5	6.49	1.57	0.07	0.0875	1.1910
6	6.81	1.72	0.11	0.1051	1.2335
7	7.16	1.88	0.15	0.1228	1.2776
8	7.53	2.06	0.21	0.1405	1.3233
9	7.92	2.25	0.28	0.1584	1.3709
10	8.34	2.47	0.37	0.1763	1.4203
11	8.80	2.71	0.47	0.1944	1.4716
12	9.28	2.97	0.60	0.2126	1.5250
13	9.81	3.26	0.74	0.2309	1.5805
14	10.37	3.59	0.92	0.2493	1.6382
15	10.98	3.94	1.13	0.2679	1.6984
16	11.63	4.33	1.37	0.2867	1.7610
17	12.34	4.77	1.66	0.3057	1.8263
18	13.10	5.26	2.00	0.3249	1.8944
19	13.93	5.80	2.40	0.3443	1.9655
20	14.83	6.40	2.87	0.3640	2.0396
21	15.82	7.07	3.42	0.3839	2.1171
22	16.88	7.82	4.07	0.4040	2.1980
23	18.05	8.66	4.82	0.4245	2.2826
24	19.32	9.60	5.72	0.4452	2.3712
25	20.72	10.66	6.77	0.4663	2.4639
26	22.25	11.85	8.00	0.4877	2.5611
27	23.94	13.20	9.46	0.5095	2.6629
28	25.80	14.72	11.19	0.5317	2.7698
29	27.86	16.44	13.24	0.5543	2.8821
30	30.14	18.40	15.67	0.5774	3.000
31	32.67	20.63	18.56	0.6009	3.1240
32	35.49	23.18	22.02	0.6249	3.2546
33	38.64	26.09	26.17	0.6494	3.3921
34	42.16	29.44	31.15	0.6745	3.5371
35	46.12	33.30	37.15	0.7002	3.6902
36	50.59	37.75	44.43	0.7265	3.8518
37	55.63	42.92	53.27	0.7536	4.0228
38	61.35	48.93	64.08	0.7813	4.2037
39	67.87	55.96	77.33	0.8098	4.3955
40	75.31	64.20	93.69	0.8391	4.5989
41	83.86	73.90	113.99	0.8693	4.8149
42	93.71	85.38	139.32	0.9004	5.0447
43	105.11	99.02	171.15	0.9325	5.2893
44	118.37	115.31	211.41	0.9657	5.5500
45	133.88	134.88	262.75	1.0000	5.8284
46	152.10	158.51	328.74	1.0355	6.1260
47	173.64	187.21	414.34	1.0724	6.4447
48	199.26	222.31	526.47	1.1106	6.7865
49	229.93	265.51	674.94	1.1504	7.1536
50	266.89	319.07	873.88	1.1918	7.5486

* (Mosher and Pace 1982).

5-6. Base Tilt Factors

These factors are used to take into account the effect of a sloping base. The base tilt factors are computed as

$$\xi_{qt} = \xi_{\gamma t} = (1 - \alpha \tan \phi)^2 \ (\alpha \text{ in radians}) \quad [5\text{-}6a]$$

$$\xi_{ct} = 1 - \left(\frac{2\beta}{\pi + 2}\right)(\alpha \text{ in radians}) \ (\text{when } \phi = 0°) \quad [5\text{-}6b]$$

$$\xi_{ct} = \xi_{qt} - \frac{1 - \xi_{qt}}{N_c \tan \phi} \quad (\text{when } \phi > 0°) \quad [5\text{-}6c]$$

where α is the angle the slip plane of the structural wedge makes with the horizontal, measured in radians. The sign of α will follow the sign convention given in Chapter 4.

5-7. Ground Slope Factors

Ground slope factors are used to correct for a sloping ground surface on the toe of the wall. The factors are computed as

$$\xi_{\gamma g} = \xi_{qg} = [1 - \tan(\beta)]^2 \quad [5\text{-}7a]$$

$$\xi_{cg} = 1 - \left[\frac{2\beta}{(\pi + 2)}\right] (\beta \text{ in radians}) \quad (\text{when } \phi = 0°) \quad [5\text{-}7b]$$

$$N_\gamma = -2 \sin \beta \quad (\text{when } \phi = 0°) \quad [5\text{-}7c]$$

$$\xi_{cg} = \xi_{qg} - \frac{1 - \xi_{qg}}{N_c \tan \phi} \quad (\text{when } \phi = > 0°) \quad [5\text{-}7d]$$

where β is the angle the ground surface makes with the horizontal, measured in radians. Beta is positive when the ground slopes down and away from the footing.

5-8. Effective Overburden Pressure

q_o is defined as the effective vertical stress due to the soil and/or surface loads above the base of the footing, on the toe side of the wall, as follows:

$$q_o = \gamma' D \quad [5\text{-}8a]$$

where

γ' = effective unit weight of the overlying soil
D = depth from the soil surface to the base of the structural wedge

For the special case of a sloping surface, compute q_o as

$$q_o = \gamma' D \cos[ABS(\beta)] \quad [5\text{-}8b]$$

5-9. Combination of Factors

As discussed in the CBEAR user's guide (Mosher and Pace 1982), the correction factors for the load inclination, base tilt, and ground slope and the adjustment for the load eccentricity should only be used in unison when all of these factors tend to produce failure in the same direction.

5-10. Example

Example problems using the general bearing capacity equation are presented in Appendix N.

1976).

Section II. Other Considerations

5-11. Settlement

A. EM 1110-2-1904. A discussion on the various factors involved in the settlement of a structure, on methods for estimating settlements, and on the limitations in the accuracy of conducting settlement analyses from laboratory tests is contained in EM 1110-2-1904. The principles and methods presented are applicable to a majority of civil works projects. Additional information for unique or special projects can be obtained from various texts on soil mechanics. The computer program CSETT (Appendix O) can assist in performing a settlement analysis.

B. ALLOWABLE SETTLEMENT. The maximum value of angular distortion (settlement/length of structure) can be tolerated without cracking of reinforced concrete retaining walls is 0.002 to 0.003 radian (Duncan and Buchignani 1976).

5-12. Deep-Seated Sliding

A deep-seated sliding analysis should be performed to check for sliding within weak layers which may exist beneath structures. The analysis should be in accordance with procedures outlined in paragraph 4-16. Active and passive wedges should be located a sufficient distance apart to allow a rotational slip surface to develop. Generally, a slip plane inscribed in an arc with a radius equal to the height of the active wedge will comply with this requirement (Figures 5-2 and 5-3). When the wall is resting on thick strata of weak soils, shallow shear failure should be investigated. This may be assumed to occur below the base of the retaining or flood wall along a cylindrical surface passing through the heel (Figure 5-4). The minimum factor of safety, which must not be less than 1.5, is determined by trial and error by changing the center of the trial circle.

5-13. Liquefaction Susceptibility

Where walls are underlain by sands below the water table in seismically active areas, an analysis should be made of the safety against foundation liquefaction. Flood walls in alluvial valleys are particularly likely to be situated over loose, saturated sands that may be liquefiable. A preliminary assessment of liquefaction susceptibility can be made using Seed's simplified method (Seed 1976, Seed and Idriss 1982) which is based on the standard penentration test. If the foundation is found to be nonliquefiable, no further analysis need be made. If liquefaction may occur, an assessment should be made of the risks and consequences of liquefaction failure and the benefits and costs of alleviating the risks. The occurrence of an earthquake during a flood is a case of the joint occurrence of independent rare events. For flood walls, the probability (risk) of an earthquake during a flood will be much smaller than the probability during a nonflood period, but the associated consequences may be much higher. For certain walls (e.g., a low retaining wall remote from other structures) the probability of liquefaction failure and the related consequences may translate into such a small risk that accepting the risk may be the preferred alternative. Possible alternatives to dealing with potentially liquefiable foundations include

A. Changing the proposed location (usually the best alternative, when feasible).

B. Removing and replacing the liquefiable materials.

C. Improving the liquefiable materials in place, by densification or grouting.

D. Accepting the risks and consequences of liquefaction.

CHAPTER 6

DESIGN AND CONSTRUCTION DETAILS AND CAUSES OF UNSATISFACTORY PERFORMANCE

6-1. Foundation Preparation

Earth foundations should be properly compacted and should be clean and damp before concrete is placed. Rock foundations should be cleaned and given any other necessary treatment to ensure proper bond of concrete to rock. Some rock foundations, primarily shales, require a protective covering such as unreinforced concrete to protect them from deterioration after being exposed and before concrete placement, unless the final excavation can be performed close enough in time to the placement of the structural base slab. When a protective coating is used, it must be such as to ensure proper bond.

6-2. Concrete Materials

Consideration should be given to the materials that are economically available for a particular project. EM 1110-2-2000 describes concrete materials requirements; all options which are applicable to the work and which include available materials should be investigated. Concrete proportions should be selected to satisfy strength and durability requirements.

6-3. Constructibility

The dimensions of the wall should be such that reinforcement and concrete can be properly placed. EM 1110-2-2000 provides guidance for concrete placement. Guide specifications CW 03301 and CW 03305 provide detail requirements for concrete placement. The top thickness of the stem for cantilever concrete walls over 8 feet high and for base slabs should be a minimum of 12 inches to facilitate concrete placement. Stems not over 8 feet high with one layer of vertical reinforcement may be 8 inches thick. The wall section should be designed for simplicity and maximum reuse of forms. Any construction constraints due to the location of the wall should be included in the design.

6-4. Joints

Walls are designed with joints to allow for expansion, contraction, and/or to divide the structure into convenient working units. The locations of all horizontal and vertical joints should be shown on the drawings.

A EXPANSION JOINTS

1. General Needs and Uses. Expansion joints are designed to prevent the crushing and distortion (including displacement, buckling, and warping) of the abutting concrete structural units that might otherwise occur due to the transmission of compressive forces. Compressive forces may be developed by expansion, applied loads, or differential movements arising from the configuration of the structure or its settlement. In general, expansion joints are needed to prevent spalling and sometimes to break continuity. In relatively thin reinforced concrete walls such joints should be located where considerable expansion or unequal settlement is anticipated, e.g., at changes in alignment or grade, at abrupt changes in section or at intermediate points when needed. In massive reinforced concrete walls and in gravity walls on rock, expansion joints usually are not provided unless required at abrupt changes in section or at angle monoliths to relieve thrust from expected expansion. Otherwise, adequate chamfers on each side of each contraction joint usually are sufficient to prevent spalling. Where temperature variations are extreme, modification of these criteria may be required. Reinforcing steel, corner protection angles, and other fixed metal embedded in or bonded to the surface of the concrete should not extend through an expansion joint. Where water tightness is needed, water stops are provided as outlined in paragraph 6-4e.

2. Joint Filler. The thickness of joint filler necessary to provide stress relief at a joint should be determined from the estimated initial contraction and subsequent expansion from maximum temperature variation. Premolded expansion joint filler and adequate chamfers should be used.

B. CONTRACTION (MONOLITH) JOINTS. These are intentional planes of weakness designed to regulate cracking that might otherwise occur due to the unavoidable, often unpredictable, contraction of concrete structural units. Contraction joints also divide the structure into convenient working units and thus also serve as construction joints. Since it is impractical and uneconomical to provide sufficient reinforcement to prevent cracks entirely, it is desirable to control their location, insofar as is practicable, by vertical contraction joints, across which reinforcement does not extend. No exact rules for the location of such joints can be made. Each job must be studied to determine where the joints should be placed, taking into account the requirements of structural design, the volume of concrete which can be placed economically in a single working unit, and the economical use of form units. Typically, contraction joints have been spaced 20 to 30 feet apart. Usually, a contraction joint has a plane surface without a key. For cantilever concrete walls, vertical contraction joints may be located only in the stem, and the footing may be a continuous placement.

C. HORIZONTAL CONSTRUCTION WALLS. These joints are provided to divide a wall into convenient working units, but they should be kept to a minimum. Keys are not permitted in horizontal construction joints as they interfere with good cleanup of the concrete surface and because a well-bonded flat surface is more dependable to transfer shear.

1. Gravity Concrete Walls. For this type of wall the horizontal construction joint locations are dictated by the height of each lift of concrete placement. Concrete for gravity walls is usually placed in lifts up to 10 feet high. The top surface of each lift is cleaned and roughened by high-pressure water jets before placing the next lift.

2. Cantilever Concrete Walls. For this type of wall a construction joint between the base and the wall stem should be provided. Additional horizontal joints in the wall stem should be provided by lifts approximately 10 feet high. The surface of each joint should be roughened to obtain as much shear strength across the joint as possible.

D. JOINT DETAILS FOR FLOOD WALLS. For expansion and contraction joint details for flood walls, see paragraph 7-14.

E. WATER STOPS. Water stops are provided across joints where water-tightness is required. Nonmetallic water stops, such as rubber or polyvinyl chloride (PVC) water stops, should be used in accordance with EM 1110-2-2102. For special flood wall water stop details, see Chapter 7, Sections II and V. Careful inspection is required for water stop installation, especially with the type 'U' water stop (Figure 7-9b), to see that special reinforcing is properly placed and that concrete is placed under the upper water stop in the base slab.

6-5. Soil Backfill

A. MATERIAL CHOICE. Many types of material can be used for backfill. It is advisable to use locally available material when possible. Unusually poor foundation material or a need to control piping may require importation of select material.

B. MATERIALS. Clean sands and gravels are the most suitable materials. They drain rapidly, are not susceptible to frost action, and remain stable. Silty sands, silts, and coarse-grained soils containing some clay are less desirable since they drain slowly, are subject to seasonal volume changes, and may lose much of their strength with time. Shrinkage cracks may develop in clay, which, when filled with water, can cause full hydrostatic pressures to act on the wall. As mentioned in paragraph 6-7, clay, as backfill or foundation material, is involved in most retaining wall failures. During winter construction, frozen backfill material should not be used under any circumstances. This material may appear satisfactory when put into place, but it can be adversely affected when it thaws.

C. PLACING AND COMPACTING. (Refer to Chapter 3 and Appendix J for additional information on compaction.) The backfill material should be carefully selected. It should be compacted to prevent large settlements due to its own weight, with the amount of compaction required depending on the material used and the purpose of the structure. Very strict control of compaction is required when the fill is a cohesive soil. When granular fill is used, the material should be placed in thin lifts with each lift being compacted before the next lift is placed (see EM 1110-2-1911). However, precautions should be taken to prevent overcompaction which will cause excessive lateral forces to be applied on the structure. If heavy compaction rollers are used near the wall, their effect on lateral earth pressures on the wall should be considered in the design. Alternatively, the allowable weight of compactors may be restricted by the specifications to control wall pressures. It is good practice to place a layer of impervious soil that is a minimum of 12 inches thick in the upper lift of the backfill to reduce infiltration of rainwater.

Backfill should be brought up equally on both sides until the lower side finished grade is reached.

6-6. Drainage

A. NEED FOR DRAINAGE SYSTEM. As mentioned in paragraph 6-7, improper drainage systems are one of the major causes of retaining wall failures. Drainage systems are necessary to eliminate excess hydrostatic pressures on the failure plane and the wall stem due to water seepage and surface infiltration of rainfall. In some cases the drainage system may be needed to prevent pressures from building up due to frost action in the backfill or to minimize pressures due to swelling of cohesive backfills. The kind of drainage system required depends upon the type of soil backfill, amount of rainfall, groundwater conditions, and potential frost action. Regardless of the drainage system used, the wall must have an adequate factor of safety assuming the drainage system is inoperative (see paragraph 3-23).

B. DRAINAGE CONTROL METHODS. All retaining walls must have adequate surface drainage to dispose of surface water. As previously mentioned, a layer of impervious soil should be placed on top of the soil backfill to reduce surface infiltration of rainfall. The most effective way to control drainage within the soil backfill is an inclined drainage blanket with longitudinal drain as shown in Figure 6-1. The inclined drainage blanket will minimize excess hydrostatic pressures on the failure plane due to groundwater seepage and surface infiltration of rainfall. A drain adjacent to the wall is less effective and will often result in higher loads against the wall (see Figure 6-2). However, for relatively low walls (typically less than 10 feet high), these higher loads may not be significant, and drains adjacent to the wall are often used. Drains adjacent to the wall may be either a drainage blanket (Figure 6-3) or a prefabricated

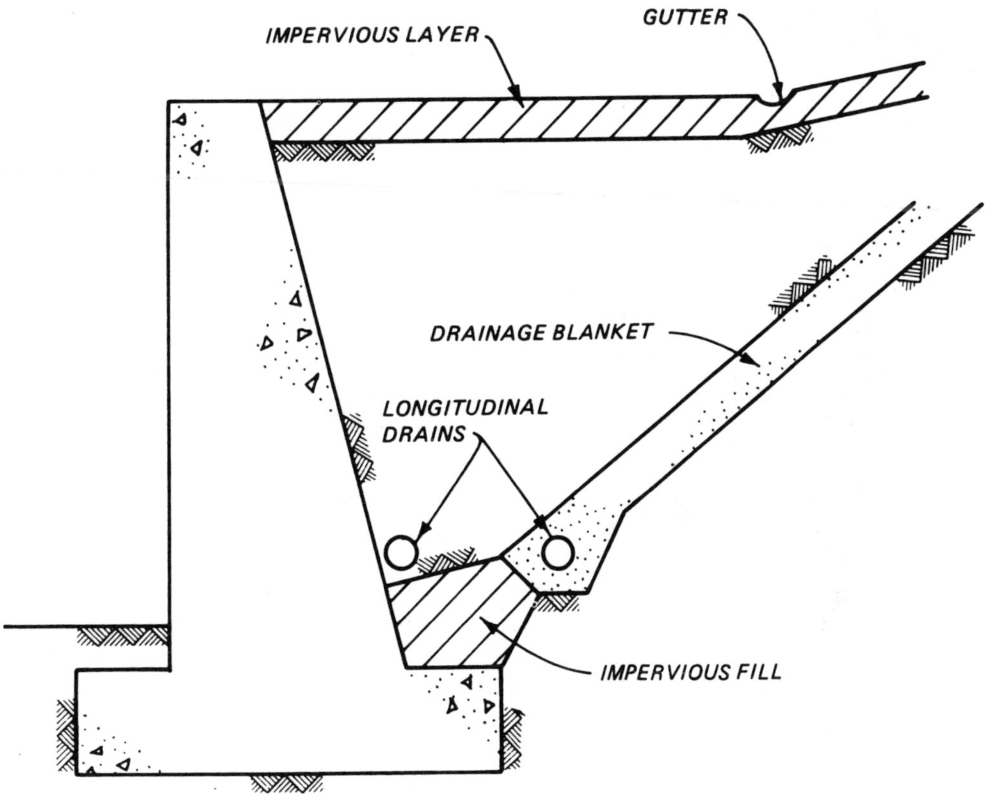

NOT TO SCALE

Fig 6-1. Inclined drainage blanket (after Department of the Navy 1982a)

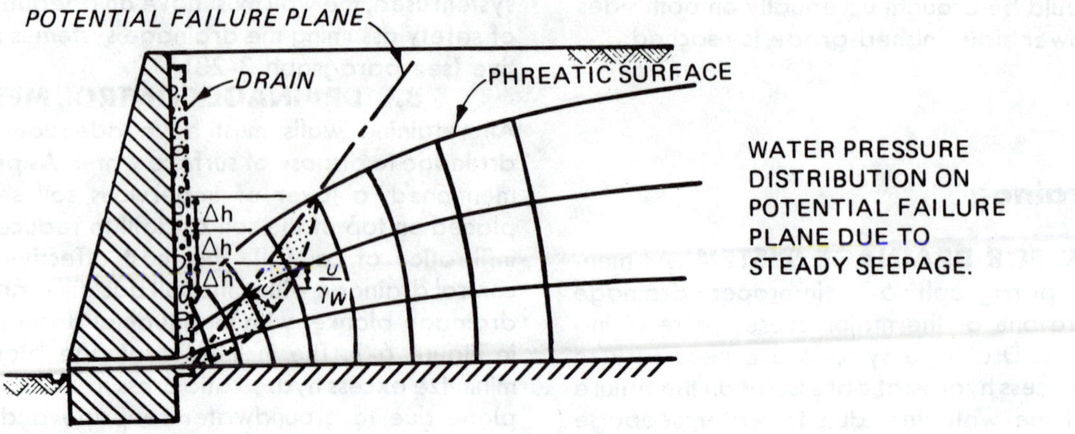

(a) NORMAL STEADY STATE SEEPAGE CONDITION (VERTICAL DRAIN)

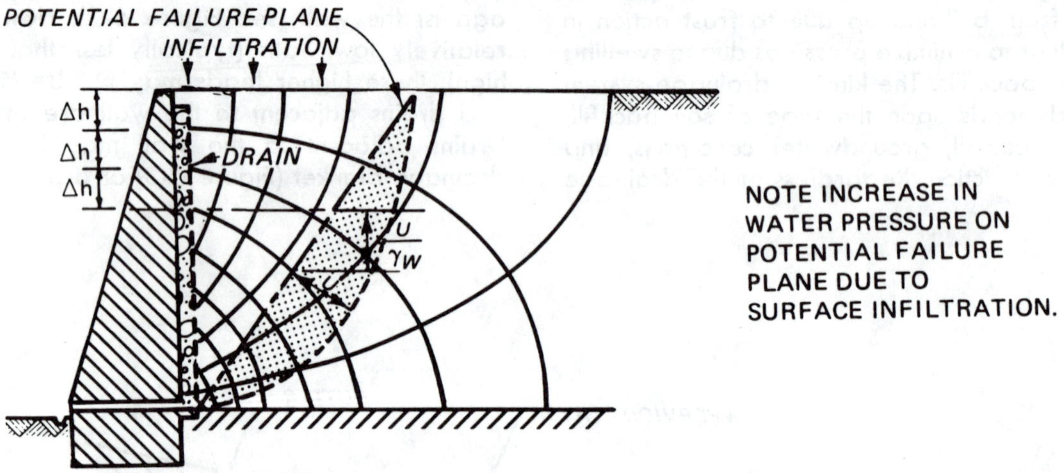

(b) SURFACE INFILTRATION (VERTICAL DRAIN)

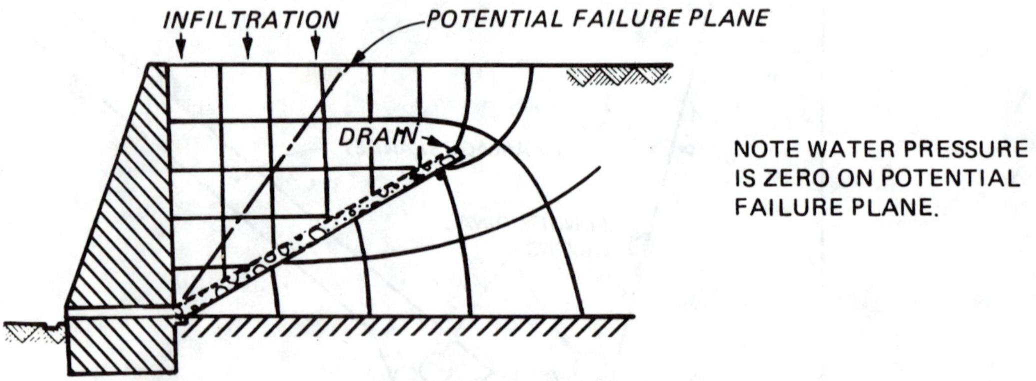

(c) SURFACE INFILTRATION (INCLINED DRAIN)

Fig 6-2. Effect of drain location on excess hydrostatic pressures on the failure plane (after Geotechnical Control Office 1982)

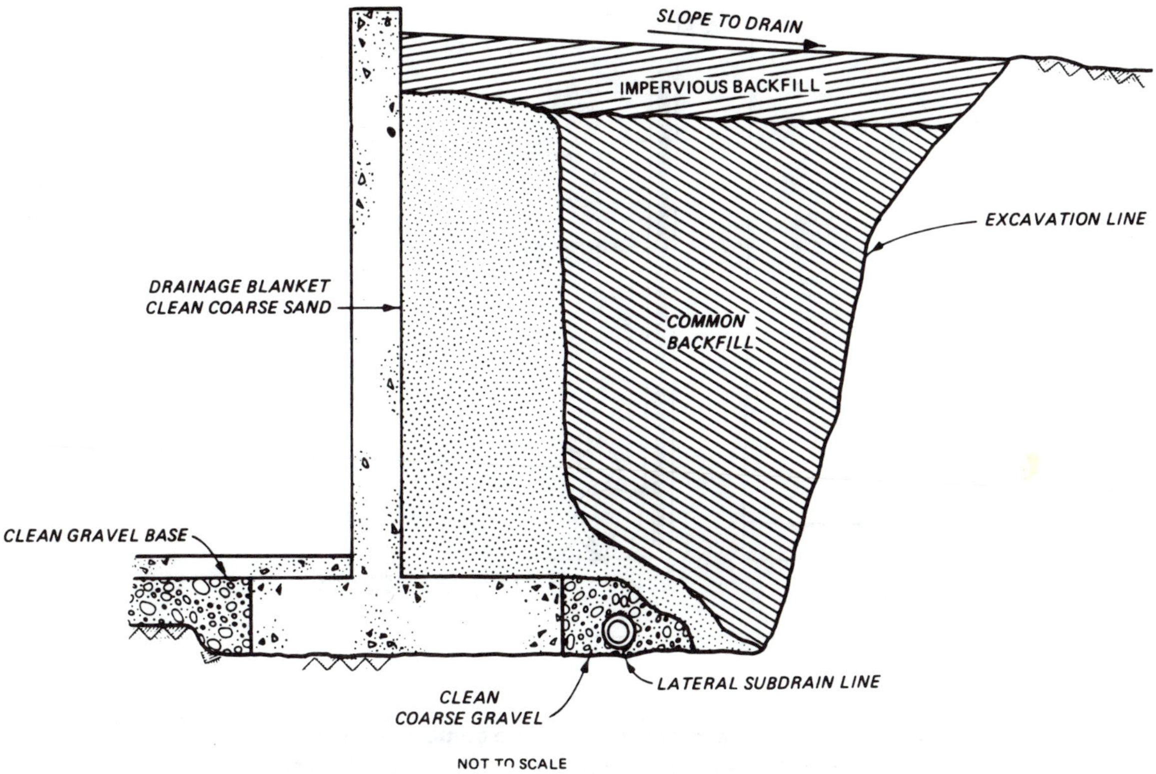

Fig 6-3. Drainage blanket located adjacent to retaining wall (after Sibley 1967)

drainage composite* (Figure 6-4). Where frost penetration is a problem, a drainage system as shown in Figure 6-5 should be used. If a cohesive soil backfill is used, a drainage system as shown in Figure 6-6 will prevent changes in moisture content of the clay and hence reduce cracking and swelling potential. Other seepage control methods are discussed in paragraph 7-4.

C. LONGITUDINAL DRAINS. Longitudinal drains within drainage blankets are used for carrying the discharge from behind the retaining wall to a ditch, manhole, or other free exit. Drains should be large enough to carry the discharge and have adequate slope to provide sufficient velocity to remove sediment from the drain.† To minimize clogging, the drain should have perforations in the bottom half of the pipe at least 22.5 degrees below the horizontal axis. Where the operation of the drains is counted on to reduce the design loadings, manholes and/or inspection holes (see Figure 6-7) should be located at sufficient intervals, and at any sharp bends in the pipe, to facilitate inspection and cleanout. The terminus of the drain should have a vertical check valve (see Figure 6-8) to prevent backflooding. The end section of pipe supporting the check valve should be secured with a coupling band which can be removed for inspection and cleaning of the pipe.

D. WEEPHOLES. Weepholes should consist of a pipe, at least 3 inches in diameter, extending through the stem of the wall. They should be protected against clogging by pockets of gravel in the soil backfill or by the use of filter fabric adjacent to the wall directly behind the weepholes. The weepholes are commonly spaced not more than 10 feet apart vertically and horizontally.

E. FILTER REQUIREMENTS. Drains should be adequately protected by filter layers so that seepage water is admitted freely but movement of the soil backfill into the drain will not occur. The piping or stability criterion is based on the grain

* Whenever a prefabricated drainage composite is used adjacent to the retaining wall, the crushing strength of the prefabricated drainage composite should be greater than three times the maximum lateral earth pressure acting on the wall. Prefabricated drainage composites are not recommended for inclined drains due to possible damage during compaction of the soil backfill and possible sliding along the plane of the drain (Smith and Kraemer 1987, Kraemer and Smith 1986).

† For a 6-inch-diameter pipe the minimum slope would be about 0.15 percent (Schwab et al. 1981).

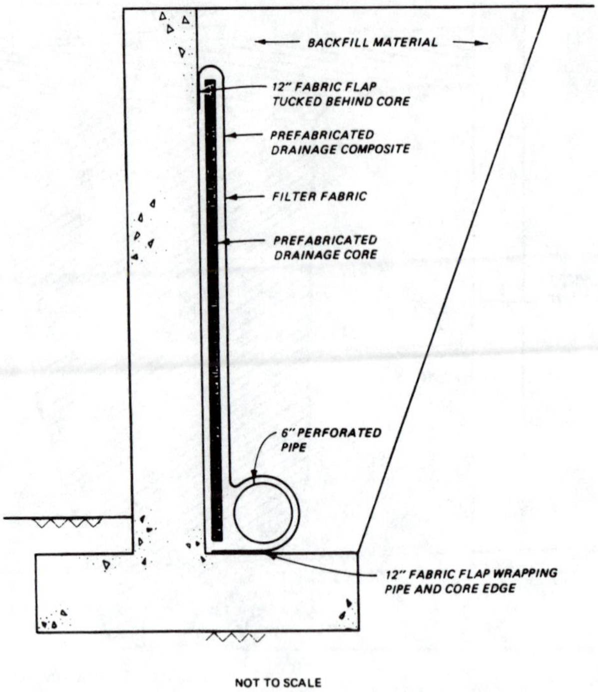

Fig 6-4. Prefabricated drainage composite used as drain adjacent to retaining wall (adapted from Carrol and Murphy 1985)

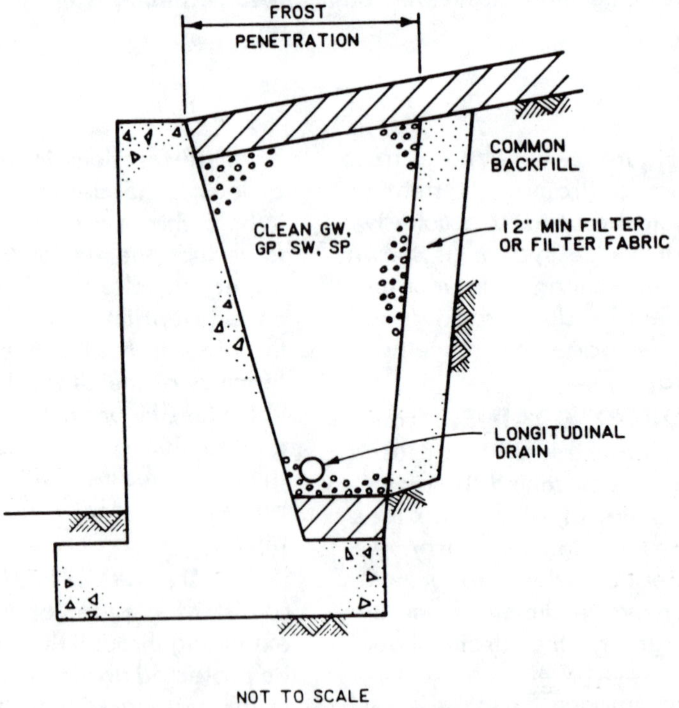

Fig 6-5. Drainage system to prevent frost penetration behind retaining wall (after Department of the Navy 1982a)

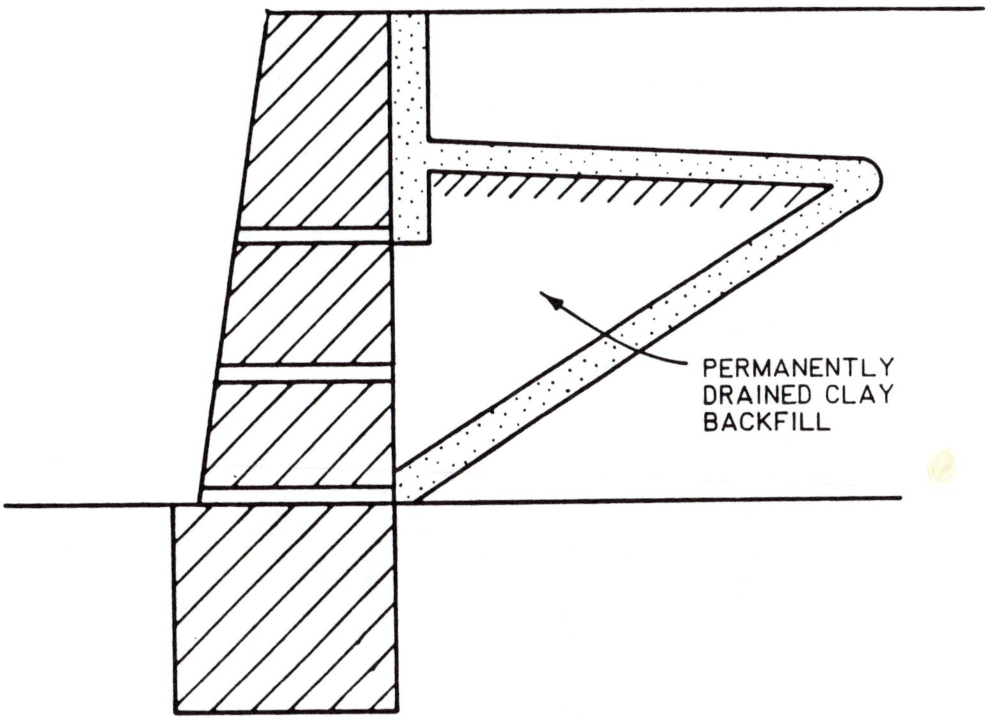

Fig 6-6. Drainage system to use with clay backfill (after Terzaghi and Peck 1948)

size relationship between the protected soild and the filter

$$\frac{D_{15_F}}{D_{85_B}} \leq 5 \qquad [6\text{-}1]$$

where

D_{15_F} = size of filter material at 15 percent passing

D_{85_B} = size of protected soil at 85 percent passing

and

$$\frac{D_{50_F}}{D_{50_B}} \leq 25 \qquad [6\text{-}2]$$

where

D_{50_F} = size of filter material at 50 percent passing

D_{50_B} = size of prtected soil at 50 percent passing

To assure that the filter material is more permeable than the material being drained, the following condition must be met:

$$\frac{D_{15_F}}{D_{15_B}} \geq 5 \qquad [6\text{-}3]$$

To prevent clogging of perforated longitudinal drains, the following requirement must be satisfied:

Circular openings

$$\frac{D_{50_F}}{\text{Hole diameter}} \geq 1.0 \qquad [6\text{-}4]$$

Slotted openings

$$\frac{D_{50_F}}{\text{Slot width}} \geq 1.2 \qquad [6\text{-}5]$$

The filter material may satisfy the criteria for stability and permeability but may be too fine to meet the criteria for circular or slotted openings. Should this happen, multilayered or graded filters are required. It may be possible to substitute filter fabric for one or more of the granular filters in a multilayered filter system. Filter cloth shall conform to the requirements of guide specification CW 02215.

F. DRAIN REQUIREMENTS. The drain must be able to carry the design flow freely without movement of soil particles. Drainage blankets may be constructed of clean sand and gravel or a prefabricated drainage composite (for certain

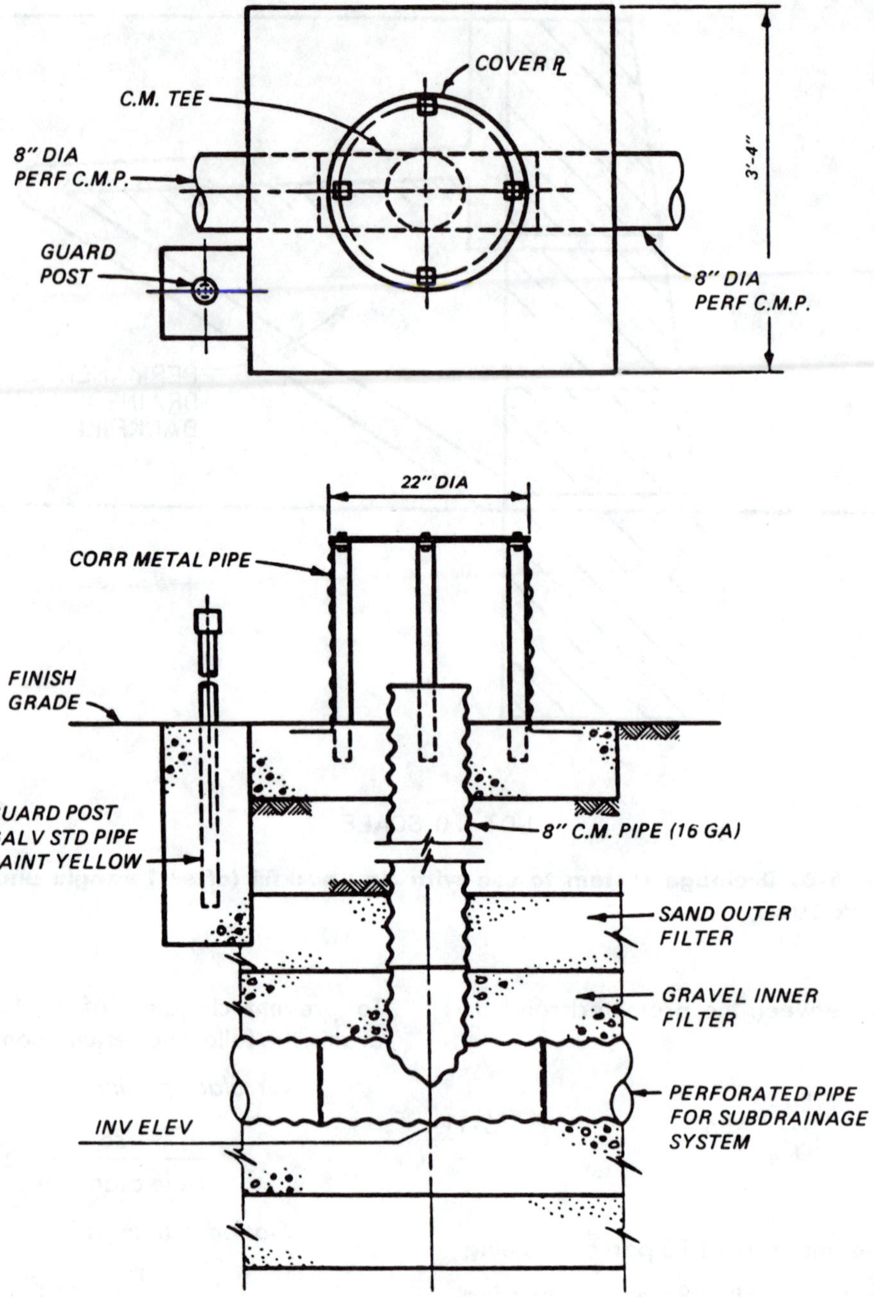

Fig 6-7. Inspection hole for longitudinal drain

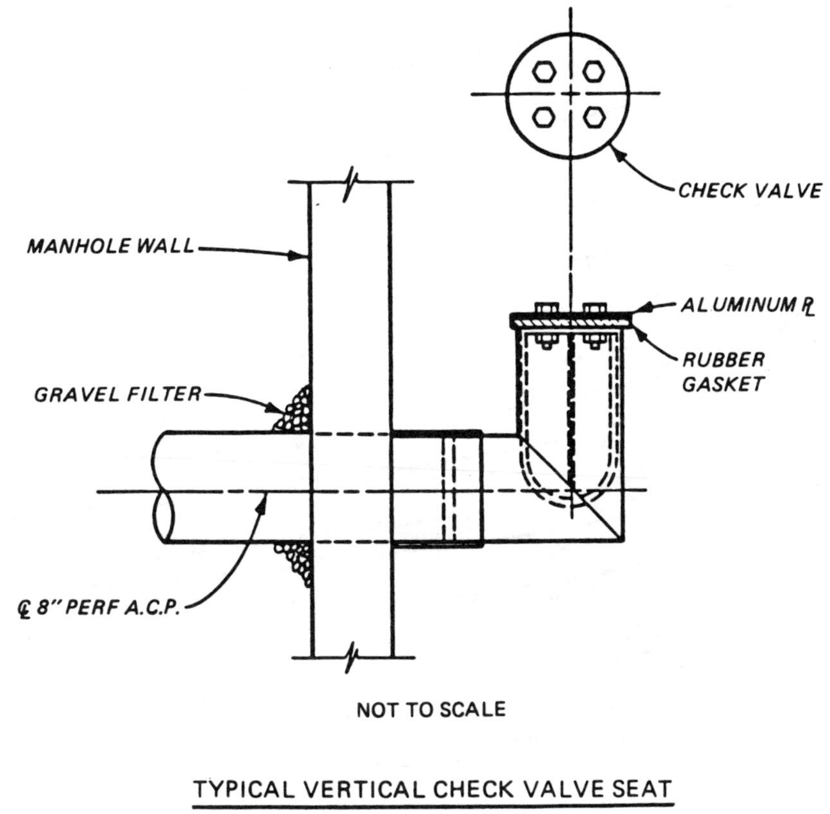

Fig 6-8. Vertical check valve at end of longitudinal drain

applications). The design flow can be determined from a flow net (Cedergren 1967). For isotropic soil conditions:

$$q_b = k_b h \frac{n_f}{n_d} \qquad [6\text{-}6]$$

where

q_b = quantity of discharge through soil backfill per linear foot of retaining wall

k_b = permeability of soil backfill

h = hydrostatic head acting on retaining wall

n_f = number of flow channels in flow net

n_d = number of equipotential drops in flow net

The minimum required permeability of the drain is

$$k_d = \frac{q_d}{i_d A_d} \qquad [6\text{-}7]$$

where

k_d = minimum required permeability of the drain

q_d = quantity of discharge through drainage blanket or prefabricated drainage composite per linear foot of retaining wall

i_d = gradient of flow in the drain (1 for vertical drain, equal to slope of drain for inclined drain)

A_d = cross-sectional flow area of drain

Seepage in coarse aggregates may be turbulent and a reduction factor should be applied to the permeability as shown in Figure 6-9. The in-plane permeability should be at least 20 times that calculated theoretically. For prefabricated drainage composites the in-plane permeability will decrease with increase in lateral pressure. Therefore, the in-plane permeability must be taken at the maximum lateral earth pressure acting on the wall.

G. CONSTRUCTION CONSIDERATIONS

1. Sand and Gravel. Sand and gravel must not become segregated or contaminated prior to, during, or after installation. Segregation will result in zones of material too fine to meet the permeability requirements and other zones too coarse to meet the stability requirements. Contamination of the filter material from muddy water, dust, etc., during construction may clog the voids in the material and prevent proper drainage. In the event that filter or drain materials are con-

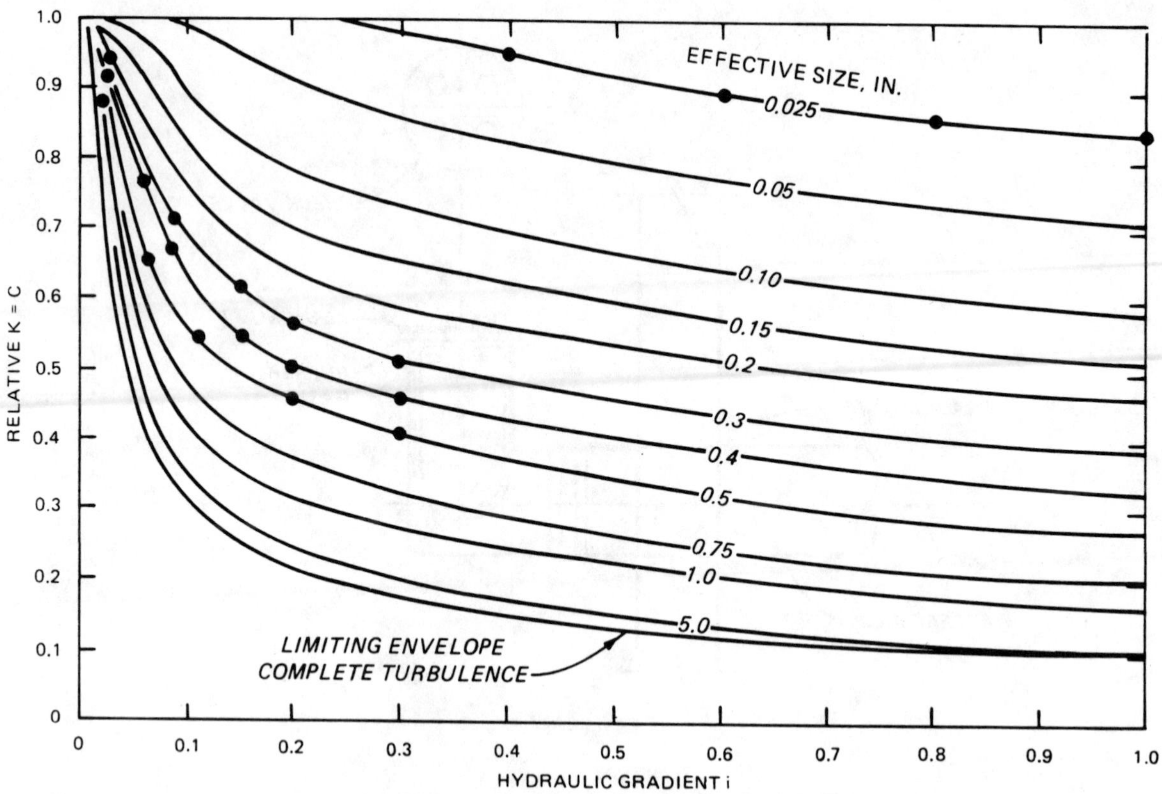

Fig 6-9. Approximation for estimating reduction in permeability of narrow size-ranged aggregate caused by turbulent flow (Cedergren 1967, courtesy of John Wiley and Sons)

taminated, they should be replaced. Filter materials subject to cementation should be rejected.

2. Prefabricated Drainage Composite. Special consideration should be given when compacting soil backfill near prefabricated drainage composites adjacent to retaining walls. Compaction adjacent to the retaining wall will induce high lateral pressures which could crush the prefabricated drainage composite and/or reduce the in-plane permeability. The drainage composite manufacturer's recommendations for backfilling and compaction near the composite should be followed. A test section may be required to determine the acceptable operating conditions of the compaction equipment. Where crushed stone is used as the backfill material, a blanket of sand should be provided against the drainage composite to protect it against damage during compaction.

3. Longitudinal Drains. One bad joint could render an entire drainage system inoperative. Care must be taken in compacting soil backfill over drains to prevent crushing of the pipe. Differential settlement can cause pipe joints to open up, permitting soil backfill to infiltrate. This should be minimized by attaining uniform adequate compaction of the underlying material.

6-7. Causes of Unsatisfactory Performance

The results of two statistical studies of retaining wall failures are given in Figure 6-10 (Tcheng and Iseux 1972, Ireland 1964). It is evident that:

A. Clay, as backfill or foundation material, is involved in most retaining wall failures.

B. Improper design of the drainage system and/or the wall base is the main cause of retaining wall failure.

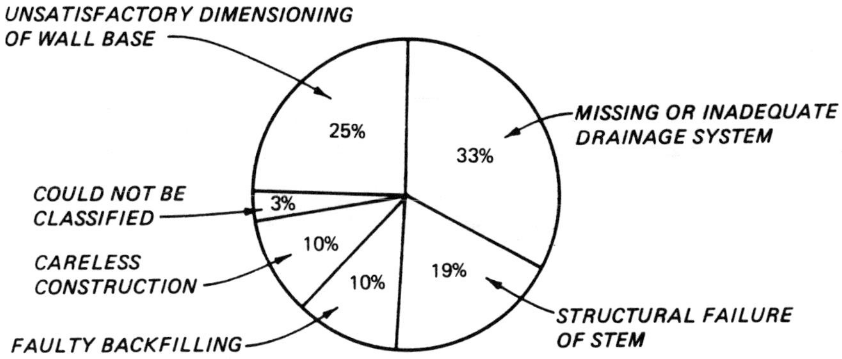

a. Causes of failure of rigid concrete retaining walls (Tcheng and Iseux 1972)

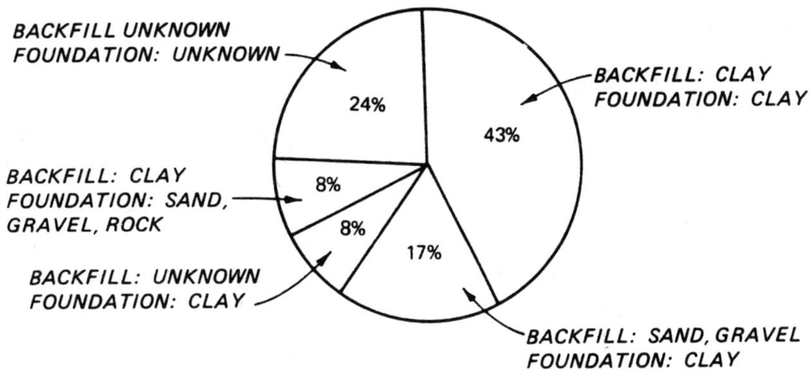

b. Foundation and backfill material of unsatisfactory retaining walls (Ireland 1964)

Fig 6-10. Summary of experience with unsatisfactory retaining walls

CHAPTER 7

SPECIAL CONSIDERATIONS FOR FLOOD WALLS

Section I. General Characteristics

7-1. Introduction

The principal function of a flood wall is to prevent flooding (inundation) of adjacent land. A flood wall is subject to hydraulic loading on one side which is resisted by little or no earth loading on the other side. The two principal types of flood walls are inland and coastal. Inland flood walls typically are installed along a riverbank and are subjected to design loadings (pool to freeboard line) for periods of hours or days (long-term loadings). Coastal flood walls are primarily subjected to short-term loadings (waves from hurricanes along with wind/tide high water surges). The wave loadings are dynamic in nature and act upon the structure for only a few seconds each. Concurrent high winds can prevent any emergency maintenance during a storm. Utility line crossings through a flood wall require careful attention to allow for independent movement of the utility lines and the wall, which requires special expansion joint details.

7-2. Rationale for Loading Cases

A. DESIGN WATER LEVEL.

1. The hydraulic data required for determining the design water level should be listed in the hydrologic/hydraulic appendix of the pertinent planning document for the project. The flow characteristics noted in historical records and indicated from detailed observation of existing conditions will usually be basic to the design of inland flood walls. Coastal flood walls will frequently require hurricane surge simulation studies and wave setup estimates. Wave overtopping can cause severe scour at or near the protected side of the stem. See paragraph 3-24 for information on surge and wave loads.

2. Factors that influence the water surface profile and level of protection, and that can reasonably be quantified, are included in the design water level; not the freeboard. Some examples of these factors are:

A. Changed conveyance, due to changing bed form, sedimentation or scour, and vegetation growth or removal.

B. Dynamic surges, and super elevation.

C. Ice, debris, and local anomalies.

D. Transverse slope due to water flowing out of or into the channel or differences in velocity head between the channel and overbank locations.

E. Profile instabilities associated with braids, meanders, etc.

F. Energy losses due to changing flow area, e.g. constrictions (bridges), abrupt expansions, and bends.

G. Future changes in flood flows due to changes in the watershed.

B. FREEBOARD. The freeboard is the marginal height of wall provided above the design water level. Freeboard is designed to accomplish design objectives and allow for uncertainty in a water surface profile.

1. Examples of design objectives are:

a. Assurance of initial overtopping at the most desirable (least hazardous) location.

b. Reduced volume of wave overtopping.

c. Extension of interval between major maintenance such as removal of sediment deposition.

2. Freeboard allowances for water surface uncertainty are allowances that are not otherwise specifically accounted for because they are considered too small to require specific determination or because they are too intractable to be quantified. Those factors that influence the water surface profile, and level of protection, and that can reasonably be quantified are included in the design water level; not the freeboard.

3. Wall settlement is identified as a separate increment added to the wall height for that purpose and is not included in the freeboard.

4. Freeboard design should be refined as a study progresses and not left entirely to a detailed design phase. The amount of effort and corresponding refinement for a given phase is a function of the importance and cost of freeboard relative to the overall plan. For an early reconnaissance phase it will generally be satisfactory to use quickly estimated freeboard values of generally accepted default values. Default values of 2 feet on agricultural and 3 feet on urban flood walls have been generally accepted. As the study progresses, these early estimated or default values will be replaced by values arrived at by a design process.

5. When large non-breaking waves are normal incident to the stem of the flood wall, the amount of freeboard will be determined by the amount of overtopping allowed. It is important to remember that such overtopping can cause significant scour on the protected (toe) side of the wall. This potential for scour can require rigid paving within a 20- to 30-foot area of the wall.

c. LOADING CASES. For determining water and soil loads acting on flood walls, refer to Chapter 3. Section I of Chapter 4 discusses loading cases.

Section II. Seepage Control

7-3. General Considerations

Water-retaining structures are subject to through-seepage, underseepage, and seepage around their sides or ends. Seepage control is a primary consideration of flood wall design. Uncontrolled seepage may result in water pressures and uplift forces on the wall base in excess of design assumptions and consequent structural instability. Excessive porewater pressures in foundation materials near the landside toe of a wall may create quick conditions evidenced by sand boils or heaving. Emerging seepage may have sufficient velocity to move cohesionless foundation materials and erode the wall foundation (piping). Seepage control entails the design of measures to ensure that seepage pressures and velocities are maintained below tolerable values. Properly controlled seepage, even if quantities are large, presents no hazard. Since flood walls are often built in congested areas, it is often necessary to pump seepage out of the protected area. While the seepage quantity is often small compared to other sources, it is occasionally appropriate to consider seepage control measures for the purpose of reducing seepage quantities. Inadequate seepage control, as shown by one example in Figure 7-1, may jeopardize the stability of a flood wall. In flood walls, control of through-seepage is provided for by water stops (paragraph 7-13). Seepage around the wall is controlled by specially designed and constructed levee wrap-around sections (paragraph 7-12). Flood walls are usually provided with a toe drain to control local underseepage along the flood wall base, as shown in Figure 7-2. As flood walls are usually founded on alluvial materials, pervious zones of significant thickness are often present at some depth below relatively impervious top stratum materials and may be hydraulically connected to the river. Because of the horizontal stratification of alluvial deposits, the horizontal permeability may be greatly in excess of the vertical permeability. The combination of these conditions may allow seepage to be readily conducted landward beneath the flood wall. Where flood walls are underlain by such pervious strata (the usual case), analysis may indicate the need for underseepage controls in addition to the toe drain. Underseepage control measures vary because the selection and design of an appropriate control scheme is highly dependent on site-specific conditions, particularly the stratification and permeability of foundation materials, availability of right-of-way, and local construction practices and costs. Various types of underseepage control measures are discussed in paragraph 7-4.

7-4. Underseepage Control

The focus of underseepage analysis is to calculate the expected exit gradient at the landside toe of a levee or flood wall and compare its value to a theoretically critical value, the critical gradient (typically 0.8 to 1.0). To provide some conservatism, underseepage controls are provided where the calculated gradient exceeds an allowable gradient, typically 0.5 to 0.8. For calculating the exit gradient, assessing the need for underseepage controls, and designing such controls, the foundation conditions are normally assumed to be a two-layer system consisting of a relatively impervious top stratum overlying a pervious substratum. Detailed analysis procedures are contained in EM 1110-2-1913 and WES Technical Memorandum 3-424 (US Army Engineer Waterways Experiment Station, 1956). In some instances, where complex problems of geometry, anisotropy, and foundation layering exist, flow nets and/or finite element seepage analyses may be neces-

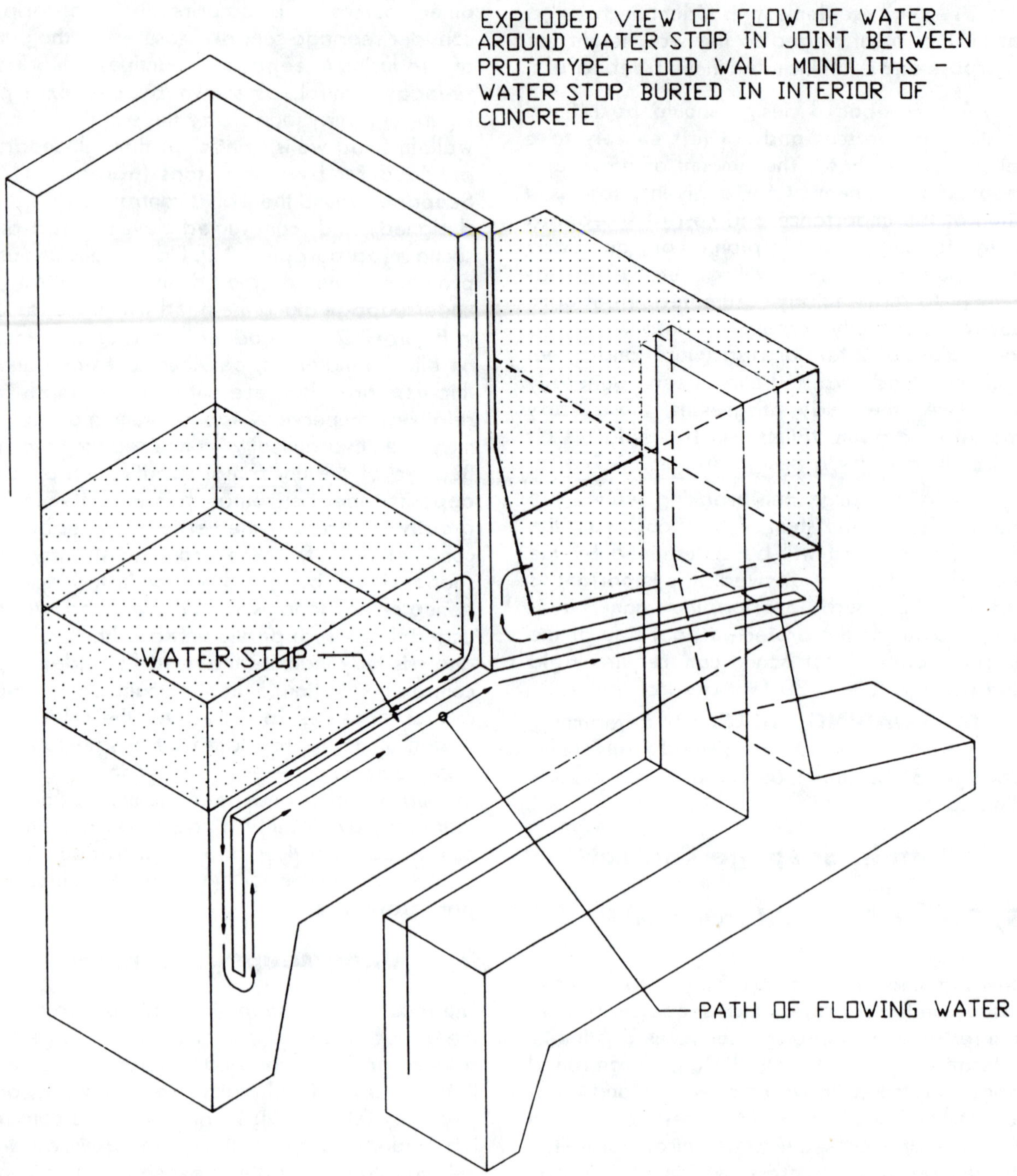

Fig 7-1. Flow around interior embedded water stop in the base

sary. Types of underseepage control measures are described in the following paragraphs. Additional discussion is given in paragraphs 3-23 and 6-6.

A. CUTOFFS. A cutoff penetrating the pervious stratum beneath the wall is the most positive means of controlling seepage. A cutoff can consist of an excavated trench backfilled with impervious compacted earth, a slurry trench, an extension of a concrete shear key, or a sheet pile wall. A cutoff is usually located at the end of the wall footing on the unprotected (heel) side. A cutoff must penetrate approximately 95 percent or more of the pervious strata before significant reductions in the quantity of flow can be realized; however, partial cutoffs can be somewhat effective in reducing uplift pressures on the wall base. Deep cutoffs will often interfere with the normal exchange of groundwater between an aquifer and

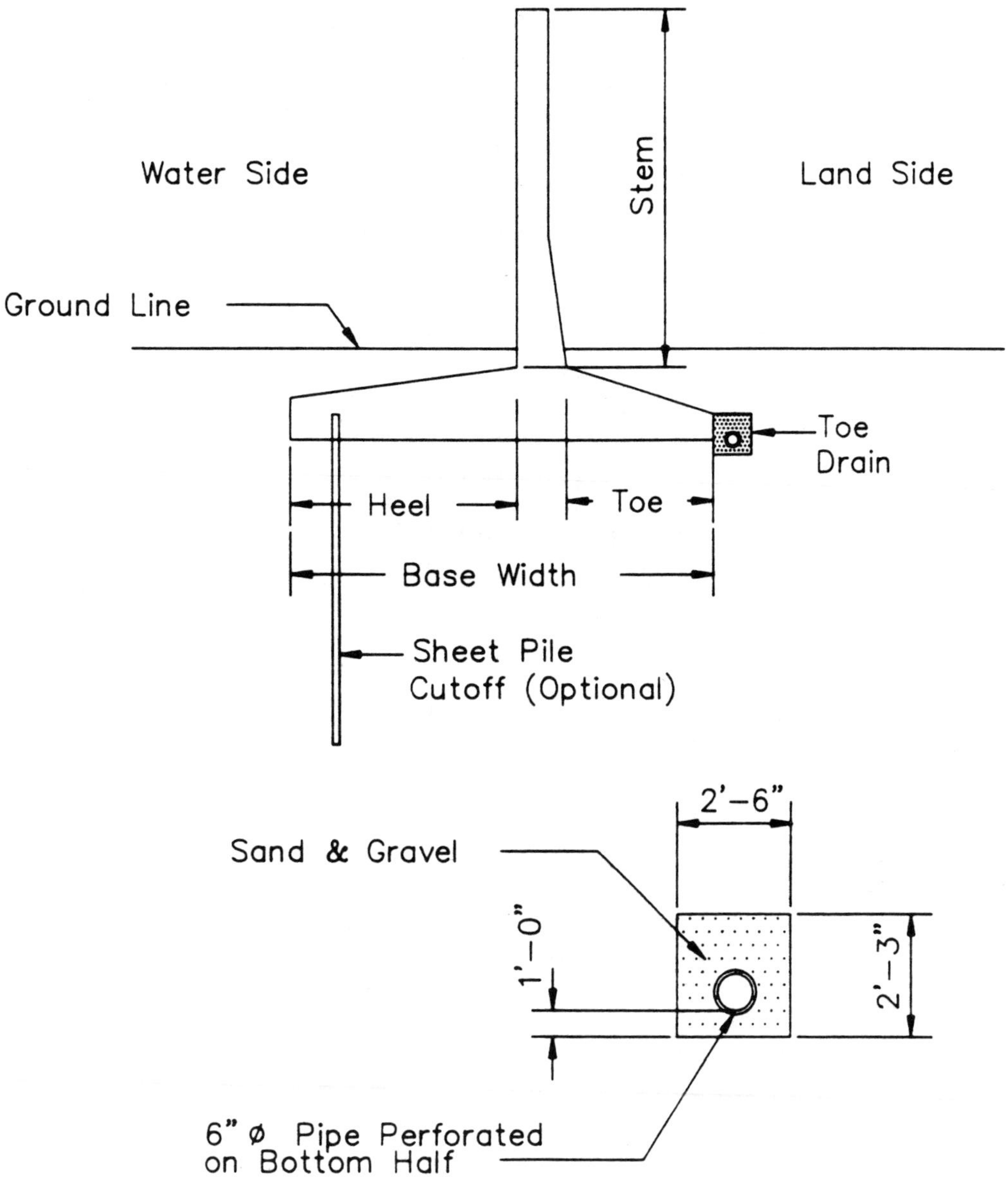

Fig 7-2. T-type flood wall—horizontal base

a river during nonflood periods and should only be considered where detailed hydrogeologic studies have been made in this regard.

The decision as to the type and depth of a cutoff should be based on an underseepage analysis considering actual site conditions. A steel sheet pile cutoff is not entirely watertight due to leakage at the interlocks but can significantly reduce the possibility of piping of coarse-grained material in the foundation. The effectiveness of a properly interlocked steel sheet pile cutoff through a coarse-grained stratum in reducing uplift can be assumed to be up to 50 percent. The design uplift diagram, as shown in Figure 7-3, should be drawn with a pressure head at point B on the unprotected side of the the cutoff equal to the full head of water on the unprotected side (neglecting any reduction in pressure due to head loss from seepage effects). The pressure head on the protected side of the cutoff at point B should equal the pressure at point B reduced by up to 50 percent of the difference between the full head value on the unprotected side and the pressure head at the end of the toe of the wall. The pressure head at the toe of the wall can be computed based on the seepage path from the cutoff wall to the saturated level on the protected side. If the effectiveness of the steel sheet pile cutoff is assumed to be greater than 50 percent, it should be based on actual experience of similar

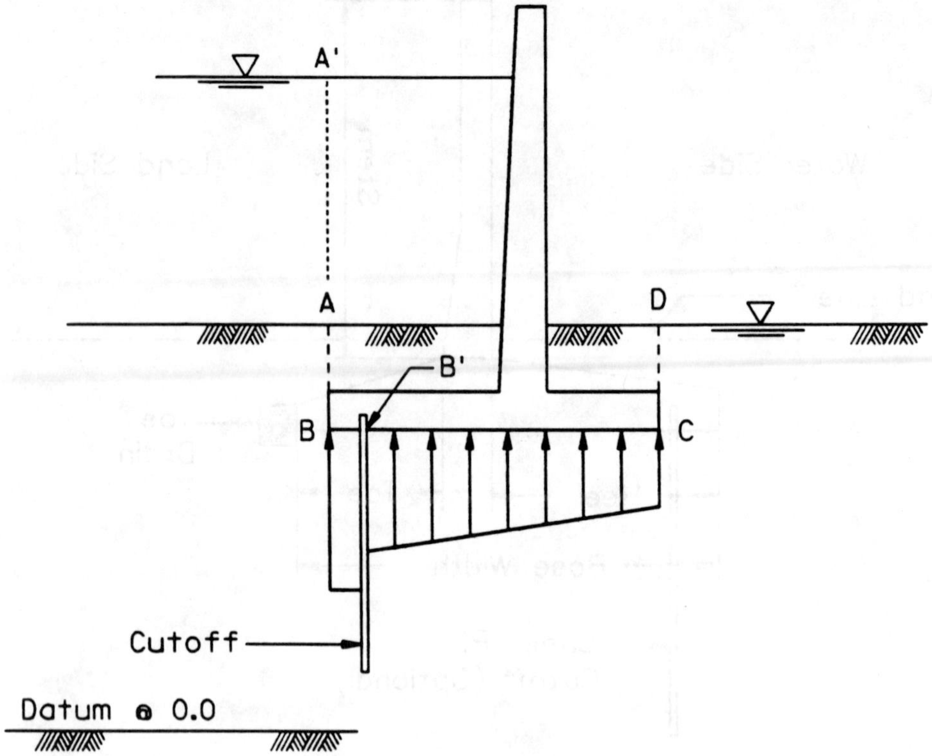

Total head at B = $El_{A'}$

Pressure head at B = $El_{A'} - El_B$

Pressure head at B' =
$(El_{A'} - El_B) - 0.5$ [(Pressure head at B) - (Pressure head at C)]

Total head at C = $El_D + \left[\frac{DC}{DCB'}\right](El_{A'} - El_D)$

Pressure head at C = Total head at C - El_C

Fig 7-3. Uplift pressures for a wall with a sheet pile cutoff

conditions and justified accordingly. An example of a flood wall with a sheet pile cutoff is shown in Example 5 of Appendix N. A sheet pile cutoff is less effective in fine-grained material than in coarse-grained material because cohesion may allow cracking and separation of the soil away from the sheet pile. Bearing value of steel sheet piling should be neglected.

B. TOE DRAINS. All inland flood walls should be provided with a landside toe drain similar to that shown in Figure 7-2. Coastal flood walls should be analyzed to determine if such drains are needed. The toe drain, which runs parallel to the wall at the landside edge of the footing, provides a positive outlet for local underseepage and a check for controlling piping and/or excessive uplift pressure beneath the base slab. For walls on impervious foundations, the toe drain may be adequate to control all underseepage; for walls on pervious foundations, additional seepage control measures will usually be required. In the case of pile-founded walls, the toe drain should be adequate to protect against 'roofing,' the loss of material from beneath the wall base. The drain should never be located under the wall footing, in order to allow maintenance access and to avoid crushing the drain. A typical toe drain design will consist of a 6- to 8-inch-diameter pipe perforated on the bottom half and surrounded in all directions with 6 to 10 inches of filter material designed by the filter criteria in paragraph 6-6e. The collected water is usually disposed of by gravity outlets into ditches, ponding areas, or pump stations. The toe drain system should provide access for inspection and maintenance at changes in alignment and at intervals not to exceed 500 feet. Discharge pipes should be provided with check valves that will prevent the entrance of surface water.

C. TRENCH DRAINS. Where the impervious top stratum is thin or non-existent, a trench drain may be used to control underseepage in the vicinity of the flood wall toe. A trench drain is an enlarged variation of a toe drain. It extends from the ground surface through shallow pervious layers or into a pervious layer underlying a shallow surface blanket. The practical depth for construction of a trench drain depends on available excavation equipment and site dewatering requirements. The excavation, pipe placement, and backfilling of the trench should always be performed in the dry. To assure adequate capacity, the collector pipe should be sized considerably larger than computations indicate to be necessary. Backfill in a trench drain should conform to the filter criteria in paragraph 6-6e. A trench drain should be provided with inspection and maintenance access and backflow protection as described for toe drains. The seepage calculations for the quantity of flow should assume the tailwater elevation equal to that of the discharge of the trench drain. However, if water can pond on the land side of the wall, the calculations for uplift pressure should check whether a more critical uplift condition can occur for the ponded case.

D. RELIEF WELLS. Pressure relief wells are used to reduce uplift pressures at depths in pervious layers which might otherwise cause sand boils and piping of foundation materials. Wells function to some extent as a controlled sand boil, relieving pressure by discharging water, but retaining materials with a screen and filter. Wells are advantageous where pervious strata are relatively thick or relatively deep. They are particularly useful in controlling large quantities of seepage in strata of pervious material having direct connections with the river. Another advantage of relief wells is the ease with which they can be constructed if piezometric pressures measured during high water indicate the need for additional underseepage control. Design of relief well systems is described in EM 1110-2-1905, EM 1110-2-1901, and WES Technical Memorandum 3-424 (US Army Engineer Waterways Experiment Station, 1956). The design entails selecting a spacing, size, and penetration for a line of wells that will result in acceptable gradients at points midway between the line of wells and at the flood wall toe. Relief wells are usually not very effective in intercepting near-surface seepage, and it is often wise to use them in combination with a toe drain. Relief wells should be pump-tested when installed. Because the efficiency of relief wells may deteriorate with time due to corrosion or bacterial incrustation, considerable monitoring and maintenance may be required to assure that the relief well system performs acceptably for the project life. To assess possible well deterioration, a representative number of wells should be periodically pump-tested, and the specific capacity (flow/drawdown) should be compared to the initial pump test results. To calculate uplift pressures on the wall, the potential head at the well line should be assumed equal to the average head in the plane of wells, a value obtained as part of the well design procedure in the cited references.

E. RIVERSIDE IMPERVIOUS BLANKETS. Impervious riverside blankets (natural or constructed) overlying a pervious foundation are effective in reducing the quantity of seepage and to some extent are effective in reducing uplift pressures and gradients landside of the flood wall. Their effects may be analyzed using seepage analysis methods found in EM 1110-2-1913 and WES Technical Memorandum 3-424 (US Army Engineer Waterways Experiment Station, 1956). Riverside blankets may be constructed over thin natural impervious blankets to improve the effects of the natural blankets or they may be constructed directly on pervious material. Excessively steep riverbanks may make blanket construction impractical. Also, it is seldom feasible to construct blankets over exposed portions of the pervious layer under water. A noncontinuous blanket has serious drawbacks, as only a small area of pervious stratum left exposed may significantly reduce the blanket's effectiveness. Riverside impervious blankets need to overlap the riverside base of the flood wall to minimize the potential for rupture of the blanket by landward deflection of the flood wall when loaded. Riverside impervious blankets may be subject to scour at high river stages when they would be most needed, or may crack open if not continuously wet. To prevent such action, blankets should be protected immediately after construction. A well-designed and well-planted vegetative cover is ordinarily sufficient along straight reaches. Along outside curves of the river, the blankets should be protected with riprap or other positive protection.

F. LAND-SIDE SEEPAGE BERMS. Land-side seepage berms function by providing an increased land-side top blanket thickness, thereby reducing the gradient. The berm also extends the seepage path by forcing the seepage exit landward. Seepage berms are typically 100 to 300 feet wide. As flood walls are usually built in areas where right-of-way cost or availability is insufficient for levee construction, seepage berms are rarely used in conjunction with flood walls. Procedures for seepage berm design are presented in WES Technical Memorandum 3-424 (US Army Engineer Waterways Experiment Station, 1956) and EM 1110-2-1913.

G. GROUTING OF OPEN ROCK JOINTS. In cases where rock is shallow enough that flood walls can be founded directly on the rock, close examination of the rock surface is necessary to determine if open joints are present. Such joints can be detrimental to underseepage control and should be cleaned out and filled with grout before the concrete base is placed. If the possibility exists for seepage flow through porous or cavernous rock in the foundation, consideration should be given to installing a grout curtain.

Section III. Foundation Considerations

7-5. Base Types

The T-wall is the most widely used flood wall type. T-walls are normally constructed with horizontal or sloped bases. The advantages of each type of base are as follows:

A. HORIZONTAL BASE (Figure 7-2)

1. The volume of foundation excavation is usually less for a horizontal base and it is simpler to construct.

2. Bearing values and base pressures for the two base types are not directly comparable. However, for equal heights, base pressures of the horizontal base generally are smaller because of its reduced earth load and slightly wider base.

B. SLOPED BASE (Figure 7-4)

1. A sloped base may allow shortening or complete elimination of a key, thereby reducing excavation difficulties. Also, a shorter key will generate less moment in the heel adjacent to the key and will generally allow for a shorter base width to maintain overturning equilibrium.

2. The deep cover or blanket over the heel of a sloped base lessens the chance of rupturing the cover as the wall moves under load.

3. The resultant of applied forces is more nearly normal to a sloped base, thereby reducing the tendency of the structure to slide along that plane.

4. A full-size flood wall test performed by the Ohio River Division (1948–1956) (U. S. Army Engineer Division, Ohio River 1958) indicated that the sloped-base wall moved consistently less than the horizontal-base wall of comparable design.

C. SELECTION. Both base types have their advantages and disadvantages. Final selection will depend upon the specific site conditions at the project under consideration.

7-6. Horizontal Water and Earth Loads on Keys

For flood walls on clay foundations, full flood head will be conservatively assumed to act at the bottom of the key and the horizontal water load acting on the riverside face of the key will be computed on this basis. The seepage path will then be assumed to begin at the bottom of the key. The land-side face of the key will normally be assumed to be in full contact with the earth-resisting movement of the wall.

7-7. Unsuitable Foundation Material and Bank Stability

Foundation material found to be unsuitable may be avoided by a change in alignment or may be removed and replaced with suitable earth fill (Figure 7-5). The wall may also be founded on piles through the unsuitable material. In some cases the removal of unsuitable foundation material involves the removal of or cutting into the existing riverbank on which the flood wall is to be placed. In other cases the right-of-way may be so restricted and confining that the flood wall may have to be placed near the top edge of the bank or even riverward of the bank. In those cases, fill placed riverward of the top bank is permitted, if proper precautionary measures are taken. Careful attention must be paid to the outlining of and removal of unsatisfactory material and to the selection of suitable replacement material. New material must be obtained, placed, and compacted to provide adequate support for the flood wall. Replacement material should undergo the

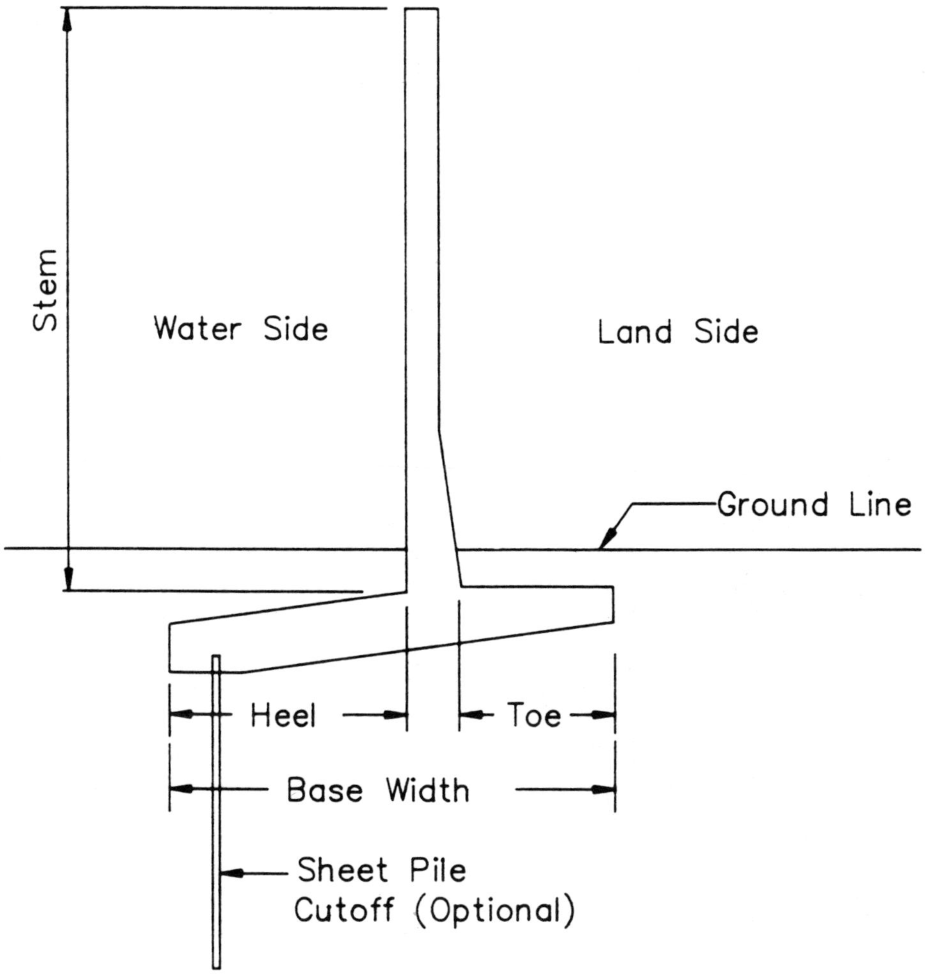

Fig 7-4. T-type flood wall—sloped base

same types of laboratory testing as existing foundation material. Placement and compaction techniques should generally be in accordance with earth dam and levee requirements. Slopes steeper than 1.0V on 1.5H and areas that require hand compaction should be minimized. Slopes on which there is evidence of past instability, or in which fill is a component, should be investigated for stability. All riverward slopes should be checked for stability if the failure of the bank would jeopardize the stability of the wall.

7-8. Scour Protection

Occasionally a flood wall is exposed to scouring because of the direction, curvature, and velocity of current or waves; characteristics of the soil; topography; etc. Scouring at the wall footing should be considered, and where anticipated, protected with riprap. Design guidance on sizing riprap is given in EM 1110-2-1601.

Section IV. Types of Monoliths

7-9. Change-of-Alignment Monoliths

Changes in alignment require special monoliths (Figure 7-6). Monoliths with less than a 10-degree change (horizontal) do not need to be analyzed as a special category. Monoliths of short length or abrupt alignment changes may require very wide bases. A 90-degree corner monolith is an indeterminate structure. Adjacent monoliths should not be considered to provide resistance in the stability analysis.

7-10. Closure and Abutment Monoliths

A number of openings must be provided in many flood walls. The openings provide access for

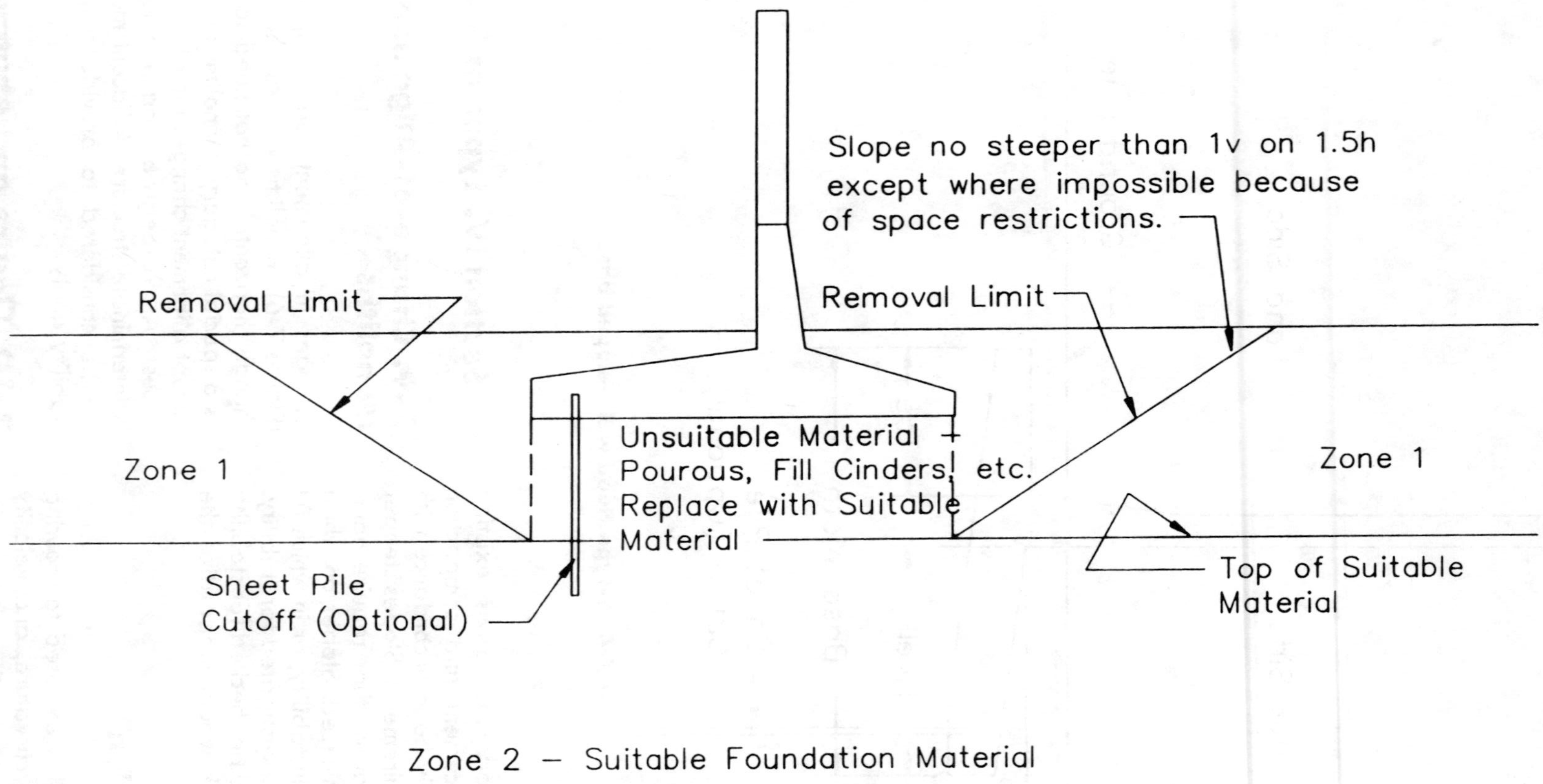

Fig 7-5. Removal limits of unsuitable foundation material

commerce, safety, and recreation during periods of low river stages. The number and size of openings depend on local requirements. Each opening must be provided with a moveable closure structure. During flood periods, the closure structure is installed on base and abutment monoliths (this combination is a special monolith). These special monoliths must be designed both for the design water load at high water and traffic loads during low-water periods.

7-11. Drainage Structure Monoliths

When topography, foundation conditions, and economics permit, it is preferable that structures housing gates and pumps be designed as integral parts of the flood wall. These special monoliths must be designed to minimize differential settlement across a monolith or between adjacent monoliths. For closure gate requirements and the need for secondary closure gates for drainage outlets, see EM 1110-2-1410.

7-12. Transition Sections Between Flood Walls and Levees

A. JUNCTURES. A junction between a T-wall and levee is not made directly or abruptly, but with a short transition concrete-capped sheet piling I-wall between the two (Figures 7-7 and 7-8). One of the primary concepts in the development of this transition is to arrange details so there will be a minimum amount of differential movement of joints of monoliths in the transition. The levee end of the transition will usually settle a considerable amount, due primarily to foundation consolidation under the added weight of the levee. The T-wall monolith immediately adjacent to the beginning of the levee adds far less superimposed weight on the foundation. Hence, there is much less settlement at this end of the transition. The I-wall can be satisfactorily adopted as a transition section between levee and T-wall because this type of construction can, and in fact must, be done after completion of the levee. A delay in inserting the I-wall allows for settling of the levee, thus lessening the differential settlement between the levee end of the transition and the T-wall. Review by hydraulic engineers is required for inland flood walls to assure that the transition geometry will not create significant flow disturbance with consequent scour.

B. I-WALL. The I-wall portion of the transition is begun where the levee slope (parallel to the protection) reaches a point 10 feet below the top of the wall. In cases where protection is already 10 feet or less above the levee, an I-wall, if used, is merely continued into the levee as shown in Figure 7-7.

C. SHEET PILING. It should be noted in Figure 7-7 that the sheet piling is continued into the levee for a specified distance beyond the last concrete cap.

Section V. Water Stops and Joints

7-13. Water Stops

As shown in Figure 7-9a, b, and c, for yielding foundations a U-shaped (type 'U') water stop should enclose almost the entire base and a center bulb (type 'Y') water stop, located in the stem, is joined to the U-shaped water stop at the bottom of the stem. Experience has shown that a center bulb or dumbbell water stop located within the base section is likely to allow excessive seepage. Between monoliths on a foundation requiring a cutoff, the type Y water stop in the stem should be extended to tie into the cutoff, and the type U water stop around the base should be deleted. The earth surface on which a type U water stop is installed must be firm and smooth, with no chips, sags, humps, clods, or loose debris that would prevent intimate contact between the water stop and soil. See Chapter 6, paragraph 6-4e, for general guidance on water stops. Because field construction problems are common for the type 'A' joints shown in Figure 7-9a with the type U water stop shown in Figure 7-9b, and because the buried base slab does not experience wide temperature changes, an optional base slab joint is allowed when the base is placed. This base slab joint uses construction joints without water stops but with the base slab longitudinal reinforcement continuous through the joint. When this option is used, longitudinal reinforcement of at least 0.4 percent of the slab cross-sectional area must be provided in the base slab, half in each face, but with not more than #9 reinforcing bars at 12-inch spacings in each face.

7-14. Contraction and Expansion Joints

Contraction and expansion joint details are illustrated in Figures 7-9a through 7-9c. Contraction joints (type 'A') should contain a bond breaker. Expansion joints (type 'B') should contain $\frac{1}{2}$-inch preformed expansion joint filler in:

A. All protruding (convex on water side) monolith bases, and in selected re-entrant monolith bases and stems as shown in Figure 7-6.

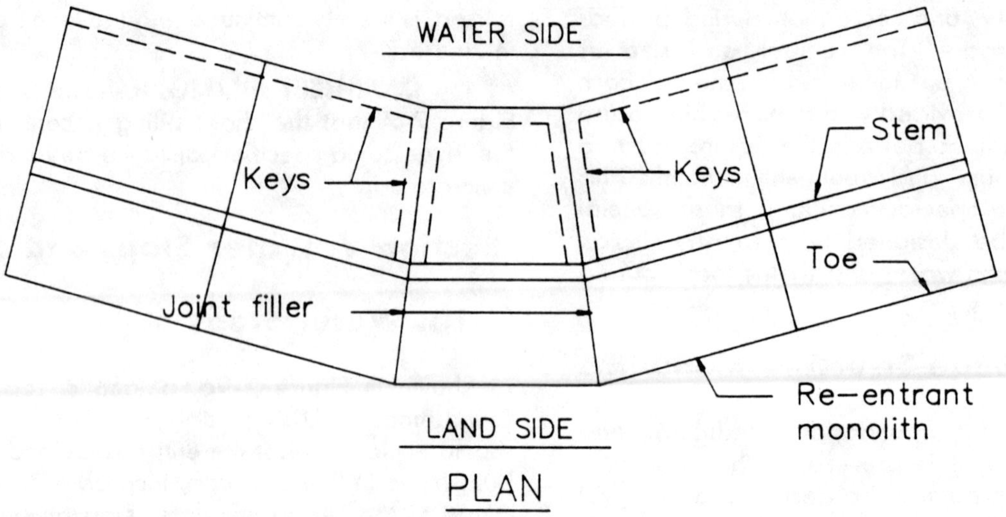

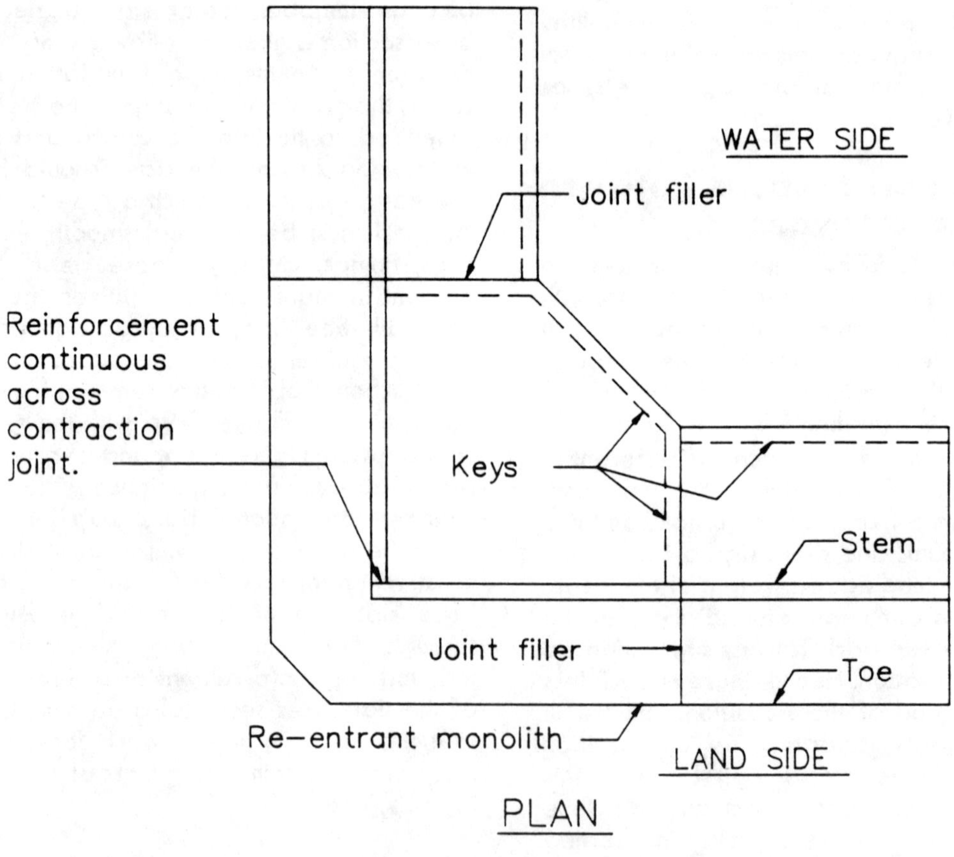

Expansion joint material (paragraph 7-14a) to be used where each end of re-entrant monolith touches the adjacent monoliths (see Figure 7-9a).

Fig 7-6. Return keys on re-entrant monolith

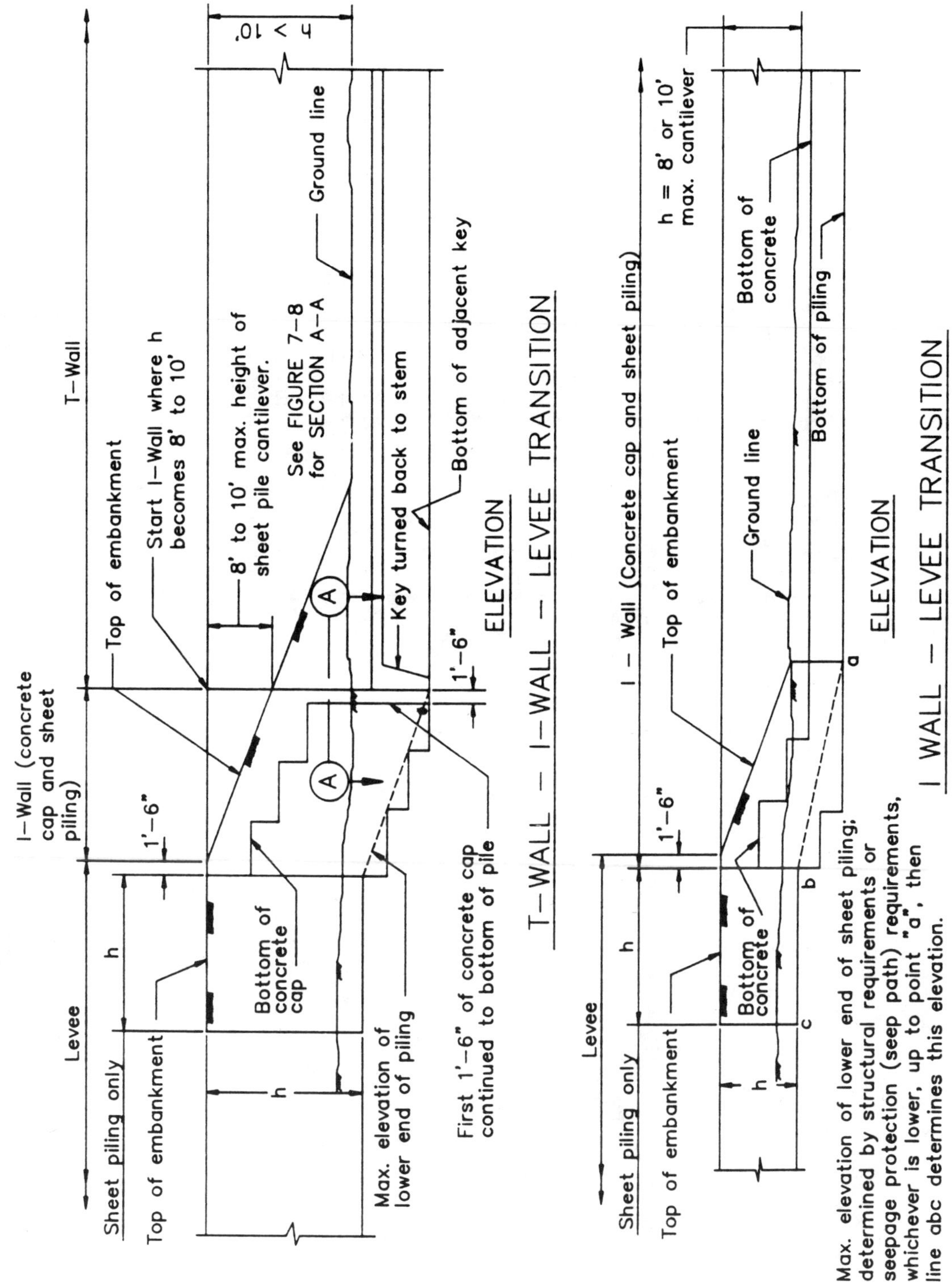

Fig 7-7. Flood wall—levee transitions

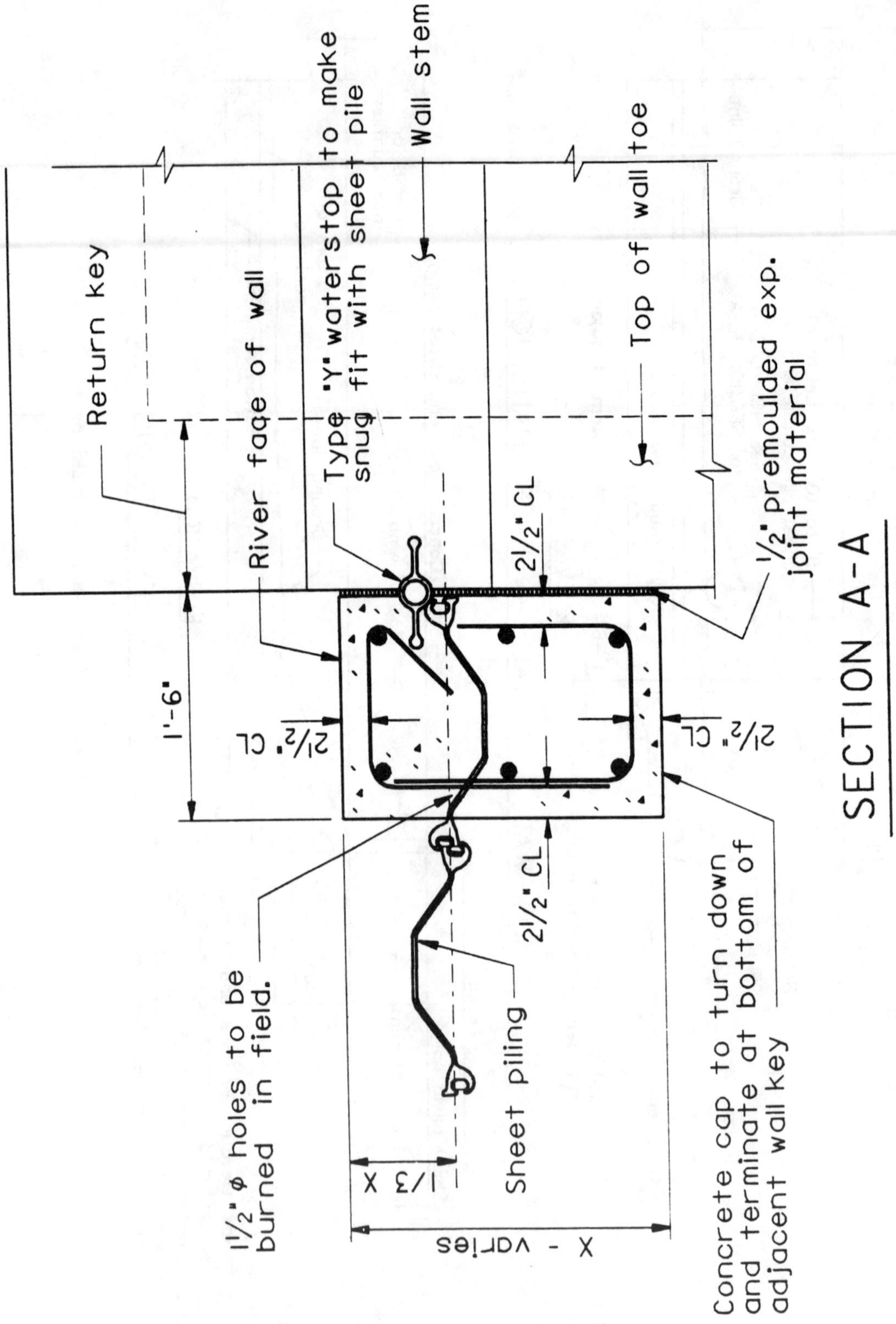

Note:
For location of section A-A, see Figure 7-7.

Fig 7-8. Typical detail of joint between I-wall and T-wall

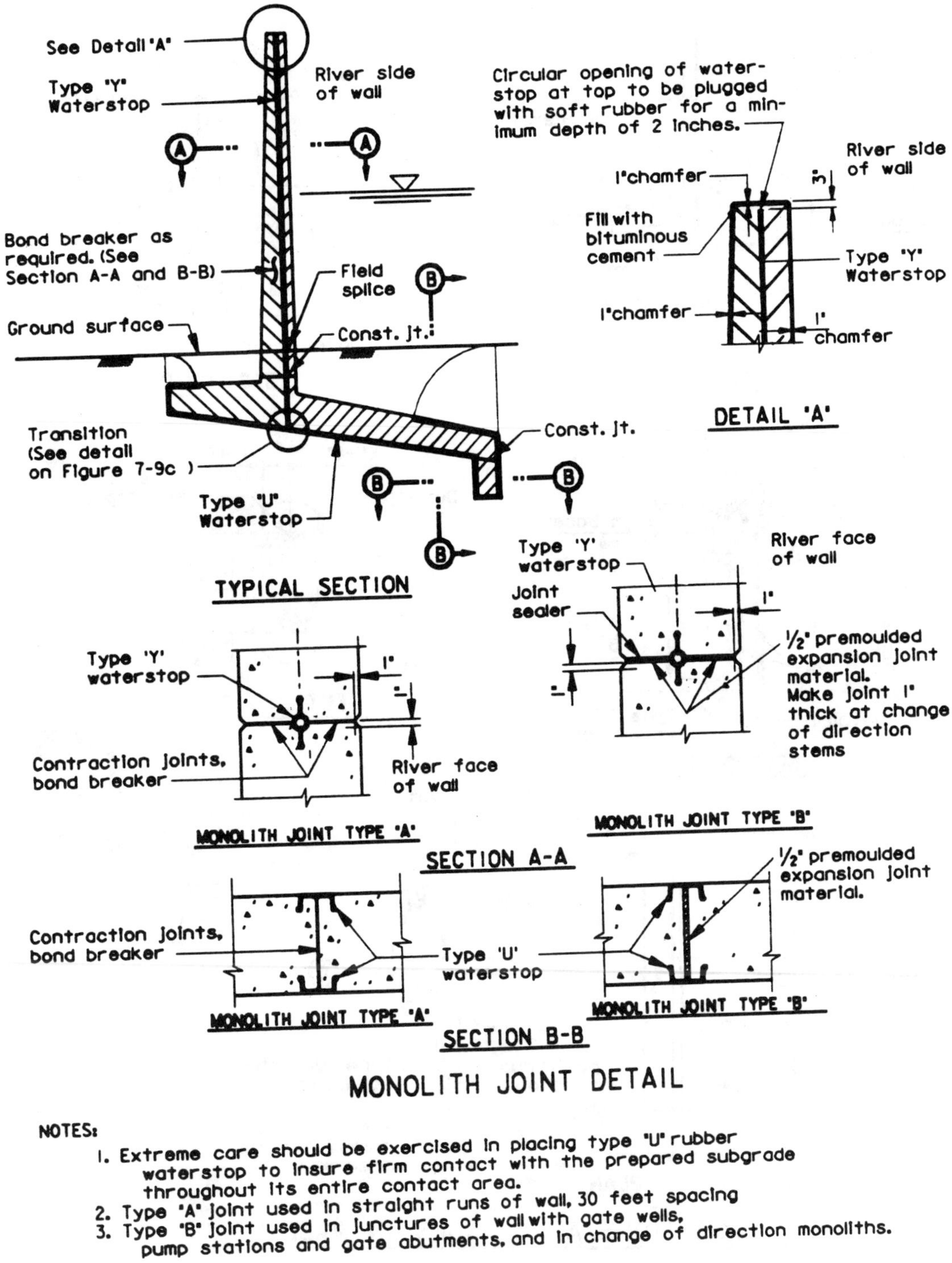

a. Monolith joint detail

Fig 7-9. Typical joint and water stop details

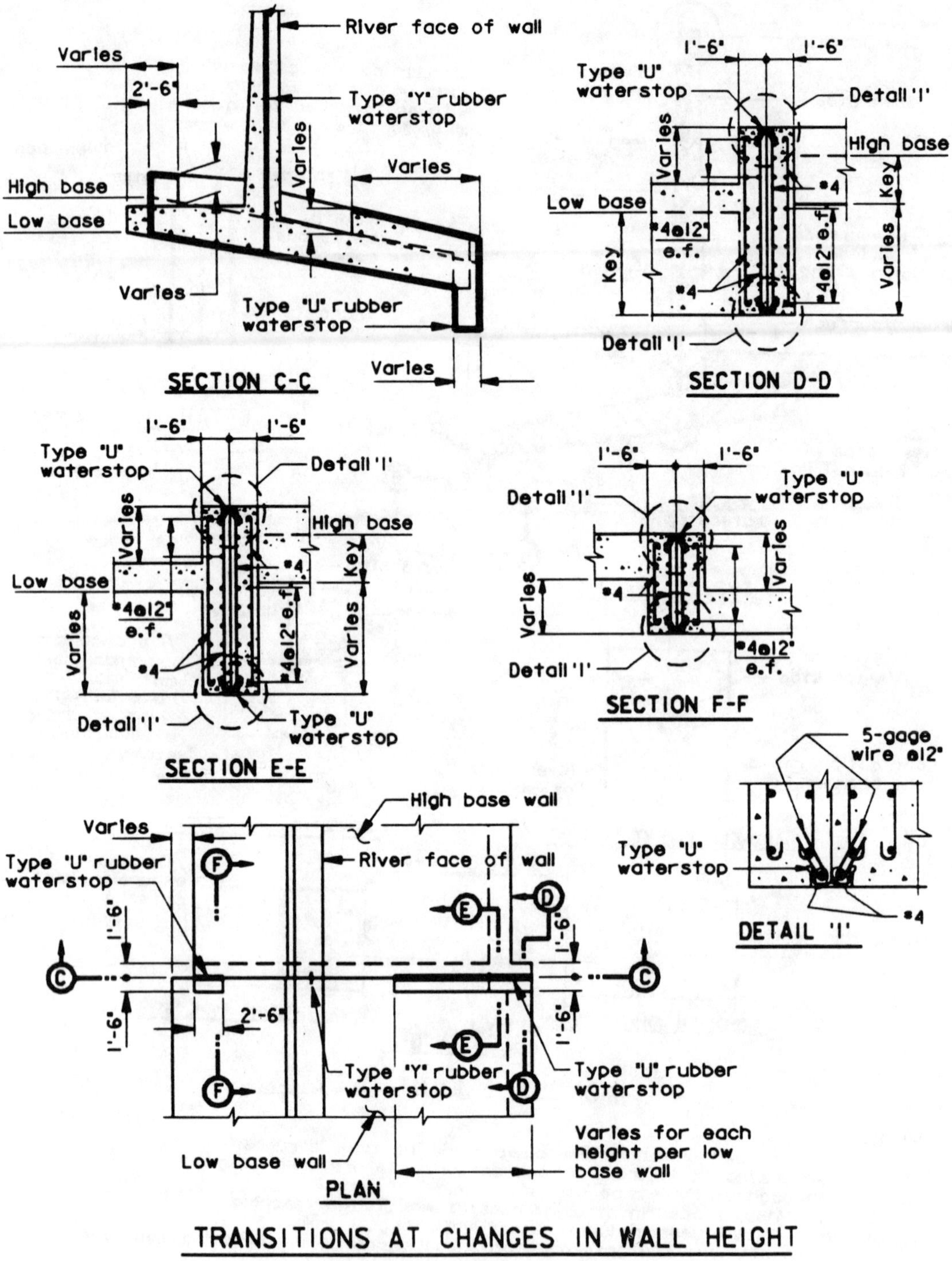

b. Transitions at changes in wall height

Fig 7-9. (Continued)

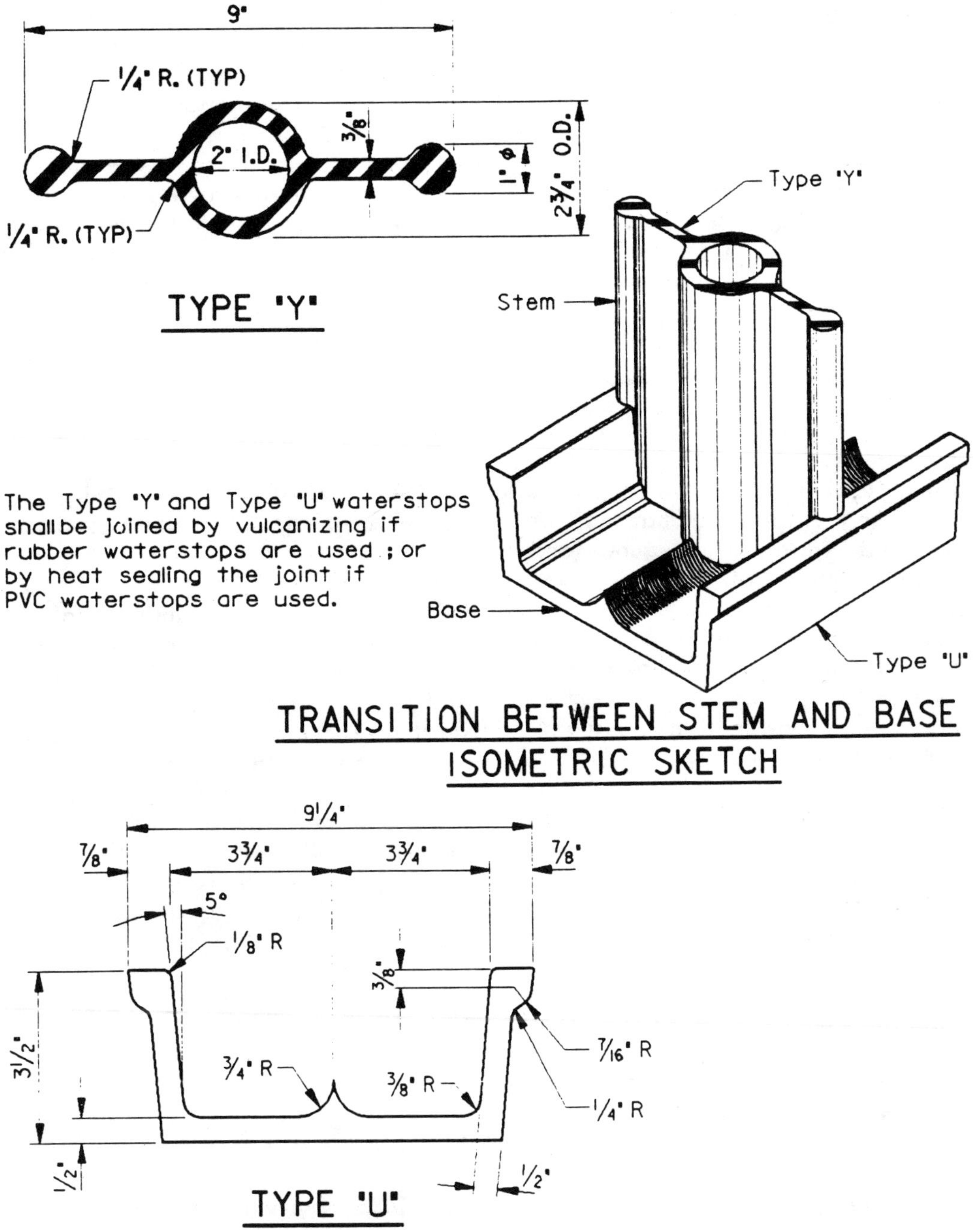

c. Water stop details

Fig 7-9. (Continued)

B. In bases and stems of alternate monolith joints in straight-line runs, if warranted by previous experience with similar foundation conditions.

C. Bases and stems of junctures of walls with gate wells, pump stations, gate abutments, and similar structures. Nonflexible material in a protruding angle joint is particularly dangerous.

D. See paragraph 6-4 for general guidance on joints.

Section VI. Site Considerations

7-15. Adjacent Structures and Rights-of-Way

Flood walls are usually built because only a narrow right-of-way is available. The presence of existing buildings or other structures is usually the reason for a narrow right-of-way. Sewer pipes with open joints, structures with basements, and excavations close to the wall may create a hazard to the safety of a flood wall. Also, new structures that are built close to existing flood walls can create the same hazards. Present right-of-way acquisition policies do not permit legal restrictions to be placed on future construction; however, local interests should be advised in writing of potential hazards, of required design and construction measures, and should be requested to closely supervise new construction close to the flood wall.

Potential hazards can be avoided by proper design and construction measures. One hazard that should be considered is seepage. A basement or other excavation on the landside of the flood wall may result in shortened seepage paths. A basement or excavation on the riverside may also create a safety hazard if it penetrates the impervious blanket or shortens the seepage path. When feasible, the basement or excavation should be backfilled with the same type of material existing in the foundation of the flood wall. If relief wells are selected to control seepage they should be located, if at all possible, between the flood wall toe and the adjacent structure. Protection of the basement area may require lowering of discharge elevations for safeguarding the wall. The location of relief wells within a basement area is not prohibited, but it leads to problems of construction, maintenance, and discharge collection. If the seepage problem is only one of quantity, sump pumping may be used during periods of high water. A second hazard that landside basements and excavations create is to lessen the resistance to sliding along a foundation failure plane. For this reason potential planes of sliding into basements or excavations should be studied. If backfilling is not possible, other measures include the addition of fill between the stem and the building or strengthening the basement to provide the needed resistance. Riverside excavations which contribute to riverward foundation instability should be backfilled, at least to the extent that stability requirements will be satisfied. For the special situation where a wall in a congested location is subjected to an unusually large horizontal force, such as the force of a breaking wave, T-type flood walls are frequently worth the extra cost over other types of construction. This situation requires an unusually wide base for sliding stability, requiring more right-of-way and, hence, more cost for construction. The relatively thin stem of the T-wall does, however, provide the most usable surface area adjacent to the stem after backfilling, in comparison with embankment, braced walls, etc., making the T-wall the preferred solution in spite of the extra construction easement right-of-way. While an I-wall also provides little intrusion on the completed surface area, its use can be precluded by the pile-driving vibration and consequent chance of damage to adjacent structures.

7-16. Architectural and Landscaping Considerations

Aesthetics should be considered in the design of flood walls, from the standpoint of blending the project with the surroundings. Whenever possible, the wall should appear to be a natural extension of the local topography. The basic design of these structures should be a coordinated effort between the design engineer, the architect, and the landscape architect. While it is seldom feasible to preserve the natural setting intact, design techniques and careful construction methods can be used to protect or even enhance the aesthetic value of the immediate project area. Landscape planting design for project structures should consider the entire area affected by the contemplated construction. Further details may be found in EM 1110-1-2009 and EM 1110-2-301.

Section VII. Instrumentation

7-17. General and Specific Considerations

Flood wall instrumentation should be considered so that performance can be monitored, particularly during periods of high water. The decision of how much, if any, instrumentation is appropriate must be based on these factors:

A. Who will monitor and evaluate the instruments—Corps of Engineers, local interests, etc.—and how meaningful their evaluations are expected to be.

B. Access to the instruments during flood conditions, especially during hurricane-flood situations where high winds may make accessibility impossible.

C. Time required for meaningful evaluation, compared with the expected duration of the flooding.

D. Other particular considerations for specific situations.

The instrumentation descriptions that follow must be implemented in light of the above decision. Specifically, areas with high walls, low embedment ratios, replaced foundation materials, overbank fills, pervious materials in the foundation, and changes in direction should be considered for instrumentation. When founding a flood wall on earth, the distance between monoliths with piezometers should not exceed 1000 feet unless warranted by site conditions. Properly installed, maintained, and observed instrumentation can forewarn of dangerous conditions that may affect the stability of the structure. All instruments should be read soon after construction is complete. Knowing the as-built conditions of the wall is essential for an accurate determination of later behavior. Initial piezometer readings should be repeated until equalization (steady state) occurs. All instrumentation readings should be made by trained survey or flood-patrol personnel. Ideally, all instruments will be read frequently during high-water stages. During design floods, the procedure may prove almost impossible because of the need for trained personnel to direct flood fights; but readings should be made at certain, previously selected, critical locations during design flood stages. During normal water stages, instruments should be read prior to district periodic inspections so that the inspection party has the necessary evaluation data. Such data also provide a history of flood wall reactions over the years, during both high and normal water. Information concerning frequency and manner of conducting periodic inspections and evaluations is contained in ER 1110-2-100, while ER 1130-2-339 covers local flood-protection projects.

7-18. Types of Instrumentation

The principal types of flood wall instrumentation monitor movements, both vertical and horizontal, and hydrostatic pressures in the foundation. The instruments selected should be simple to install and observe, and efficient in performance and functional reliability. The monitoring of the movements provides an indication of possible sliding instability or possible water stop rupture. The piezometers provide a record of hydrostatic pressures in the foundation which can indicate uplift and possible excessive seepage pressures. Instrumentation systems, installations, and devices are discussed in detail in EM 1110-2-4300.

A. MOVEMENT MONITORING. All reference points to monitor movements should be tied in to a permanent baseline located so that it is unaffected by movements of the wall. When establishment of a baseline is not feasible, the relative movements observed between monoliths or by means of triangulation can provide valuable data on behavior of the wall. Reference points to monitor the wall movements need to be installed during construction. Noncorrosive metal plugs should be installed in the top surfaces of the stems within 6 inches of each end of each monolith. The reference marks in the plugs of four to six successive monoliths should be placed in a straight line with theodolite or stretched wire. At changes in alignment, the straight line should be continued until it intersects the far side of the next monolith and a reference point for alignment control is placed. Each plug's changes in horizontal movement and elevation should be measured to 0.001 foot. Stations to be read with electronic optical reading devices need to be established at locations near the ground surface level on the land-side of the stem. Selection of electronic-optical station locations for the stem should be based on factors such as changes of direction, areas of overbank fill, foundation replacement, high walls, low embedment ratios, and junctures of flood walls with drainage structures. The monitoring system selected should be vandalproof. In many cases the monitoring system can be tied into the same baselines established for the reference markers on top of the wall. Tilting of stems can be measured by a tiltmeter.

B. FOUNDATION PIEZOMETERS. Design, installation, and observations of piezometers are described in EM 1110-2-1908, Part I. The simplest, most reliable method of measuring porewater pressures is the open tube piezometer. For impervious soils, the Casagrande type of piezometer with 24-inch-long porous stone is recommended. In order to measure the piezometric pressure at the porous tip, the boring for installation of the Casagrande piezometer must be effectively sealed against migration of seepage along the piezometer riser. For semipervious to pervious soils, a driven wellpoint type of piezometer is recommended. Where possible, the wellpoint should be driven into undersized, prebored holes. More piezometers can be added if foundation conditions warrant.

Section VIII. Operation and Maintenance Manual Requirements

7-19. General Coverage

General coverage of the requirements of local cooperation is contained in EM 1120-2-109. As written, the regulations are general in nature and obviously cannot give detailed instructions for the maintenance and operation of a specific project. Therefore, it is necessary for the district office having jurisdiction over the specific project to issue an adequate operation and maintenance manual for the guidance of local interests.

Section IX. Review of Existing Flood Walls

7-20. Inspection

Flood walls should be examined during scheduled periodic inspections, after major periods of high water, and when special events warrant an inspection (building or excavating near the wall, etc.). A determination of areas which may be weak or critical from the standpoint of leakage and stability should be made. Criteria for this determination are described below. Areas deficient in any of the criteria will be considered weak or critical, depending on the degree of deficiency.

A. HORIZONTAL MOVEMENT. Areas in which movement of a straight section of monoliths or differential movement between any two monoliths is greater than expected will be considered critical.

B. JOINT OPENING OR SPREADING. Joints referred to in this paragraph are those having a water stop embedded in the interior of the section. Using the results of the full-size flood wall test performed by the Ohio River Division, (ORD) in 1955, expected spreading of joints at 90-degree re-entrant corner monoliths (concave on the riverside) will be 42 percent of the expected movement of the straight run walls. Not only may joints at corner monoliths become critical upon application of load, but open joints below ground should be considered critical. Any joint can become open through loss of joint filler or through unequal settlement between adjacent monoliths or structures such as levees, pump houses, gate wells, and gate abutments. Some joints below ground may need to be excavated to determine the adequacy of joint filler. If the expected joint opening is greater than the allowable, the area should be considered critical.

C. FOREIGN MATERIAL IN JOINTS. The presence of inflexible foreign material, such as grout and pieces of aggregate, in expansion joints is dangerous from two standpoints. Grout, particularly if located within the fold of the water stop, destroys the flexibility of the water stop and, upon the occurrence of differential movements, allows the water stop to be torn. Grout and pieces of aggregate anywhere in the joint prevent the joint from fulfilling its expansion function. This condition becomes particularly dangerous at protruding angle locations; i.e., where the wall appears convex when viewed from the river. Here, the wall may be tilted waterward by a wedging action upon expansion of adjacent monoliths in hot weather. This wedging of adjacent monoliths at changes in alignment is likely to force excessive flexure in the stem, sufficient to cause failure. The same tilting can occur at re-entrant monoliths (Figure 7-6), but there the tilting is landward and the reinforcing is more adequate to resist the stress. For angle monoliths protruding toward the river, the land-side temperature steel can be quickly overstressed.

D. WATER STOPS. Joints with torn or parted water stops should be considered critical. Torn water stops may not be noticed during an inspection, particularly if the joint has not spread open. If sufficient differential movement has occurred, it should be assumed that the water stop is torn. The amount of tearing to be allowed should be based on factors causing piping; however, this is very difficult to predict. In the above cases, if a total differential movement (transverse and longitudinal combined) of $\frac{1}{2}$ inch or more has occurred, the water stop should be considered torn unless shown otherwise.

E. FOUNDATION VOIDS. All unequal settlements should be viewed with suspicion. In particular, unequal settlements adjacent to structures such as pump houses and gate wells should be the subject of rigid examination. Usually one or two monoliths (or a portion of one monolith) are constructed on compacted fill in these areas. Initial unequal settlement may cause the first monolith to bridge or wedge between the second monolith and the other structure. Further consolidation of the fill then leaves a dangerous void or voids under this base. Only underground examination will reveal the presence of these voids.

F. STABILITY ANALYSES. Original seepage assumptions or patterns should be reviewed for realistic representation of actual foundation conditions. Particular attention should be paid to foundations having pervious strata which connect directly with the river. Where

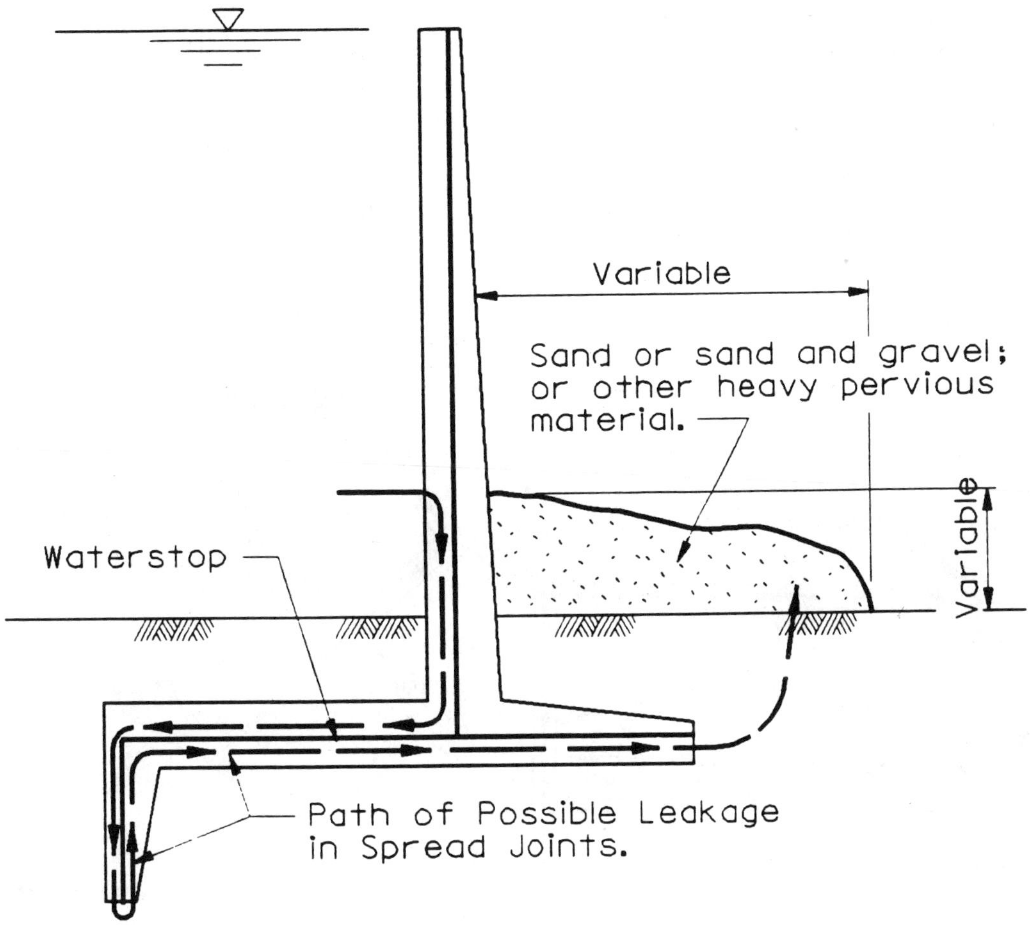

Fig 7-10. Emergency measures to control piping

indicated, seepage and/or stability analyses should be recomputed as described in Chapters 3, 4, and 5. In addition to a recomputation of uplift, the shear strengths used in the original analyses should be reevaluated on the basis of a study of types of soil and their drainage and consolidation characteristics. In cases where there is a lack of sufficient foundation information in areas suspected to be weak, new soil samples should be obtained as close to the existing wall as is feasible. Areas found to have questionable stability should be closely observed during high floods.

G. BASEMENTS AND OTHER EXCAVATIONS. The seepage aspects and the foundation stability of walls which have had basements excavated on either side of and adjacent to the wall since the original design and construction were completed should be investigated.

H. SEEPAGE CONDITIONS LANDSIDE OF FLOOD WALLS. These areas should be investigated thoroughly and seepage control of pressure relief provided, if needed.

7-21. Repair Measures

A. GENERAL. The following repair measures are only suggestions. Their use is not mandatory if more feasible or economical measures can be devised for the individual problems involved.

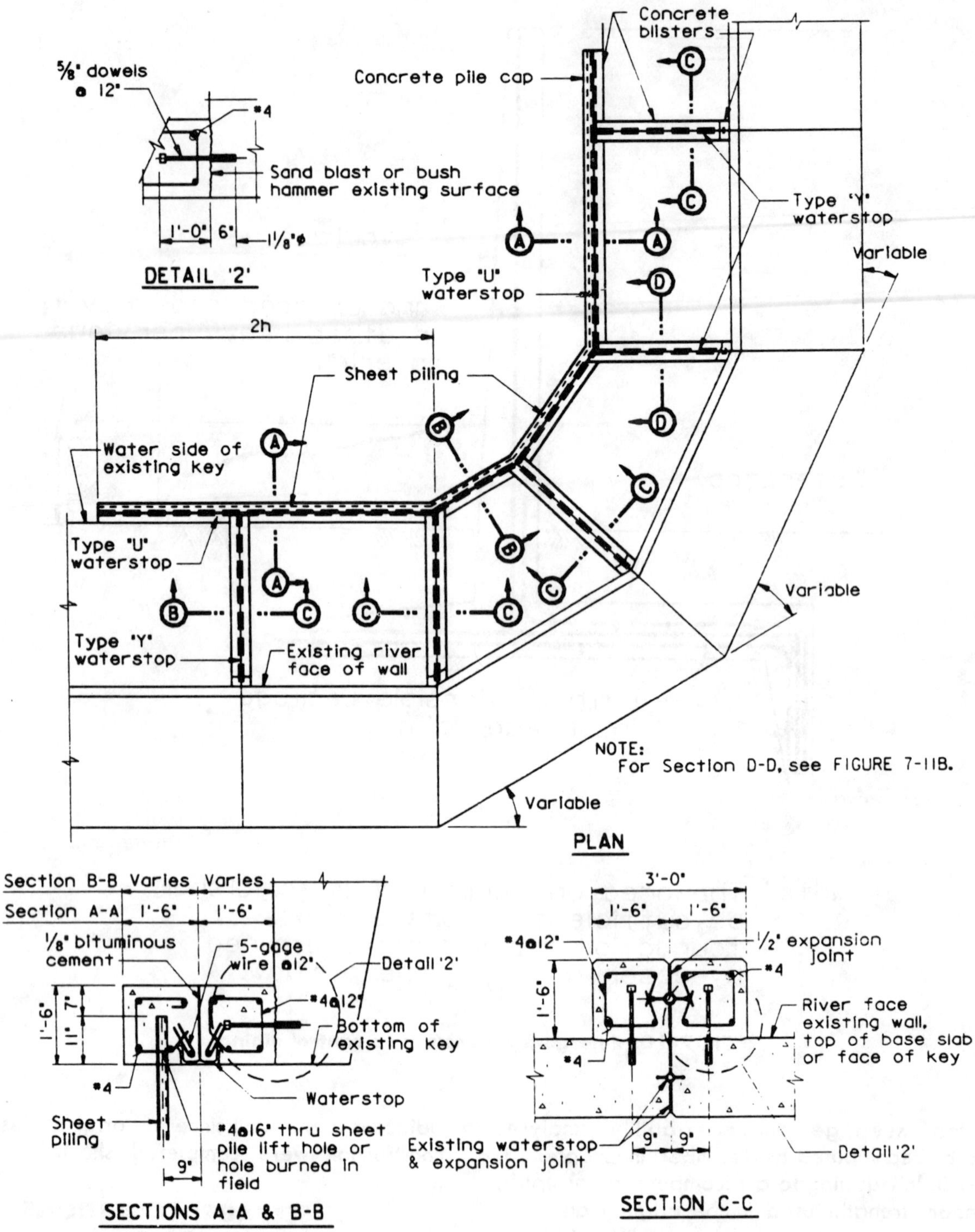

a.

Fig 7-11. Permanent water stop repair measures

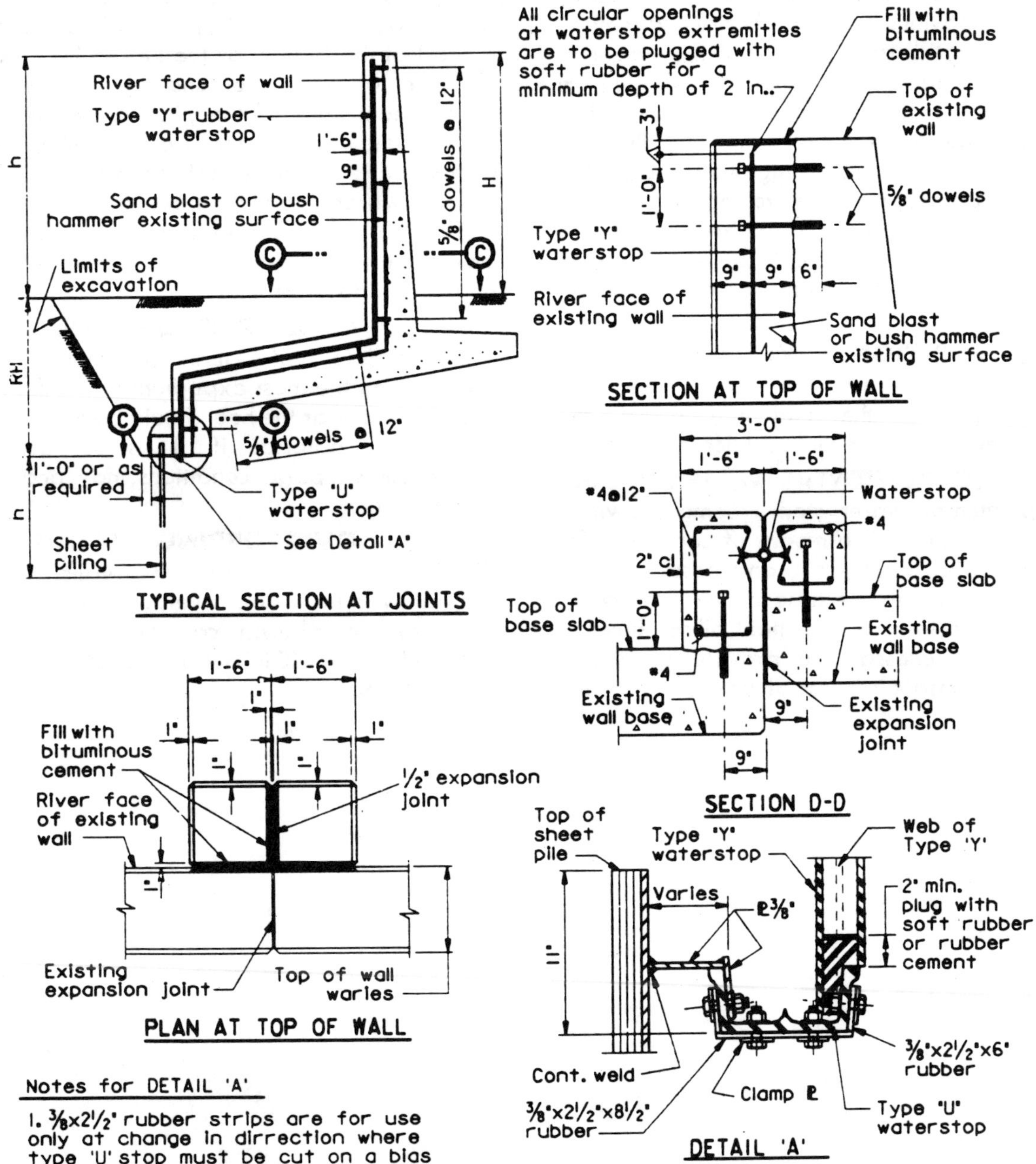

Notes for DETAIL 'A'

1. 3/8x2 1/2" rubber strips are for use only at change in dirrection where type 'U' stop must be cut on a bias and re-joined. On straight runs these strips need not be used.
2. Bulbs of type 'Y' stop are shaved down to web to provide a flat surface for bolting to inside of type 'U' stop.
3. All rubber surfaces in contact with each other are coated with rubber cement.

NOTES:
1. For details of "Y" & "U" type rubber waterstop, see FIGURE 7-9C
2. All steel in blisters 2" clear.
3. For Section C-C, see FIGURE 7-11A
4. For Section D-D location, see FIGURE 7-11A

b.

Fig 7-11. (Continued)

B. ADDITIONAL LAND-SIDE COVER. The most obvious and straightforward method for reducing anticipated horizontal movement or increasing sliding stability is the addition of land-side cover or fill to the wall (see Figure 7-10). At locations where additional land-side fill is not feasible or possible due to highways, railroads, and other structures, measures to reduce seepage pressure (such as those described below in paragraph 7-21d) will have to be employed to decrease landward movement or increase sliding stability.

C. ADDITIONAL WATER-SIDE COVER. In areas where earth cover over the waterward end of the heel is deficient, the recommended remedy is the addition of cover.

D. SUPPLEMENTAL WATER STOPS. The supplemental water stop scheme shown in Figure 7-11 a and b is a means of correcting for torn water stops, open joints, and possible earth cracking over the key because of thin heel cover or excessive movements. The sheet piling shown in the scheme is necessary to provide additional cutoff to compensate for loss of part or all of the normal seep path between earth and the water-side face of the key. The pile cap should be placed at the bottom of the key to limit excessive leakage of water around the upstream and downstream ends of the pile curtain as the wall moves landward under load. Another possible method of repair is to seal the opening below the existing water stop in the base by injecting cement grout. The opening above the water stop in the base could be sealed with an elastic sealant such as polysulfide elastomer.

E. OTHER PROBLEM AREAS. Foreign incompressible material in the joints should be removed by the most expedient method. Riverside excavations near the heel should be backfilled with impervious material if it is suspected that dangerous seepage conditions may occur during high water.

F. OVERTOPPING SCOUR CONTROL. For coastal walls or other walls where scour has removed land-side cover, consideration should be given to placing concrete slabs over the restored cover within a distance of 20 feet from the wall stem.

CHAPTER 8

CONCRETE GRAVITY WALLS

8-1. General Factors

Factors favoring concrete gravity retaining walls are shallow depth of overburden, a competent foundation, and an adequate source of fine and coarse aggregate for the required volume of concrete. See Chapter 2, Section I for additional comments on gravity walls.

8-2. Foundation Investigation

The requirements for the foundation investigation are discussed in Chapter 2, Section V.

8-3. Materials

A concrete compressive strength of 2,000 to 2,500 psi will usually meet the requirements for the gravity type wall. Where the environment requires durability, such at as the outer surface of the wall, the higher strength should be achieved with the appropriate water-cement ratio from EM 1110-2-2000.

8-4. Design

A. MAGNITUDE AND DISTRIBUTION OF FORCES

1. Dead Load. The unit weight of concrete is usually assumed to be 150 lb/cu ft. This value may vary, depending on the aggregate. Other dead loads that should be considered are superimposed backfill and the weights of any equipment or other structures supported by the wall.

2. External Water Pressure. The pressure exerted by water above ground and water in the ground should be determined as described in Section III of Chapter 3.

3. Internal Water Pressure (uplift). The uplift on a lift (horizontal construction joint) within the body of a concrete gravity wall for long-term water levels should be taken as 50 percent of the value obtained by assuming a straight line variation between the full hydrostatic pressures acting on each side of the wall. Uplift pressures on the base of the wall should be determined by the methods described in Section III of Chapter 3.

4. Lateral Earth Pressures. Lateral earth pressures should be determined by the methods presented in Section II of Chapter 3. Computation examples, as applied to gravity walls, are shown in Examples 5, 6, 8, 10, and 11 of Appendix M.

5. Wind and Earthquake Forces. These supplemental forces should be determined by the methods presented in Section IV of Chapter 3.

B. LOAD CASES. The load cases should be those described in Section I of Chapter 4.

C. EXTERNAL STABILITY. Sliding and overturning stability should be determined by the methods and criteria discussed in Chapter 4. Computer program 3DSAD will significantly assist in performing stability analyses. An example of a complete stability analysis of a gravity wall section is shown in Example 2 of Appendix N.

D. INTERNAL STABILITY. The resultant of all forces acting on any horizontal section should fall within the kern or sufficiently close to the kern of the section to keep the tensile stresses low. See EM 1110-1-2101 for allowable concrete stresses.

E. FOUNDATION ANALYSES. Foundation analyses should be performed in accordance with the methods described in Chapters 4 and 5.

CHAPTER 9

CANTILEVER REINFORCED CONCRETE WALLS

9-1. General Characteristics

The cantilever reinforced concrete wall is a special type of gravity wall in which part of the stabilizing weight is supplied by the weight of the backfill resting on the base slab. The structural members are designed for stresses due to bending and shear. Chapter 2, Section I, offers additional general comments on cantilever concrete walls.

9-2. Foundation Investigation

The requirements for the foundation investigation are discussed in Chapter 2, Section V.

9-3. Materials

Concrete materials and mixture proportioning, with appropriate water-cement ratios for durability, should follow guide specification CW 03301 and EM 1110-2-2000. Typically, a concrete compressive strength of 3000 psi is used for retaining walls. The age at which the specified strength is to be obtained should be decided by the designer depending on the loading conditions anticipated. Steel reinforcement bars should follow the specifications in the American Concrete Institute (ACI) Building Code (ACI 318), with the exception that for hydraulic structures the grade of steel will be limited to ASTM Grade 60 without special approval.

9-4. Reinforcement Cover

For hydraulic structures the minimum reinforcement cover should comply with EM 1110-2-2103. For structures not subject to hydraulic action the minimum reinforcement cover should comply with the ACI Building Code requirements.

9-5. Load Cases

The load cases should be those described in Section I of Chapter 4. The magnitude and distribution of the loads should be determined as described in Chapter 3.

9-6. Structural Stability

Sliding and overturning stability should be determined by the methods and criteria discussed in Chapter 4. Forces and moments for structural design should be based on external forces allocated according to paragraphs 3-7 through 3-9 and calculated as described in Section III of Chapter 4 for overturning stability. Sample stability calculations are shown in Appendix N.

9-7. Structural Design

A. GENERAL. Reinforced concrete walls should be designed for the loading cases given in Section I of Chapter 4 and the foundation pressures obtained from the overturning stability analysis described in Section III of Chapter 4. Wall components should be analyzed as cantilever beams. Compression reinforcement is not normally used. Temperature and shrinkage reinforcement should conform with EM 1110-2-2103. Example calculations are shown in Appendix N. When the top surface of backfill is sloping upward, a shear force in addition to the horizontal earth force should be considered acting on the structural wedge (see Figure 9-1).

B. STEM. Axial loads are usually small and may be neglected in design.

C. TOE. The toe should be designed with loads imposed by soil, water, concrete, bearing pressures, etc. The effects of axial loads are not ordinarily substantial enough to be taken into account.

D. HEEL. The loads for calculating design moments are the weight of soil, water, and concrete acting downward, along with uplift and bearing pressure acting upward. The bearing pressure should be determined using the horizontal earth force and shear when the backfill surface is sloping upward (see paragraphs 9-7a and 4-8c).

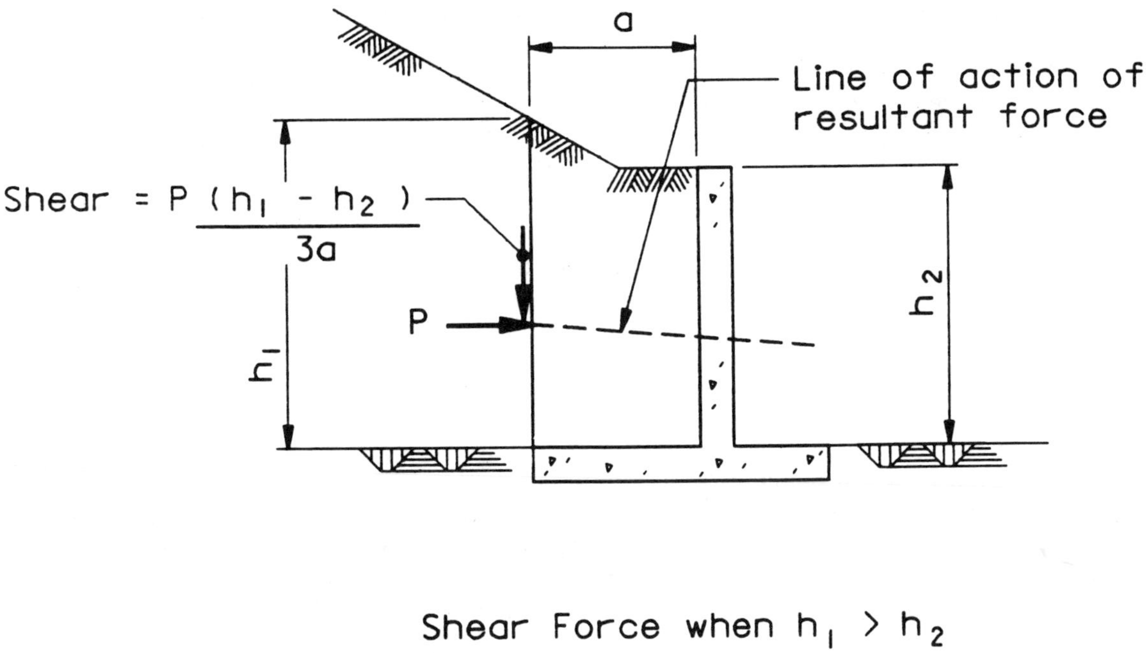

Fig 9-1. Shear force for upward-sloping backfill

With no key, the base shear should be neglected when computing reinforcement, as illustrated in Appendix N, Example 1.

E. SPECIAL CONSIDERATIONS FOR WALLS WITH KEYS. The overturning stability criteria for walls with keys include an assumed uniform distribution of earth pressure on the resisting side of the key that may result in unconservative design for reinforcement in the top face of the wall heel at and near the face of the stem. A portion of this force may actually act along the plane at the base slab of the wall and not on the key. The designer is cautioned to consider this in developing a reinforcing design. A conservative approach for design of the heel top steel at the stem would result from the use of foundation pressures obtained from a stability analysis assuming that all of the earth resistance acts along the plane at the base of the wall. See Section III of Chapter 4, especially paragraph 4-8b. Stability calculations for walls with keys are shown in Examples 3 and 6 of Appendix N.

9-8. Reinforced Concrete Design

A. GENERAL. Reinforced concrete walls should be designed with the strength design method in accordance with the current ACI Building Code, except as herein specified. Notations used are the same as those in the ACI Code, except those defined herein. (Appendix D lists the notations used in Chapter 9.) WES Technical Report SL-80-4 (Liu and Gleason 1981) contains design aids consistent with the information presented in paragraph 9-8b of this chapter. Retaining walls and flood walls may be designed using the same load factor for concrete weight as that selected for earth and water loads, as explained in paragraph 9-8b(1), Equations 9-5 and 9-6.

B. HYDRAULIC STRUCTURES—STRENGTH AND SERVICEABILITY.

1. Required Strength. Reinforced concrete hydraulic structures should be designed to have strengths in all sections equal at least to those calculated for the factored loads and forces in the following combinations that are applicable.

a. For usual loading cases R1, I1, C1, C2a, and C2c as described in Chapter 4:

$$U = 1.5D + 1.9L \text{ (if D and L have the same sign)} \quad [9\text{-}1]$$

or

$$U = 0.9D + 1.9L \text{ (if D opposes L)} \quad [9\text{-}2]$$

where

D = internal forces and moments from dead load of the concrete members only

L = internal forces and moments from live loads (loads other than the dead load of concrete members)

b. For unusual or extreme loading conditions such as cases R2, R3, I2, I3, I4, C2b, C3,

C4, and C5, earthquakes, and short-term loadings

$$U = 0.75(1.5D + 1.9L) \quad \text{(if D and L have the same sign)} \quad [9\text{-}3]$$

or

$$U = 0.75(0.9D + 1.9L) \quad \text{(if D opposes L)} \quad [9\text{-}4]$$

c. In most retaining walls and flood walls, dead loads represent a small percentage of total loads and the additional effort to recompute another stability analysis using the above two factors may not be warranted. Therefore, a single load factor as defined by Equation 9-5 may be substituted for Equations 9-1 and 9-2 to avoid having to recompute an alternate stability analysis with a different set of loadings. Likewise, Equation 9-6 may be substituted for Equations 9-3 and 9-4.

$$U = 1.9(D + L) \quad [9\text{-}5]$$

$$U = 0.75[1.9(D + L)] \quad [9\text{-}6]$$

Note that the ACI definition of D is modified so that

D = dead load of the concrete members only or related axial forces, shears, and moments

L = all loads other than dead load of concrete, or related axial forces, shears, and moments

d. When multiple load factors are used and the reactions (i.e., base reactions, pile reactions, resisting earth pressures, etc.) are computed using the applied factored loads, the following combinations should be considered:

From Equation 9-1: $U = 1.5D + 1.9L + R_f$ [9-7]

From Equation 9-2: $U = 0.9D + 1.9L + R_f$ [9-8]

From Equation 9-3: $U = 0.75(1.5D + 1.9L + R_f)$ [9-9]

From Equation 9-4: $U = 0.75(0.9D + 1.9L + R_f)$ [9-10]

where R_f equals internal forces and moments resulting from reactions induced by the applied factored dead and live loads.

e. When the single load factor is used and the reactions (i.e., base reactions, pile reactions, resisting earth pressures, etc.) are computed using the applied *unfactored* loads, the following combinations should be considered: (See paragraphs j and k, Example 1, Appendix N):

From Equation 9-5: $U = 1.9(D + L + R)$ [9-11]

From Equation 9-6: $U = 0.75[1.9(D + L + R)]$ [9-12]

where R equals internal forces and moments resulting from reactions induced by applied *unfactored* dead and live loads.

2. Design Strength of Reinforcement. The design should be based on yield strengths of reinforcement of 40,000 psi and 48,000 psi for ASTM Grades 40 and 60 steels, respectively, except for calculating development lengths. The development length for Grades 40 and 60 steels should be based on yield strengths of 40,000 psi and 60,000 psi, respectively. Reinforcement with a yield strength in excess of Grade 60 should not be used unless a detailed investigation of ductility and serviceability requirements is conducted in consultation with and approved by Headquarters, US Army Corps of Engineers (HQUSACE) (CECW-ED).

3. Maximum Tension Reinforcement. For flexural members and for members subject to combined flexure and compressive axial load when the design load strength ϕP_n is less than the smaller of $0.10f'_cA_g$ or ϕP_b, the ratio of tension reinforcement provided generally should not exceed $0.25\rho_b$. Reinforcement ratios greater than $0.25\rho_b$ but less than $0.50\rho_b$ may be used in retaining walls if excessive deflections are not predicted when using the method specified in the ACI Building Code. Reinforcement ratios in excess of 0.50 ρb should not be used unless a detailed investigation of serviceability requirements, including computation of deflections, is conducted in consultation with and approved by HQUSACE (CECW-ED).

4. Minimum Reinforcement of Flexural Members. At any section of a flexural member where reinforcement is required by analysis, the minimum reinforcement requirements specified in the ACI Building Code, should apply, except that f_y should be in accordance with paragraph 9-8b(2).

5. Control of Deflections and Cracking. Cracking and deflections due to service loads need not be investigated if the limits on design strength specified in paragraph 9-8b(2) and a reinforcement ratio of $0.25\rho_b$ are not exceeded. Where these limitations are exceeded, extensive investigation of deformation and cracking due to service loads should be made in consultation with higher authority.

6. Distribution of Flexural Reinforcement. The spacing of flexural tension reinforcement should not generally exceed 18 inches for Grade 40 steel, or 12 inches for Grade 60 steel.

7. Extreme Loadings. For extreme loadings which are highly improbable, such as from earthquakes which have a frequency of occurrence that greatly exceeds the economic life of the structure, selection of less conservative load factors than given in Equations 9-3, 9-4, and 9-6 and less conservative strength criteria than given above may be justified. For extreme loadings, requests and the justification for varying from the guidance should be submitted to HQUSACE (CECW-E) for approval.

C. HYDRAULIC STRUCTURES—REINFORCED CONCRETE DESIGN

1. Design Assumptions

a. Strain. The assumed maximum usable strain at the extreme concrete compression fiber should be equal to 0.003. The design strain ε_m at the extreme concrete compression fiber should be limited to 0.5 of the maximum usable strain for hydraulic structures.

b. Balanced Conditions. Balanced conditions exist at a cross section when the tension reinforcement reaches the strain corresponding to its specified yield strength f_y just as the concrete in compression reaches its design strain ε_m. T-wall members should be designed for a ductile failure on the tensile side of balance, as described in paragraphs 9-7a, 9-8b(3), and 9-8b(4).

c. Concrete Stress. A concrete stress of $0.85f'_c$ should be assumed uniformly distributed over an equivalent compression zone bounded by the edges of the section and a straight line lying parallel to the neutral axis at a distance $a = \beta_m c$ from the extreme compression fiber. The factor β_m should be taken as 0.55 for values of f'_c up to 4000 psi. For values of f'_c greater than 4000 psi, β_m should be 0.50.

2. Design Equations. Equations for design and investigation of reinforced concrete sections are given in Figures 9-2 through 9-5. These will be the only equations required to determine flexural adequacy for sections of retaining and flood walls in practically all cases.

a. The minimum effective depth (d) needed to provide the amount of ductility required by criteria may be determined from the following equation:

$$d_{min} = \sqrt{\frac{M_u/\phi}{0.85f'_c k_m b\left(1 - \dfrac{k_m}{2}\right)}} \qquad [9\text{-}13]$$

where

$$k_m = \frac{f_y \rho_{max}}{0.85f'_c}, \; \rho_{max} = \lambda \rho_b$$

and λ is 0.25 for hydraulic structures, compared to a value of 0.75 allowed by the ACI Building Code. Equation 9-13 is valid only for flexure.

b. Design aids that will provide essentially the same results as the equations given in Figures 9-2 through 9-5 may be found in ACI publication SP-17. These will be valid for hydraulic structures so long as λ does not exceed 0.25 and the allowable capacity of the cross section is limited by flexural tension. Computer program CSTR (X0066) can assist in the design or investigation of strength of members in hydraulic structures (Appendix O).

D. STRUCTURES NOT SUBJECT TO HYDRAULIC ACTION—STRENGTH AND SERVICEABILITY. The strength and serviceability requirements for structures not subject to hydraulic action should be in accordance with the current ACI Building Code. Computer program CASTR (X0067) can assist in the design or investigation of strength of members in walls not subject to hydraulic action (Appendix O).

E. STRUCTURES NOT SUBJECT TO HYDRAULIC ACTION—REINFORCED CONCRETE DESIGN. Limits on strain, reinforcement, and concrete stress should be in accordance with the current ACI Building Code.

F. SHEAR STRENGTH. The shear strength V_c provided by concrete should be computed in accordance with the ACI Building Code requirements. For cantilever retaining walls the maximum factored shear force should be computed at a distance d from the base of the stem for stem design, at a distance d from the stem for toe design, at the face of the stem for heel design, and at the top of the key for key design. Wherever an L-shaped wall without a toe is used, the shear force should be computed at the base of the stem for stem design and at the face of the stem for heel design.

9-9. Foundation Analyses

Foundation analysis should be performed in accordance with the methods described in Chapters 4 and 5 and illustrated in Appendix N. Concrete design should be for earth pressures corresponding to loading conditions which produce maximum tension in the respective elements of the foundation slab based on factored ultimate loads.

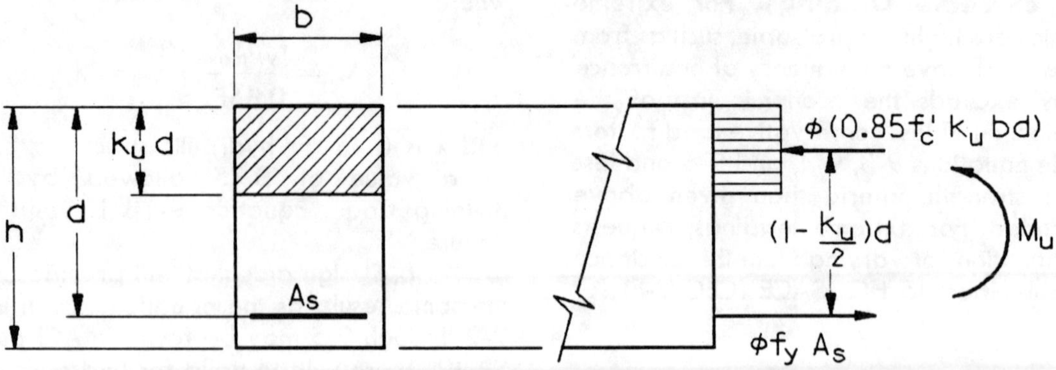

DESIGN

GIVEN : M_u, b, d, h, f_y, f'_c, ϕ=0.9

FIND: k_u, A_s AND ρ

$$k_u = 1-\sqrt{1-\frac{M_u}{0.425\,\phi\, f'_c\, bd^2}} \quad , \quad A_s = \frac{0.85 f'_c\, k_u\, bd}{f_y}$$

$$\rho = \frac{A_s}{bd} \leq \rho_{MAX} \quad , \quad \rho = \frac{0.85 f'_c\, k_u}{f_y}$$

INVESTIGATION

GIVEN : b, d, h, f_y, f'_c, ϕ=0.9, A_s, $\rho = \frac{A_s}{bd} \leq \rho_{MAX}$

FIND: k_u AND M_u

$$k_u = \frac{f_y \rho}{0.85 f'_c} \quad , \quad M_u = \phi f_y \rho \left(1-\frac{f_y \rho}{1.7 f'_c}\right) bd^2$$

Fig 9-2. Rectangular beam, simple bending with no compression reinforcement

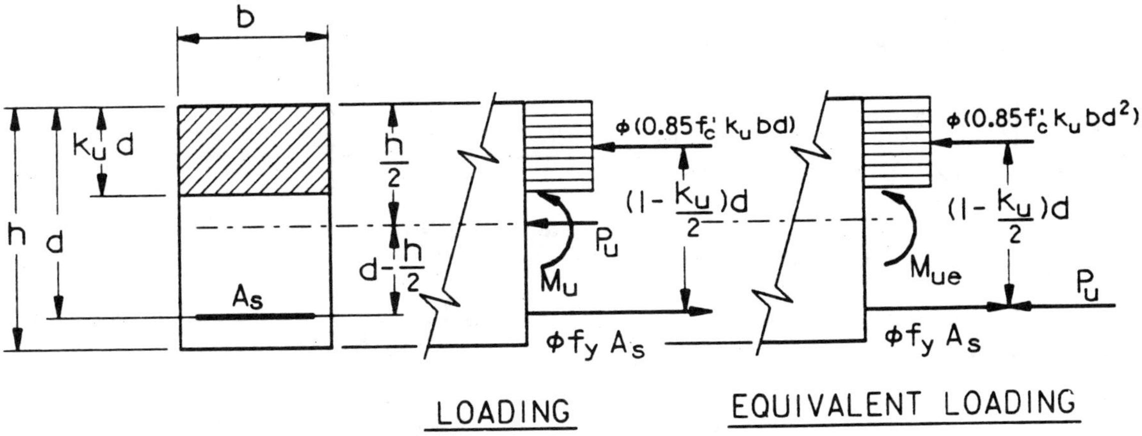

FOR EQUIVALENT LOADING: $M_{ue} = M_u + P_u(d - \frac{h}{2})$

DESIGN

GIVEN : P_u, M_u, b, d, h, f_y, f'_c,

(ϕ SHALL NOT BE LESS THAN 0.7) $\phi = 0.9 - (\frac{P_u}{0.1 f'_c bh})(0.2)$

FIND: k_u, A_s AND ρ

$$k_u = 1 - \sqrt{1 - \frac{M_{ue}}{0.425\,\phi\, f'_c\, bd^2}}$$

$$A_s = \frac{0.85 f'_c k_u bd - \frac{P_u}{\phi}}{f_y}$$

$$\rho = \frac{0.85 f'_c k_u - \frac{P_u}{\phi bd}}{f_y} \leq \rho_{MAX}$$

INVESTIGATION

GIVEN : P_u, b, d, h, f_y, f'_c, ϕ (SEE DESIGN), A_s, $\rho \leq \rho_{MAX}$

FIND: k_u, M_{ue} AND M_u

$$k_u = \frac{\frac{P_u}{\phi bd} + f_y \rho}{0.85 f'_c}$$

$$M_{ue} = [0.85 f'_c k_u (1 - \frac{k_u}{2})\, bd^2]\phi$$

$$M_u = M_{ue} - P_u(d - \frac{h}{2})$$

Fig 9-3. Rectangular member, bending with small axial compression load, no compression reinforcement

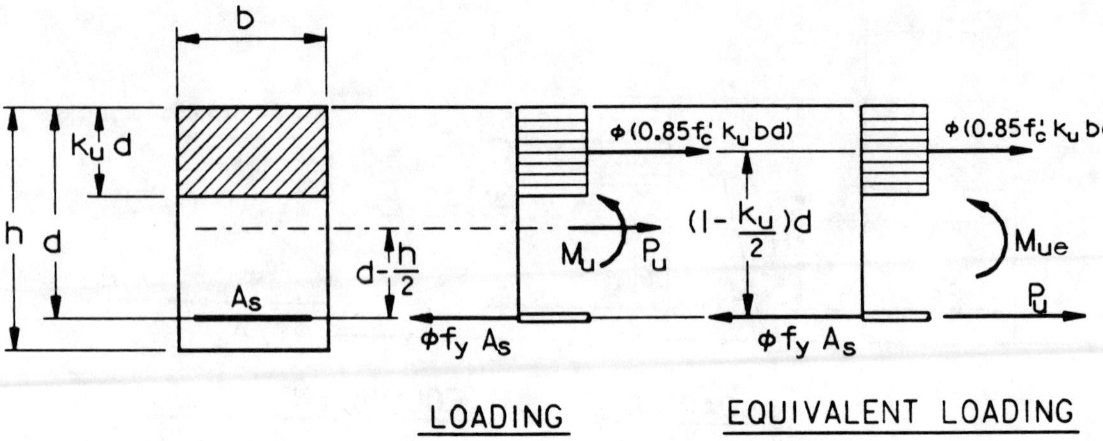

FOR EQUIVALENT LOADING: $M_{ue} = M_u - P_u(d - \frac{h}{2})$

DESIGN

GIVEN : $P_u, M_u, b, d, h, f_y, f'_c, \phi = 0.9$

FIND: k_u, A_s, AND ρ

$$k_u = 1 - \sqrt{1 - \frac{M_{ue}}{0.425 \phi f'_c bd^2}}$$

$$A_s = \frac{0.85 f'_c k_u bd + \frac{P_u}{\phi}}{f_y}$$

$$\rho = \frac{0.85 f'_c k_u + \frac{P_u}{\phi bd}}{f_y} \leq \rho_{MAX}$$

INVESTIGATION

GIVEN : $P_u, b, d, h, f_y, f'_c, \phi, A_s, \rho < \rho_{MAX}$

FIND: k_u, M_{ue}, AND M_u

$$k_u = \frac{f_y \rho - \frac{P_u}{\phi bd}}{0.85 f'_c}, \quad M_{ue} = [0.85 f'_c k_u (1 - \frac{k_u}{2}) bd^2]\phi$$

$$M_u = M_{ue} + P_u(d - \frac{h}{2})$$

Fig 9-4. Rectangular member, bending with axial tensile load, where $M_u/P_u \geq (d - h/2)$

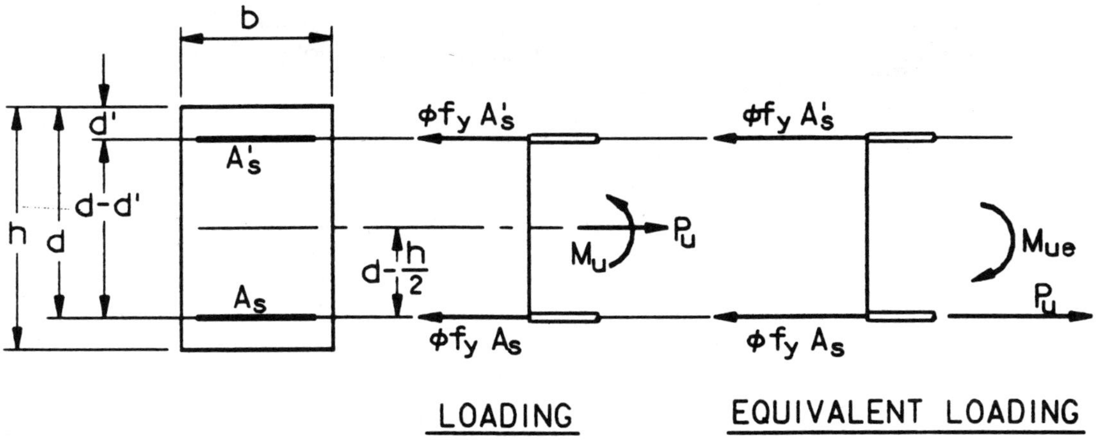

FOR EQUIVALENT LOADING: $M_{ue} = -M_u + P_u\,(d - \frac{h}{2})$

DESIGN

GIVEN : P_u, M_u, b, d, h, f_y, f'_c, $\phi = 0.9$

FIND: A_s AND A'_s

$$A_s = \frac{\frac{P_u}{\phi} - f_y A'_s}{f_y} \quad , \quad A'_s = \frac{M_{ue}}{\phi f_y (d-d')}$$

INVESTIGATION

GIVEN : P_u, b, d, h, f_y, f'_c, $\phi = 0.9$, A_s AND A'_s

FIND: M_{ue}, M_u AND f'_s (STRESS IN A'_s)

$$M_{ue} = (P_u - \phi f_y A_s)(d-d')$$

$$M_u = P_u\,(d - \frac{h}{2}) - M_{ue}$$

$$f'_s = \frac{\frac{P_u}{\phi} - f_y A_s}{A'_s}$$

(NOT NECESSARILY EQUAL TO f_y. $f'_s \leq f_y$ IS CRITERIA.)

Fig 9-5. Rectangular member, bending with axial tensile load, where $M_u/P_u < (d - h/2)$

The loading conditions corresponding to SMF = 2/3 should be considered as a minimum for single wedge analysis. This does not preclude the use of any other rational method of analysis that will produce an equivalent design.

CHAPTER 10

ALTERNATE TYPES OF RETAINING WALLS

Section I. Introduction

10-1. Classes of Retaining Walls

The four basic classes of retaining walls are gravity, cantilever, anchored, and mechanically stabilized backfill. Gravity walls rely on the weight of the wall system to resist overturning. The cantilever wall is fully reinforced to resist applied moments and shears. Anchored walls resist lateral forces primarily by the use of tieback anchors. Mechanically stabilized backfill involves the inclusion of reinforcement in the soil to form a coherent mass (Godfrey 1984; Mitchell, Villet, and DiMillio 1984; and Jones 1985).

10-2. Alternate Types of Retaining Walls

As discussed previously in Chapter 2, the most common types of retaining walls are gravity and cantilever walls constructed of cast-in-place concrete. Recently, however, a number of wall systems utilizing mechanically stabilized backfill as well as new types of gravity walls have been developed (Godfrey 1984). This chapter briefly describes mechanically stabilized backfill systems and precast concrete modular systems. The mention of any specific wall system does not constitute an endorsement or approval. Numerous wall systems are available and should be considered when appropriate. This manual does not attempt to provide complete design and/or construction procedures for the types of walls described in this chapter. Normally, design and construction procedures are provided by the manufacturer. However, the manufacturer normally provides only part of the design. The design engineer must ensure the overall adequacy of the design.

Section II. Mechanically Stabilized Backfill Systems

10-3. General Background

Reinforced soil is a construction material composed primarily of soil with a performance that has been improved by the introduction of small quantities of other materials. These materials are in the form of strips, grids, sheets, rods, or fibers which strengthen the soil to resist tensile forces that soil alone is unable to withstand (Al-Hussani and Perry 1976, Collin 1986).

10-4. Available Systems

Several mechanically stabilized backfill systems are available for retaining walls (Mitchell and Villet 1986).

A. BASIC COMPONENTS. Mechanically stabilized backfill systems have three major components: reinforcements, soil backfill, and facing elements. Both metallic and nonmetallic (geotextile, plastic) materials have been used for reinforcement. Granular material is normally used for soil backfill to meet stress transfer, durability, and drainage requirements. Facing elements are used to retain backfill material at the face of the wall, to prevent erosion of steep faces, and for aesthetic reasons. The facings are designed to resist only small horizontal earth pressures. Facing materials commonly used include precast concrete panels, prefabricated metal sheets and plates, welded wire mesh, inclusion of intermediate reinforcements between main reinforcement layers at the face, and seeding of the exposed soil.

B. BASIC MECHANISMS AND BEHAVIOR. The two primary mechanisms of stress transfer between the reinforcement and soil are friction between plane contact surface areas and passive soil-bearing resistance on reinforcement surfaces oriented transverse to the direction of movement. Strip, sheet, and rod reinforcements transfer stresses to the soil by friction, while grid reinforcements transfer stresses primarily by passive resistance. Geogrid reinforcements develop both frictional and passive soil resistance.

C. STRIP REINFORCEMENT. With strip reinforcement, a mechanically stabilized backfill is created by placing strips in horizontal planes between successive lifts of soil backfill. Reinforced earth, shown schematically in Figure 10-1, is a strip reinforcement system.

D. GRID REINFORCEMENT. Grid reinforcement systems are formed by placing metallic or polymeric tensile resistant elements in horizontal planes in the soil backfill. Retaining walls using bar-mesh reinforcement have been constructed by the California Department of Transportation, Hilfiker Retaining Walls; VSL Corporation, and the Georgia State Highway Department (see Figures 10-2 and 10-3). Grid reinforcements are also made of polymer materials, such as Tensar Geogrid (see Figure 10-4).

10-5. Advantages and Disadvantages

The advantages and disadvantages of mechanically stabilized backfill systems are outlined below (Mitchell and Villet 1986).

A. ADVANTAGES

1. Mechanically stabilized backfill systems are economical when compared to conventional retaining walls.

2. Construction of mechanically stabilized backfill systems usually is easy and rapid. It does not require skilled labor or specialized equipment. Many of the components are prefabricated allowing relatively quick construction.

3. Regardless of the height or length of the wall, the structure remains stable during construction.

4. When compared to conventional retaining walls, mechanically stabilized backfill systems are relatively flexible and can tolerate large lateral deformations and large differential vertical settlements (when this is anticipated, vertical sliding joints can be installed at intervals to compensate for movement). The flexibility of mechanically stabilized backfill systems allows the use of a lower factor of safety for bearing capacity design than conventional more rigid structures.

5. Mechanically stabilized backfill systems are potentially better suited for earthquake loading than conventional retaining walls because of the flexibility and inherent energy absorption capacity of the coherent earth mass. In designing mechanically stabilized backfill systems for earthquake regions, provision should be made for slippage of reinforcement elements rather than tension failure of the elements, resulting in a ductile structure (McKittrick 1979).

6. Mechanically stabilized backfill systems, because of their flexibility and mass, are capable of withstanding dynamic loads imposed by wheel loads, wave action, and impact of small boats.

7. Polymeric reinforcements are stable under chemical and biological conditions normally occurring in soils.

8. Since facing elements play only a secondary structural role, a greater flexibility is available to meet aesthetic requirements than for conventional retaining walls. Facing arrangements range from concrete panels of various shapes, textures, and colors to provision of vegetation at the exposed face of the soil.

B. DISADVANTAGES

1. Corrosion of metallic reinforcement occurs and must be assessed on a project basis by determining the potential aggressiveness of the soil. Special coatings such as galvanized zinc and resin-bonded epoxy are used with a sacrificial thickness of steel added in the design to give the required service life.

2. Although polymeric reinforcement is a robust material, some allowance must be made for decrease in strength due to abrasion during construction. This will vary with the type of reinforcement material.

3. Different polymers have different creep characteristics. Allowable loads in the grid should be selected based on allowable deformations, as well as the results of creep tests (10,000 hour). See McGown et al. (1985), for load-strain-time behavior of Tensar Geogrids.

4. The construction of mechanically stabilized embankments in cut regions requires a wider excavation than conventional retaining walls.

5. Excavation behind the mechanically stabilized wall is restricted.

10-6. Cost Considerations

Mechanically stabilized backfill systems are particularly economical when compared to conventional retaining walls for earth fill situations where the retaining wall has a total surface area greater than 2000 sq ft and an average wall height greater than 10 feet, or where a rigid conventional wall requires a deep foundation for support. Precast concrete modular systems are cost-effective compared to conventional retaining walls for cut sides of excavations where the wall surface area is greater than 500 sq ft and average wall height is greater than 8 feet. The cost effectiveness of mechanically stabilized backfill systems will probably be reduced by high-cost backfill,

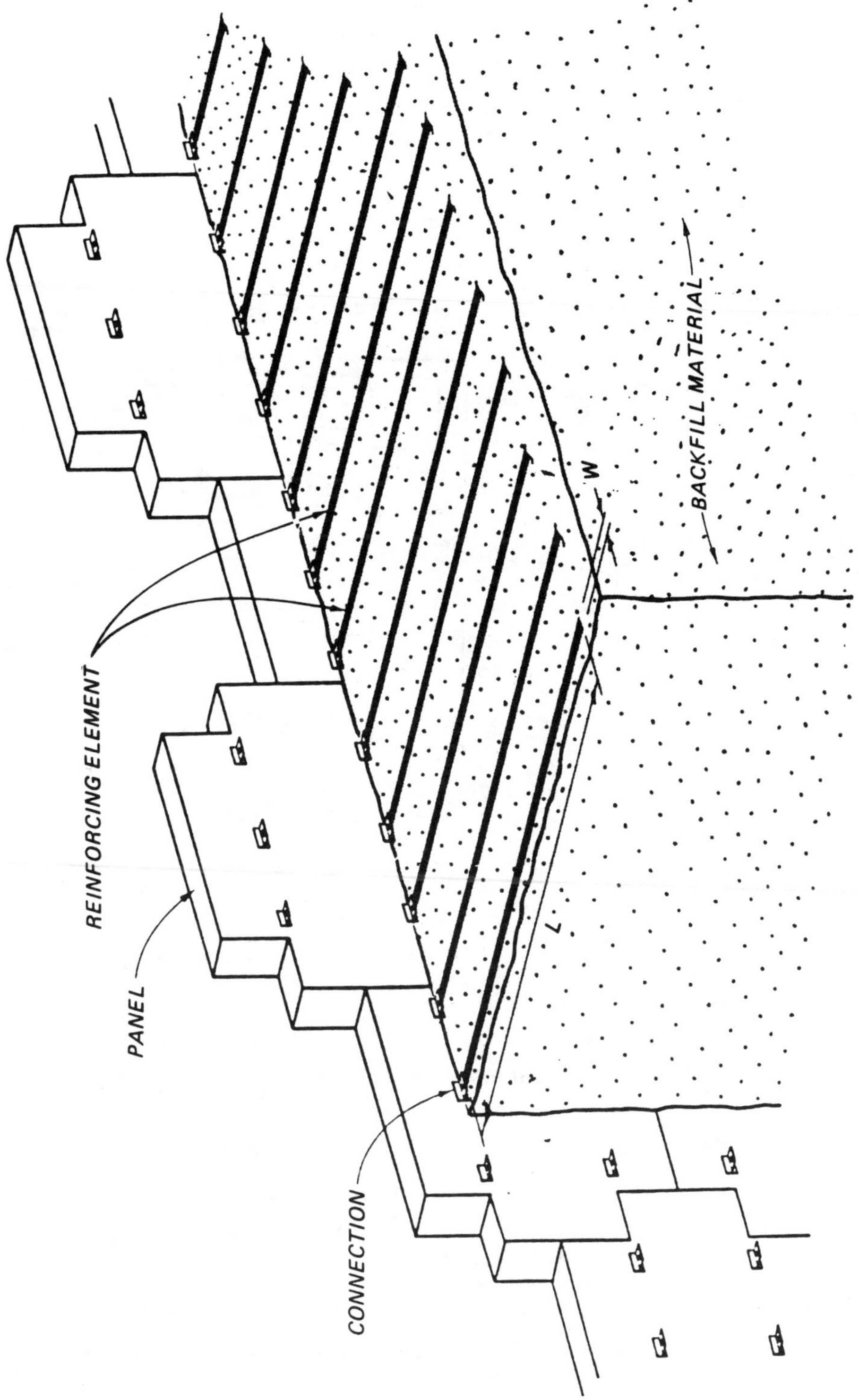

Fig 10-1. Schematic diagram of reinforced earth retaining wall

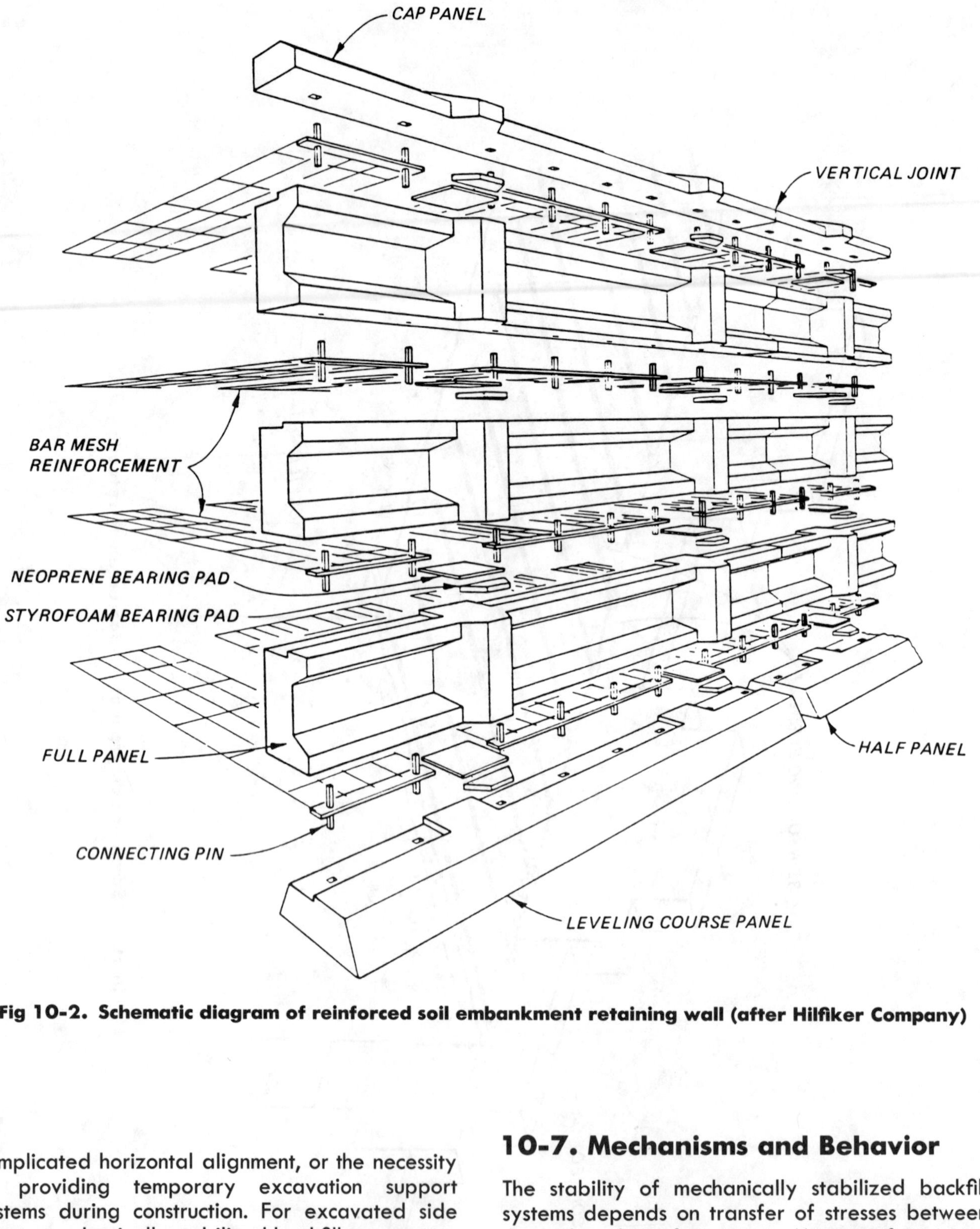

Fig 10-2. Schematic diagram of reinforced soil embankment retaining wall (after Hilfiker Company)

complicated horizontal alignment, or the necessity of providing temporary excavation support systems during construction. For excavated side slopes, mechanically stabilized backfill systems can be constructed for 30 to 50 percent less than conventional retaining walls. However, a short life, serious consequences of failure, or high repair or replacement costs could offset a lower first cost. Similar savings in construction time are obtained using mechanically stabilized backfill systems, according to Leary and Klinedinst (1984).

10-7. Mechanisms and Behavior

The stability of mechanically stabilized backfill systems depends on transfer of stresses between the soil and reinforcements. Most reinforcements are inextensible in that they rupture at strains much less than those required to cause soil failure* (Mitchell and Villet 1986; Mitchell, Villet, and DiMillio 1984).

* Some geotextiles, which require large deformations to cause failure, are the exception.

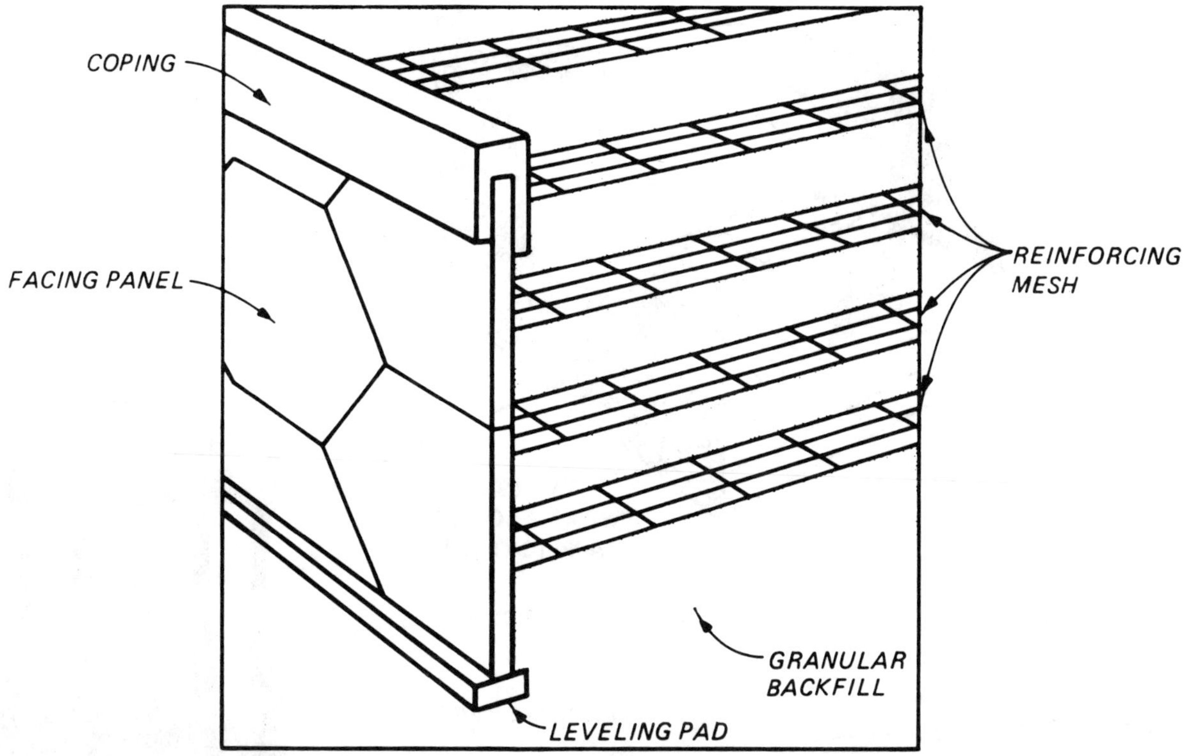

Fig 10-3. VSL retained earth retaining wall (adapted from VSL Corporation 1984)

A. MECHANISMS. The transfer of stress between soil and reinforcement is by friction and/or passive soil resistance when the reinforcement is loaded in tension. In many reinforcement systems both mechanisms are present, and the relative contribution of each is indeterminate.

1. Friction. The load transferred by friction per unit area of reinforcement depends on the interface characteristics of the soil and reinforcement, and on the normal stress between them, which in turn depends on the stress-deformation behavior of the soil. This latter behavior is itself stress-dependent. Therefore, the effective friction coefficient cannot be estimated by analytical procedures. The results of experiments such as pullout tests, direct shear tests between soil and reinforcements, and instrumented model and full-scale tests, are often used to select friction coefficients. The coefficient of friction is defined as the average mobilized shear stress along the reinforcement divided by the normal stress from the overburden pressure. Empirical data from pullout tests on strip reinforcements show a decrease in this coefficient with depth regardless of the type of reinforcement (smooth or ribbed). This occurs because the effective normal stress is altered by the soil to reinforcement interaction. As shear strains are imposed on a dense granular soil, the soil tends to dilate. If the tendency to dilate is partially restrained by boundary conditions, local confining stresses will increase with the tendency to dilate decreasing as the confining stress increases. Hence, the influence of dilatancy on friction coefficients computed from pullout tests decreases with depth. Therefore, since the influence of dilatancy decreases with depth, the coefficient of friction also decreases with depth. Also, recent experience in construction with granular soils of low uniformity coefficients† (less than 4) indicates a relatively low friction coefficient ($\simeq 1.0$) for these types of granular soils.

2. Passive Soil Resistance. Passive soil resistance to pullout of reinforcement develops

---

† $C_u = \dfrac{D_{60}}{D_{10}}$

where C_u = coefficient of uniformity

D_{60} = grain diameter at 60 percent passing

D_{10} = grain diameter at 10 percent passing

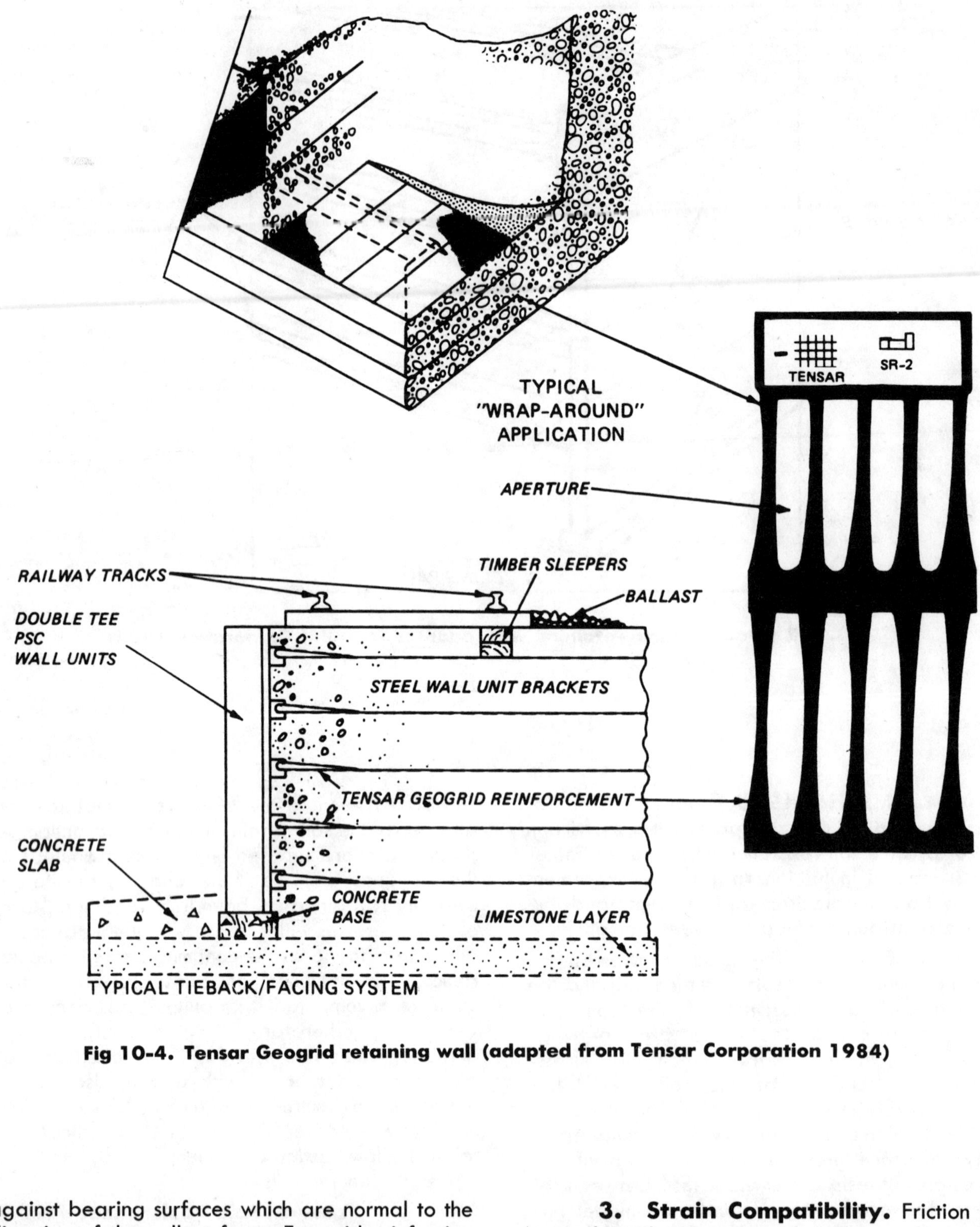

Fig 10-4. Tensar Geogrid retaining wall (adapted from Tensar Corporation 1984)

against bearing surfaces which are normal to the direction of the pullout force. For grid reinforcing systems with the spacing of cross bars parallel to the wall equal to or greater than 6 inches, the major portion of the resistance (approximately 90 percent for bar-mesh in a sandy gravel) is obtained by passive soil resistance or bearing capacity on the front face of elements oriented transverse to the pullout force direction.

3. Strain Compatibility. Friction between the soil and a smooth reinforcement requires a small displacement of about 0.05 inch. Passive soil resistance against surfaces normal to displacement requires relative soil-to-reinforcement displacements as large as 4 inches for complete mobilization. However, a significant portion (greater than 50 percent) of the maximum value is mobilized at deflections of about 0.25 inch (Elias

1986). For bar mat grid reinforcement systems, the small beneficial effects of friction are neglected in view of possible strain incompatibility between frictional behavior and passive soil resistance.

B. BEHAVIOR. The distribution of lateral earth pressure within reinforced soil depends on the extensibility of the reinforcements, the construction methods used, and the type of reinforced structure. The active horizontal stress state is used for systems which are able to undergo relatively large lateral deformations, such as geotextiles. Higher lateral stresses, such as at-rest pressures, are associated with less extensible reinforcements, e.g., steel strips, bar meshes, welded wire mesh, and relatively low confining pressures, e.g., at shallow depths in the soil backfill where dilatancy is most effective. Under low confining stresses a reinforcement system may fail by pullout between the reinforcement and soil. Under high confining stresses the same system may fail by breakage of the reinforcements.

10-8. Materials

As previously mentioned, the three basic components of mechanically stabilized backfill systems are reinforcements, soil backfill, and facing elements (Mitchell and Villet 1986).

A. REINFORCEMENT. The reinforcements may be characterized by the type of material (metallic and nonmetallic) and geometry (strips, grids, and sheets). Important material properties for reinforcements are strength and stability (low tendency to creep), high coefficient of friction with soil backfill, and durability. Depending on the electrochemical properties of the soil backfill and structure environment (marine or freshwater, presence of stray electrical currents in the ground, etc.) galvanized zinc-coated steel, resin-bonded epoxy-coated steel, or polymeric reinforcements are used. Polymeric reinforcements are not subject to corrosion but do exhibit creep characteristics (decrease in strength with time at constant load and soil temperature).

B. SOIL BACKFILL. Most mechanically stabilized embankment systems have used cohesionless soil backfill. However, since grid reinforcements have a much greater pullout resistance than strip reinforcements, it is possible to construct mechanically stabilized embankment systems using silty or clayey material as backfill (Forsyth 1979 and Jackura 1984). The advantages of cohesionless soil backfill are that it is stable (will not creep), free-draining, not susceptible to frost, and relatively noncorrosive to reinforcement. The main disadvantage, where cohesionless soil has to be imported, is cost. The main advantage of cohesive soils is availability and hence lower cost. The disadvantages are long-term durability problems (corrosion and/or frost susceptibility) and distortion of the structure (due to creep of the soil backfill). When cohesionless soil backfill is readily available it should be used. When it is not readily available, the costs of importing cohesionless soil backfill should be weighed against the potentially poorer performance of using the lower-cost locally available cohesive soil backfill.

C. FACING ELEMENTS. Since facing elements play only a secondary structural role, a greater flexibility in choice of materials is available to meet aesthetic requirements than is the case for conventional retaining walls. A wide variety of materials, shapes, architectural finishes, and colors are available for facing elements. Selecting among these depends on the function of the structure, type of reinforcement, and aesthetics.

10-9. Design Considerations

The various engineering companies involved in a project provide site-specific designs for their proprietary system. Mechanically stabilized embankment systems must be designed for both external and internal stability. External stability is evaluated in a manner similar to a conventional gravity retaining wall. Internal stability depends on there being neither pullout nor breakage of the reinforcement (Mitchell and Villet 1986, Collin 1986).

A. EXTERNAL STABILITY. The mechanically stabilized backfill system must be stable against sliding along the base of the structure, overturning about the toe of the wall, bearing capacity failure of the foundation soil, overall slope stability, and differential settlement along the structure. For external stability calculations the mechanically stabilized backfill system is assumed to behave as a coherent block.

1. Sliding Along the Base of the Structure. The mechanically stabilized backfill system must be stable against sliding due to the lateral pressure of the soil retained by the system. The minimum factor of safety against sliding should be 1.5. Sliding considerations may govern the design for high structures (greater than 30 feet) or structures with sloping backfills.

2. Overturning About the Toe of the Wall. The mechanically stabilized backfill system must be stable against overturning about the toe of the wall. Since in reality the structure is flexible, it would probably never fail by overturning. One hundred percent of the base should always be in contact with the subgrade for all loading conditions (Elias 1986). Overturning considerations seldom govern the design of structures when the minimum reinforcement length is 70 percent of the wall height.

3. Bearing Capacity Failure and Settlement. The mechanically stabilized backfill system must be stable against bearing capacity failure of the foundation soil. The minimum factor of safety against bearing capacity failure should be 2.0. This is lower than that used for conventional retaining walls (see Table 4-1) because of the flexibility of the mechanically stabilized backfill system and its ability to function satisfactorily after experiencing large differential settlements. If the foundation does not meet stability requirements, consideration should be given to ground improvement techniques such as stone columns, vibroflotation, and dynamic compaction to improve foundation stability. The maximum allowable differential settlement of mechanically stabilized backfill systems is limited by the longitudinal deformability of the facing and the purpose of the structure. For precast concrete panels, without built-in vertical joints,* the limiting tolerable differential settlement is 1 foot per 100 feet of wall length.

4. Overall Slope Stability. The mechanically stabilized backfill systems, retained soil, and foundation should be stable against slope failure. All potential slip surfaces should be investigated including those passing through the reinforcement and deep-seated sliding. The minimum factor of safety for slope stability should be 1.5.

B. INTERNAL STABILITY. The mechanically stabilized backfill system must be stable against reinforcement pullout and reinforcement breaking.

1. Reinforcement Pullout. In determining the reinforcement pullout capacity, the effective length of reinforcement behind the theoretical failure surface must be great enough to assure the transfer of stress from the reinforcement to the backfill soil without reinforcement pullout. The resistance to pullout may be frictional (strip reinforcement), passive (bar-mesh reinforcement), or frictional-passive (Geogrid). Using data from laboratory pullout tests at a maximum of 0.75 inch of deformation, the structure should be designed with a minimum factor of safety against reinforcement pullout of 1.5 at each reinforcement level.

2. Reinforcement Breaking. To assure a sufficient reinforcement breaking capacity, the effective cross-sectional area of the reinforcement (corrected for corrosion effects over the design life of the structure) must be great enough to allow for the transfer of stress from the reinforcement to the backfill soil without the reinforcement breaking. The design stress in the reinforcement should be taken as 55 percent of the yield stress (Elias 1986).

3. Durability of Reinforcements. The durability of reinforcements, over the design life of the structure, is an important design consideration. Deterioration of polymeric reinforcements may occur due to abrasion during construction and decrease in strength with time at constant load and soil temperature. Corrosion of metallic reinforcement occurs due to exposure to air, water, and chemicals in the soil backfill. Galvanized zinc-coated steel (with a sacrificial thickness of steel added to give the required service life) is often used for reinforcing mild to moderately corrosive soil backfill with the following properties† (Frondistou-Yannas 1985):

Resistivity $>$ 3000 ohm-centimeters

pH 5–10

Chlorides $<$ 200 parts per million

Sulfates $<$ 1000 parts per million

For mild to moderately corrosive soil backfills the corrosion loss rates are

Galvanization 6 μm/year for first 2 years

2.5 μm/year for subsequent years

Steel 9 μm/year after all zinc is lost

The use of aluminum and stainless steel reinforcement is not recommended as several failures have occurred using these materials (McGee 1985). For structures exposed to marine environments, stray electrical currents in the ground, or with soil backfill properties outside the electrochemical guidelines previously given, resin-bonded epoxy-coated metallic reinforcements or a conventional or precast concrete modular gravity wall should be used. A minimum epoxy coating thickness of 18 mils is necessary to survive transportation and installation, and to provide an acceptable level of design confidence. When epoxy-coated metallic reinforcement is used, the soil backfill should consist

* Hilfiker Reinforced Soil Embankment retaining walls have vertical joints built into the wall every 12.5 feet and can tolerate large differential settlements.

† Galvanized zinc-coated steel should not be used if the soil backfill contains illite clay because zinc is sensitive to illite.

of rounded stone with a maximum particle size of 1 inch. For design purposes, the life of the epoxy coating should be assumed to be the same as a galvanized zinc coating of 2 oz/sq ft, or 30 years. A sacrificial thickness of steel should be added to provide the epoxy-coated reinforcement an adequate factor of safety at the design life of the structure (Frondistou-Yannas 1985, Jones 1985).

c. DRAINAGE. Drainage measures must be considered for all mechanically stabilized backfill systems to prevent saturation of the soil backfill and to intercept surface flows containing aggressive elements such as deicing chemicals. When mechanically stabilized backfill systems support roadways which are chemically deiced in the winter, an impervious membrane should be placed between the pavement and the first row of metallic reinforcements to intercept any surface flows containing aggressive chemicals.

10-10. Construction Considerations

The construction of mechanically stabilized embankment systems does not require specialized contractors, skilled labor, or special equipment. Many of the components are prefabricated, providing ease of handling and forming and relatively quick construction. A small crane is used to handle and erect precast concrete facing panels. Front end loaders are used for loading dump trucks and spreading the soil backfill. Vibratory rollers are used to compact the soil backfill while small hand-operated compactors are used for compaction near the wall face. Preparation of the construction area consists of clearing vegetation, debris, and other deleterious material from the site. A concrete leveling pad, which is not a structural member, is constructed to facilitate the erection of the concrete panels. The first layer of soil backfill is placed and compacted and the reinforcement is laid on the surface of the compacted fill and covered with the next layer of fill. Construction equipment must not run on top of the reinforcement. Concrete panels are battered to the inside to compensate for the small outward movement to mobilize the resistance of the reinforcement. Filler material (cork, Styrofoam, neoprene, etc.) is used between all horizontal joints to provide a uniform bearing surface between adjacent panels. A geotextile is placed over all joints on the fill side of the concrete panels to prevent fines from migrating from behind the wall (Mitchell and Villet 1986).

10-11. Instrumentation and Monitoring

The history of mechanically stabilized embankment systems is relatively short compared to the design life of the structure.* Therefore, continued accumulation of field data on a full-scale structure is necessary to verify design assumptions. Structures should be instrumented and monitored whenever atypical conditions exist such as cohesive soil backfill, epoxy-coated metallic reinforcement, or adverse groundwater conditions (outside the range specified in paragraph 10-9b[3]). Measurements should be made of horizontal and vertical displacements of the wall facing; soil pressures on the facing or on a vertical plane near the facing, the base of the wall, and perpendicular planes (horizontal and vertical) along the anticipated maximum tensile force line; tensile forces in the reinforcement including near the locus of maximum tensile force and near the wall facing; and pullout tests on short reinforcements. All mechanically stabilized embankment structures should be monitored once they are placed in operation to ensure stability. External stability of the mechanically stabilized embankment structure could be threatened by the same factors as a conventional retaining wall; e.g., clogging of the drainage system, erosion at the toe of the wall, etc. However, the mechanically stabilized embankment system could also fail due to changes in conditions which adversely influence the internal stability of the system. These include excavation within the soil backfill, changes in the groundwater conditions (outside the range specified in paragraph 10-9b[3]), and possible damage to the stabilizing ties because of vandalism to the exposed structure (Mitchell and Villet 1986, Al-Hussani and Perry 1978).

10-12. Maintenance and Repair

Since mechanically stabilized embankment systems are relatively new there is very limited field experience regarding maintenance and repair. Maintenance problems arising with facing elements could be repaired by conventional methods since the facing elements play a secondary role and resist only small horizontal earth pressures (Long et al. 1984). However, problems with the reinforcements, such as corrosion of metallic reinforcements, would be difficult to repair. One possible solution would be to use soil nailing to stabilize the structure (Jones 1985). Another method would be to place a stone buttress in front of the structure (Mitchell and Villet 1986).

* Strip reinforcement was first utilized in U. S. construction in 1972, bar-mesh reinforcement in 1975, and Geogrid reinforcement in 1984.

Section III. Precast Concrete Modular Systems

10-13. Background

Precast concrete modular systems consist of interlocking soil-filled reinforced concrete modules which form a gravity retaining wall. They can be erected rapidly and are cost-competitive with mechanically stabilized backfill systems.*

10-14. Basic Components

The basic components of precast concrete modular systems are interlocking precast reinforced concrete modular elements filled with soil and resting on natural soil or a concrete foundation (see Figures 10-5 to 10-8). Some systems have resets or an open-face structure at the wall face (see Figures 10-7 and 10-8) with evergreen vegetation to reduce noise levels and eliminate the problem of graffiti.

10-15. Advantages and Disadvantage

The advantages and disadvantage of precast concrete modular systems are listed below.

A. ADVANTAGES.

1. Modular systems are economical when compared to conventional retaining walls in cut situations, particularly where the retaining wall has a total surface area greater than 500 sq ft and average wall heights greater than 8 feet.

2. Assembly of the wall components requires no fasteners and the modules may be reused easily and economically.

3. The precast concrete modular retaining wall does not utilize reinforcing elements and therefore is not subject to corrosion damage.

4. Excavation behind the precast concrete modular retaining wall does not adversely influence the stability of the system as might occur for the mechanically stabilized wall.

B. DISADVANTAGE. The precast concrete modular retaining wall could sustain cracking of interior connecting members due to relatively small (0.5 foot per 100 feet of wall length) longitudinal differential settlement.

* For certain applications, such as where large differential vertical settlements are anticipated, consideration should be given to steel-bin type retaining walls, which are generally more expensive than mechanically stabilized backfill or precast concrete modular systems but less expensive than conventional retaining walls.

10-16. Design Considerations

Various engineering companies involved will provide site-specific plans and limited designs for their proprietary system. Stability is evaluated in a manner similar to a conventional gravity retaining wall. For stability calculations the interlocking precast concrete modular system is assumed to behave as a coherent block. The system must be stable against sliding along the base of the structure, overturning about the toe of the wall, bearing capacity failure of the foundation soil, differential settlement, and overall slope stability.

A. SLIDING ALONG THE BASE OF THE STRUCTURE. The precast concrete modular system must be stable against sliding due to the lateral pressure of the soil retained by the system. The minimum factor of safety against sliding should be 1.5.

B. OVERTURNING ABOUT THE TOE OF THE WALL. The precast concrete modular system must be stable against overturning about the toe of the wall. Since the concrete modular units are not tied together vertically, the stability against overturning must be checked at each concrete module level for a given width. One hundred percent of the base should always be in contact with the subgrade for all loading conditions (Elias 1986). Normally overturning (not sliding) criteria govern the design.

C. BEARING CAPACITY FAILURE AND SETTLEMENT. The precast concrete modular system must be stable against bearing capacity failure of the foundation soil. The minimum factor of safety against bearing capacity failure should be 3.0 (Elias 1986). If the foundation does not meet stability requirements, consideration should be given to use of a mechanically stabilized backfill system or ground improvement techniques such as stone columns, vibroflotation, and dynamic compaction to improve foundation stability. As previously stated, the precast concrete modular retaining wall could sustain cracking of interior connecting members due to relatively small (0.5 foot per 100 feet of wall length) longitudinal differential settlement. Precast concrete modular retaining walls are also susceptible to damage from differential settlement perpendicular to the wall face, particularly on high walls where the bottom wall units may be as wide as 20 feet.

D. OVERALL SLOPE STABILITY. The precast concrete modular system, retained soil, and foundation should be stable against slope failure. All potential slip surfaces should be investigated including deep-seated sliding. The minimum factor of safety for slope stability should be 1.5.

E. DRAINAGE. Drainage measures must be considered for all precast concrete modular systems to prevent saturation of the soil backfill. Also, for closed-face modular systems (see Figure 10-5), a geotextile is placed over all joints on the back side of the front face of the wall to prevent fines from migrating from behind the wall.

10-17. Construction Considerations

The construction of precast concrete modular systems does not require specialized contractors, skilled labor, or special equipment. The components are prefabricated providing ease of handling and forming and relatively quick construction. Soil backfill within the precast modular units should receive adequate compaction to minimize post-construction settlements.

10-18. Instrumentation and Monitoring

The history of precast concrete modular systems is relatively short compared to the design life of the structure.* Therefore, continued accumulation of field data is necessary to verify design assumptions. Structures should be instrumented and monitored whenever atypical conditions exist such as anticipated large differential vertical settlement. Measurements should be made of horizontal and vertical displacements of the front face of the wall and soil pressures on the rear face of the wall. All precast concrete modular structures should be monitored once they are placed in operation to ensure stability. Stability of the precast concrete modular structure could be threatened by the same factors as a conventional retaining wall; e.g., clogging of the drainage system, erosion at the toe of the wall, etc. Precast concrete modular structures should also be monitored for possible damage from differential settlements.

10-19. Maintenance and Repair

Since precast concrete modular structures are relatively new, there is very limited field experience regarding maintenance and repair. Possible methods of repair to a section of the structure which has sustained damage from differential settlement include replacing the section with a wall more tolerant to differential settlement, such as a mechanically stabilized embankment system with vertical joints (see Figure 10-2) or a steel-bin type wall, or placing a stone buttress in front of the structure.

* Criblock was first used in US construction in 1978, Doublewal in 1979, and Evergreen in 1986.

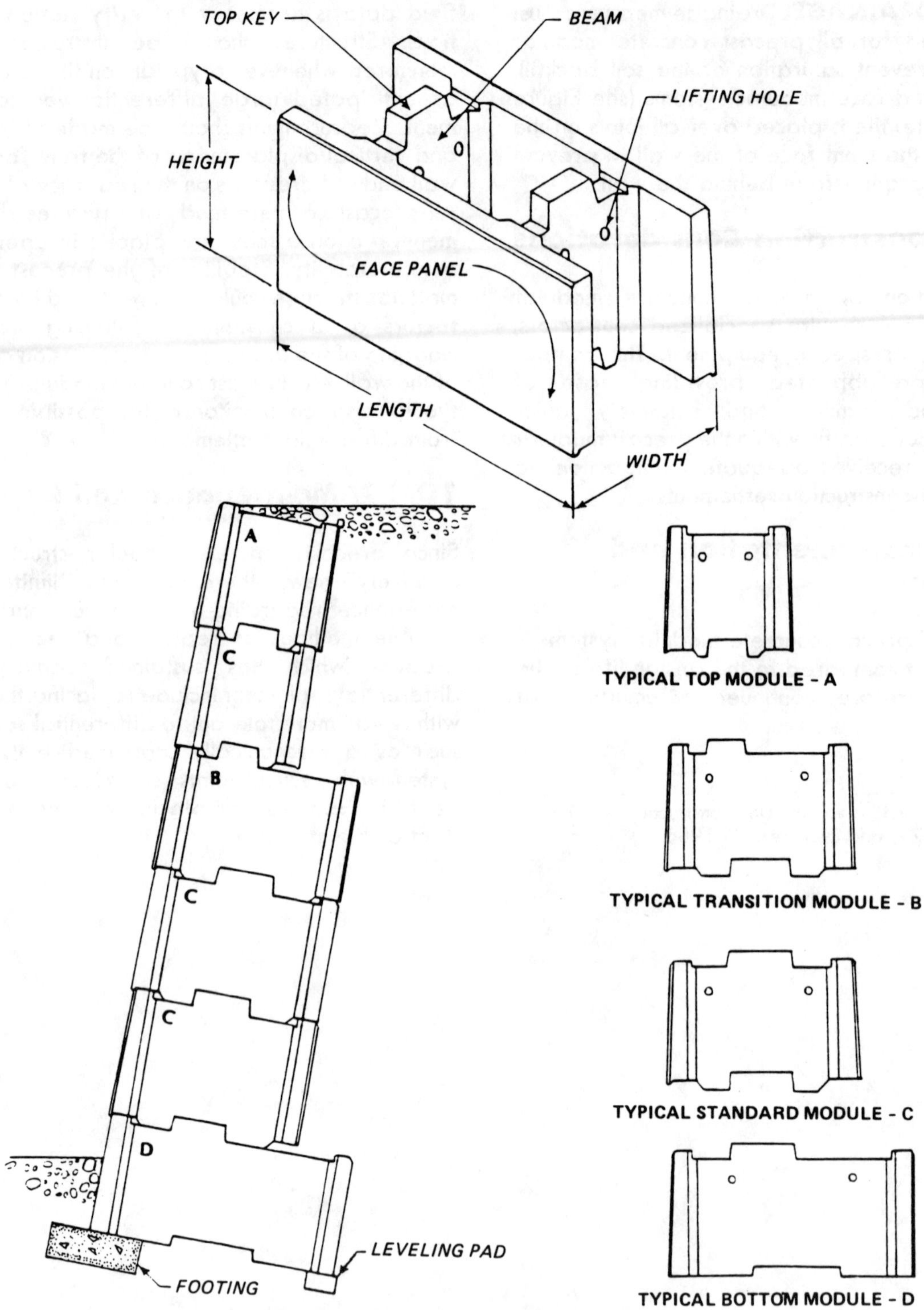

Fig 10-5. Schematic diagram of Doublewal retaining wall (after Doublewal Corporation 1984)

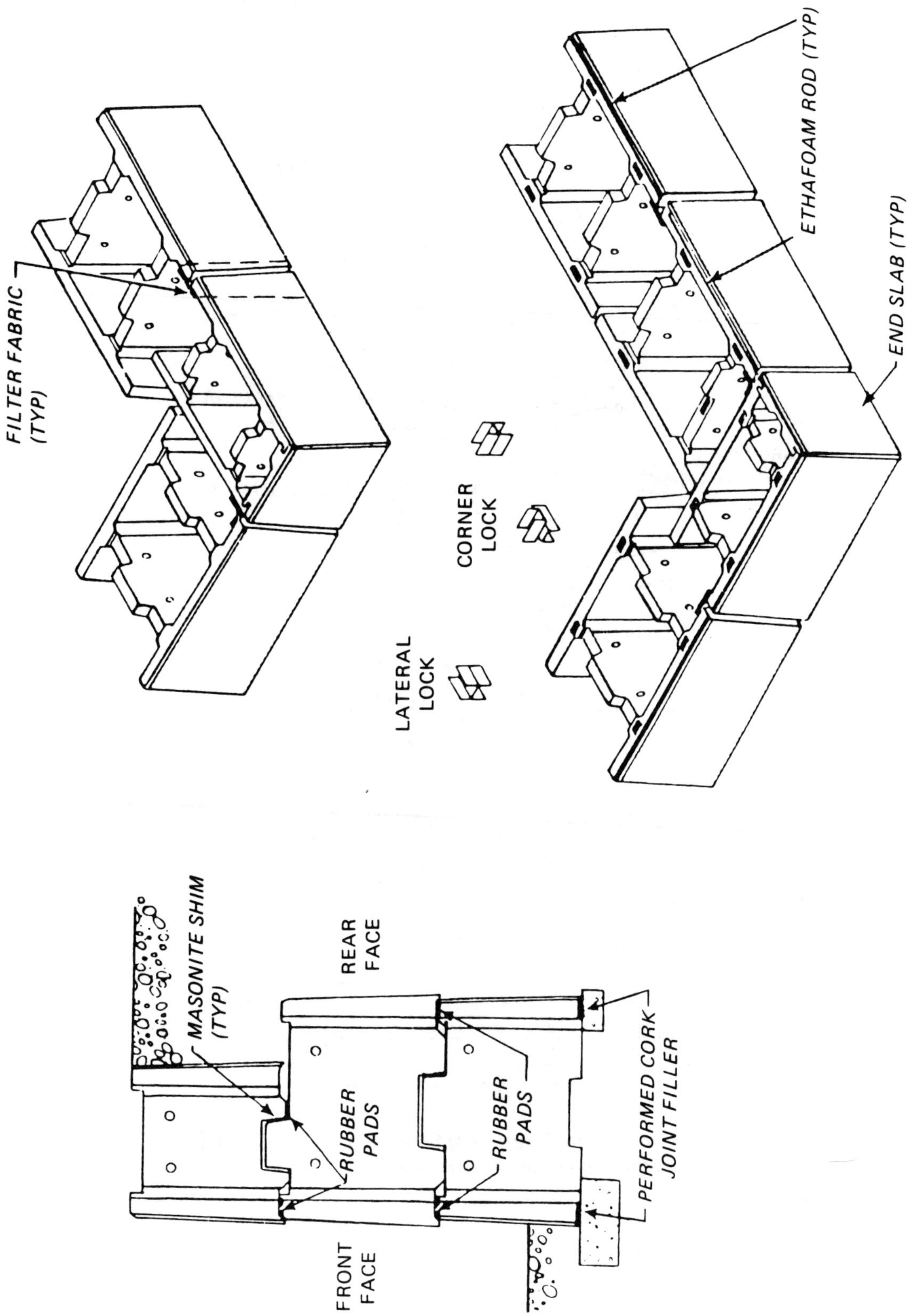

Fig 10-6. Materials used to construct Doublewal retaining wall (after Doublewal Corporation 1984)

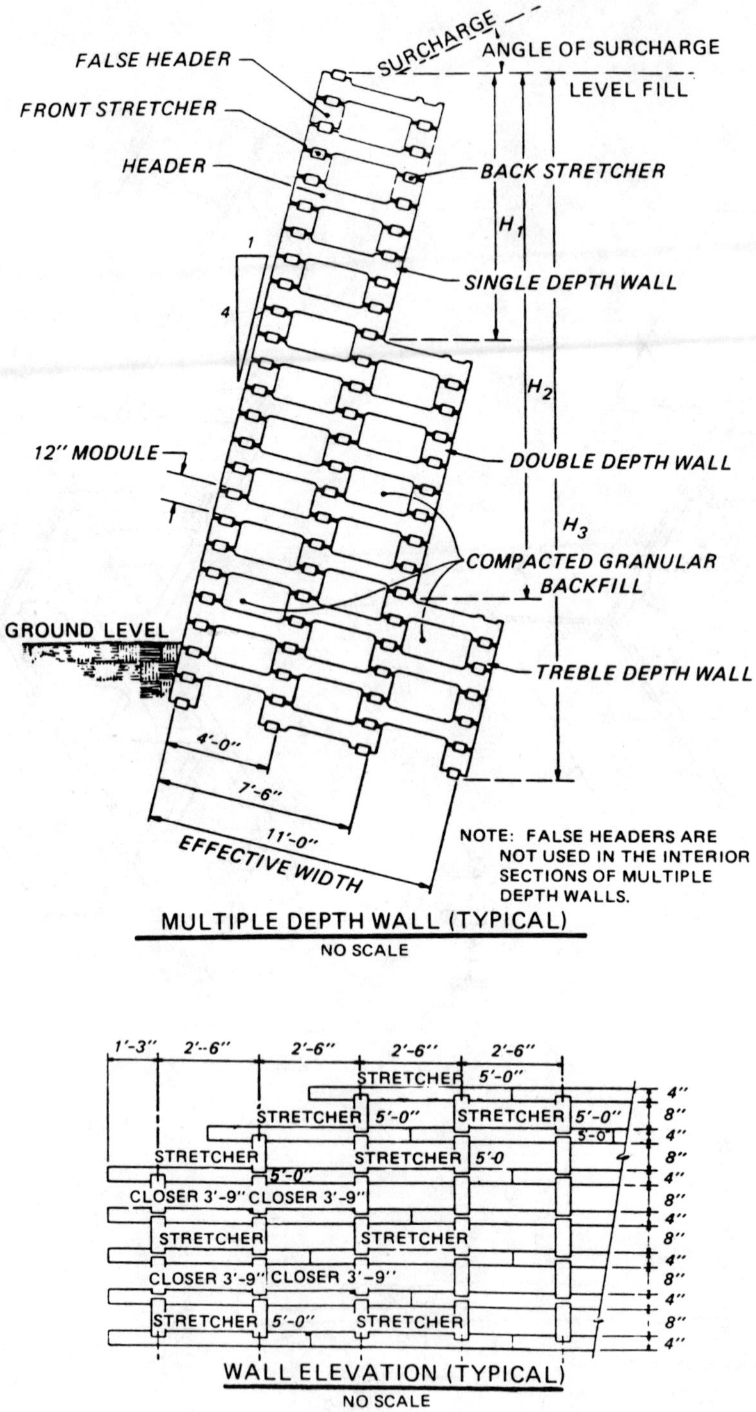

Fig 10-7. Schematic diagram of Criblock retaining wall (after Criblock Retaining Walls of America)

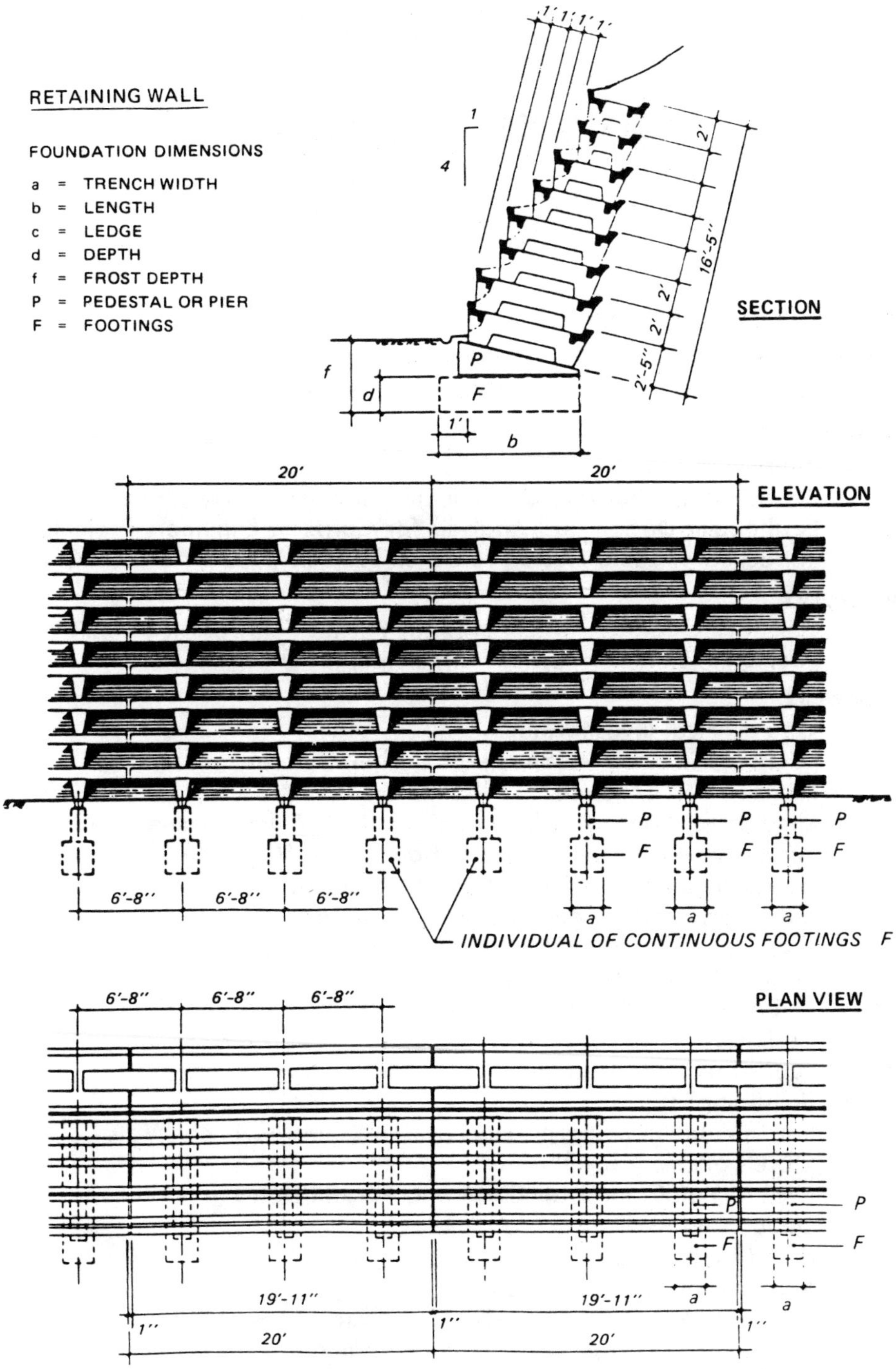

Fig 10-8. Schematic diagram of Evergreen retaining wall (after Evergreen Systems, Inc.)

APPENDIX A

REFERENCES

A-1. Government Publications

Department of the Army, Corps of Engineers. These publications are available from the Corps of Engineers Publication Depot, 2803 52nd Avenue, Hyattsville, MD 20781.

1. Engineer Regulations

ER 1110-1-803 Constructibility
ER 1110-2-100 Periodic Inspection and Continuing Evaluation of Completed Civil Works Structures
ER 1110-2-1806 Earthquake Design and Analysis for Corps of Engineers Projects
ER 1130-2-339 Inspection of Local Flood Protection Projects (RC5 DAEN-CWC-40)

2. Technical Manuals

TM 5-818-1 Soils and Geology: Procedures for Foundation Design of Buildings and Other Structures (Except Hydraulic Structures)

3. Engineer Manuals

EM 1110-1-1804 Geotechnical Investigations
EM 1110-1-2101 Working Stresses for Structural Design
EM 1110-1-2009 Architectural Concrete
EM 1110-2-301 Landscape Planting at Floodwalls, Levees & Embankment Dams
EM 1110-2-1410 Interior Drainage of Leveed Urban Areas: Hydrology
EM 1110-2-1601 Hydraulic Design of Flood Control Channels
EM 1110-2-1901 Soil Mechanics Design—Seepage Control
EM 1110-2-1902 Stability of Earth and Rock-Fill Dams
EM 1110-2-1903 Bearing Capacity of Soils
EM 1110-2-1904 Soil Mechanics Design—Settlement Analysis
EM 1110-2-1905 Design of Finite Relief Well Systems
EM 1110-2-1906 Laboratory Soil Testing
EM1110-2-1907 Soil Sampling
EM 1110-2-1908 Instrumentation of Earth and Rock-Fill Dams (Ground-Water & Pore Pressure Observations) (Part 1)
EM 1110-2-1911 Construction Control for Earth and Rockfill Dams
EM 1110-2-1913 Design and Construction of Levees
EM 1110-2-2000 Standard Practice for Concrete
EM 1110-2-2102 Waterstops and Other Joint Materials
EM 1110-2-2103 Details of Reinforcement—Hydraulic Structures
EM 1110-2-2906 Design of Pile Structures and Foundations
EM 1110-2-4300 Instrumentation for Concrete Structures
EM 1120-2-109 Federal Participation in Major Drainage Improvements

4. Guide Specifications

CW 02215 Plastic Filter Fabric
CW 03301 Cast-in-Place Structural Concrete
CW 03305 Mass Concrete

A-2. Other Publications

Agostinelli, V. M., et al. 1981 (Feb). 'Program Criteria Specifications Document for Computer Program TWDA for Design and Analysis of Inverted-T Retaining Walls and Floodwalls.' Available from: U. S. Army Engineer Waterways Experiment Station, P. O. Box 631, Vicksburg, MS 39180.

Al-Hussani, M. M., and Perry, E. B. 1976 (Sept). 'Effect of Horizontal Reinforcement on Stability of Earth Masses,' Technical Report S-76-11. U. S. Army Engineer Waterways Experiment Station, P. O. Box 631, Vicksburg, MS 39180.

Al-Hussani, M. M., and Perry, E. B. 1978 (Mar). 'Field Experiment of Reinforcement Earth Wall,' *Journal of the Geotechnical Engineering Division,* American Society of Civil Engineers, Vol 104, No. GT3, pp 307–322. Available from U. S. Army Engineer Waterways Experiment Station, P. O. Box 631, Vicksburg, MS 39180.

American Concrete Institute. 1967. "Ultimate Strength Design Handbook," ACI SP-17. Available from American Concrete Institute, P. O. Box 19150, Redford Station, Detroit, MI 48219.

American Concrete Institute. 1983. "Building Code Requirements for Reinforced Concrete," ACI 318, Available from American Concrete Institute, P. O. Box 19150, Redford Station, Detroit, MI 48219.

ANSI A58.1 "Minimum Design Loads for Buildings and Other Structures," American National Structures Institutes. Available from American National Structures Institute, Inc., 1430 Broadway, New York, NY 10018.

ASTM D1586 "Standard Method for Penetration Test and Split-Barrel Sampling of Soils." Available from American Society for Testing and Materials, 1916 Race St., Philadelphia, PA 19103.

ASTN D3441-79 "Standard Method for Deep Quasi-Static, Cone and Friction-Cone Penetration Tests of Soil," Available from American Society for Testing and Materials, 1916 Race St., Philadelphia, PA 19103.

Applied Technology Council (ATC). 1981 (Oct). "Seismic Design Guidelines for Highway Bridges," FHWA Report No. FHWA/RD-81/081. Available from NTIS, Springfield, VA 22161.

Baguelin, F., Jezequel, J. F., and Shields, J. F. 1978. *The Pressuremeter and Foundation Engineering,* Trans-Tech Publications, Rockport, MA.

Bowles, J. E. 1982. *Foundation Analysis and Design,* 3rd ed., McGraw-Hill, New York.

Brooker, E. W., and Ireland, H. O. 1965 (Feb). 'Earth Pressures At-Rest Related to Stress History,' *Canadian Geotechnical Journal,* Vol 2, No. 1, pp 1–15.

Caquot, A., and Kerisel, J. 1948. *Tables for the Calculation of Passive Pressure, Active Pressure, and Bearing Capacity of Foundations* (translated by M. A. Bec, London), Gauthier-Villars, Paris.

Carroll, R. G., Jr., and Murphy, J. C. 1985 (May–June). "Drainage Objective: Prefabricated Drainage Composites," *Geotechnical Fabrics Report,* Vol 3, No. 3, pp 14–18.

Casagrande, A. 1937. ''Seepage Through Dams,'' *Contributions to Soil Mechanics, 1925–1949,* Boston Society of Civil Engineers, Boston, MA, pp 295–336.

Casagrande, L. 1973 (Feb). "Comments of Conventional Design of Retaining Structures," *Journal, Soil Mechanics and Foundations Division, American Society of Civil Engineers,* Vol 99, No. SM 2, pp 181–197. Available from ASCE Publication Fulfillment, 345 East 47th St., New York, NY 10017.

Cedergren, H. R. 1967. *Seepage, Drainage, and Flow Nets,* John Wiley & Sons, New York.

Collin, J. C. 1986 (May). "Earth Wall Design," Ph.D. dissertation, University of California, Berkeley, CA. Available from University Microfilms International, 300 North Zeeb Road, Ann Arbor, MI 48106.

Criblock Retaining Walls of America, Inc. "Criblock Interlocking Concrete Criblock." Available from: Criblock Retaining Walls of America, Inc., P. O. Box 849, Sierra Madre, CA 91024.

Danish Geotechnical Institute. 1978. "Code of Practice for Foundation Engineering," Bulletin No. 32, p 52, Danish Geotechnical Institute, Copenhagen, Denmark.

Das, B. M. 1984. *Principles of Foundation Engineering,* Brooks/Cole Engineering Division, Monterey, CA.

Department of the Navy. 1982a (May). "Soil Mechanics," NAVFAC DM-7.1. Available from Department of the Navy, Naval Facilities Engineering Command, 200 Stovall St., Alexandria, VA 22332.

Department of the Navy. 1982b (May). "Foundations and Earth Structures," NAVFAC DM-7.2. Available from Department of the Navy, Naval Facilities Engineering Command, 200 Stovall St., Alexandria, VA 22332.

Doublewal Corporation 1984. "Doublewal Interlocking Precast Retaining Wall System Field Manual." Available from Doublewal Corporation, 59 East Main St., Plainville, CT 06062.

Driscoll, D. D. 1979 (Dec). *Retaining Wall Design Guide,* Foundation Services, Inc., Portland, OR. Prepared for USDA Forest Service. Available from: USDA Forest Service Region 6, 319 S.W.

Pine St., Portland, OR 97208.

Duncan, J. M., and Buchignani, A. L. 1976 (June). *An Engineering Manual for Settlement Studies,* Department of Civil Engineering, University of California, Berkeley, CA.

Elias, V. 1986. "Retaining Walls," *Draft Bridge Design Manual,* Part 4, Structures, Section 5, Pennsylvania Department of Transportation. Available from V. Elias and Associates, 6809 Carlynn Court, Bethesda, MD 20817.

Evergreen Systems, Inc. "Retaining Walls." Available from Evergreen Systems, Inc., P. O. Box 345, Kings Point, NY 11754.

Forsyth, R. A. 1979. "Alternate Earth Reinforcements," *Proceedings, Symposium on Earth Reinforcement, American Society of Civil Engineers,* pp 358–370. Available from U. S. Army Engineer Waterways Experiment Station, P. O. Box 631, Vicksburg, MS 39180.

Frondistou-Yannas, S. 1985 (Jan). "Corrosion Susceptibility of Internally Reinforced Soil Retaining Structures," Report No. FHWA-RD-83-105, U. S. Department of Transportation, Federal Highway Administration. Available from Federal Highway Administration, Turner-Fairbank Highway Research Center, 6300 Georgetown Pike, McLean, VA 22101.

Geotechnical Control Office. 1982 (June). *Guide to Retaining Wall Design,* Geoguide 1, Engineering Development Department, Hong Kong. Available from U. S. Army Engineer Waterways Experiment Station, P. O. Box 631, Vicksburg, MS 39180.

Godfrey, K. A., Jr. 1984 (Dec). "Retaining Walls: Competition or Anarchy?", *Civil Engineering,* Vol 54, No. 12, pp 48–52. Available from U. S. Army Engineer Waterways Experiment Station, P. O. Box 631, Vicksburg, MS 39180.

Harr, M. E. 1962. Groundwater and Seepage, McGraw-Hill, New York.

Harr, M. E. 1977. Mechanics of Particulate Media, McGraw-Hill, New York.

Hilfiker Company. "R. S. E. Construction Guide." Available from Hilfiker Company, 3900 Broadway, Eureka, CA 95501.

Ingold, T. S. 1979a. "Retaining Wall Performance During Backfilling," *Journal, Geotechnical Engineering Division,* American Society of Civil Engineers, Vol 105, No. GT5, pp 613–626.

Ingold, T. S. 1979b. "The Effect of Compaction on Retaining Walls," *Geotechnique,* Vol 29, No. 3, pp 265–283.

Ireland, H. O. 1964 (Aug). "Design and Construction of Retaining Walls," *Design of Structures to Resist Earth Pressure,* Soil Mechanics Lecture Series, Department of Civil Engineering, Illinois Institute of Technology, Chicago, IL. Available from U. S. Army Engineer Waterways Experiment Station, P. O. Box 631, Vicksburg, MS 39180.

Jackura, K. A. 1984 (May). "Results of Minor Research on Bar-Mat Pullout Tests," CA/TL-84-08, California Department of Transportation, Sacramento, CA.

Jaky, J. 1944. "The Coefficient of Earth Pressure At-Rest," *Journal, Society of Hungarian Architects and Engineers,* Budapest, Hungary, pp 355–358.

Jones, C. J. F. P. 1985. *Earth Reinforcement and Soil Structures,* Butterworths, London. Available from U. S. Army Engineer Waterways Experiment Station, P. O. Box 631, Vicksburg, MS 39180.

Kenney, T. C. 1959. "Discussion of 'Geotechnical Properties of Glacial Clays,' by T. H. Wu," *Journal of the Soil Mechanics and Foundations Division,* American Society of Civil Engineers, Vol 85, No. SM3, pp 67–79.

Kraemer, S. R., and Smith, A. D. 1986 (Oct). "Geocomposite Drains," Vol I, *Engineering Assessment and Preliminary Guidelines,* Report No. FHWA-RD-86-171, U. S. Department of Transportation, Federal Highway Administration. Available from Federal Highway Administration, 6300 Georgetown Pike, McLean, VA 22101.

Leary, R. M., and Klinedinst, G. L. 1984 (May). "Retaining Wall Alternates," *Proceedings, 34th Annual Highway Geology Symposium, Atlanta, GA.* Available from Federal Highway Administration, Geotechnical and Materials Branch, 400 7th St., S.W., Washington, DC 20590.

Liu, T. C., and Gleason, S. 1981 (Dec). "Strength Design of Reinforced Concrete Hydraulic Structures," Technical Report SL-80-4. Available from U. S. Army Engineer Waterways Experiment Station, P. O. Box 631, Vicksburg, MS 39180.

Long, N. T., et al. 1984 (May). "Repair of a Reinforced Earth Wall," *Proceedings, International Conference on Case Histories in Geotechnical Engineering,* Vol 1, pp 335–339, University of Missouri-Rolla, Rolla, MO. Available from: U. S. Army Engineer Waterways Experiment Station, P. O. Box 631, Vicksburg, MS 39180.

McGee, P. E. 1985 (Nov). "Reinforced Earth Wall Strip Serviceability Study," Special Research Study No. 8405, Geotechnical Engineering Bureau, Office of Materials and Research, Georgia Department of Transportation, Forest Park, GA. Available from Engineering Design Unit, Georgia Department of Transportation, Office of Materials and Research, 15 Kennedy Drive, Forest Park, GA 30050.

McGown, A., et al. 1985. "The Load-Strain-Time Behavior of Tensor Geogrids," *Pro-*

ceedings, Conference on Polymer Grid Reinforcement, pp 11–17, Thomas Telford Ltd., London. Available from U. S. Army Engineer Waterways Experiment Station, P. O. Box 631, Vicksburg, MS 39180.

McKittrick, D. 1979. "Design Construction, Technology, and Performance of Reinforced Earth Structures," *Proceedings, Symposium on Earth Reinforcement,* pp. 596–616, American Society of Civil Engineers. Available from U. S. Army Engineer Waterways Experiment Station, P. O. Box 631, Vicksburg, MS 39180.

Marcuson, W. F., and Bieganousky, W. A. 1977 (Nov). "SPT and Relative Density in Coarse Sands," *Journal of the Geotechnical Engineering Division,* American Society of Civil Engineers, Vol 103, No. GT11, pp 1295–1309.

Massarsch, K. R. 1979. "Lateral Earth Pressure in Normally Consolidated Clays," *Proceedings of the Seventh European Conference on Soil Mechanics and Foundation Engineering, Brighton, England,* Vol 2, pp 245–249.

Matsuo, M., Kenmochi, S., and Yagi, H. 1978 (Sept). "Experimental Study on Earth Pressure of Retaining Wall by Field Tests," *Soils and Foundations,* Japanese Society of Soil Mechanics and Foundation Engineering, Vol 18, No. 3, pp 27–41. Available from Japanese Society of Soil Mechanics and Foundation Engineering, Suga-yama Building-4F, Kanda Awaji-cho 2-23 Chiyoda-ku, Tokyo 101, Japan.

Meyerhof, G. G. 1956. "Penetration Tests and Bearing Capacity of Cohesionless Soils," *Journal of the Soil Mechanics and Foundations Divisions,* American Society of Civil Engineers, Vol 82, No. SM1, pp 1–19.

Mitchell, J. K., and Villet, W. C. B., eds. 1986. *Reinforcement of Earth Slopes and Embankments,* 4 Vols, National Cooperative Highway Research Program Project 24-2, National Research Council, Transportation Research Board, Washington, DC. Available from National Cooperative Highway Research Program, 2101 Constitution Avenue, N.W., Washington, DC 20418.

Mitchell, J. K., Villet, W. C. B., and DiMillio, A. F. 1984 (Dec). "Soil Reinforcement for Stabilization of Earth Slopes and Embankments," *Public Roads,* U. S. Department of Transportation, Vol 48, No. 3, pp 88–95. Available from U. S. Army Engineer Waterways Experiment Station, P. O. Box 631, Vicksburg, MS 39180.

Mosher, R. L., and Pace, M. E. 1982 (June). "User's Guide: Computer Program for Bearing Capacity Analyses of Shallow Foundations (CBEAR)," Instruction Report K-82-7. Available from U. S. Army Engineer Waterways Experiment Station, P. O. Box 631, Vicksburg, MS 39180

Pace, M. E., and Noddin, V. R. 1987 (Oct). "Sliding Stability of Concrete Structures (CSLIDE)," Instruction Report ITL-87-5. Available from U. S. Army Engineer Waterways Experiment Station, P. O. Box 631, Vicksburg, MS 39180.

Pace, M. E., Williams, D. R., Wolff, T. F., and Mosher, R. 1984 (Sept). "Seepage Analysis of Confined Flow Problems by the Method of Fragments (CFRAG)," Instruction Report K-84-8. Available from U. S. Army Engineer Waterways Experiment Station, P. O. Box 631, Vicksburg, MS 39180.

Peck, R. B., and Bazaraa, A. S. 1969. "Discussion on Settlement of Spread Footings on Sand," *Journal of the Soil Mechanics and Foundations Division,* American Society of Civil Engineering, Vol 95, No. SM3, pp 905–909.

Peck, R. B., Hanson, W. E., and Thornburn, T. H. 1974. *Foundation Engineering,* 2nd. ed., John Wiley & Sons, New York.

Robertson, P. K., and Campanella, R. G. 1983. "Interpretation of Cone Penetration Tests; Parts I and II," *Canadian Geotechnical Journal,* Vol 20, No. 4, pp 718–745.

Schmertmann, J. H. 1975. "Measurement of In-Situ Strength," *Proceedings of the Conference on In-Situ Measurement of Soil Properties,* pp 55–138, American Society of Civil Engineers.

Schwab, G. O., et al. 1981. *Soil and Water Conservation Engineering,* 3rd eds., John Wiley & Sons, New York, Available from U. S. Army Engineer Waterways Experiment Station, P. O. Box 631, Vicksburg, MS 39180.

Seed, H. B. 1976 (Sept). "Evaluation of Soil Liquefaction Effects on Level Ground During Earthquakes," *Liquefaction Problems in Geotechnical Engineering,* ASCE annual convention and exposition, Philadelphia, PA, preprint 2752, pp 1–104.

Seed, H. B., Arango, I., and Chan, C. K. 1975. "Evaluation of Soil Liquefaction Potential During Earthquakes" Report No. EERC 75-28, Earthquake Engineering Research Center, College of Engineering, University of California, Berkeley, CA.

Seed, H. B., and Idriss, I. M. 1982. "Ground Motions and Soil Liquefaction During Earthquakes," Earthquake Engineering Research Institute, University of California, Berkeley, CA.

Seed, H. B., and Whitman, R. V. 1970. "Design of Earth Retaining Structures for Dynamic Loads," *ASCE Specialty Conference, Lateral Stresses in the Ground and Design of Earth Retaining Structures,* pp 103–147.

Shields, D. H., and Tolunay, A. Z. 1973. "Passive Pressure Coefficients by Method of Slices," *Journal of the Soil Mechanics and Foundations Division,* American Society of Civil

Engineers, Vol 99, No. SM12, pp 1043–1053.

Sibley, E. A. 1967. "Backfill Adjacent to Structures," *Proceedings of the Montana Conference on Soil Mechanics and Foundation Engineering,* Department of Civil Engineering, Montana State University, Bozeman, MT. Available from U. S. Army Engineer Waterways Experiment Station, P. O. Box 631, Vicksburg, MS 39180.

Smith, A. D., and Kraemer, S. R. 1987 (Feb). "Creep of Geocomposite Drains," *Proceedings, Geosynthetic '87 Conference, New Orleans, LA,* Vol 2, pp 422–433. Available from U. S. Army Engineer Waterways Experiment Station, P. O. Box 631, Vicksburg, MS 39180.

Spangler, M. G., and Mickle, J. 1956. "Lateral Pressure on Retaining Walls Due to Backfill Surface Loads," Highway Research Board Bulletin 141, pp 1–18.

Tcheng, Y., and Iseux, J. 1972. "Essais de Butee en Vraie Grandeur et Constraintes Engendrees par une Surcharge Rectangulaire sur un Mur Vertical," *Proceedings, 5th European Conference on Soil Mechanics and Foundation Engineering, Madrid., Spain.*

Tensar Corporation, 1984. "Tensar." Available from The Tensar Corporation, 1210 Citizens Parkway, Morrow, GA 30260.

Terzaghi, K. 1934. "Large Retaining Wall Tests," *Engineering News—Record,* Vol 85, February 1–April 19.

Terzaghi, K. 1943. *Theoretical Soil Mechanics,* John Wiley & Sons, New York.

Terzaghi, K., and Peck, R. B. 1948. *Soil Mechanics in Engineering Practice,* John Wiley & Sons, New York. Available from U. S. Army Engineer Waterways Experiment Station, P. O. Box 631, Vicksburg, MS 39180.

Tracy, F. T. 1977a (Aug). "An Interactive Graphics Finite Element Method Grid Generator for Two-Dimensional Problems," Miscellaneous Paper K-77-5. Available from U. S. Army Engineer Waterways Experiment Station, P. O. Box 631, Vicksburg, MS 39180.

Tracy, F. T. 1977b (Aug). "An Interactive Graphics Post-Processor for Finite Element Method Static Analyses," Miscellaneous Paper K-77-4. Available from U. S. Army Engineer Waterways Experiment Station, P. O. Box 631, Vicksburg, MS 39180.

Tracy, F. T. 1983 (Sept). "User's Guide for a Plane and Axisymmetric Finite Element Program for Steady-State Seepage Problems," Instruction Report K-83-4. Available from U. S. Army Engineer Waterways Experiment Station, P. O. Box 631, Vicksburg, MS 39180.

Tschebatarioff, G. 1949. "Large Scale Earth Pressure Tests with Model Flexible Bulkheads," Final Report to U. S. Navy, Bureau of Yards and Docks, Princeton University, Princeton, NJ.

U. S. Army Engineer Division, Ohio River. 1958 (June). "Final Report on Full-size Floodwall (T-type) Stability Test." Available from HQUSACE Library. U. S. Army Engineer Waterways Experiment Station. 1956 (Oct). "Investigation of Underseepage and Its Control," Technical Memorandum No. 3-424. Available from U. S. Army Engineer Waterways Experiment Station, P. O. Box 631, Vicksburg, MS 39180.

U. S. Army Engineer Waterways Experiment Station. 1960 (Apr). 'The Unified Soil Classification System,' Technical Memorandum No. 3-357. Available from U. S. Army Engineer Waterways Experiment Station, P. O. Box 631, Vicksburg, MS 39180.

U. S. Army Engineer Waterways Experiment Station. 1962 (June). "Engineering Properties of Fine-Grained Mississippi Valley Alluvial Soils Meander Belt and Blackswamp Deposits." Technical Report TR3-604. Available from U. S. Army Engineer Waterways Experiment Station, P. O. Box 631, Vicksburg, MS 39180.

U. S. Army Engineer Waterways Experiment Station. 1980 (Aug). *Rock Testing Handbook, Standard and Recommended Methods.* Available from U. S. Army Engineer Waterways Experiment Station, P. O. Box 631, Vicksburg, MS 39180.

U. S. Army Engineer Waterways Experiment Station. 1984. *Shore Protection Manual,* Vol 2. Available from Superintendent of Documents, U. S. Government Printing Office, Washington, DC 20402.

Vesic, A. S. 1975. "Bearing Capacity of Shallow Foundations," *Foundation Engineering Handbook,* H. F. Winterkorn and H-Y. Fang, eds., Van Nostrand Reinhold, New York.

VSL Corporation. 1984. "Retaining Earth." Available from VSL Corporation,101 Albright Way, Los Gatos, CA 95030.

Westergaard, H. M. 1933. "Water Pressures on Dams During Earthquakes," *Transaction, American Society of Civil Engineers,* Vol 98, pp 418–433. Available from ASCE Publication Fulfillment, 345 East 4th St., New York, NY 10017.

Whitman, R. V., and Liao, S. 1985 (Jan). "Seismic Design of Gravity Retaining Walls," Miscellaneous Paper GL-85-1. Available from U. S. Army Engineer Waterways Experiment Station, P. O. Box 631, Vicksburg, MS 39180.

Winterkorn, H. F., and Fang, H-Y., eds. 1975. *Foundation Engineering Handbook,* Van Nostrand Reinhold, New York.

Wu, T. H. 1966. *Soil Mechanics,* Allyn and Bacon, Inc., Boston, MA.

APPENDIX B

BIBLIOGRAPHY

Al-Yassin, Z. 1980 (Nov). "Laboratory Pullout Test Results on VSL Retained Earth," Available from VSL Corporation, 101 Albright Way, Los Gatos, CA 95030.

Bang, S. 1985 (Mar). "Active Earth Pressure Behind Retaining Walls," *Journal, Geotechnical Engineering,* American Society of Civil Engineers, Vol 111, No. 3.

Bishop, J. A., and Anderson, L. A. 1979 (Dec). "Performance of a Welded Wire Retaining Wall," Report to Hilfiker Pipe Company, Department of Civil Engineering, University of Utah, Logan, UT. Available from U. S. Army Engineer Waterways Experiment Station, P. O. Box 631, Vicksburg, MS 39180.

Boyd, M. S. 1978 (Oct). "Practical Application of Reinforced Earth Theory," *Proceedings,Symposium on Soil Reinforcing and Stabilizing Techniques in Engineering Practice*, Vol 1, pp 83–110. Available from U. S. Army Engineer Waterways Experiment Station, P. O. Box 631, Vicksburg, MS 39180.

Dawkins, W. P. 1981 (Feb). "User's Guide: Computer Program for Design and Analysis of Sheet Pile Walls by Classical Methods (CSHTWAL)," Instruction Report K-81-2. Available from U. S. Army Engineer Waterways Experiment Station, P. O. Box 631, Vicksburg, MS 39180.

Department of the Army, Corps of Engineers. "Expansion, Contraction, and Construction Joints in Concrete," Guide Specifications CW 03150. Available from Corps of Engineers Publication Depot, 890 South Pickett St., Alexandria, VA 22304.

Doublewal Corporation. "Doublewal Interlocking Precast Retaining Wall System." Available from Doublewal Corporation, 59 East Main St., Plainville, CT 06062.

Doublewal Corporation. 1982. "Design Manual Doublewall Interlocking Reinforced Concrete Retaining Wall Systems." Available from Doublewal Corp., 59 East Main St., Plainville, CT 06062.

Egan, P. D. 1984 (June) "Design and Performance of Reinforced Earth Bridge Abutments," Paper presented at First International Bridge Conference, Pittsburgh, PA, Available from Reinforced Earth Company, Rosslyn Center, 1700 North Moore St., Arlington, VA 22209.

Goldberg-Zoino & Associates, Inc. 1983 (May) "Report on Full Scale Field Test of Twenty-Two Foot High Doublewal," Report to Doublewal Corporation, Newton Upper Falls, MA. Available from Doublewal Corporation, 59 East Main St., Plainville, CT 06062.

Hilfiker Company. 1984. "Reinforced Soil Embankment." Available from Hilfiker Company, 3900 Broadway, Eureka, CA 95501.

International Society for Rock Mechanics, Commission on Standardization of Laboratory and Field Tests. 1974 (Feb). "Suggested Methods for Determining Shear Strengths," Document No. 1. Available from Printing and Publishing Office, National Academy of Sciences—National Research Council, 2101 Constitution Ave., N.W., Washington, DC 20418.

McKittrick, D. P. 1978 (Oct). "Reinforced Earth: Application of Theory and Research to Practice," *Proceedings, Symposium on Soil Reinforcing and Stabilizing Techniques in Engineering Practice*, Vol 2, pp 1–44. Available from U. S. Army Engineer Waterways Experiment Station, P. O. Box 631, Vicksburg, MS 39180.

Neilson, M. R. 1984. "Pullout Resistance of Welded Wire Mats Embedded in Soil," M.S. thesis, Department of Civil Engineering, University of Utah, Logan, UT. Available from U. S. Army Engineer Waterways Experiment Station, P. O. Box 631, Vicksburg, MS 39180.

Netlon Limited. 1982. "Guidelines for the Design and Construction of Reinforced Soil Retaining Walls Using Tensar Geogrids," Available from The Tensar Corporation, 1210 Citizens Parkway, Morrow, GA 30260.

Netlon Limited. 1984. "Tensar Test Methods and Physical Properties of Tensar Geogrids," Available from The Tensar Corporation., 1210 Citizens Parkway, Morrow, GA 30260.

Peterson, L. M., and Anderson, L. R. 1980 "Pullout Resistance of Welded Wire Mesh Embedded in Soil," Department of Civil Engineering, University of Utah, Logan, UT. Available from U. S. Army Engineer Waterways Experiment Station, P. O. Box 631, Vicksburg, MS 39180.

Price, William A., et al. 1980 (Dec). "User's Guide: Computer Program for Design and Analysis of Inverted-T Retaining Walls and Flood Walls (TWDA)," Instruction Report K-80-6 and K-80-7. Available from U. S. Army Engineer Waterways Experiment Station, P. O. Box 631, Vicksburg, MS 39180.

Reinforced Earth Company. 1980. "Reinforced Earth Retaining Walls," Brochure 2.18b/RE. Available from Reinforced Earth Company, Rosslyn Center, 1700 North Moore St., Arlington, VA 22209.

Selvage, Heber, Nelson and Associates, Inc. 1984 (Apr). "Engineering Analysis of the Gridcote System." Available from Hilfiker Company, 3900 Broadway, Eureka, CA 95501.

VSL Corporation. 1983 (Mar). "Retained Earth, Field Erection Manual," Available from VSL Corporation, 101 Albright Way, Los Gatos, CA 95030.

VSL Corporation. 1984 (Mar). "Retained Earth Walls, Design Procedure for Standard Walls. Available from VSL Corporation, 101 Albright Way, Los Gatos, CA 95030.

VSL Corporation. 1984 (July). "General Specification for Retaining Earth Walls," Available from VSL Corporation, 101 Albright Way, Los Gatos, CA 95030.

Welti, C. 1982 (Oct). "Soil-Concrete Interaction Tests of 12′ Wide Doublewal," Glastonbury, CT. Available from Doublewal Corporation, 59 East Main St., Plainville, CT 06062.

APPENDIX C

NOTATIONS FOR

CHAPTERS 3, 4, AND 5

Symbol	Term
b	Distance over which base is in compression
B	Width of base of structure
$\overline{B}$	Effective width of base
c	Cohesion on slip plane of wedge
c_d	Developed cohesion on slip plane of wedge
c_1	Variable used to compute critical slip plane angle
c_2	Variable used to compute critical slip plane angle
C_E	Hydrodynamic factor
C_f	Coefficient used to compute wind loads
d_b	Depth of water at breaking wave occurrence
d_c	Depth of tension crack in cohesive backfill
d_s	Water depth
D	Depth of material in front of wall to base of structural wedge
e	Eccentricity of resultant at base of structural wedge
F	Inertia force of wall
FS	Factor of safety
F_T	Magnification factor
h	Wall height; height of fill against gravity wall or height of fill against "structural wedge"
H	Design wave height
H_b	Height of wave which breaks in water depth d_b
H_L	Horizontal force, acting to the right, applied to wedge
H_{Li}	Any horizontal force applied above the top or below the bottom of the left side adjacent wedge
H_R	Horizontal force, acting to the left, applied to wedge
H_{Ri}	Any horizontal force applied above the top or below the bottom of the right side adjacent wedge
H_s	Average of highest one-third of all waves or depth of saturation in wedge
H_1	Average of highest 1 percent of all waves
i	Number of wedge being analyzed
k_h	Horizontal acceleration in g's
k_v	Vertical acceleration in g's
K	Lateral earth pressure coefficient
K_A	Active earth pressure coefficient
K_{AE}	Mononobe-Okabe active earth pressure coefficient
K_i	Horizontal earth pressure coefficient for the ith layer
K_o	At-rest earth pressure coefficient
$K_{o\beta}$	At-rest earth pressure coefficient calculated by Danish Code equation
K_P	Passive earth pressure coefficient
K_{PE}	Mononobe-Okabe passive earth pressure coefficient
L	Length along slip plane of wedge
L_i	Length along slip plane of the ith wedge
N'	Resultant of forces normal to the assumed sliding plane
N_c, N_q, N_r	Bearing capacity factors for a strip load
P_E	Hydrodynamic pressure at depth y below top surface of water
p'_{hm}	Pressure due to compaction at depth z_{cr}
p'_{hz}	Effective horizontal pressure at depth z
p'_{vi}	Vertical effective earth pressure at the top of the ith layer
p_{AH}	Horizontal active earth pressure at depth z
p_{PH}	Horizontal passive earth pressure at depth z
P	Lateral (horizontal) force produced by wedge
P_A	Total active force on a unit length of wall backfilled with a cohesionless material; static component for a driving wedge
P_{AE}	Static and dynamic forces due to driving wedge
P_{AH}	Horizontal component of active earth force
P_D	Absolute value of total horizontal force from driving wedges
P_E	Hydrodynamic force given by Westergaard's equation

P_{EE}	Effective horizontal earth force contributed by wedge or wedge segment
$(P_{i-1} - P_i)$	Summation of applied forces acting horizontally on ith wedge
P_P	Static component for a resisting wedge
P_{PE}	Static and dynamic forces due to resisting wedge
P_{PH}	Horizontal component of passive earth force
P_R	Absolute value of total horizontal force from resisting wedges
P_W	Internal water force acting on the side of the wedge free body
P_{ws}	Static component of water force for partially saturated wedge
q'	Foundation pressure at base of structural wedge
q_o	Effective overburden pressure
Q	Vertical component of ultimate bearing capacity
SMF	Strength mobilization factor
T	Force tangential to slip plane of wedge
T	Earthquake period of vibration in seconds
T_F	Shear force that will cause failure along slip plane
U	Uplift force normal to slip plane of wedge
U_i	Uplift force exerted normal to slip plane of the ith angle
V	Vertical force applied to wedge
V_i	Any vertical force applied above the top of the ith wedge
$\sum V$	Summation of vertical forces for structural wedge
V_{max}	Maximum value of V for which the equations for c_1 and c_2 are valid
W	Total weight of material in wedge
W_i	Total weight of water, soil, rock, or concrete in the ith wedge
x_R	Location of resultant force from toe of structure
Y_{AE}	Line of action of P_{AE}
Y_{PE}	Line of action of P_{PE}
z	Depth from the top of the wall
z_i	Vertical distance measured down from the top of the ith layer
z_{cr}	Critical depth where pressure due to compaction equals p'_{hm}
z_w	Depth to water table

z_w	Angle between slip plane of wedge
α	and horizontal
	Angle between slip plane of the ith
α_i	wedge and the horizontal
	Angle between top of wedge and
β	horizontal
	Unit weight of material
γ	Angle of wall friction
δ	Effective unit weight of material
γ'	Effective unit weight of the ith layer
$\gamma_{sat}, \hat{\gamma}_s$	Saturated unit weight of fill
γ_w	Unit weight of water
γ_b	Buoyant unit weight of fill
	Angle of internal friction on slip
ϕ	plane of wedge
	Drained friction angle
ϕ'	Developed angle of internal friction
ϕ_d	on slip plane of wedge
	Seismic inertia angle
ψ	Stress normal to slip plane
σ	Effective normal stress
σ'_n	Applied shear stress on slip plane of
τ	wedge
τ_f	Shear strength of wedge material
θ	Angle of the wall face from horizontal or inclination of wall with respect to vertical
ξ	Correction factors for bearing capacity
$\tan \phi$	Unfactored shear strength parameter of the foundation material through which sliding plane passes
$\tan \phi_d$	Portion of shear strength considered to be mobilized or developed along the slip plane(s)
tan c	Unfactored shear strength parameter of the foundation material through which sliding plane passes
Δp_{HX}	Increase in horizontal pressure at distance x due to surcharge
Δp_{HZ}	Increase in horizontal pressure at depth z due to surcharge
ΔP_{AE}	Dynamic component of total driving force
ΔP_H	Increase in horizontal force due to surcharge
ΔP_{PE} wedge	Dynamic component of total driving force

APPENDIX D

NOTATIONS FOR CHAPTER 9

Symbol	Term
a	Depth of equivalent rectangular stress block
A_g	Gross area of section, sq. in.
A_s	Area of tension reinforcement, sq. in.
A'_s	Area of compression reinforcement, sq. in.
b	Width of compression face of member, in.
c	Distance from extreme compression fiber to neutral axis, in.
d	Distance from extreme compression fiber to centroid of tension reinforcement, in.
d'	Distance from extreme compression fiber to centroid of compression reinforcement, in.
D	Dead load of the concrete members only or related internal forces and moments
e'	Eccentricity of axial load measured from the centroid of the tension reinforcement $= d - h/2$
E_s	Modulus of elasticity of reinforcement, psi
f'_c	Specified compressive strength of concrete, psi
f_s	Calculated stress at centroid of tension reinforcement when compression controls
f'_s	Calculated stress at centroid of compression reinforcement
f_y	Specified yield strength of reinforcement, psi
h	Total depth of section, in.
k_m	Ratio of stress block depth (a) to the effective depth (d), at balanced strain conditions for hydraulic structures
k_u	Ratio of stress block depth (a) to the effective depth (d)
L	Live loads or related internal moments and forces
M_u	Internal moment due to factored axial load, Pu
P_b	Axial load at given eccentricity at the balanced strain condition
P_n	Nominal axial load strength at given eccentricity
P_u	Factored axial load at given eccentricity
R	Internal forces and moments resulting from reactions induced by applied *unfactored* dead and live loads.
R_f	Internal forces and moments resulting from reaction induced by the applied *factored* dead and live loads
U	Required strength to resist factored loads or related internal moments and forces
V_c	Shear strength
W	Wind load or related internal moments and forces
β_m	A coefficient which accounts for the difference between the distribution of actual compression stresses and the assumed rectangular distribution of stresses
β_1	Defined the same as β_m but the value is in accordance with ACI 318
ε_c	Maximum usable strain at extreme concrete compression fiber = 0.003
ε_m	Design strain at the extreme concrete compression fiber $= 0.5\ \varepsilon_c$
ε_y	Strain at the centroid of the tension reinforcements $\leq f_y/E_s$
λ	Maximum percentage of ρ_b allowed for hydraulic structures
ϕ	Strength reduction factor
ρ	Ratio of tension reinforcement, A_s/bd
ρ_b	Reinforcement ratio producing balanced strain conditions
ρ_m	Reinforcement ratio corresponding to balanced strain conditions for hydraulic structures

APPENDIX E

COMPARISON OF DANISH CODE AND JAKY EQUATIONS FOR AT-REST COEFFICIENT WITH COULOMB COEFFICIENT FOUND USING REDUCED SHEAR STRENGTH PARAMETER ϕ_d

$$\phi_d = \tan^{-1}(\tfrac{1}{2}\tan\phi)$$

Jaky, Equation 3-4 Danish Code, Equation 3-6 Coulomb, Equation 3-14

(Table E-1 appears on next page.)

Table E-1. At-rest earth pressure coefficients for comparison of Coulomb's equation with an SMF of 2/3 to the Danish Code and Jaky equations

		Horizontal Backfill		Inclined Backfill							
				$\beta = -10^6$		$\beta = 10^6$		$\beta = 10^6$		$\beta = 20^6$	
ϕ (deg)	ϕ (rad)	K_o (Jaky)	K_o (SMF = 2/3)	$K_{o\beta}$ (Danish)	K_o (SMF = 2/3)	$K_{o\beta}$ (Danish)	K_o (SMF = 2/3)	$K_{o\beta}$ (Danish)	K_o (SMF = 2/3)	$K_{o\beta}$ (Danish)	$\varnothing_o$ (SMF = 2/3)
10	0.175	0.826	0.791	0.683	0.703	0.826	0.791	0.970			
11	0.192	0.809	0.772	0.669	0.686	0.809	0.772	0.950			
12	0.209	0.792	0.754	0.655	0.669	0.792	0.754	0.930			
13	0.227	0.775	0.736	0.640	0.653	0.775	0.736	0.910			
14	0.244	0.758	0.718	0.626	0.638	0.758	0.718	0.890			
15	0.262	0.741	0.701	0.612	0.622	0.741	0.701	0.870	0.931		
16	0.279	0.724	0.684	0.599	0.607	0.724	0.684	0.850	0.871		
17	0.297	0.708	0.667	0.585	0.593	0.708	0.667	0.831	0.833		
18	0.314	0.691	0.651	0.571	0.578	0.691	0.651	0.811	0.802		
19	0.332	0.674	0.634	0.557	0.564	0.674	0.634	0.792	0.774		
20	0.349	0.658	0.618	0.544	0.550	0.658	0.618	0.772	0.748	0.883	
21	0.367	0.642	0.603	0.530	0.537	0.642	0.603	0.753	0.724	0.861	
22	0.384	0.625	0.587	0.517	0.523	0.625	0.587	0.734	0.702	0.839	
23	0.401	0.609	0.572	0.503	0.510	0.609	0.572	0.715	0.680	0.818	
24	0.419	0.593	0.557	0.490	0.497	0.593	0.557	0.696	0.659	0.796	
25	0.436	0.577	0.542	0.477	0.485	0.577	0.542	0.678	0.638	0.775	
26	0.454	0.562	0.528	0.464	0.472	0.562	0.528	0.659	0.619	0.754	
27	0.471	0.546	0.513	0.451	0.460	0.546	0.513	0.641	0.599	0.733	
28	0.489	0.531	0.499	0.438	0.448	0.531	0.499	0.623	0.581	0.712	
29	0.506	0.515	0.485	0.426	0.436	0.515	0.485	0.605	0.563	0.691	0.809
30	0.524	0.500	0.471	0.413	0.424	0.500	0.471	0.587	0.545	0.671	0.742
31	0.541	0.485	0.458	0.401	0.412	0.485	0.458	0.569	0.527	0.651	0.696
32	0.559	0.470	0.445	0.388	0.400	0.470	0.445	0.552	0.510	0.611	0.660
33	0.576	0.455	0.431	0.376	0.389	0.455	0.431	0.534	0.494	0.611	0.627
34	0.593	0.441	0.418	0.364	0.378	0.441	0.418	0.517	0.478	0.592	0.598
35	0.611	0.426	0.405	0.352	0.367	0.426	0.405	0.500	0.462	0.572	0.572
36	0.628	0.412	0.393	0.341	0.356	0.412	0.393	0.484	0.446	0.553	0.547
37	0.646	0.398	0.380	0.329	0.345	0.398	0.380	0.467	0.431	0.534	0.523
38	0.663	0.384	0.368	0.318	0.334	0.384	0.368	0.451	0.416	0.516	0.501
39	0.681	0.371	0.356	0.306	0.323	0.371	0.356	0.435	0.401	0.497	0.479
40	0.698	0.357	0.344	0.295	0.313	0.357	0.344	0.419	0.386	0.479	0.459
41	0.716	0.344	0.332	0.284	0.303	0.344	0.332	0.404	0.372	0.462	0.439
42	0.733	0.331	0.320	0.273	0.292	0.331	0.320	0.388	0.358	0.444	0.420
43	0.750	0.318	0.309	0.263	0.282	0.318	0.309	0.373	0.345	0.427	0.402
44	0.768	0.305	0.298	0.252	0.272	0.305	0.298	0.358	0.331	0.410	0.384
45	0.785	0.293	0.286	0.242	0.263	0.293	0.286	0.344	0.318	0.393	0.367

Note: A blank entry in the table denotes where $\beta > \phi_d$

APPENDIX F

DERIVATION OF GENERAL WEDGE EQUATION FOR SINGLE WEDGE ANALYSIS (EQUATION 3-23)

F-1. Effective horizontal earth force

Given the following driving wedge, an equation for P_{EE}, the effective horizontal earth force, will be derived.

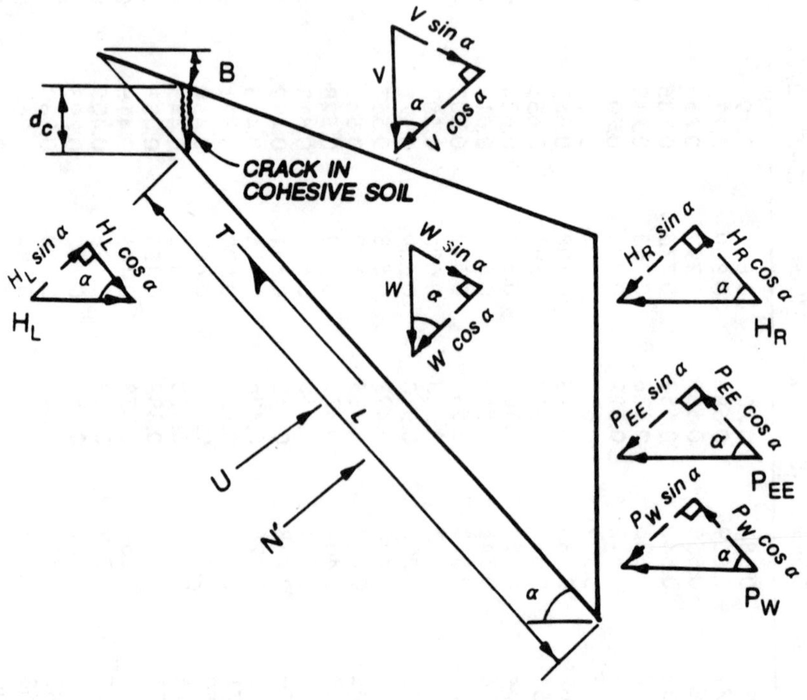

Summing forces normal to the slip plane yields,

$$H_L \sin\alpha + U + N' - V\cos\alpha - H_R\sin\alpha - P_{EE}\sin\alpha - P_W\sin\alpha - W\cos\alpha = 0$$

Solving for N′ yields,

$$N' = (-H_L + H_R + P_{EE} + P_W)\sin\alpha - U + (V + W)\cos\alpha \qquad \text{[F-1]}$$

According to the Mohr-Coulomb failure criterion,

$$T = N'\tan\phi + cL \qquad \text{[F-2]}$$

Inserting Equation F-1 into Equation F-2 yields,

$$T = [(-H_L + H_R + P_{EE} + P_W)\sin\alpha - U + (V + W)\cos\alpha]\tan\phi + cL$$

Setting the summation of forces parallel to the slip plane equal to zero yields

$$H_L\cos\alpha + V\sin\alpha + W\sin\alpha - H_R\cos\alpha - P_{EE}\cos\alpha - P_W\cos\alpha - T = 0 \qquad \text{[F-3]}$$

Solving for T from Equation F-3 and substituting this expression into Equation F-2 yields,

$$(H_L - H_R - P_{EE} - P_W)\cos\alpha + (V + W)\sin\alpha + [(H_L - H_R - P_{EE} - P_W)\sin\alpha + U - (V + W)\cos\alpha] \tan\phi - cL = 0$$

Simplifying and solving for P_{EE} yields,

$$P_{EE} = (V + W)\frac{(\sin\alpha - \cos\alpha\tan\phi)}{(\cos\alpha + \sin\alpha\tan\phi)} + \frac{U\tan\phi - c}{\cos\alpha + \sin\alpha\tan\phi} + H_L - H_R - P_W$$

$$P_{EE} = (V + W)\frac{(1 - \cot\alpha\tan\phi)\tan\alpha}{(1 + \tan\alpha\tan\phi)} + \frac{U\tan\phi - cL}{(1 + \tan\alpha\tan\phi)\cos\alpha} + H_L - H_R - P_W \quad [F-4]$$

F-2. Soil Parameters

For a particular SMF, the corresponding factored soil parameters ϕ_d and c_d can be inserted into Equation F-4 to yield

$$P_{EE} = (V + W)\frac{(1 - \cot\alpha\tan\phi_d)\tan\alpha}{(1 + \tan\alpha\,\phi_d)} + \frac{U\tan\phi_d - c_dL}{(1 + \tan\alpha\tan\phi_d)\cos\alpha} + H_L - H_R - P_W$$

F-3. Resisting wedge

The same procedure can be applied to a resisting wedge to yield the following equation for P_{EE}:

$$P_{EE} = (W + V)\frac{(1 + \tan\phi_d\cot\alpha)\tan\alpha}{(1 - \tan\phi_d\tan\alpha)} - \frac{U\tan\phi_d - c_dL}{\cos\alpha(1 - \tan\phi_d\,\alpha)} - H_L + H_R - P_W$$

If more than one driving or resisting wedge exists, the value of P_{EE} will equal the difference of the earth forces applied to the wedge.

APPENDIX G

DERIVATION OF EQUATIONS FOR CRITICAL SLIP-PLANE ANGLE (α) FOR DRIVING AND RESISTING WEDGES

G-1. Derivation of Critical Slip-Plane Angle (α) for Driving Wedges for the Static Condition

In the following paragraphs the equations for the critical slip-plane angle (α) will be derived for the static case for a driving wedge.

A. INTRODUCTION. The wedge equation (Equation 3-23) with the uplift term omitted is:

$$P_{EE} = \frac{(W + V)(1 - \tan\phi \cot\alpha)\tan\alpha}{1 + \tan\phi \tan\alpha} - \frac{cL}{\cos\alpha(1 + \tan\phi\tan\alpha)} \qquad \text{[G-1]}$$

where

$$L = \text{length of slip plane} = \frac{h - d_c}{\cos\alpha(\tan\alpha - \tan\beta)}$$

$$W = \frac{\gamma(h^2 - d_c^2)}{2(\tan\alpha - \tan\beta)}$$

as shown in the following figure.

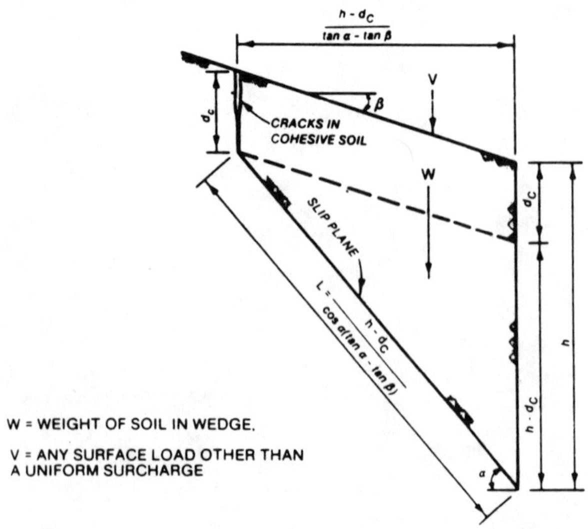

Inserting the terms for L and W into Equation G-1 yields,

$$P_{EE} = \frac{\gamma(h^2 - d_c^2)}{2(\tan\alpha - \tan\beta)} \cdot \frac{(1 - \tan\phi\cot\alpha)\tan\alpha}{1 + \tan\phi\tan\alpha} + \frac{V(1 - \tan\phi\cot\alpha)\tan\alpha}{1 + \tan\phi\tan\alpha} - \frac{c(h - d_c)}{\cos^2\alpha(\tan\alpha - \tan\beta)(1 + \tan\phi\tan\alpha)}$$

Using the trigonometric identity $\frac{1}{\cos^2\alpha} = \sec^2\alpha = \tan^2\alpha$ and putting all terms over a common denominator

$$P_{EE} = [\gamma(h^2 - d_c^2)(\tan\alpha - \tan\phi) + 2V(\tan\alpha - \tan\phi)(\tan\alpha - \tan\beta) - 2c(h - d_c)(1 + \tan^2\alpha)] \div [2(\tan\alpha - \tan\beta)(1 + \tan\phi\tan\alpha)] \qquad \text{[G-2]}$$

The first derivative of P_{EE} with respect to α may now be taken, and set equal to zero. The resulting expression is solved for the critical value of tan α. For simplification, the calculus is done in parts.

B. FIRST DERIVATIVE OF γ TERM. Grouping the terms from Equation G-2 that are associated with γ yields,

$$P_{EE\gamma} = \frac{\gamma(h^2 - d_c^2)}{2} \cdot \frac{\tan \quad - \tan\phi}{(\tan\alpha - \tan\beta)(1 + \tan\phi\tan\alpha)}$$

$$P_{EE\gamma} = \frac{\gamma(h^2 - d_c^2)}{2} \cdot \frac{\tan\alpha - \tan\phi}{\tan\phi\tan^2\alpha + (1 - \tan\phi\tan\beta)\tan\alpha - \tan\beta}$$

$$= \frac{m_\gamma}{n} \qquad [G\text{-}3]$$

The first derivative of Equation G-3 is equation

$$\frac{dP_{EE\gamma}}{d\alpha} = \frac{\dfrac{ndm_\gamma}{d\alpha} - \dfrac{m_\gamma dn}{d\alpha}}{n^2}$$

The denominator may be neglected because it is the same for all terms in Equation G-2.

$$\frac{dm_\gamma}{d\alpha} = a\sec^2\alpha,\ \text{let}\ \frac{\gamma(h^2 - d_c^2)}{2} = a$$

$$\frac{dn}{d\alpha} = [2\tan\phi\tan\alpha + (1 - \tan\phi\tan\beta)]\sec^2\alpha$$

$$\frac{ndm_\gamma}{d\alpha} = a[\tan\phi\tan^2\alpha + (1 - \tan\phi\tan\beta)\tan\alpha - \tan\beta]\sec^2\alpha$$

$$\frac{m_\gamma dn}{d\alpha} = a[2\tan\phi\tan^2\alpha - (2\tan^2\phi + 1 - \tan\phi\tan\beta)\tan\alpha + \tan^2\phi\tan\beta - \tan\phi(1 - \tan\phi\tan\beta)]\sec^2\alpha$$

$$\frac{ndm_\gamma}{d\alpha} - \frac{m_\gamma dn}{d\alpha} = \frac{\gamma(h^2 - d_c^2)}{2} \times [-\tan\phi\tan^2\alpha + 2\tan^2\phi\tan\alpha + \tan\phi(1 - \tan\phi\tan\beta) - \tan\beta]\sec^2\alpha \qquad [G\text{-}4]$$

C. FIRST DERIVATIVE OF c TERM.

Grouping the terms from Equation G-2 that are associated with c yields,

$$P_{EEc} = -c(h - d_c) \cdot \frac{1 + \tan^2\alpha}{\tan\phi\tan^2\alpha + (1 - \tan\phi\tan\beta)\tan\alpha - \tan\beta}$$

$$= \frac{m_c}{n}$$

Let $-c(h - d_c) = a$

$$\frac{dm_c}{d\alpha} = a(2\tan\alpha)\sec^2\alpha$$

$$\frac{dn}{d\alpha} = [2\tan\phi\tan\alpha + (1 - \tan\phi\tan\beta)]\sec^2\alpha$$

$$\frac{ndm_c}{d\alpha} = a[2\tan\phi\tan^3\alpha + 2(1 - \tan\phi\tan\beta)\tan^2\alpha - 2\tan\beta\tan\alpha]\sec^2\alpha$$

$$\frac{m_c dn}{d\alpha} = a[2\tan\phi\tan^3\alpha + (1 - \tan\phi\tan\beta)\tan^2\alpha + 2\tan\phi\tan\alpha + (1 - \tan\phi\tan\beta)]\sec^2\alpha$$

$$\frac{ndm_c}{d\alpha} - \frac{m_c dn}{d\alpha} = -c(h - d_c)[(1 - \tan\phi\tan\beta)\tan^2\alpha - 2(\tan\phi + \tan\beta)\tan\alpha - (1 - \tan\phi\tan\beta)]\sec^2\alpha$$

D. FIRST DERIVATIVE OF V TERM.

Grouping the terms from Equation G-2 that are associated with the V term yields,

$$P_{EEV} = \frac{V[\tan^2\alpha - (\tan\phi + \tan\beta)\tan\alpha + \tan\phi\tan\beta}{\tan\phi\tan^2\alpha + (1 - \tan\phi\tan\beta)\tan\alpha - \tan\beta}$$

$$= \frac{m_v}{n}$$

$$\frac{dm_v}{d\alpha} = V[2\tan\alpha - (\tan\phi\tan\beta)]\sec^2\alpha$$

$$\frac{dn}{d\alpha} = [2\tan\phi\tan\alpha + (1 - \tan\phi\tan\beta)]\sec^2\alpha$$

$$\frac{ndm_v}{d\alpha} = V\{[2\tan\phi\tan^3\alpha + (2 - 3\tan\phi\tan\beta - \tan^2\phi)]\tan^2\alpha - [3\tan\beta + \tan\phi - \tan\phi\tan\beta(\tan\phi + \tan\beta)]\tan\alpha + \tan\beta(\tan\phi + \tan\beta)\}\sec^2\alpha$$

$$\frac{m_v dn}{d\alpha} = V\{[2\tan\phi\tan^3\alpha + (1 - 3\tan\phi\tan\beta - 2\tan^2\phi)]\tan^2\alpha + [2\tan^2\phi\tan\beta - (1 - \tan\phi\tan\beta)]\tan\alpha + \tan\phi\tan\beta(1 - \tan\phi\tan\beta)\}\sec^2\alpha$$

$$\frac{ndm_v}{d\alpha} - \frac{m_v dn}{d\alpha} = V[(1 + \tan^2\phi)\tan^2\alpha$$

$$- 2\tan\beta(1+\tan^2\phi)\tan\alpha + \tan^2\beta(1+\tan^2\phi)]\sec^2\alpha \quad \text{[G-6]}$$

E. Summation of Terms. From Equations G-4, G-5, and G-6, set the sum of

$$\left(\frac{ndm_\gamma}{d\alpha} - \frac{m_\gamma dn}{d\alpha}\right) + \left(\frac{ndm_c}{d\alpha} - \frac{m_c dn}{d\alpha}\right) + \left(\frac{ndm_v}{d\alpha} - \frac{m_v dn}{d\alpha}\right)$$

equal to zero, and divide by $\sec^2\alpha$.

$$\left[-\frac{\gamma(h^2-d_c^2)\tan\phi}{2} - c(h-d_c)(1-\tan\phi\tan\beta) + V(1+\tan^2\phi)\right]\tan^2\alpha$$
$$+\left[\frac{\gamma(h^2-d_c^2)(2\tan^2\phi)}{2} + 2c(h-d_c)(\tan\phi+\tan\beta) - 2V\tan\beta(1+\tan^2\phi)\right]\tan\alpha$$
$$+\left\{\frac{\gamma(h^2-d_c^2)}{2}[\tan\phi(1-\tan\phi\tan\beta)-\tan\beta] + c(h-d_c)(1-\tan\phi\tan\beta) + V\tan^2\beta(1+\tan^2\phi)\right\} = 0$$

Divide all terms by $-\dfrac{\gamma(h^2-d_c^2)}{2}$.

$$\left[\tan\phi + \frac{2c(1-\tan\phi\tan\beta)}{\gamma(h+d_c)} - \frac{2V(1+\tan^2\phi)}{\gamma(h^2-d_c^2)}\right]\tan^2\alpha$$
$$-\left[2\tan^2\phi + \frac{4c(\tan\phi+\tan\beta)}{\gamma(h+d_c)} - \frac{4V\tan\beta(1+\tan^2\phi)}{\gamma(h^2-d_c^2)}\right]\tan\alpha$$
$$-\left[\tan\phi(1-\tan\phi\tan\phi) - \tan\beta + \frac{2c(1-\tan\phi\tan\beta)}{\gamma(h+d_c)} + \frac{2V\tan^2\beta(1+\tan^2\phi)}{\gamma(h^2-d_c^2)}\right] = 0 \quad \text{[G-7]}$$

In Equation G-7, let the coefficient of $\tan^2\alpha$ equal A, let the coefficient of $\tan\alpha$ equal $-Ac_1$, and let the third coefficient equal $-Ac_2$. Equation G-7 yields

$$A\tan^2\alpha - Ac_1\tan\alpha - Ac_2 = 0$$
$$\tan^2\alpha - c_1\tan\alpha - c_2 = 0$$

and

$$\tan\alpha_{crit.} = \frac{c_1 + \sqrt{c_1^2 + 4c_2}}{2}$$

or

$$\alpha_{crit.} = \tan^{-1}\left(\frac{c_1 + \sqrt{c_1^2 + 4c_2}}{2}\right) \quad \text{[G-8]}$$

The coefficients c_1 and c_2 are defined as

$$c_1 = \frac{2\tan^2\phi + \dfrac{4c(\tan\phi+\tan\beta)}{\gamma(h+d_c)} - \dfrac{4V\tan\beta(1+\tan^2\phi)}{\gamma(h^2-d_c^2)}}{A} \quad \text{[G-9]}$$

$$c_2 = \frac{\tan\phi(1-\tan\phi\tan\beta) - \tan\beta + \dfrac{2c(1-\tan\phi\tan\beta)}{\gamma(h+d_c)} + \dfrac{2V\tan^2\beta(1+\tan^2\phi)}{\gamma(h^2-d_c^2)}}{A} \quad \text{[G-10]}$$

where

$$A = \tan\phi + \frac{2c(1-\tan\phi\tan\beta)}{\gamma(h+d_c)} - \frac{2V(1+\tan^2\phi)}{\gamma(h^2-d_c^2)} \quad \text{[G-11]}$$

G-2. Derivation of Equation for V_{max}

The equations for calculating the critical value of α are not valid when the value of V is too large. When V is equal to or greater than a certain maximum value (V_{max}), the value of α should be set in accordance with Figure 3-15. The value of V_{max} is found by setting Equation 3-28 equal to zero and solving.

$$\tan\phi + \frac{2c(1-\tan\phi\tan\beta)}{\gamma(h+d_c)} - \frac{2V_{max}(1+\tan^2\phi)}{\gamma(h^2-d_c^2)} = 0$$

$$\frac{2V_{max}(1+\tan^2\phi)}{\gamma(h^2-d_c^2)} = \frac{\gamma(h+d_c)\tan\phi + 2c(1-\tan\phi\tan\beta)}{\gamma(h-d_c)}$$

$$V_{max} = \frac{\gamma(h^2-d_c^2)\tan\phi + 2c(h-d_c) \times (1-\tan\phi\tan\beta)}{2(1+\tan^2\phi)} \quad \text{[G-12]}$$

G-3. Equations for Critical Slip-Plane Angles for Resisting Wedges for the Static Condition

The following equations, for resisting wedges, where derived from Equation 3-33 in the same manner used for deriving the driving wedge equations for the static condition.

$$A = \tan\phi + \frac{2c(1+\tan\phi\tan\beta)}{\gamma h} + \frac{2V(1+\tan^2\phi)}{\gamma h^2} \quad \text{[G-13]}$$

$$c_1 = \frac{2\tan^2\phi + \dfrac{4c(\tan\phi-\tan\beta)}{\gamma h} - \dfrac{4V\tan\beta(1+\tan^2\phi)}{\gamma h^2}}{A} \quad \text{[G-14]}$$

$$c_2 = \frac{\tan\phi(1+\tan\phi\tan\beta)+\tan\beta + \dfrac{2c(1+\tan\phi\tan\beta)}{\gamma h} - \dfrac{2V\tan^2\beta(1+\tan^2\phi)}{\gamma h^2}}{A} \quad \text{[G-15]}$$

$$\alpha = \tan^{-1}\left(\frac{-c_1+\sqrt{c_1^2+4c_2}}{2}\right) \quad \text{[G-16]}$$

Note: $d_c = 0$, in cohesive soil on resisting side of walls.

G-4. Effect of Water on Critical Slip-Plane Angle

The unit weight of soil (γ) to use in equations for calculating the slip-plane angle should be the average unit weight of soil in the wedge. The average unit weight is determined by using the moist unit weight of soil above the water table and the buoyant unit weight below the water table. The effect of seepage should be taken into account in determining the buoyant unit weight. See Appendices M and N for example calculations.

G-5. Derivation of Equations for Critical Slip-Plane Angle (α) for Driving Wedges for Earthquake Condition

The equations for the critical slip-plane angle (α) for a driving wedge as shown in the figure following for the dynamic case will be derived.

A. SUMMATION OF FORCES. Summing forces normal to the sliding plane yields

$$N = W\cos\alpha + P_{AE}\sin\alpha - k_h W\sin\alpha$$

The shearing force T is defined by the Mohr-Coulomb failure criterion as

$$T = N\tan\phi + cL$$

$$= (W\cos\alpha + P_{AE}\sin\alpha - k_h W\sin\alpha)\tan\phi + cL$$

Summing forces parallel to the slip plane and inserting the expression for T yields

$$P_{AE}\cos\alpha - k_h W\cos\alpha - W\sin\alpha$$

$$+ (W\cos\alpha + P_{AE}\sin\alpha - k_h W\sin\alpha)\tan\phi + cL = 0$$

$$P_{AE}(\cos\alpha + \tan\phi\sin\alpha) = W(\sin\alpha - \tan\phi\cos\alpha)$$

$$+ k_h W(\cos\alpha + \tan\phi\sin\alpha) - cL$$

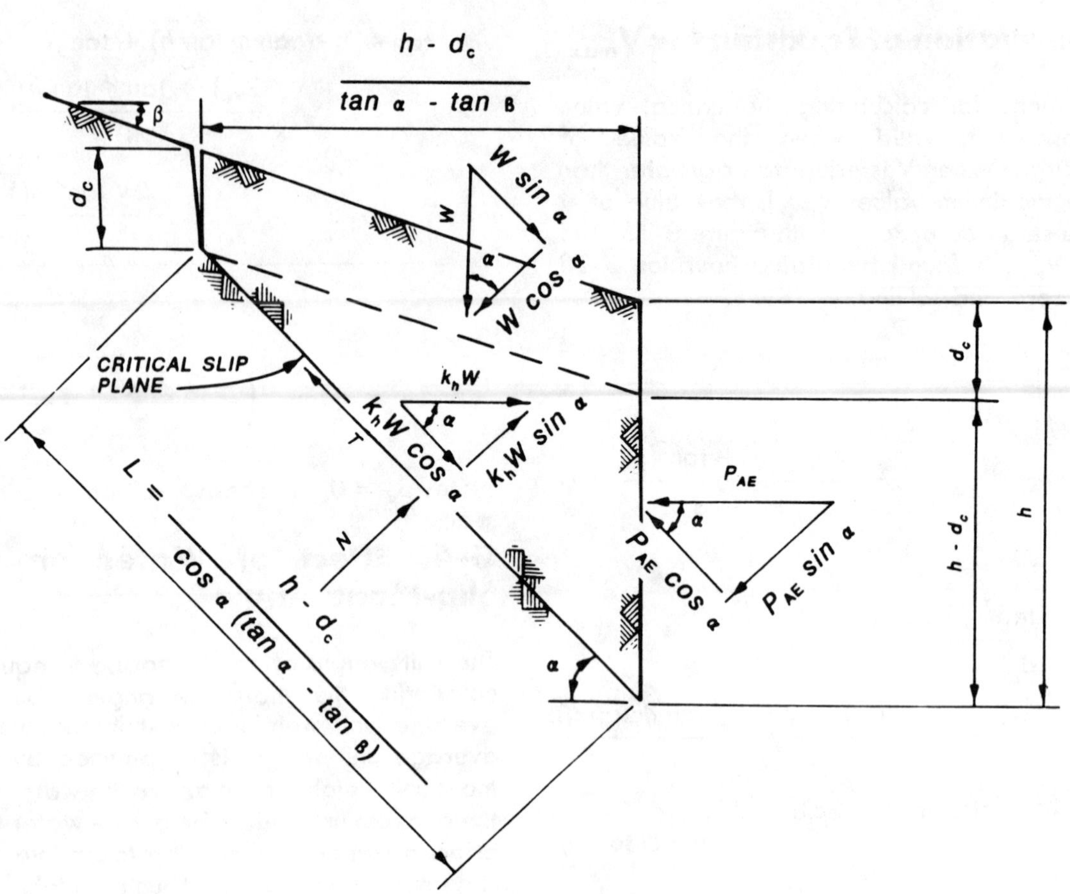

Solving for P_{AE} yields

$$P_{AE} = \frac{W[(1 + k_h \tan\phi)\sin\alpha - (\tan\phi - k_h)\cos\alpha] - cL}{\cos\alpha + \tan\phi \sin\alpha}$$

Referring to the previous figure,

$$W = \frac{\gamma(h^2 - d_c^2)}{2(\tan\alpha - \tan\beta)},$$

$$L = \frac{h - d_c}{\cos\alpha(\tan\alpha - \tan\beta)}$$

Inserting the terms for W and L into the equation for P_{AE} yields

$$P_{AE} = \frac{\gamma(h^2 - d_c^2)[(1 + k_h \tan\phi)\tan\alpha - (\tan\phi - k_h)]}{2(\tan\alpha - \tan\beta)(1 + \tan\phi\tan\alpha)} - \frac{c(h - d_c)}{\cos\alpha(\tan\alpha - \tan\beta)(\cos\alpha + \tan\phi\sin\alpha)}$$

Using the trigonometric relationship $\sec^2\alpha = \frac{1}{\cos^2\alpha}$ yields

$$P_{AE} = \frac{\gamma(h^2 - d_c^2)[(1 + k_h \tan\phi)\tan\alpha - (\tan\phi - k_h)]}{2(\tan\alpha - \tan\beta)(1 + \tan\phi\tan\alpha)} - \frac{2c(h - d_c)\sec^2\alpha}{2(\tan\alpha - \tan\beta)(1 + \tan\phi\tan\alpha)}$$

Now using the trigonometric relationship $\sec^2\alpha = 1 + \tan^2\alpha$ and rearranging yields

$$\frac{2P_{AE}}{\gamma(h^2 - d_c^2)} = \frac{(1 + k_h\tan\phi)\tan\alpha - (\tan\phi - k_h) - \frac{2c}{h + d_c}(1 + \tan^2\alpha)}{\tan\phi\tan^2\alpha + (1 - \tan\phi\tan\beta)\tan\alpha - \tan\beta} = \frac{m}{n}$$

B. FIRST DERIVATIVE OF P_{AE} WITH RESPECT TO α. The first derivative of PAE with respect to α may now be taken and set equal to zero. The resulting expression is solved for the critical value of α. The first derivative is equal to

$$\frac{d\left[\left(\frac{2P_{AE}}{\gamma(h^2-d_c^2)}\right)\right]}{d\alpha} - = \frac{n\frac{dm}{d\alpha} - m\frac{dn}{d\alpha}}{n^2}$$

For this derivative to be equal to zero it is only necessary for $\left(n\frac{dm}{d\alpha} - m\frac{dn}{d\alpha}\right)$ to be equal to zero.

In order to simplify operations let:

$$1 + k_h \tan\phi = a, \quad \tan\phi - k_h = b,$$

and

$$1 - \tan\phi\,\beta = d$$

Then

$$m = a\tan\alpha - b - \left(\frac{2c}{h+d_c}\right)(1+\tan^2\alpha)$$

$$n = \tan\phi\tan^2\alpha + d\tan\alpha - \tan\beta$$

$$\frac{dm}{d\alpha} = \left[a - \left(\frac{4c}{h+d_c}\right)\tan\alpha\right]\sec^2\alpha$$

$$\frac{dn}{d\alpha} = (2\tan\phi\tan\alpha + d)\sec^2\alpha$$

$$n\frac{dm}{d\alpha} = \left[a\tan\phi\tan^2\alpha + ad\tan\alpha - a\tan\beta - \frac{4c}{\gamma(h+d_c)}\tan\phi\tan^3\alpha - \frac{4c}{\gamma(h+d_c)}d\tan^2\alpha + \frac{4c}{\gamma(h+d_c)}\tan\beta\tan\alpha\right]\sec^2\alpha$$

$$m\frac{dn}{d\alpha} = \left[2a\tan\phi\tan^2\alpha - 2b\tan\alpha\tan\alpha - \frac{4c}{\gamma(h+d_c)}\tan\phi\tan\alpha - \frac{4c}{\gamma(h+d_c)}\tan\phi\tan^3\alpha + ad\tan\alpha - bd - \frac{2c}{\gamma(h+d_c)}d - \frac{2c}{\gamma(h+d_c)}d\tan^2\alpha\right]\sec^2\alpha$$

$$\frac{m\frac{dn}{d\alpha} - n\frac{dm}{d\alpha}}{\sec^2\alpha} = \left[-a\tan\phi - \frac{2c}{\gamma(h+d_c)}d\right]\tan^2\alpha + 2b\left[\tan\phi\frac{4c}{\gamma(h+d_c)}(\tan\beta+\phi)\right]\tan\alpha + \left[-a\tan\beta + bd + \frac{2c}{\gamma(h+d_c)}d\right] = 0$$

Substitute $1 + k_h\tan\phi$ for a, $\tan\phi - k_h$ for b, and $1 - \tan\phi\tan\beta$ for d

$$\left[-(1+k_h\tan\phi)\tan\phi - \frac{2c}{\gamma(h+d_c)}(1-\tan\phi\tan\beta)\right]\tan^2\alpha + \left[2(\tan\phi-k_h)\tan\phi + \frac{4c}{\gamma(h+d_c)}(\tan\beta+\tan\phi)\right]\tan\alpha + \left[-(1+k_h\tan\phi)\tan\beta + (\tan\phi-k_h)(1-\tan\phi\tan\beta) + \frac{2c}{\gamma(h+d_c)}(1-\tan\phi\tan\beta)\right] = 0$$

$$\tan^2\alpha - \left[\frac{2\tan\phi(\tan\phi-k_h) + \frac{4c}{\gamma(h+d_c)}(\tan\phi+\tan\beta)}{(1+k_h\tan\phi)\tan\phi + \frac{2c}{\gamma(h+d_c)}(1-\tan\phi\tan\beta)}\right]\tan\alpha$$

$$-\left[\frac{\tan\phi(1-\tan\phi\tan\beta)-(k_h+\tan\beta)+\dfrac{2c}{\gamma(h+d_c)}(1-\tan\phi\tan\beta)}{(1+k_h\tan\phi)\tan\phi+\dfrac{2c}{\gamma(h+d_c)}(1-\tan\phi\tan\beta)}\right]=0$$

then

$$\alpha=\tan^{-1}\left[\frac{c_1+\sqrt{c_1^2+4c_2}}{2}\right] \quad \text{[G-17]}$$

where

$$c_1=\frac{2\tan\phi(\tan\phi-k_h)+\dfrac{4c}{\gamma(h+d_c)}(\tan\phi+\tan\beta)}{(1+k_h\tan\phi)\tan\phi+\dfrac{2c}{\gamma(h+d_c)}(1-\tan\phi\tan\beta)} \quad \text{[G-18]}$$

$$c_2=\frac{\tan\phi(1-\tan\phi\tan\beta)-(k_h+\tan\beta)+\dfrac{2c}{\gamma(h+d_c)}(1-\tan\phi\tan\beta)}{(1+k_h\tan\phi)\tan\phi+\dfrac{2c}{\gamma(h+d_c)}(1-\tan\phi\tan\beta)} \quad \text{[G-19]}$$

$$d_c=\frac{c}{(\sin\alpha\cos\alpha-\tan\phi\cos^2\alpha)\gamma} \quad \text{[G-20]}$$

G-6. Equations for Critical Slip-Plane Angles for Resisting Wedges for Earthquake Condition

The equations for the critical slip-plane angle (α) for a resisting wedge are derived in the same manner used for the driving wedges. The resulting equations are

$$\alpha=\tan^{-1}\left[\frac{-c_1+\sqrt{c_1^2+4c_2}}{2}\right] \quad \text{[G-21]}$$

$$A=(1+k_h\tan\phi)\tan\phi+\frac{2c(1+\tan\phi\tan\beta)}{\gamma h} \quad \text{[G-22]}$$

$$c_1=\frac{2(\tan\phi-k_h)\tan\phi+\dfrac{4c(\tan\phi-\tan\beta)}{\gamma h}}{A} \quad \text{[G-23]}$$

$$c_2=\frac{\tan\phi(1+\tan\phi\tan\beta)+(\tan\beta-k_h)+\dfrac{2c(1+\tan\phi\tan\beta)}{\gamma h}}{A} \quad \text{[G-24]}$$

G-7. Procedure for Finding the Critical Slip-Plane Angle, for a Wedge in a Layer Below the Top Layer, of a Stratified Backfill

A layered soil system is shown in the next figure. In order to find the critical slip-plane angle for layer 2, the weight of soil in layer 1 will be considered a surcharge supported by layer 2. See the following figure on page 175.

The weight of soil in parallelogram ABCD of the last figure is a uniformly distributed surcharge that varies with α_2. The uniformly distributed surcharge does not affect the angle α_2, but does have the effect of increasing the unit weight of soil in wedge 2. The increased soil unit weight is calculated as follows.

γ' = increased unit weight

$$\frac{\gamma' h_2^2}{2(\tan\alpha_2-\tan\beta_2)}=\frac{\gamma_2 h_2^2}{2(\tan\alpha_2-\tan\beta_2)}+\gamma_1 h_2\left[\frac{h_1+\dfrac{h_2(\tan\beta_1-\beta_2)}{\tan\alpha_2-\tan\beta_2}}{\tan\alpha_2-\tan\beta_2}\right]$$

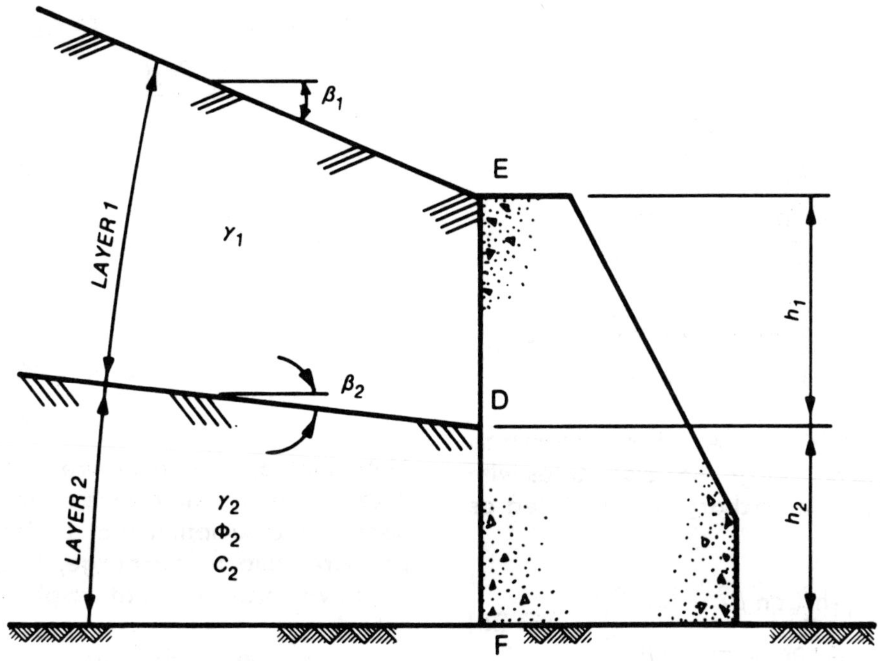
β1
E
γ1
LAYER 1
h1
β2
D
LAYER 2
γ2
Φ2
C2
h2
F

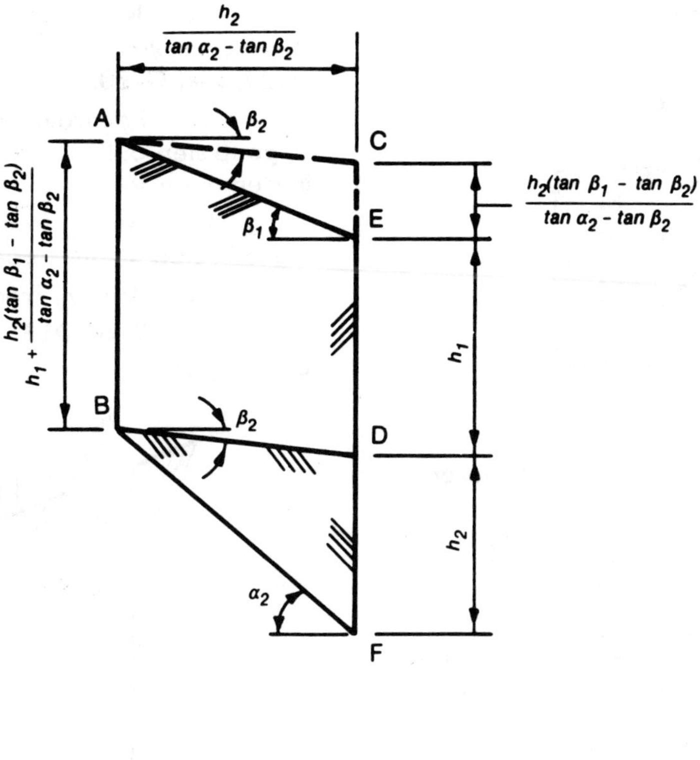
h2 / tan α2 - tan β2
A
β2
C
h2(tan β1 - tan β2) / tan α2 - tan β2
β1
E
h1 + h2(tan β1 - tan β2) / tan α2 - tan β2
h1
B
β2
D
h2
α2
F

$$\frac{\gamma' h_2^2}{2(\tan\alpha - \tan\beta_2} = \frac{2\gamma_1 h_1 h_2 + \gamma_2 h_2^2}{2(\tan\alpha_2 - \tan\beta_2)} + \frac{\gamma_1 h_2^2(\tan\beta_1 - \tan\beta_2)}{(\tan\alpha_2 - \tan\beta_2)^2}$$

$$\gamma' = \frac{2\gamma_1 h_1}{h_2} + \gamma_2 + \frac{2\gamma_1(\tan\beta_1 - \tan\beta_2)}{\tan\alpha_2 - \tan\beta_2} \qquad \text{[G-25]}$$

The weight of soil in triangle ACE is considered to be a negative strip surcharge that also varies with α. The value of this strip surcharge is calculated as follows.

$$V_\alpha = -\frac{\gamma_1 h_2^2(\tan\beta_1 - \tan\beta_2)}{2(\tan\alpha_2 - \tan\beta_2)^2} \qquad \text{[G-26]}$$

Using the terms γ' and V_α, we have a wedge whose critical slip-plane angle may be determined from Equations 3-25, 3-28, 3-29, and 3-30. See the next figure.

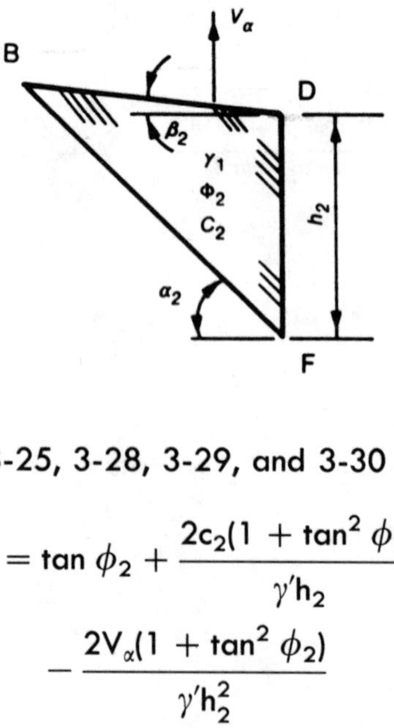

Equations 3-25, 3-28, 3-29, and 3-30 become

$$A' = \tan\phi_2 + \frac{2c_2(1 + \tan^2\phi_2)}{\gamma' h_2} - \frac{2V_\alpha(1 + \tan^2\phi_2)}{\gamma' h_2^2} \qquad \text{[G-27]}$$

$$c_1' = \frac{2\tan^2\phi_2 + \dfrac{4c_2(\tan\phi_2 + \tan\beta_2)}{\gamma' h_2} - \dfrac{4V_\alpha \tan\beta_2(1 + \tan^2\phi_2)}{\gamma' h_2^2}}{A'} \qquad \text{[G-28]}$$

$$c_2' = \frac{\tan\phi_2(1 - \tan\phi_2\tan\beta_2) - \tan\beta_2 + \dfrac{2c_2(1 - \tan\phi_2\tan\beta_2)}{\gamma' h_2} + \dfrac{2V_\alpha \tan^2\beta_2(1 + \tan^2\phi_2)}{\gamma' h_2^2}}{A'} \qquad \text{[G-29]}$$

$$\alpha_2 = \tan^{-1}\left(\frac{c_1' + \sqrt{c_1'^2 + 4c_2'}}{2}\right) \qquad \text{[G-30]}$$

Note that d_c (depth of crack in cohesive soil) is taken as zero in the above equations. This is based upon the assumption that horizontal compressive pressure, due to surcharge, is greater than any negative pressure that might develop due to cohesion.

A. PROCEDURE

1. Choose a trial value for α_2.

2. Using the trial value of α_2, calculate γ' (Equation G-25) and (Equation G-26) V_α.

3. Using the values of γ' and V_α from step 2, calculate α_2 from Equations G-27, G-28, G-29, and G-30.

4. If α_2 from step 3 is equal to α_2 in step 1, go to step 6. If α_2 from step 3 is not equal to α_2 in step 1 go to step 5.

5. Use α_2 from step 3 as a new trial value in step 1, and repeat the procedure.

6. Stop. α_2 from step 3 is critical value.

B. EXAMPLE. Find α_2 critical to nearest 0.5 degree.

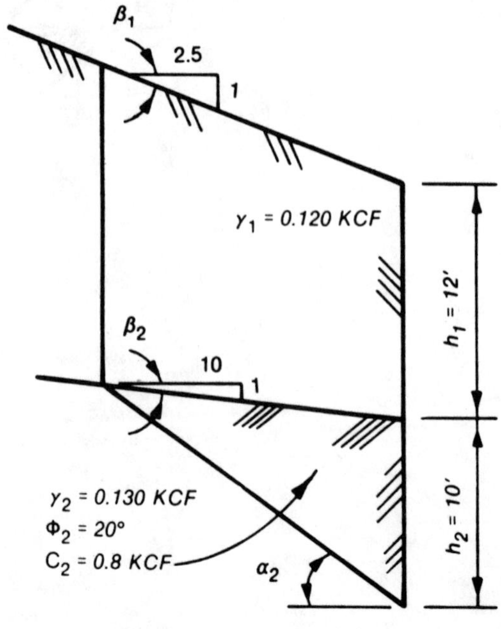

From Equation G-25,

$$\gamma' = \frac{2(0.12)(12)}{10} + 0.13 + \frac{2(0.12)(0.4-0.1)}{\tan\alpha_2 - 0.1}$$

$$\gamma' = 0.418 + \frac{0.072}{\tan\alpha_2 - 0.1}$$

From Equation G-26

$$V_\alpha = -\frac{0.12(10)^2(0.4-0.1)}{2(\tan\alpha_2 - 0.1)^2} = -\frac{1.80}{(\tan\alpha_2 - 0.1)^2}$$

1. α_2 for Long-Term Loading ($\phi_2 = 20°$, $c_2 = 0$).

a. First Trial: $\alpha_2 = 45°$, $\gamma' = 0.4980$, $V_\alpha = -2.2222$

Using Equations G-27, G-28, G-29, and G-30 yields

$$A' = 0.36398 - \frac{2(-2.2222)(1.132474)}{0.4980(10)^2}$$

$$= 0.465038$$

$$c_1' = \frac{0.26498 - \dfrac{4(-2.2222)(0.1)(1.132474)}{0.4980(10)^2}}{0.465038}$$

$$= 0.613200$$

$$c_2' = \frac{0.36397(1 - 0.36397 \times 0.1) - 0.1 + \dfrac{2(-2.2222)(0.01)(1.132474)}{0.4980(10)^2}}{0.465038}$$

$c_2' = 0.536971$, $\alpha_2 = 47.75° \approx 48° \neq 45°$

b. Second Trial: $\alpha_2 = 48.0°$, $\gamma' = 0.4892$, $V_\alpha = -1.7624$

$$A' = 0.3697 - \frac{0.022649(-1.7624)}{0.4892} = 0.445566$$

$$c_1' = \frac{0.264948 - \dfrac{0.004530(-1.7684)}{0.4892}}{0.445566}$$

$$= 0.631260$$

$$c_2' = \frac{0.250723 + \dfrac{0.0002265(-1.7624)}{0.4892}}{0.445566}$$

$$= 0.560875$$

$$\alpha_2 = 48.45° \approx 48.5° \neq 48°$$

c. Third Trial. $\alpha_2 = 48.5°$, $\gamma' = 0.4879$, $V_\alpha = 1.6957$

$$A' = 0.36397 - \frac{0.022649(-1.6957)}{0.4879}$$

$$= 0.442687$$

$$c_1' = \frac{0.264948 - \dfrac{0.004530(-1.6957)}{0.4879}}{0.443687}$$

$$= 0.634064$$

$$c_2'^2 = \frac{0.250723 + \dfrac{0.0002265(-1.6957)}{0.4879}}{0.443687}$$

$$= 0.564588$$

$$\alpha_2 = 48.56° \approx 48.5°$$

$\underline{\underline{\alpha_2 = 48.5°}}$

2. α_2 for Short-Term Loading ($\phi_2 = 0$, $c_2 = 0.8$ ksf).

a. First Trial. $\alpha_2 = 45°$, $\gamma' = 0.4980$, $V_\alpha = -2.2222$

$$A' = \frac{2(0.8)}{0.498(10)} - \frac{2(-2.2222)}{0.498(10)^2} = 0.410530$$

$$c_1' = 2\tan\beta_2 = 0.2$$

$$c_2' = \frac{-0.1 + \dfrac{2(0.8)}{0.498(10)} + \dfrac{2(-2.2222)(0.01)}{0.498(10)^2}}{0.410530}$$

$$= 0.537936$$

$$\alpha_2 = 40.04° \approx 40°$$

b. Second Trial. $\alpha_2 = 40°$, $\gamma' = 0.5154$, $V_\alpha = -3.2951$

$$A' = \frac{0.16}{0.5154} = \frac{0.02(-3.7218)}{0.5154} = 0.438304$$

$$c_1' = 0.2,$$

$$c_2' = \frac{-0.1 + \dfrac{0.16}{0.5154} + \dfrac{0.0002(-3.2951)}{0.5154}}{0.438304}$$

$$= 0.477203$$

$$\alpha_2 = 38.59° \approx 38.5° \neq 40°$$

c. Third Trial. $\alpha_2 = 38.5°$, $\gamma' = 0.5215$, $V_\alpha = -3.7218$

$$A' = \frac{0.16 - 0.02(-3.7218)}{0.5215} = 0.448542$$

$$c_1' = 0.2,$$

$$c_2' = \frac{-0.1 + \dfrac{0.16 + 0.0002(-3.7218)}{0.5215}}{0.449542}$$

$$= 0.456865$$

$$\alpha_2 = 38.07° \approx 38° \neq 38.5°$$

d. Fourth Trial. $\alpha_2 = 38°$, $\gamma' = 0.5237$, $V_\alpha = -3.8781$

$$A' = \frac{0.16 - 0.02(-3.8781)}{0.5237} = 0.453622$$

$$c_1' = 0.2,$$

$$c_2' = \frac{-0.1 + \dfrac{0.16 + 0.0002(-3.8781)}{0.5237}}{0.453622}$$

$$= 0.449796$$

$$\alpha_2 = 37.89° \approx 38°$$

$$\underline{\underline{\alpha_2 = 38°}}$$

3. Check $\alpha_3 = 48.5°$ (for Long-Term Loading) by Using Equation 3-23.

$$W = \frac{0.13(10)^2}{2(\tan\alpha_2 - 0.1)} + \frac{0.12(12)(10)}{\tan\alpha_2 - 0.1} + \frac{0.12(10)^2(\tan\beta_1 - \tan\beta_2)}{2(\tan\alpha_2 - 0.1)^2}$$

$$W = \frac{20.9}{\tan\alpha_2 - 0.1} + \frac{1.8}{(\tan\alpha_2 - 0.1)^2}$$

$$P_{EE} = \frac{W(\tan\alpha_2 - \tan\phi_2)}{1 + \tan\phi_2 \tan\alpha_2}$$

$\alpha_2 = 47.5°$: $W = 22.9149$ kips

$$P_{EE} = \frac{22.9149(1.091309 - 0.36397)}{1 + 0.36397(1.091309)}$$

$$= 11.9288 \text{ kips}$$

$\alpha_2 = 48.5°$: $W = 21.9812$ kips

$$P_{EE} = \frac{21.9812(1.130294 - 0.36397)}{1 + 0.36397(1.130294)}$$

$$= 11.9348 \text{ kips} > 11.9288 \text{ kips}$$

$\alpha_2 = 49.5°$: $W = 21.0869$ kips

$$P_{EE} = \frac{21.0869(1.170850 - 0.36397)}{1 + 0.36397(1.170850)}$$

$$= 11.9304 \text{ kips} < 11.9348$$

The critical α_2 equals 48.5°, which checks with the value computed in paragraph G-7b(1)(c).

4. Check $\alpha_2 = 38°$ (for Short-Term Loading) by Using equation 3-23.

$$W = \frac{20.9}{\tan\alpha_2 - 0.1} + \frac{1.8}{(\tan\alpha_2 - 0.1)^2},$$

$$L = \frac{10}{\cos\alpha_2\,(\tan\alpha_2 - 0.1)}$$

$$P_{EE} = W\tan\alpha_2 - c2L/\cos\alpha$$

$\alpha_2 = 37°$: $W = 36.1931$ kips, $L = 19.1589$ ft

$$P_{EE} = 36.1931(0.753554) - \frac{0.8(19.1589)}{0.798636}$$

$$= 8.0818 \text{ kips}$$

$\alpha_2 = 38°$: W = 34.5554 kips, L = 18.6268 ft

$$P_{EE} = 34.5554(0.781286) - \frac{0.8(18.6268)}{0.788011}$$

$$= 8.0875 \text{ kips} > 0.0818$$

$\alpha_2 = 39°$: W = 33.0185 kips, L = 18.1289 ft

$$P_{EE} = 33.0185(0.809784) - \frac{0.8(18.1289)}{0.777146}$$

$$= 8.0758 \text{ kips} < 8.0875$$

The critical α_2 equals 38°, which checks with the value computed in paragraph G-7b(2)(d).

APPENDIX H

DERIVATION OF PRESSURE COEFFICIENTS FOR SOLUTION OF LATERAL EARTH PRESSURE PROBLEMS

H-1. Introduction

A. Engineers are familiar with earth pressure coefficients and their use in determining pressures and forces acting on retaining walls. The most familiar and most often used are:

$$K_A = \tan^2\left(45° - \frac{\phi}{2}\right), \text{ for driving pressure}$$

$$K_P = \tan^2\left(45° + \frac{\phi}{2}\right), \text{ for resisting pressure}$$

The above represent special cases of Coulomb's equations for earth pressure coefficients.

B. These simple coefficients are proper to use only when

1. Backfill surface is a level plane.

2. Any surcharge applied to the surface is uniform and covers the entire surface of the backfill wedge.

C. The general equations from which these simple coefficients are obtained are subject to the following limitations:

1. Backfill must be cohesionless, unless the top surface is horizontal.

2. Fill must be completely saturated or completely unsaturated, unless the top surface is horizontal.

3. Top can be a constant slope but must be an unbroken plane.

4. Any surcharge must be uniform and cover the entire surface of the backfill wedge.

D. The correct lateral earth force due to any backfill wedge may be obtained from the general wedge equation, and the general wedge equation is subject to none of the above limitations. This equation may be used to solve the most complicated problems of wedge geometry and surface loading.

E. If lateral earth pressure coefficients are derived from the general wedge equation, these coefficients may be used in a rather simple manner to solve complex earth pressure problems. The general wedge equation will now be derived and the pressure coefficients obtained from the derived equation.

H-2. Driving-Side Earth Pressure

A. A typical driving wedge is shown in the following figure.

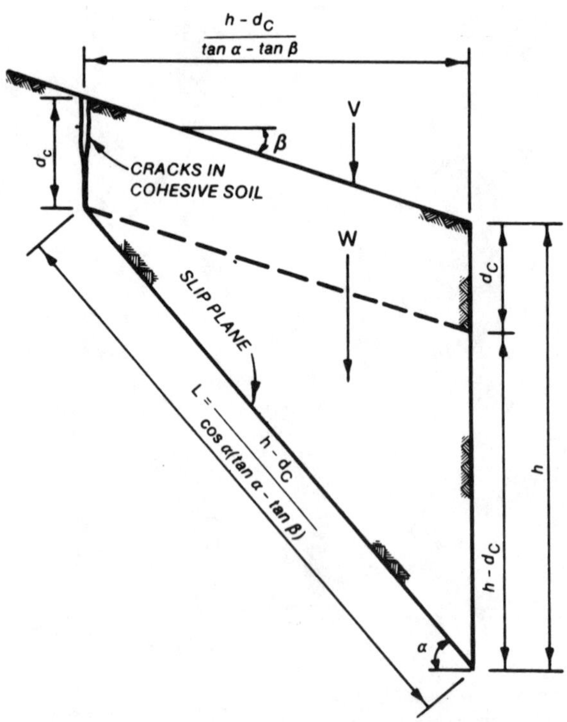

W = WEIGHT OF SOIL IN WEDGE,

V = ANY SURFACE LOAD OTHER THAN A UNIFORM SURCHARGE

In the derivation on the following pages it will be assumed that shear on the vertical face of the wedge is zero.

B. In the next figure, the forces acting on the wedge are shown along with their components parallel and normal to the slip plane.

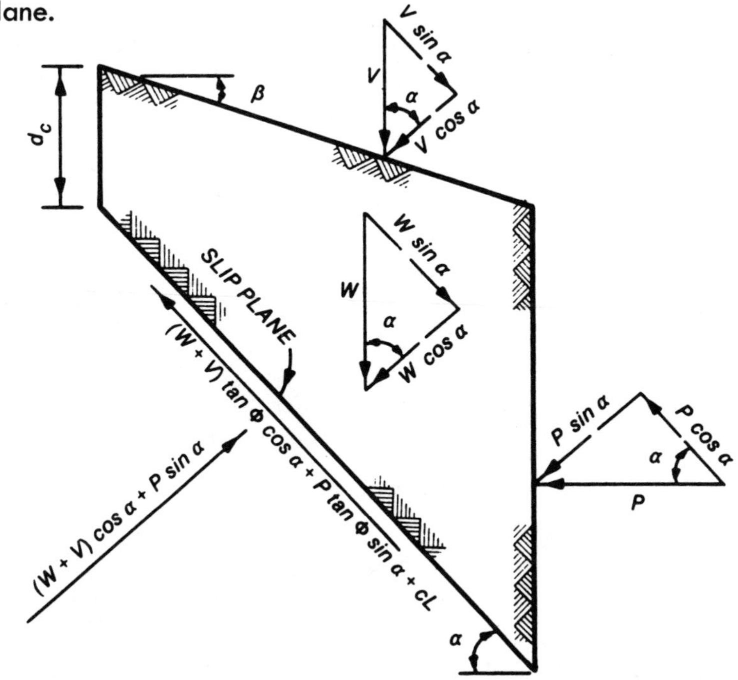

Setting the summation of forces parallel to the slip plane equal to zero

$$P(\cos\alpha + \tan\phi \sin\alpha) + (W + V)(\tan\phi\cos\alpha - \sin\alpha) + cL = 0$$

Rearranging and solving for P

$$P(\cos\alpha + \tan\phi \sin\alpha) + (W + V)(\tan\phi\cos\alpha - \sin\alpha) + cL = 0$$

From the previous figure

$$L = \frac{h - d_c}{\cos\alpha\,(\tan\alpha \tan\beta)} \quad \text{and} \quad W = \frac{\gamma(h^2 - d_c^2)}{2(\tan\alpha - \tan\beta)}$$

where

γ = unit weight of wedge material

c = cohesion

The equation for P will now be rewritten with the individual terms for W, V, and c separated.

$$P = \frac{\gamma(h^2 - d_c^2)}{2} \cdot \frac{\sin\alpha - \tan\phi\cos\alpha}{\cos\alpha + \tan\phi\sin\alpha} \cdot \frac{1}{\tan\alpha - \tan\beta} + \frac{V(\sin\alpha - \tan\phi\cos\alpha)}{\cos\alpha + \tan\phi\sin\alpha} - \frac{c(h - d_c)}{\cos\alpha\,(\tan\alpha - \tan\beta)(\cos\alpha + \tan\phi\sin\alpha)}$$

This may be simplified and rewritten as

$$P = \frac{\gamma(h^2 - d_c^2)}{2} \cdot \frac{1 - \tan\phi\cot\alpha}{1 + \tan\phi\tan\alpha} \cdot \frac{\tan\alpha}{\tan\alpha - \tan\beta} + \frac{V(1 - \tan\phi\cot\alpha)\tan\alpha}{1 + \tan\phi\tan\alpha} - \frac{2c(h - d_c)\tan\alpha}{2\sin\alpha\cos\alpha\,(1 + \tan\phi\tan\alpha)(\tan\alpha - \tan\beta)} \quad \text{[H-1]}$$

C. Everything in the first term except $\gamma(h^2 - d_c^2)/2$ is the lateral earth coefficient for γ and will be called K_1. Everything in the second term except V is the lateral coefficient for the strip surcharge and will be called K_v. Everything in the third term except $2c(h - d_c)$ is the lateral coefficient for cohesion and will be called K_c.

D. Note that K_1 and K_v both contain the term $\left(\frac{1 - \tan\phi \cot\alpha}{1 + \tan\phi \tan\alpha}\right)$. This is the lateral coefficient for γ when the top surface is a level unbroken plane. This term will be called K. Therefore

$$K_1 = K\left(\frac{\tan\alpha}{\tan\alpha - \tan\beta}\right)$$

$$K_v = K \tan\alpha$$

E. Note also that K_1 and K_c both contain this term:

$$\frac{\tan\alpha}{\tan\alpha - \tan\beta}$$

G. The individual forces and pressure distributions are shown in the next figure.

H. The lateral coefficient (K_b) to apply to the effective weight of fill (γ_b) below the saturation level in a partially saturated wedge will now be determined. If the water table rises in a wedge that was previously unsaturated, the lateral earth force will be reduced by the same amount on both sides of the wedge. The slope of the earth pressure diagram, acting on the vertical projection of the wedge slip plane, is $K\gamma$ above the saturation level and $K\gamma_b$ below the saturation level. The slope of the earth pressure diagram, acting on the vertical face of the wedge, is $K_1\gamma$ above the saturation level and $K_b\gamma b$ below the saturation

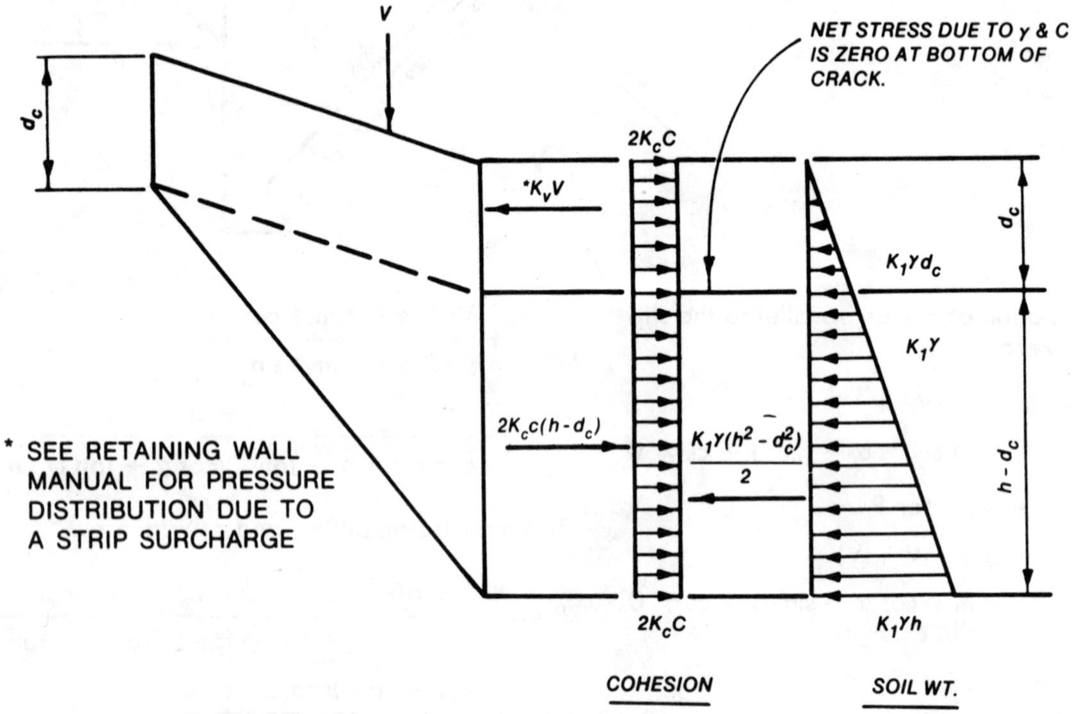

This term modifies K_1 and K_c for the effect of surface slope. It becomes unity when $\beta = 0$.

F. The equation for the total lateral force (P) produced by a driving wedge can now be written, where the wedge material possesses both cohesion and internal friction, where the top surface is a sloping plane, where the top surface supports a strip surcharge, and where the wedge is either completely saturated or completely unsaturated. The equation is

$$P = \frac{K_1\gamma(h^2 - d_c^2)}{2} + K_vV - 2K_cc(h - d_c) \qquad \text{[H-2]}$$

level. See the next figure. The force reduction on each side of the wedge (−P) is represented by the shaded parts of the pressure diagrams.

$$\frac{K(\gamma - \gamma_b)h_s^2}{2} = \frac{(K_1\gamma - K_b\gamma_b)h_s^2}{2},$$

$$K_b\gamma_b = K_1\gamma - K\gamma + K\gamma_b$$

$$K_b = (K_1 - K)\left(\frac{\gamma}{\gamma_b}\right) + K$$

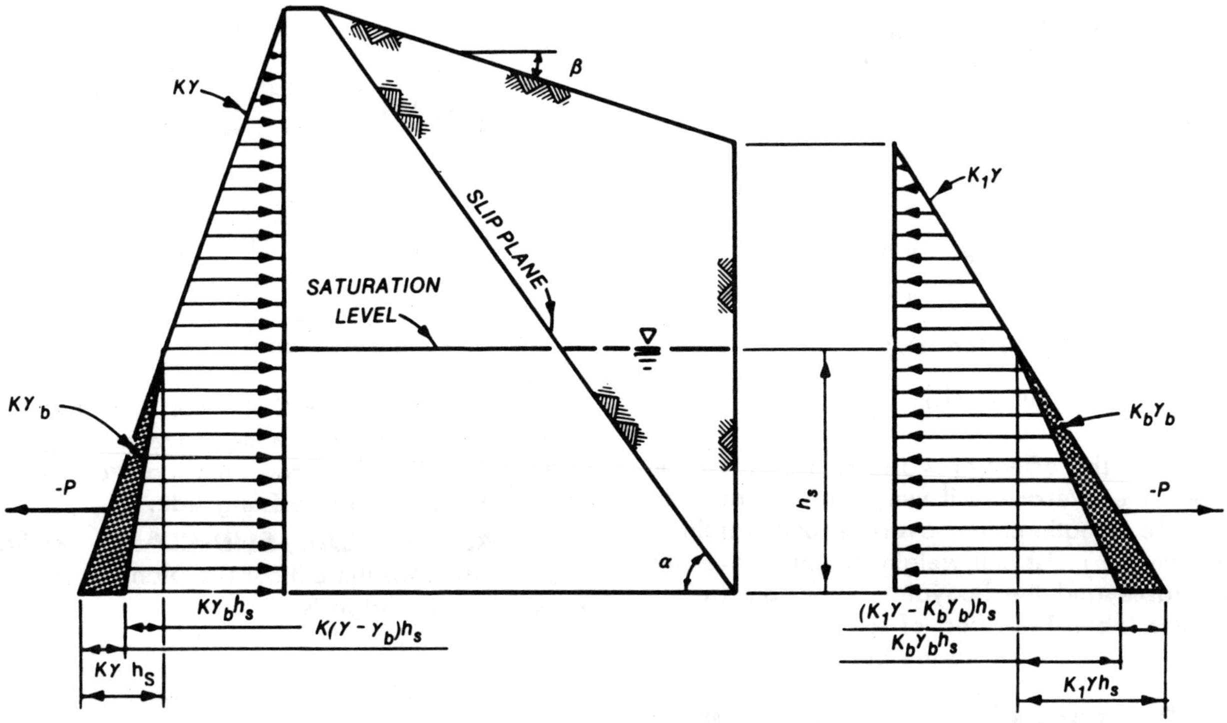

$$K_b = K\left[1 + \left(\frac{\tan\alpha}{\tan\alpha - \tan\beta} - 1\right)\left(\frac{\gamma}{\gamma_b}\right)\right]$$

I. In summary, the coefficients for driving wedges are

$$K = \frac{1 - \tan\phi\cot\alpha}{1 + \tan\phi\tan\alpha}, \text{ basic value}$$

$$K_1 = K\left(\frac{\tan\alpha}{\tan\alpha - \tan\beta}\right),$$

apply to moist fill above saturation level

$$K_b = K\left[1 + \left(\frac{\tan\alpha}{\tan\alpha - \tan\beta} - 1\right)\left(\frac{\gamma}{\gamma_b}\right)\right],$$

apply to buoyant weight of fill below the saturation level

$K_v = K\tan\alpha$, multiply times V to obtain lateral force due to strip surcharge

$$K_c = \frac{1}{2\sin\alpha\cos\alpha\,(1 + \tan\phi\tan\alpha)} \cdot \frac{\tan\alpha}{\tan\alpha - \tan\beta},$$

multiply times 2c to obtain negative pressure due to cohesion

J. A problem remains that must be solved before the coefficients can be used for calculating driving pressures and forces, and that problem is the value of the critical slip-plane angle. The critical angle is the angle which produces the maximum driving force. The equations for calculating this angle are derived in Appendix G and presented in Chapter 3 of the manual and are as follows:

Equation 3-25:

$$\alpha = \tan^{-1}\left(\frac{c_1 + \sqrt{c_1^2 + 4c_2}}{2}\right)$$

where

Equation 3-28:

$$c_1 = \left[2\tan^2\phi_d + \frac{4c_d(\tan\phi_d + \tan\beta)}{\gamma(h + d_c)} - \frac{4V\tan\beta(1 + \tan^2\phi_d)}{\gamma(h^2 - d_c^2)}\right] \div A$$

Equation 3-29:

$$c_2 = \left[\tan\phi_d\,(1 - \tan\phi_d\tan\beta) - \tan\beta + \frac{2c_d(1 - \tan\phi_d\tan\beta)}{\gamma(h + d_c)}\right.$$

$$+ \frac{2V \tan^2 \beta(1 + \tan^2 \phi_d)}{\gamma(h^2 - d_c^2)}\Bigg] \div A$$

and

Equation 3-30:

$$A = \tan \phi_d + \frac{2c_d(1 - \tan \phi_d \tan \beta)}{\gamma(h + d_c)} - \frac{2V(1 + \tan^2 \phi_d)}{\gamma(h^2 - d_c^2)}$$

K. The effect of water is accounted for by using the average unit weight of soil for γ in the above equations. The average unit weight is based on the moist unit weight of soil above the water table and the buoyant unit weight below the water table. The effect of seepage should be considered in determining the buoyant weight.

H-3. Resisting-Side Earth Pressure

A. PRESSURE COEFFICIENTS. The pressure coefficients for the resisting side are

$$K = \frac{1 + \tan \phi \cot \alpha}{1 - \tan \phi \tan \alpha}, K_1 = K\left(\frac{\tan \alpha}{\tan \alpha - \tan \beta}\right)$$

$$K_v = K \tan \alpha$$

$$K_b = K\left[1 + \left(\frac{\tan \alpha}{\tan \alpha - \tan \beta} - 1\right)\left(\frac{\gamma}{\gamma_b}\right)\right]$$

$$K_c = \frac{1}{2 \sin \alpha \cos \alpha\, (1 - \tan \phi \tan \alpha)} \cdot \frac{\tan \alpha}{\tan \alpha - \tan \beta}$$

The term for cohesion ($2K_c c$) is positive for resisting pressure calculations, not negative as it was for driving pressure. Therefore, the crack of depth d_c does not exist on the resisting side.

B. CRITICAL SLIP-PLANE ANGLE. The equation for the critical slip-plane angle for a resisting side wedge is

$$\alpha = \tan^{-1}\left(\frac{-c_1 + \sqrt{c_1^2 + 4c_2}}{2}\right)$$

Values for c_1 and c_2 are determined from the equations in Appendix G.

C. With the above changes, the method for determining resisting pressures and forces is the same as for the driving case.

APPENDIX I

DERIVATION OF EQUATION FOR DEPTH OF CRACK d_c IN COHESIVE SOIL

I-1. Definition

By definition, the bottom of the crack will lie at the depth where the net horizontal stress in the soil is zero.

$$P_h = K_1 \gamma d_c - 2K_c c = 0 \qquad \text{[I-1]}$$

where

$$K_1 = \frac{1 - \tan\phi \cot\alpha}{1 + \tan\phi \tan\alpha} \cdot \frac{\tan\alpha}{\tan\alpha - \tan\beta)}$$

$$K_c = \frac{\tan\alpha}{2 \sin\alpha \cos\alpha\,(1 + \tan\phi \tan\alpha)(\tan\alpha - \tan\beta)}$$

The coefficients K_1 and K_c are derived in Appendix H. Rearranging Equation I-1 and solving for d_c yields

$$d_c = \frac{2K_c c}{K_1 \gamma} \qquad \text{[I-2]}$$

Equation I-2 is valid for a sloping soil layer.

I-2. Horizontal Top Surface

When the top surface is horizontal,

$$K_1 = K_A \text{ (Coulomb)}$$

$$K_c = \sqrt{K_A}$$

and

$$d_c = \frac{2c}{\sqrt{K_A}\gamma}$$

APPENDIX J

LATERAL PRESSURES DUE TO COMPACTION

J-1. DESIGN PRESSURE ENVELOPE

The design pressure envelope for nonyielding walls with compaction effects will be derived. The lateral pressure due to at-rest conditions is shown in Figure J-1. The lateral pressure induced by a compaction roller line load is shown in Figure J-2.

In Figure J-2,

$$\Delta P'_v = \frac{2P}{\pi z}$$

where P is the roller load, lb/linear ft

$$\Delta P'_h = \frac{2K_oP}{\pi z}$$

where P is the roller load, lb/linear ft

The maximum lateral pressure occurs at z_{cr} and the passive pressure is

$$P'_p = K_p\gamma'z \qquad [J\text{-}1]$$

Taking

$$K_p = \frac{1}{K_A}$$

and inserting this into Equation J-1 yields

$$P'_p = \frac{\gamma'z}{K_A} \qquad [J\text{-}2]$$

The horizontal pressure due to the earth and roller is

$$P'_h = K_o\gamma'z + \frac{2K_oP}{\pi z} \qquad [J\text{-}3]$$

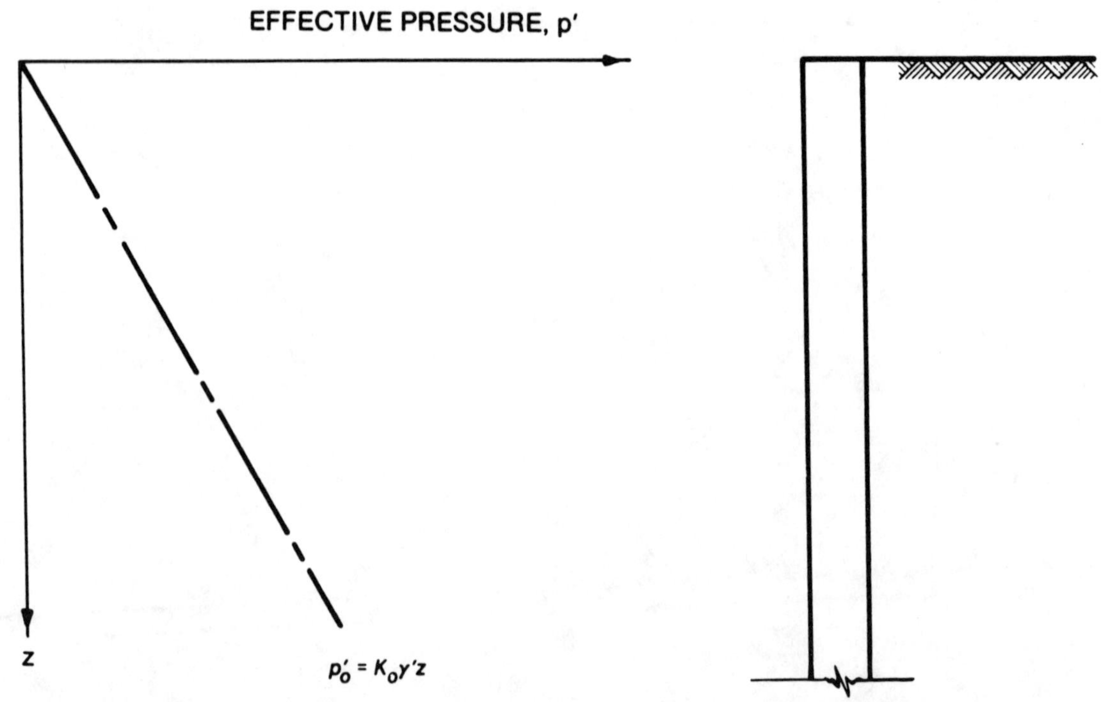

Fig J-1. Lateral pressure due to at-rest conditions ($\gamma' = \gamma$ above water table)

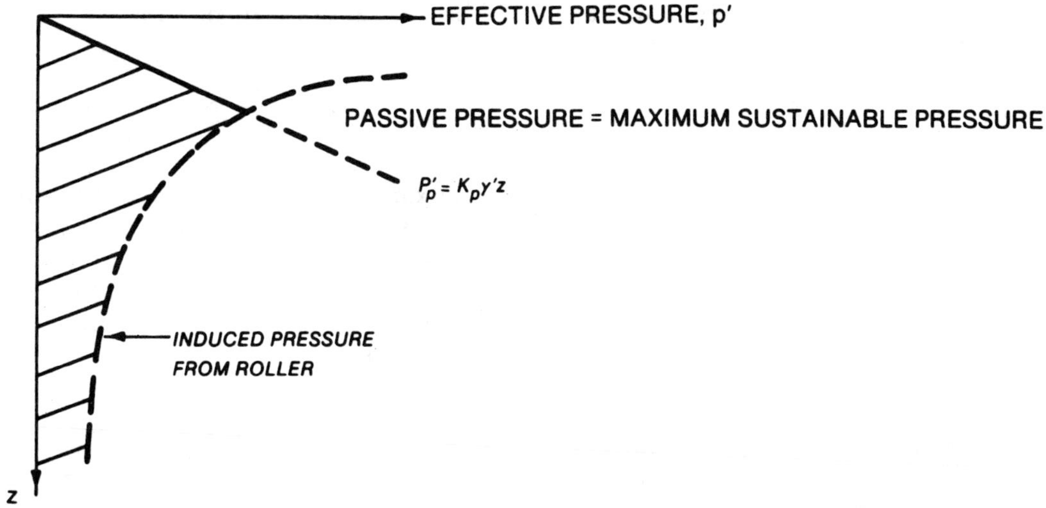

Fig J-2. Lateral pressure induced by compaction roller line load

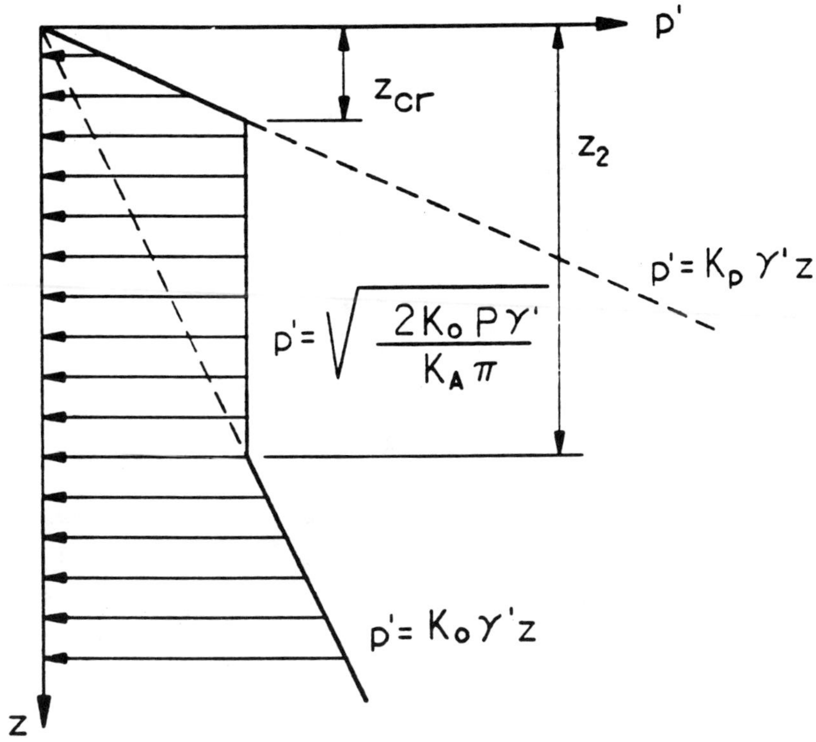

$$z_{cr} = \sqrt{\frac{2K_A K_o P}{\pi \gamma}} \qquad z_2 = \sqrt{\frac{2P}{K_A K_o \pi \gamma'}}$$

Fig J-3. Design pressure envelope for nonyielding walls with compaction effects

Using Equations J-2 and J-3 and solving for the critical depth z_{cr} yields

$$\frac{\gamma' z_{cr}}{K_A} = K_o \gamma' z_{cr} + \frac{2K_o P}{\pi z_{cr}}$$

$$\gamma'^2_{cr} = K_A K_o \gamma' z^2_{cr} + \frac{2K_A K_o P}{\pi}$$

$$\gamma' z^2_{cr} (1 - K_A K_o) = \frac{2K_A K_o P}{\pi}$$

Assuming that $(1 - K_A K_o) \approx 1$

$$z_{cr} = \sqrt{\frac{2K_A K_o P}{\pi \gamma'}}$$

The horizontal pressure at z_{cr} is

$$P'_h = \frac{\gamma' z_{cr}}{K_A} = \sqrt{\frac{2K_o P \gamma'}{K_A \pi}}$$

The maximum pressure is constant below z_{cr} until it is exceeded by the at-rest pressure because the foregoing analysis represents each successive top lift. The design pressure envelope for nonyielding walls including the effects of compaction is shown in Figure J-3.

APPENDIX K

DERIVATION OF VERTICAL SHEAR FORCE FOR UPWARD-SLOPING BACKFILL

K-1. Derivation for Shear Required on Vertical Faces of Earth Wedges

When the vertical face of an earth wedge does not lie on the same plane as the face of the structural wedge, a shear force is required in order that the horizontal earth force may be transferred between wedges. A general situation where such a force is required is shown in Figure K-1.

The earth force calculated for wedge 1, using the provisions of the text, is a force acting on Surface AB. In order for the same force to act on Surface CD, a shear force must exist on vertical planes in Block ABCD. A free body of Block ABCD is shown in Figure K-2.

Taking moments about D

$$P_{AB}\left[\left(\frac{h_{AB}}{3} + a \tan \beta_2\right) - \frac{h_{CD}}{3}\right] - va = 0$$

where

$$h_{CD} = h_{AB} - a(\tan \beta_1 - \tan \beta_2)$$

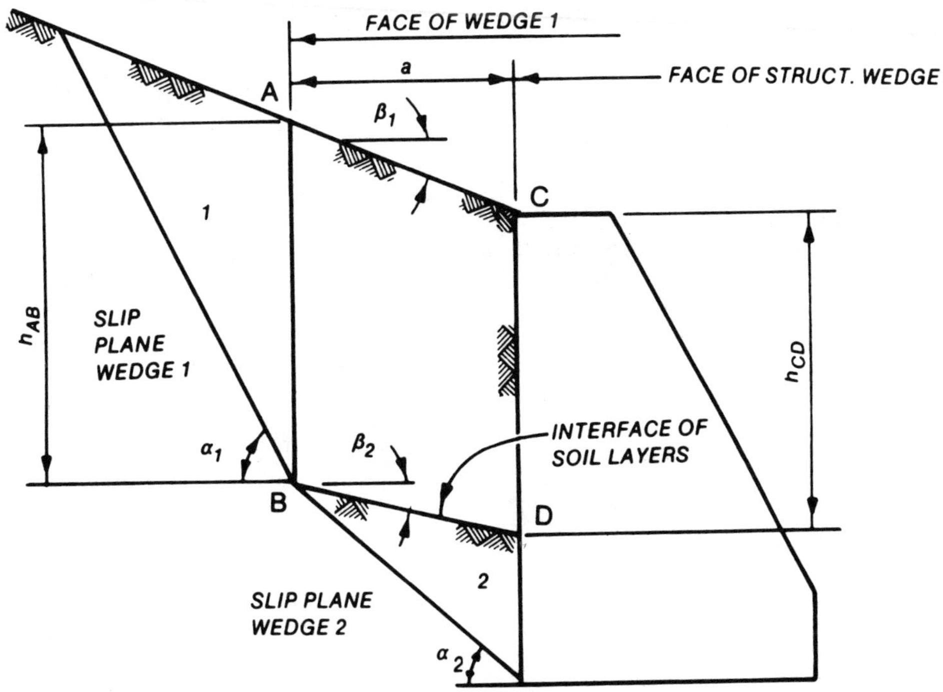

Fig K-1. General condition where shear force is required

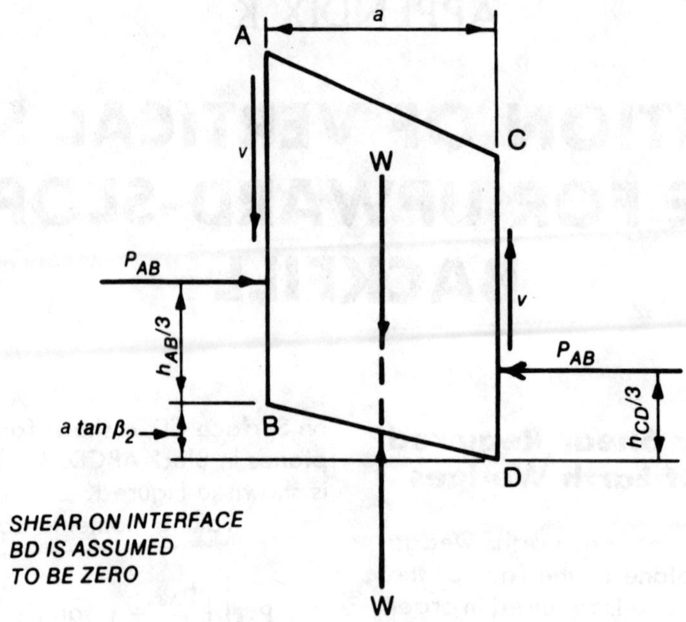

Fig K-2. Free body of Block ABCD

then

$$P_{AB}\left(\frac{h_{AB}}{3} + a \tan \beta_2 - \frac{h_{AB} - a \tan \beta_1 + a \tan \beta_2}{3}\right) = va$$

$$P_{AB}\left(a \tan \beta_2 + \frac{a \tan \beta_1}{3} - \frac{a \tan \beta_2}{3}\right) = va$$

$$v = \frac{P_{AB}(\tan \beta_1 + 2 \tan \beta_2)}{3}$$

See Example 6 in Appendix M and Example 1 in Appendix N for applications of the above equation.

APPENDIX L

DERIVATION OF GENERAL WEDGE EQUATION FOR MULTIPLE WEDGE ANALYSIS

L-1. Free-Body Diagram

A free-body diagram of an ith wedge is shown below

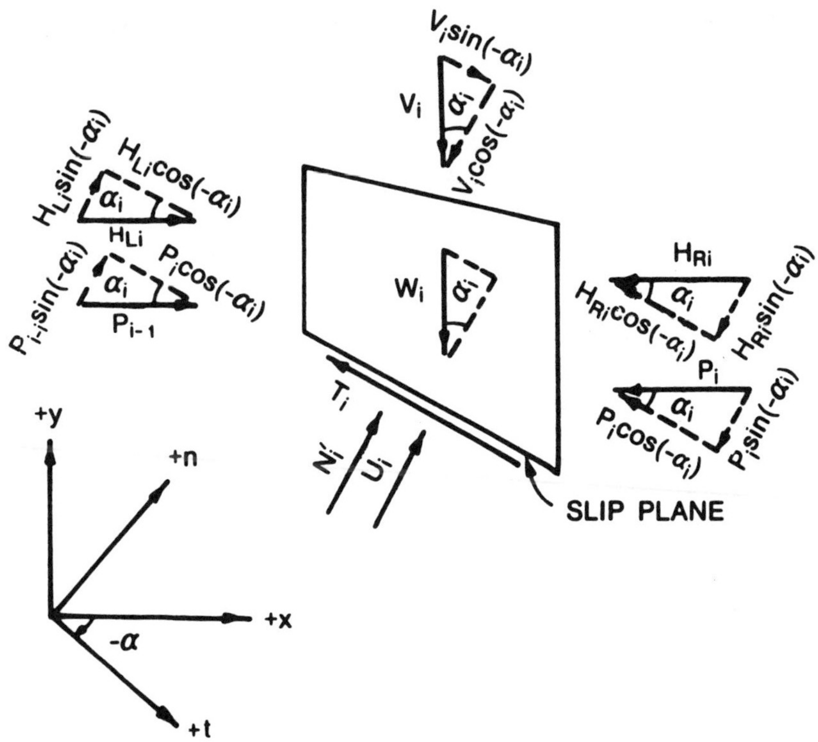

Writing equilibrium equations normal and parallel to slip plane yields,

$$\sum F_n = 0$$

$$0 = N_i + U_i - W\cos(-\alpha_i) - V_i\cos(-\alpha_i) + H_{Li}\sin(-\alpha_i) - H_{Ri}\sin(-\alpha_i) + P_{i-1}\sin(-\alpha_i) - P_i\sin(-\alpha_i)$$

Using the trigonometric relationships,

$$\cos(-\alpha_i) = \cos\alpha$$

and

$$\sin(-\alpha) = -\sin\alpha$$

yields

$$N_i = (W_i + V_i)\cos(\alpha_i) - (H_{Li} - H_{Ri})(-\sin\alpha_i - (P_{i-1} - P_i)(-\sin\alpha_i) - U_i$$

$$\sum F_t = 0:$$

$$0 = -T_i + W_i \sin(-\alpha_i) + V_i \sin(-\alpha_i) + H_{Li} \cos(-\alpha_i) - H_{Ri} \cos(-\alpha_i) + P_{i-1} \cos(-\alpha_i) - P_i \cos(-\alpha_i)$$

$$T_i = (W_i + V_i)(-\sin \alpha_i) + (H_{Li} - H_{Ri})\cos \alpha_i + (Pi-1 - P_i)\cos \alpha_i$$

According to the Mohr-Coulomb failure criterion,

$$T_F = N_i \tan \phi_i + c_i L_i$$

The equation for the sliding factor of safety is

$$FS_i = \frac{T_F}{T_i} = \frac{N_i \tan \phi_i + c_i L_i}{T_i}$$

where FS_i = factor of safety.

Substituting the expressions for T_i and N_i into the equation for FSi yields

$$FS_i = \frac{[(W_i + V_i)\cos \alpha_i + (H_{Li} - H_{Ri})\sin \alpha_i + (P_{i-1} - P_i)\sin \alpha_i - U_i]\tan \phi_i + c_i L_i}{-(W_i - V_i)\sin \alpha_i + (H_{Li} - H_{Ri})\cos \alpha_i + (P_{i-1} - P_i)\cos \alpha_i}$$

Solving for (Pi-1 - Pi),

$$(P_{i-1} - P_i) = [(W_i + V_i)(\tan \phi_{di} \cos \alpha_i + \sin \alpha_i) - U_i \tan \phi_{di} + (H_{Li} - H_{Ri}) \times (\tan \phi_{di} \sin \alpha_i - \cos \alpha_i) + c_{di} L_i] \div [\cos \alpha_i - \tan \phi_{di} \sin \alpha_i] \quad [L\text{-}1]$$

Equation L-1 is the general wedge equation.

APPENDIX M

LATERAL EARTH PRESSURE COMPUTATIONS, EXAMPLES

Summary of Examples

Example 1, page 195

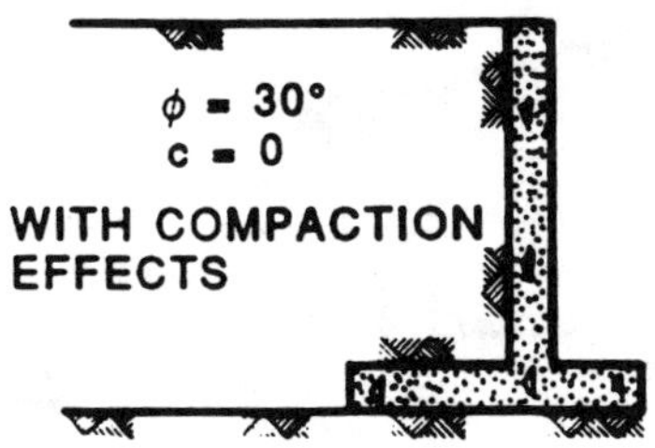

Example 4, page 200

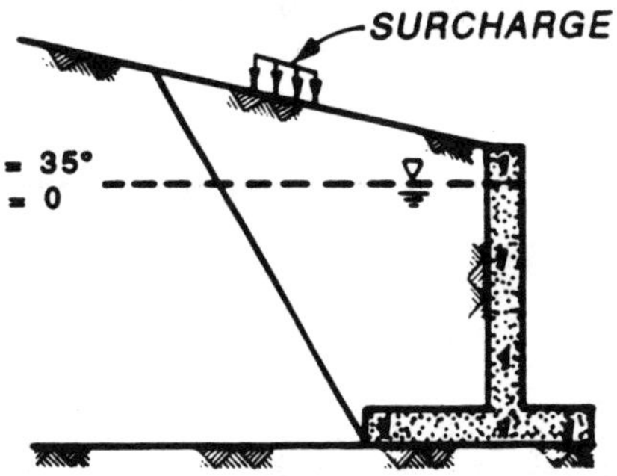

Example 2, page 196

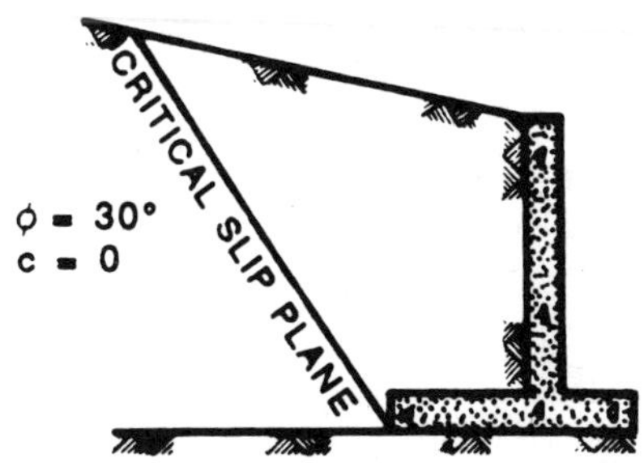

Example 5, page 203

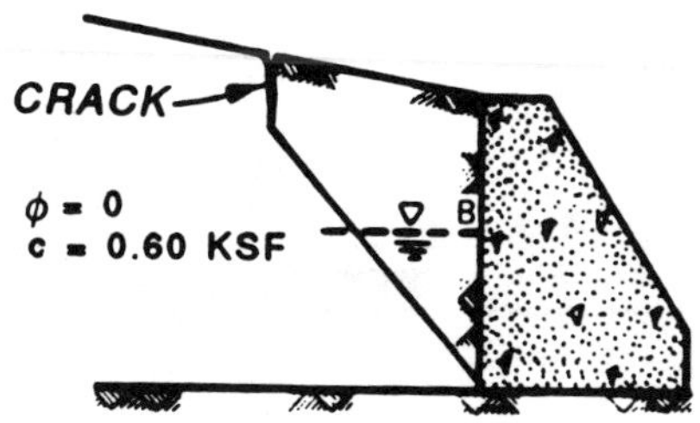

Example 3, page 198

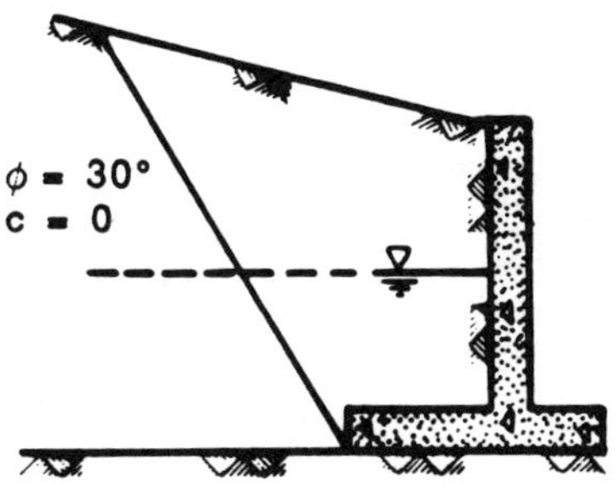

Example 6, page 208

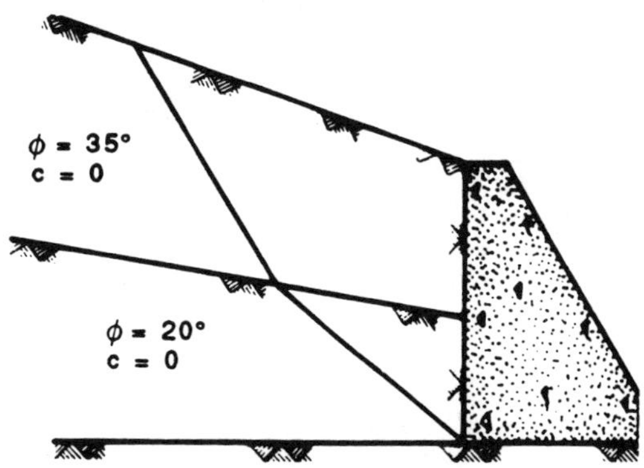

Example 7, page 212

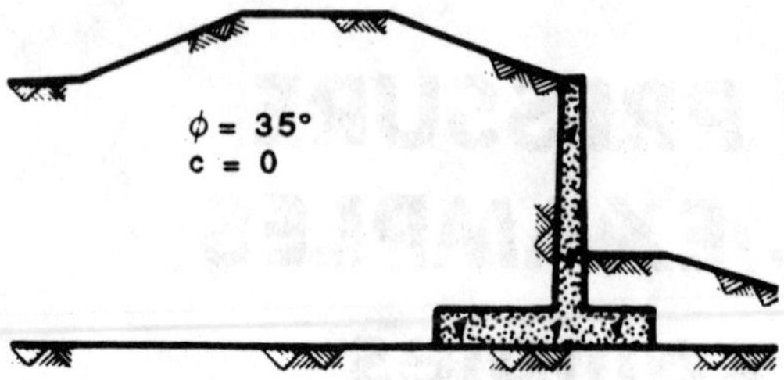

Example 8, page 215

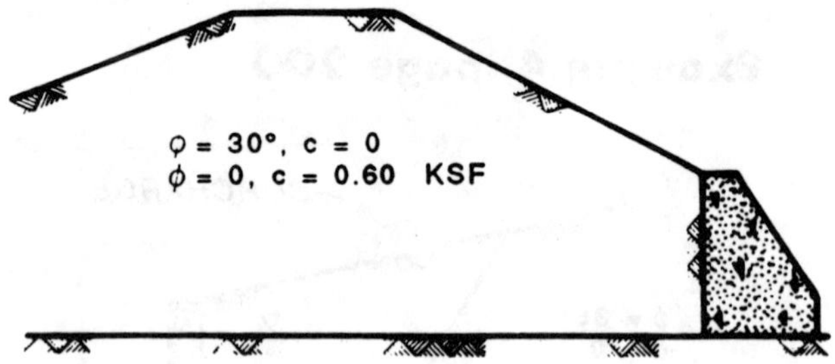

Example 9, page 220

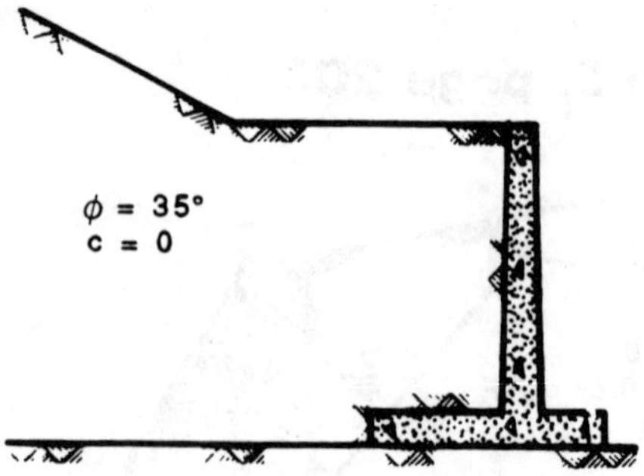

Example 10, page 223

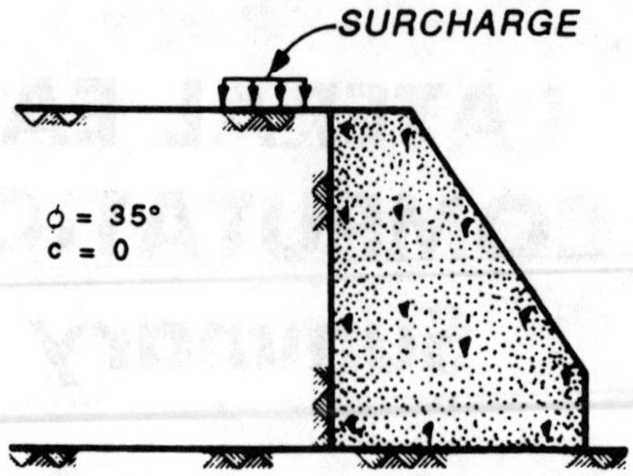

Example 11, page 227

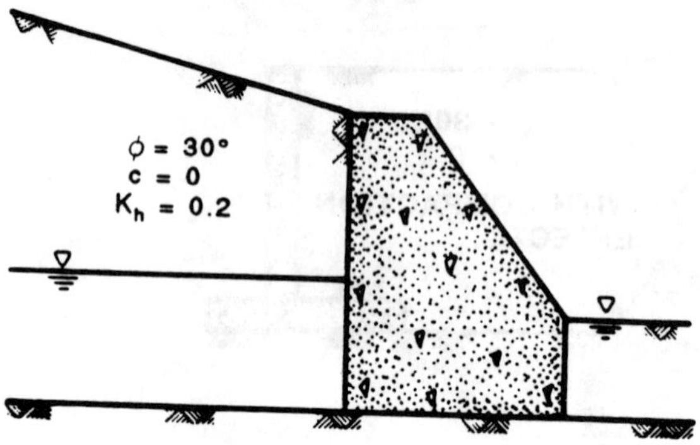

M-1. Example 1

Find the lateral earth force and its pressure distribution on Surface AB. Consider the effects of compaction in accordance with paragraph 3-17.

P = Line load for compaction roller = 5 k/ft

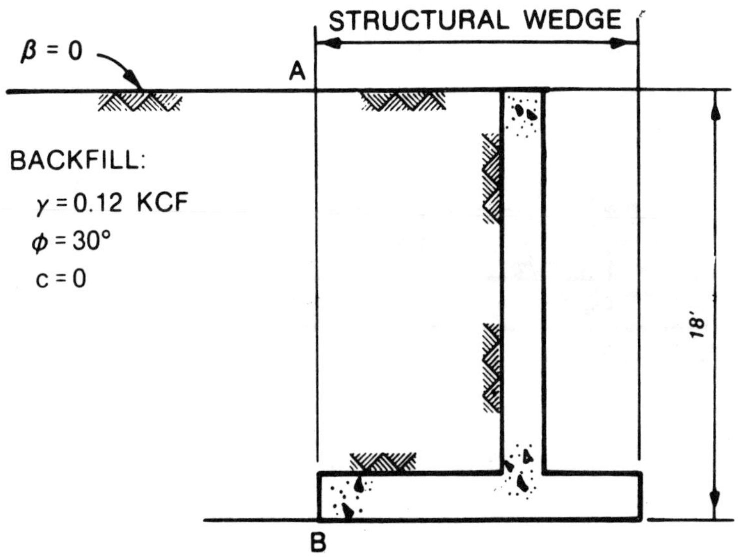

A. ACTIVE AND PASSIVE PRESSURE COEFFICIENTS

$$SMF = 1.00,\ \phi_d = \phi = 30^\circ$$

From Equation 3-15

$$K_A = \tan^2(45^\circ - \phi_d/2) = 1/3$$

Equation 3-20

$$K_P = \tan^2(45^\circ + \phi_d/2) = 3$$

B. AT-REST EARTH PRESSURE COEFFICIENT

From paragraph 3-11c

$$SMF = 2/3,\ \phi_d = \tan^{-1}(2/3 \tan \phi) = 21^\circ$$

$$K_o = \tan^2(45^\circ - \phi_d/2) = 0.47$$

C. CALCULATION OF EARTH PRESSURES. From Figure 3-30:

$$z_{cr} = \sqrt{\frac{2K_AK_oP}{\pi\gamma}} = \sqrt{\frac{2/3(0.47)(5)}{\pi(0.12)}} = 2.04 \text{ ft}$$

$$z_2 = \sqrt{\frac{2P}{K_AK_o\pi\gamma}} = \sqrt{\frac{2(5)}{1/3(0.47)(\pi)(0.12)}}$$

$$= 13.01 \text{ ft}$$

$$p'_{hm} = \sqrt{\frac{2K_oP\gamma}{K_A\pi}} = \sqrt{\frac{2(0.47)(5)(0.12)}{(1/3)\pi}}$$

$$= 0.734 \text{ ksf}$$

The force and pressure distribution are shown next.

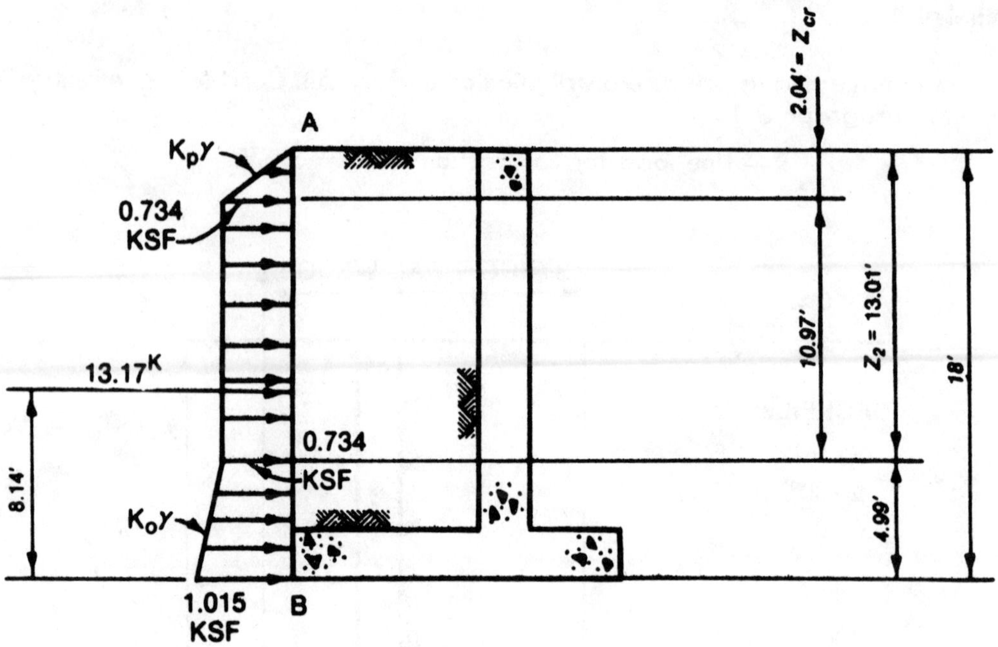

M-2. Example 2

Find the lateral earth force and pressure distribution acting on Surface AB.

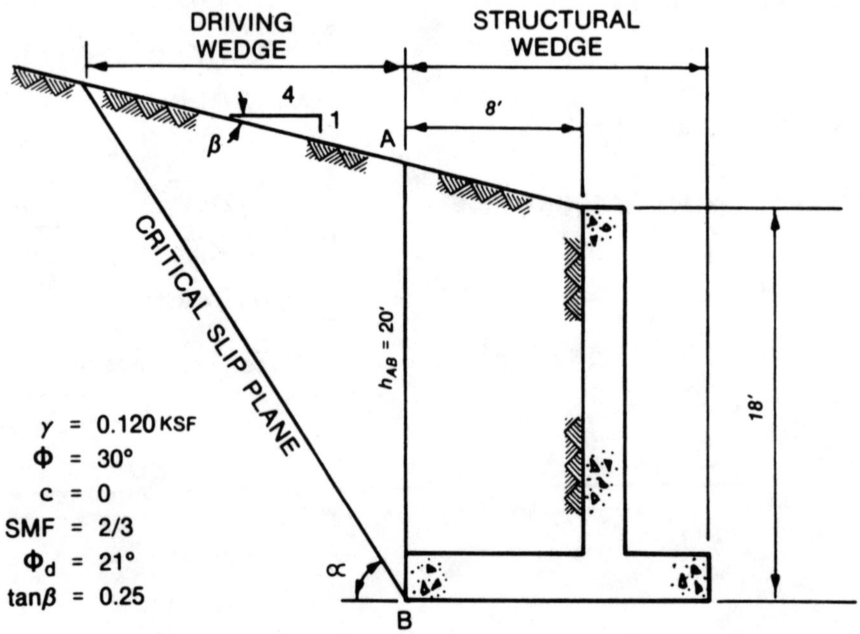

A. CALCULATE α

From Equation 3-26

$$c_1 = 2 \tan \phi_d = 2(0.383864) = 0.767728$$

From Equation 3-27

$$c_2 = 1 - \tan \phi_d \tan \beta - \left(\frac{\tan \beta}{\tan \phi_d}\right)$$

$$c_2 = 1 - 0.383864(0.25) - \frac{0.25}{0.383864}$$

$$= 0.252762$$

From Equation 3-25

$$\alpha = \tan^{-1}\left(\frac{c_1 + \sqrt{c_1^2 + 4c_2}}{2}\right) = 45.466^\circ$$

B. LATERAL EARTH PRESSURE COEFFICIENT (see Appendix H).

From the equations contained in Appendix H

$$K_1 = \left(\frac{1 - \tan \phi_d \cot \alpha}{1 + \tan \phi_d \tan \alpha}\right)\left(\frac{\tan \alpha}{\tan \alpha - \tan \beta}\right)$$

$$K_1 = \left(\frac{1 - 0.383864 \times 0.983864}{1 + 0.383864 \times 1.016400}\right)\left(\frac{1.016400}{0.766400}\right)$$

$$= 0.5937$$

Alternatively K_1 may be calculated by using Equation 3-14:

$$K_1 = \frac{\cos^2 \phi_d}{\left[1 + \sqrt{\dfrac{\sin \phi_d \sin (\phi_d - \beta)}{\cos \beta}}\right]^2}$$

$$\beta = \tan^{-1}(0.25) = 24.0362^\circ,$$

$$\phi_d - \beta = 6.9638^\circ$$

$$K_1 = \frac{(0.933580)^2}{\left[1 + \sqrt{\dfrac{0.358368(0.121242)}{0.970143}}\right]^2} = 0.5937$$

C. LATERAL FORCE AND PRESSURE DISTRIBUTION.

The lateral force and pressure distribution are shown in the next figure.

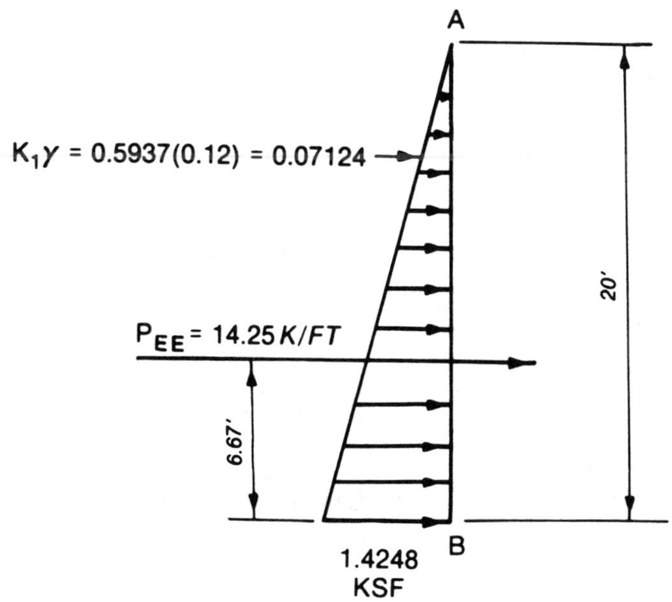

M-3. Example 3

Find lateral earth and water forces acting on Surface ABC. Use earth pressure coefficients and check by iteration of Equation 3-23.

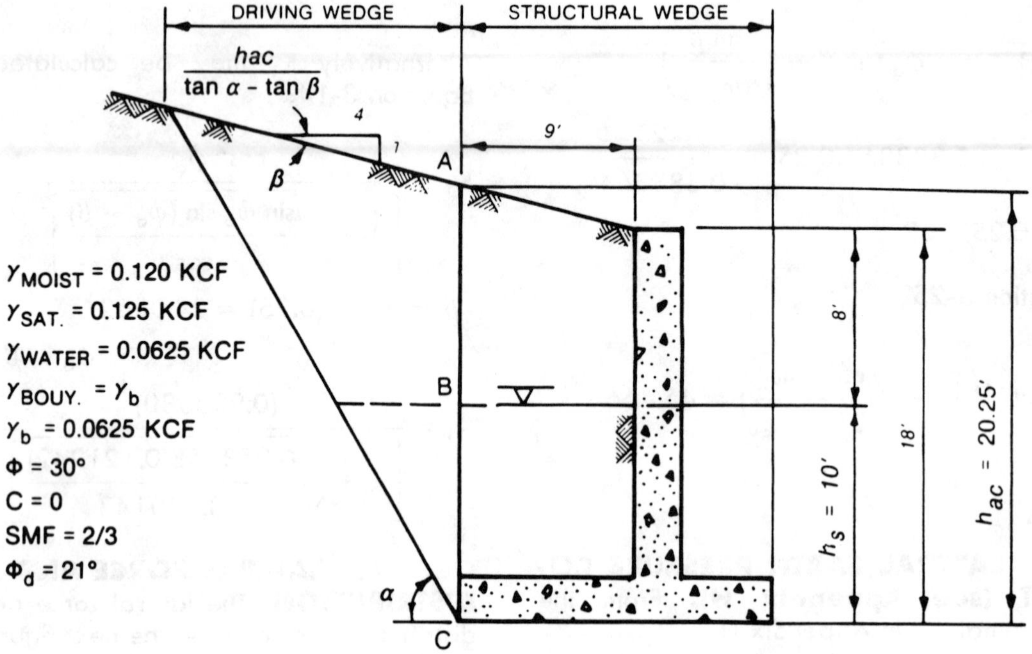

A. LATERAL EARTH PRESSURE COEFFICIENTS (see Appendix H). The critical slip-plane angle $\alpha = 45.466°$ (from Example 2)

$$K = \frac{1 - \tan \phi_d \cot \alpha}{1 + \tan \phi_d \tan \alpha}$$

$$= \frac{1 - 0.383864(0.983864)}{1 + 0.383864(1.016400)} = 0.447668$$

$$K_1 = K\left(\frac{\tan \alpha}{\tan \alpha - \tan \beta}\right)$$

$$= 0.447668\left(\frac{1.016400}{0.766400}\right) = 0.5937$$

$$K_b = K\left[1 + \left(\frac{\tan \alpha}{\tan \alpha - \tan \beta} - 1\right)\left(\frac{\gamma_m}{\gamma_b}\right)\right]$$

$$= 0.447668\left[1 + \left(\frac{1.0164}{0.7664} - 1\right)\left(\frac{0.120}{0.0625}\right)\right]$$

$$K_b = 0.7280$$

B. LATERAL FORCES AND PRESSURE DISTRIBUTION. The lateral forces and pressure distribution are shown in the figure on page 199.

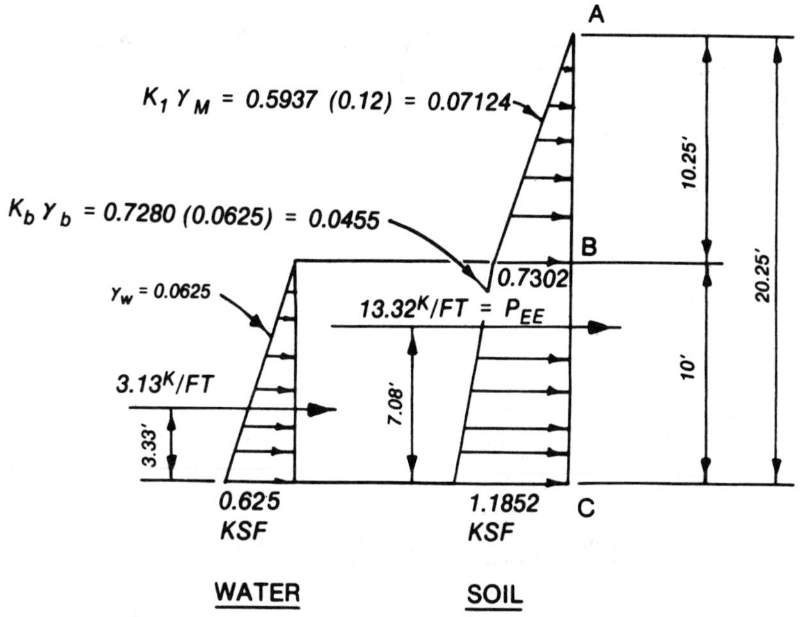

C. CHECK SOIL FORCE BY ITERATION EQUATION 3-23

W = total wedge weight, including water

$$W = \frac{\gamma_m h_{AC}^2}{2(\tan\alpha - \tan\beta)} + \frac{(\gamma_{sat} - \gamma_m)h_m^2}{2\tan\alpha}$$

$$W = \frac{0.12(20.25)^2}{2(\tan\alpha - \tan\beta)} + \frac{(0.005)(10)^2}{2\tan\alpha}$$

$$= \frac{24.604}{\tan\alpha - \tan\beta} + \frac{0.25}{\tan\alpha}$$

$$U = \text{uplift} = \frac{\gamma_{water} h_s^2}{2\sin\alpha} = \frac{0.0625(10)^2}{2\sin\alpha} = \frac{3.125}{\sin\alpha}$$

$$P_W = \frac{\gamma_{water} h_s^2}{2} = \frac{0.0625(10)^2}{2} = 3.125 \text{ k/ft}$$

$$P_{EE} = \frac{W(\tan\alpha - \tan\phi_d + U\tan\phi_d/\cos\alpha}{1 + \tan\phi_d \tan\alpha} - P_W$$

Let $\alpha = 45.466°$:

$$W = \frac{24.604}{0.766400} + \frac{0.25}{1.016400} = 32.3493 \text{ k/ft}$$

$$U = \frac{3.125}{0.712834} = 4.3839 \text{ k/ft}$$

$$P_{EE} = \frac{32.3493(1.016400 - 0.383864) + 4.3839(0.383864)/0.701332}{1 + 0.383864(1.016400)} - 3.125 = 13.3203 \text{ k/ft}$$

Let $\alpha = 44.466°$:

$$W = \frac{24.604}{0.731531} + \frac{0.25}{0.981531} = 33.8883 \text{ k/ft}$$

$$U = \frac{3.125}{0.700486} = 4.4612 \text{ k/ft}$$

$$P_{EE} = \frac{33.8883(0.981531 - 0.383864) + 4.4612(0.383864)/0.713666}{1 + 0.383864(0.981531)} - 3.125 = 13.3290 \text{ k/ft} > 13.3203 \text{ k/ft}$$

Let $\alpha = 43.466°$:

$$W = \frac{24.604}{0.697837} + \frac{0.25}{0.947837} = 35.5213 \text{ k/ft}$$

$$U = \frac{3.125}{0.687924} = 4.5427 \text{ k/ft}$$

$$P_{EE} = \frac{35.5213(0.947837 - 0.383864) + 4.5427(0.383864)/0.725783}{1 + 0.383864(0.947837)} - 3.125 = 13.3254 \text{ k/ft} < 13.3290 \text{ k/ft}$$

α critical = 44.466° instead of 45.466°.
$P_{max} = P_{EE}$ = 13.33 k/ft instead of 13.32 k/ft

D. CONCLUSION. These small differences in α and P_{EE} are due to the fact that the effect of water on the critical slip-plane angle is neglected in Equations 3-26 and 3-27. These differences are well within the permissible range of error required for soil pressure calculations. It can be concluded that for a cohensionless soil without a finite surcharge (V), it is not necessary to consider the effect of water when calculating the value of α.

M-4. Example 4

Find lateral soil force (P_{EE}) that will act on Surface ABC. Show the effect of water on slip-plane angle (α), and earth force.

A. CALCULATE α—NEGLECTING EFFECT OF WATER. From Equation 3-30 (omitting the c term)

$$A = \tan \phi_d - \frac{2V(1 + \tan^2 \phi_d)}{\gamma_m h^2}$$

$$= 0.466308 - \frac{2(6)(1.217443)}{0.12(20.25)^2}$$

$$A = 0.169416$$

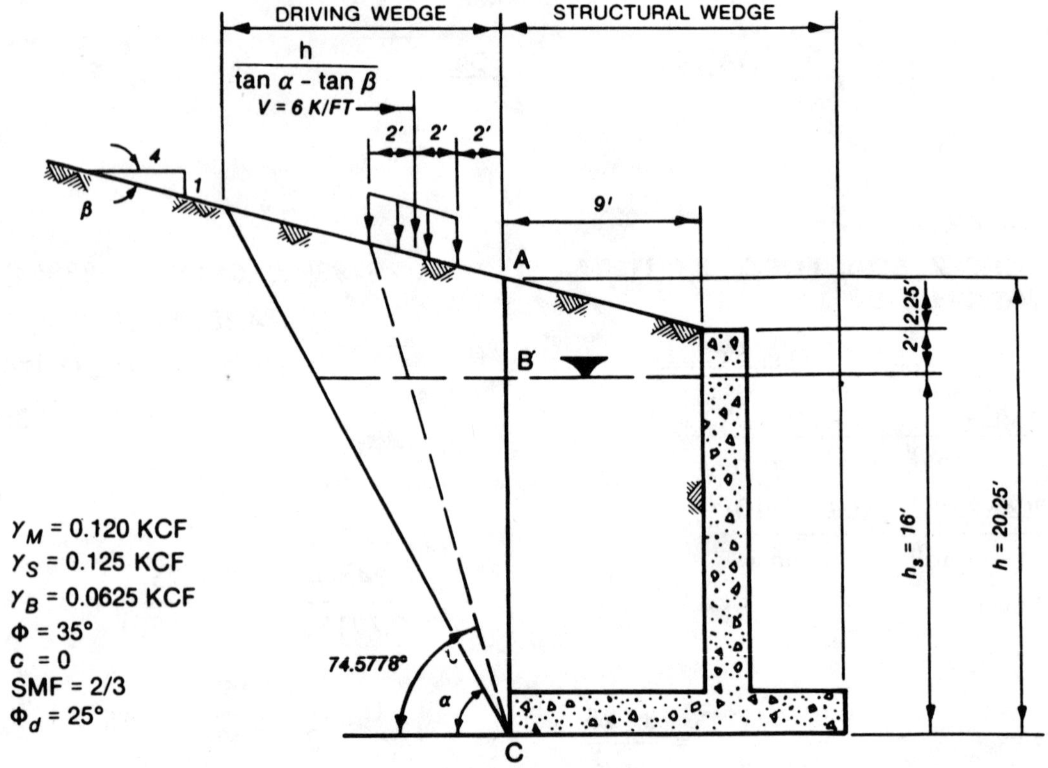

From Equation 3-28 (omitting the c term)

$$c_1 = \frac{2 \tan^2 \phi_d - \dfrac{4V \tan \beta (1 + \tan^2 \phi_d)}{\gamma_m h^2}}{A}$$

$$= \frac{2(0.466308)^2 - \dfrac{4(6)(1/4)(1.217443)}{0.12(20.25)^2}}{0.169416}$$

$$c_1 = 1.690751$$

From Equation 3-29 (omitting the c term)

$$c_2 = \frac{\tan \phi_d(1 - \tan \phi_d \tan \beta) - \tan \beta + \dfrac{2V \tan^2 \beta\,(1 + \tan^2 \phi_d)}{\gamma_m h^2}}{A}$$

$$c_2 = \frac{0.466308[(1 - 0.466308(1/4)] - \dfrac{1}{4} + \dfrac{2(6)(1/16)(1.217443)}{0.12(20.25)^2}}{0.169416}$$

$$= 1.065442$$

From Equation 3-25

$$\alpha = \tan^{-1}\left(\frac{c_1 + \sqrt{c_1^2 + 4c_2}}{2}\right) = \underline{\underline{65.354^\circ}}$$

B. CALCULATE α—INCLUDE EFFECT OF WATER. Use γ_{avg} instead of γ_m, where

$$\gamma_{avg} = \left[\frac{\gamma_m h^2}{2(\tan \alpha - \tan \beta)} - \frac{(\gamma_m - \gamma_b)h_s^2}{2 \tan \alpha}\right] \div \left[\frac{h^2}{2(\tan \alpha - \tan \beta)}\right]$$

Let $\alpha = 74.5778^\circ$:

$$\gamma_{avg} = \left[\frac{0.12(20.25)^2}{2(3.374990)} - \frac{0.0575(16)^2}{2(3.624990)}\right] \div \left[\frac{(20.25)^2}{2(3.374990)}\right]$$

$$\gamma_{avg} = 0.087 \text{ kcf}$$

$$A = 0.466308 - \frac{2(6)(1.217443)}{0.087(20.25)^2} = 0.056802$$

$$c_1 = \frac{2(0.466308)^2 - \dfrac{4(6)(1/4)(1.217443)}{0.087(20.25)^2}}{0.056802}$$

$$= 4.051497$$

$$c_2 = \frac{0.466308[1 - 0.466308(1/4)] - \dfrac{1}{4} + \dfrac{2(6)(1/16)(1.217443)}{0.087(20.25)^2}}{0.056802}$$

$$c_2 = 3.301668$$

$$\alpha = \tan^{-1}\left(\frac{c_1 + \sqrt{c_1^2 + 4c_2}}{2}\right)$$

$$= \underline{\underline{78.104^\circ}} > 74.5778^\circ$$

Use $\alpha = 74.5778^\circ$; the entire surcharge will not lie on the top surface of the wedge when the angle is greater.

C. CALCULATE EARTH PRESSURE COEFFICIENTS (see Appendix H). For $\alpha = 65.354^\circ$ (neglecting effect of water)

$$K = \frac{1 + \tan \phi_d \cot \alpha}{1 - \tan \phi_d \tan \alpha}$$

$$K = \frac{1 - 0.466308(0.458807)}{1 + 0.466308(2.179565)} = 0.389841$$

$$K_1 = K\left(\frac{\tan \alpha}{\tan \alpha - \tan \beta}\right)$$

$$K_1 = 0.389841\left(\frac{2.179565}{1.929565}\right) = 0.44035$$

$$K_b = K\left[1 + \left(\frac{\tan \alpha}{\tan \alpha - \tan \beta} - 1\right)\left(\frac{\gamma_m}{\gamma_b}\right)\right]$$

$$K_b = 0.389841 \times \left[1 + \left(\frac{2.179565}{1.929565} - 1\right)\left(\frac{0.120}{0.0625}\right)\right]$$

$$= 0.48682$$

$$K_v = K \tan \alpha$$

$$K_v = 0.389841(2.179565) = 0.84968$$

For $\alpha = 74.5778^\circ$ (including effect of water)

$$K = \frac{1 - 0.466308(0.275863)}{1 + 0.466308(3.624990)} = 0.323883$$

$$K_1 = 0.323883\left(\frac{3.624990}{3.374990}\right) = 0.34787$$

$$K_b = 0.323883 \times \left[1 + \left(\frac{3.624990}{3.374990} - 1\right)\left(\frac{0.120}{0.0625}\right)\right]$$

$$= 0.36995$$

$$K_v = 0.323883(3.624990) = 1.17407$$

D. CALCULATE LATERAL SOIL SOURCE (USING COEFFICIENTS). For $\alpha = 65.354°$

$$P_\gamma = (\tfrac{1}{2})K_1\gamma_m(h - h_s)^2 + (\tfrac{1}{2})(h_s)[2K_1\gamma_m(h - h_s) + K_b\gamma_b h_s]$$

$$P_\gamma = (\tfrac{1}{2})(0.44035)(0.12)(4.25)^2 + (\tfrac{1}{2})(16)[2(0.44035)(0.12)(4.25) + 0.48682(0.0625)(16)]$$

$$P_\gamma = 7.965 \text{ k/ft}$$

$$P_v = K_v V = 0.84968(6) = 5.098 \text{ k/ft}$$

$$P_{EE} = 7.965 + 5.098 = \underline{\underline{13.063 \text{ k/ft}}}$$

For $\alpha = 74.5778°$

$$P_\gamma = (\tfrac{1}{2})(0.34787)(0.12)(4.25)^2 + (\tfrac{1}{2})(16)[2(0.34787)(0.12)(4.25) + 0.36995(0.0625)(16)]$$

$$P_\gamma = 6.175 \text{ k/ft}$$

$$P_v = 1.17407(6) = 7.044 \text{ k/ft}$$

$$P_{EE} = 6.175 + 7.044 = \underline{\underline{13.219 \text{ k/ft}}}$$

E. FIND LATERAL SOIL FORCE BY ITERATION OF EQUATION 3-23. Simplifying Equation 3-23

$$P_{EE} = \frac{(W+V)(\tan\alpha - \tan\phi_d) + U\tan\phi_d/\cos\alpha}{1 + \tan\phi_d\tan\alpha} - P_W$$

$$W = \frac{\gamma_m h^2}{2(\tan\alpha - \tan\beta)} + \frac{(\gamma_s - \gamma_m)h_s^2}{2\tan\alpha}$$

$$W = \frac{0.12(20.25)^2}{2(\tan\alpha - \tan\beta)} + \frac{(0.005)(16)^2}{2\tan\alpha}$$

$$= \frac{24.6038}{\tan\alpha - \tan\beta} + \frac{0.64}{\tan\alpha}$$

$$U = \frac{\gamma_w h_s^2}{2\sin\alpha} = \frac{0.0625(16)^2}{2\sin\alpha} = \frac{8}{\sin\alpha}$$

$$P_W = \frac{\gamma_w h_s^2}{2} = \frac{0.0625(16)^2}{2} = 8.000 \text{ k/ft}$$

Let $\alpha = 73.5778°$:

$$W = \frac{24.6038}{3.142854} + \frac{0.64}{3.392854} = 8.0171 \text{ k/ft}$$

$$U = \frac{8.00}{0.959205} = 8.3402 \text{ k/ft,}$$

$$W + V = 14.0171 \text{ k/ft}$$

$$P_{EE} = \frac{14.017(3.392854 - 0.466308) + 8.3402(0.466308)/0.282713}{1 + 0.466308(3.392854)} - 8.000 = \underline{\underline{13.214 \text{ k/ft}}}$$

Let $\alpha = 74.5778°$:

$$W = \frac{24.6038}{3.374990} + \frac{0.64}{3.624990} = 7.4666 \text{ k/ft}$$

$$W + V = 13.4666 \text{ k/ft,}$$

$$U = \frac{8.00}{0.963992} = 8.2988 \text{ k/ft}$$

$$P_{EE} = \frac{13.4666(3.624990 - 0.466308) + 8.2988(0.466308)/0.265930}{1 + 0.466308(3.624990)} - 8.000 = \underline{13.220 \text{ k/ft}} > 13.214$$

Let $\alpha = 75.5778°$ (all of V does not lie on top of wedge):

$$\frac{h}{\tan\alpha - \tan\beta} = \frac{20.25}{3.638487} = 5.5655 \text{ ft}$$

$$V = \left(\frac{5.5655 - 2}{4}\right)6 \text{ k/ft} = 5.3483 \text{ k/ft}$$

$$W = \frac{24.6038}{3.638487} + \frac{0.64}{3.888487} = 6.9267 \text{ k/ft}$$

$$W + V = 12.2750 \text{ k/ft,}$$

$$U = 8/0.96847 = 8.2603$$

$$P_{EE} = \frac{12.275(3.888487 - 0.466308) + 8.2603(0.466308)/0.249065}{1 + 0.466308(3.888487)} - 8.000 = \underline{\underline{12.429 \text{ k/ft}}} < 13.220$$

P_{EE} = 13.220 k/ft (agrees with coefficient solution where α is found considering the effect of water)

F. DETERMINE LINE OF ACTION FOR LATERAL FORCE DUE TO V BY USING THE APPROXIMATE METHOD OF FIGURE 3-29. Find α', the slip-plane angle without surcharge: Use SMF = 1.00 in Equations 3-26 and 3-27.

$$c_1 = 2 \tan \phi = 2(0.700208) = 1.400416$$

$$c_2 = 1 - \tan \phi \tan \beta - \frac{\tan \beta}{\tan \phi}$$

$$c_2 = 1 - 0.700208(\tfrac{1}{4}) - \frac{(\tfrac{1}{4})}{0.700208} = 0.467911$$

From Equation 3.25

$$\alpha' = \tan^{-1}\left(\frac{c_1 + \sqrt{c_1^2 + 4c_2}}{2}\right) = 59.22°$$

The location of P_v is determined in the following figure.

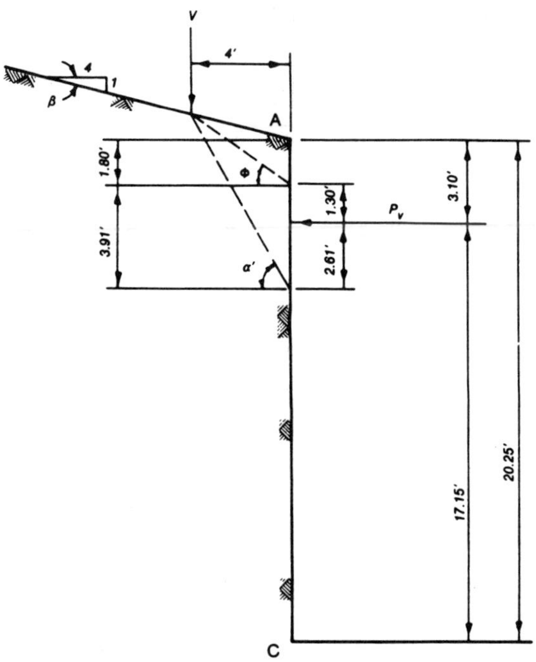

G. PRESSURE DISTRIBUTION AND LINE OF ACTION OF EARTH FORCE P_{EE} FOR α = 65.354°. The pressure distribution and line of action of earth force P_{EE} for $\alpha = 65.354°$ are shown in the next figure. A pressure distribution must be assumed for the lateral force due to V since only the location of the resultant force is known. The pressure distribution will be defined by an isosceles triangle with the apex located at the point of application of the resultant force due to V. See Example 10 of this appendix for the computation of the lateral force due to a surcharge by the elastic method.

H. PRESSURE DISTRIBUTION AND LINE OF ACTION OF EARTH FORCE P_{EE} for α = 74.5778°. The pressure distribution and line of action of earth force P_{EE} for $\alpha = 74.5778°$ are shown in the next figure.

I. CONCLUSION. For a cohesionless soil when calculating α from Equations 3-28, 3-29, and 3-30, the unit weight of soil (γ) in the strip surcharge (V) term should be the average effective unit weight (γ_{avg}). γ_{avg} should be calculated using the moist unit weight above the water table and the buoyant unit weight below the water table.

M-5. Example 5

Find the lateral earth force acting on Surface ABC. Use the moist unit weight of soil to calculate α. Check the solution by iteration of Equation 3-23.

A. CALCULATE SLIP-PLANE ANGLE α. Estimate $\alpha = 40°$. Using Equation I-2 with $\phi = 0$ yields

$$d_c = \frac{c_d/\gamma_m}{\sin \alpha \cos \alpha} = \frac{0.4/0.12}{0.642788(0.766044)} = 6.77 \text{ ft}$$

From Equation 3-30

$$A = \frac{2c_d}{\gamma_m(h + d_c)} = \frac{2(0.4)}{0.12(24.77)} = 0.269143$$

From Equation 3-28

$$c_1 = \frac{\dfrac{4c_d \tan \beta}{\gamma_m(h + d_c)}}{A} = \frac{\dfrac{4(0.4)(1/4)}{0.12(24.77)}}{0.269143} = 0.50$$

From Equation 3-29

$$c_2 = \frac{-\tan \beta + \dfrac{2c_d}{\gamma_m(h + d_c)}}{A}$$

$$= \frac{-\dfrac{1}{4} + \dfrac{2(0.4)}{0.12(24.77)}}{0.269143} = 0.071125$$

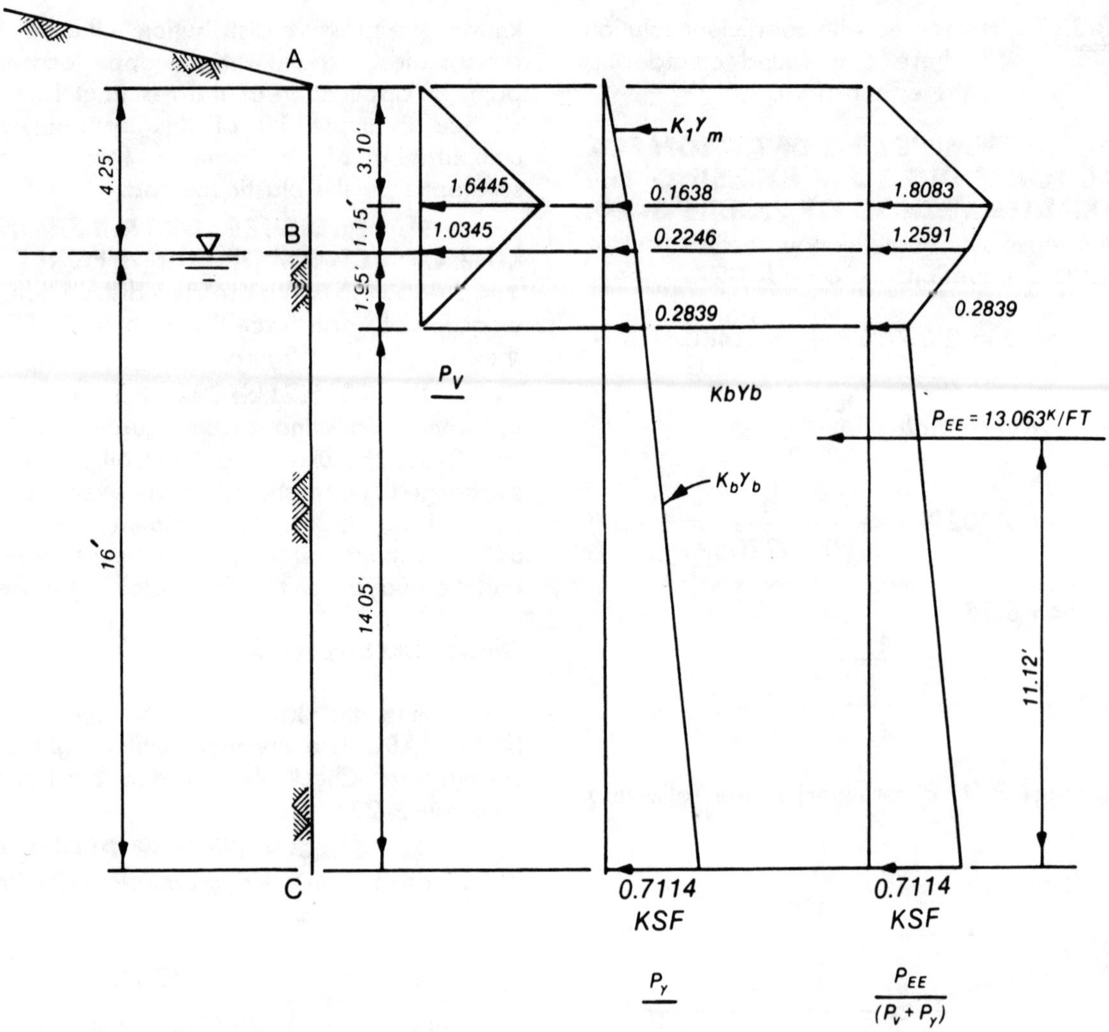
A
B
C
4.25'
16'
3.10'
1.15'
1.95'
14.05'
1.6445
1.0345
P_V
$K_1\gamma_m$
0.1638
0.2246
0.2839
KbYb
$K_b\gamma_b$
0.7114
KSF
P_γ
1.8083
1.2591
0.2839
$P_{EE} = 13.063^K/FT$
11.12'
0.7114
KSF
P_{EE}
$(P_v + P_\gamma)$

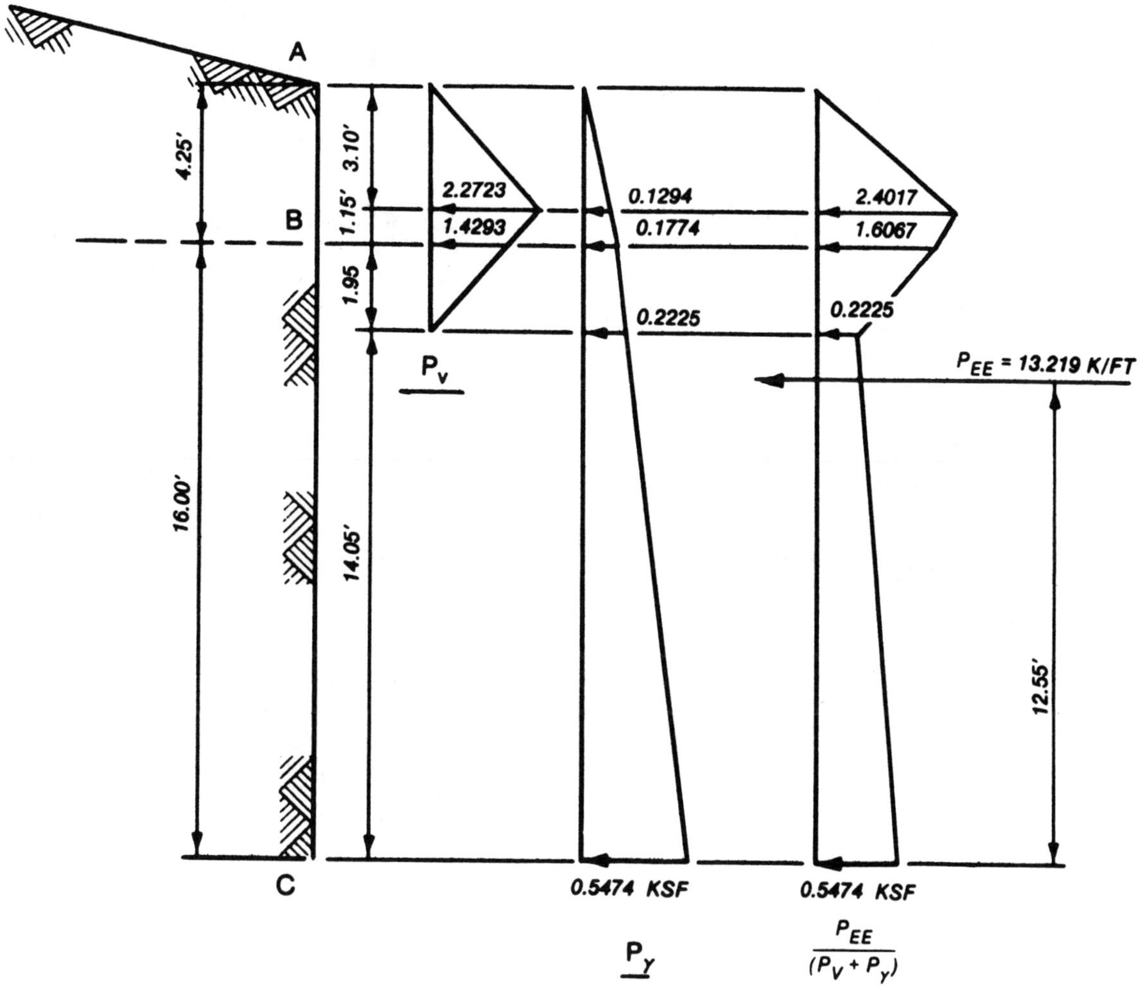
A
B
C
4.25'
16.00'
3.10'
1.15'
1.95
14.05'
2.2723
1.4293
Pv
0.1294
0.1774
0.2225
0.5474 KSF
Py
2.4017
1.6067
0.2225
PEE = 13.219 K/FT
12.55'
0.5474 KSF
PEE / (PV + PY)

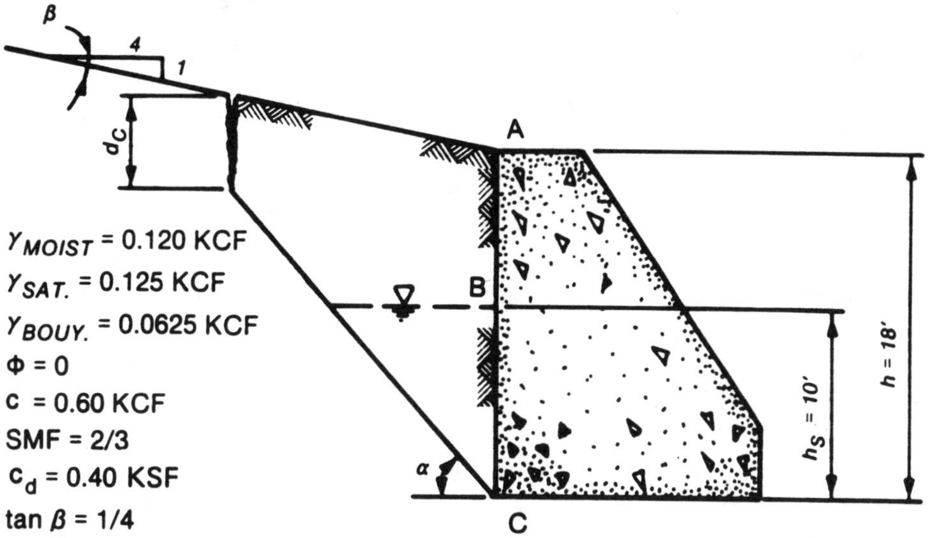
β
4
1
dc
A
B
C
α
hs = 10'
h = 18'
γMOIST = 0.120 KCF
γSAT. = 0.125 KCF
γBOUY. = 0.0625 KCF
Φ = 0
c = 0.60 KCF
SMF = 2/3
cd = 0.40 KSF
tan β = 1/4

From Equation 3-25

$$\alpha = \tan^{-1}\left(\frac{c_1 + \sqrt{^2_1 + 4c_2}}{2}\right) = 31.61^\circ \neq 40^\circ$$

Let $\alpha = 31.61^\circ$:

$$d_c = \frac{0.4/0.12}{0.524135(0.851635)} = 7.47 \text{ ft}$$

$$A = \frac{2(0.4)}{0.12(25.47)} = 0.261746$$

$$c_1 = \frac{\dfrac{4(0.4)(1/4)}{0.12(25.47)}}{0.261746} = 0.50$$

$$c_2 = \frac{-\dfrac{1}{4} + \dfrac{2(0.4)}{0.12(25.47)}}{0.261746} = 0.044875$$

$$\alpha = \tan^{-1}\left(\frac{c_1 + \sqrt{c_1^2 + 4c_2}}{2}\right) = 30.01^\circ \neq 31.61^\circ$$

Let $\alpha = 30^\circ$:

$$d_c = \frac{0.4/0.12}{0.5(0.866025)} = 7.70 \text{ ft}$$

$$A = \frac{2(0.4)}{0.12(25.70)} = 0.259403$$

$$c_1 = 0.50,\ c_2 = \frac{-\frac{1}{4} + 0.259403}{0.259403} = 0.036249$$

$$\alpha = \tan^{-1}\left(\frac{c_1 + \sqrt{c_1^2 + 4c_2}}{2}\right) = 29.43^\circ \neq 30^\circ$$

Let $\alpha = 29^\circ$:

$$d_c = \frac{0.4/0.12}{0.484810(0.874620)} = 7.86 \text{ ft}$$

$$A = \frac{2(0.4)}{0.12(25.86)} = 0.257798$$

$$c_1 = 0.5,\ c_2 = \frac{-\frac{1}{4} + 0.257798}{0.257798} = 0.030248$$

$$\alpha = \tan^{-1}\left(\frac{c_1 + \sqrt{c_1^2 + 4c_2}}{2}\right) = 29.01^\circ$$

$\underline{\underline{\alpha = 29^\circ}}$ (nearest degree), $d_c = \underline{\underline{7.86 \text{ ft}}}$

B. EARTH PRESSURE COEFFICIENTS AND EARTH FORCE FOR $\alpha = 29^\circ$ (see Appendix H).

$$K = \frac{1 - \tan\phi_d \cot\alpha}{1 + \tan\phi_d \tan\alpha} = 1.00$$

$$K_1 = K\left(\frac{\tan\alpha}{\tan\alpha - \tan\beta}\right) = 1.00\left(\frac{0.554309}{0.304309}\right)$$
$$= 1.8215$$

$$K_b = K\left[1 + \left(\frac{\tan\alpha}{\tan\alpha - \tan\beta} - 1\right)\left(\frac{\gamma_m}{\gamma_b}\right)\right]$$

$$K_b = 1.00\left[1 + \left(\frac{0.554309}{0.304309} - 1\right)\left(\frac{0.120}{0.0625}\right)\right]$$
$$= 2.5773$$

$$P_{EE} = (1/2)K_1\gamma_m(h - h_s - d_c)^2 + (1/2)h_s[2K_1\gamma_m(h - h_s - d_c) + K_b\gamma_b h_s]$$

$$h = 18 \text{ ft},\ h_s = 10 \text{ ft},\ d_c = 7.86 \text{ ft}$$

$$P_{EE} = (1/2)(1.8215)(0.12)(0.14)^2 + (1/2)(10)[2(1.8215)(0.12)(0.14) + 2.5773(0.0625)(10)]$$
$$= \underline{\underline{8.362 \text{ k/ft}}}$$

C. CHECK SOLUTION BY ITERATION OF EQUATION 3-23.

$$W = \frac{\gamma_m(h^2 - d_c^2)}{2(\tan\alpha - \tan\beta)} + \frac{(\gamma_s - \gamma_m)h_s^2}{2\tan\alpha}$$

$$W = \frac{0.06(h^2 - d_c^2)}{\tan\alpha - \tan\beta} + \frac{0.25}{\tan\alpha}$$

$$d_c = \frac{c_d/\gamma_m}{\sin\alpha\cos\alpha} = \frac{3.3333}{\sin\alpha\cos\alpha}$$

$$L = \frac{h - d_c}{\cos\alpha\,(\tan\alpha - \tan\beta)},$$

$$P_W = \tfrac{1}{2}\gamma_m h_s^2 = 3.125 \text{ k/ft}$$

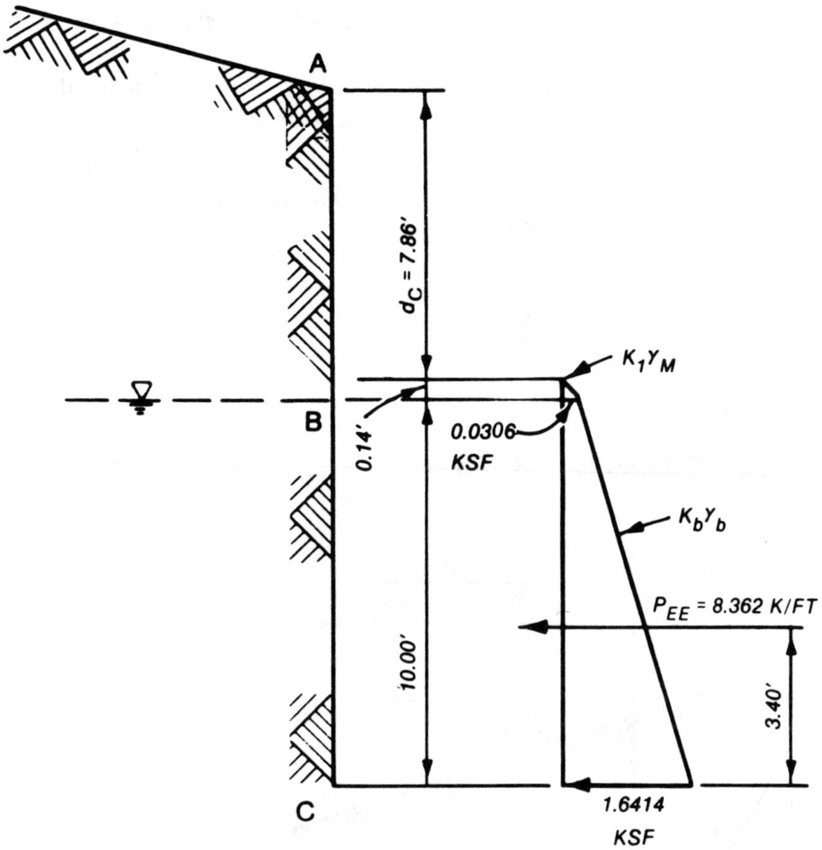

From Equation 3-23

$$P_{EE} = W \tan \alpha - c_d L/\cos \alpha - P_W$$

Let $\alpha = 28°$:

$$d_c = \frac{3.3333}{0.469472(0.882948)} = 8.04 \text{ ft},$$

$$h^2 - d_c^2 = 259.3584$$

$$L = \frac{9.96}{0.882948(0.281709)} = 40.0427 \text{ ft}$$

$$W = \frac{0.06(259.3584)}{0.281709} + \frac{0.25}{0.531709}$$
$$= 55.7098 \text{ k/ft}$$

$$P_{EE} = 55.7098(0.531709) - 0.4(40.0427)/0.882948 - 3.125$$

$$P_{EE} = 8.356 \text{ k/ft}$$

Let $\alpha = 29°$:

$$d_c = \frac{3.3333}{0.484810(0.874620)} = 7.86 \text{ ft}$$

$$h^2 - d_c^2 = 262.2204,$$

$$L = \frac{10.14}{0.87462(0.304309)} = 38.0981 \text{ ft}$$

$$W = \frac{0.06(262.2204)}{0.304309} + \frac{0.25}{0.554309}$$
$$= 52.1424 \text{ k/ft}$$

$$P_{EE} = 52.1525(0.554309) - 0.4(38.0981)/0.87462 - 3.125$$

$$P_{EE} = \underline{\underline{8.360 \text{ k/ft}}} > 8.356$$

Let $\alpha = 30°$:

$$d_c = \frac{3.3333}{0.5(0.866025)} = 7.70 \text{ ft}, \; h^2 - d_c^2 = 264.71$$

$$L = \frac{10.30}{0.866025(0.327350)} = 36.3324 \text{ ft}$$

$$W = \frac{0.06(264.71)}{0.327350} + \frac{0.25}{0.57735} = 48.9517 \text{ k/ft}$$

$$P_{EE} = 48.9517(0.57735) - 0.4(36.3324)/0.866025 - 3.125$$

$$P_{EE} = 8.356 \text{ k/ft} < 8.360$$

Iteration of Equation 3-23 shows that:

$\alpha_{crit} = 29°$ (to nearest degree), $P_{EE} = 8.360$ k/ft

which agrees with the pressure coefficient solution.

D. CONCLUSION. Use the moist unit weight of soil (γ_m) in the cohesion terms of Equations 3-28, 3-29, and 3-24, even if the soil is partially saturated.

M-6. Example 6

Find the lateral earth forces on Surfaces CD and DE (stratified soil).

A. CALCULATE α_2 (see Appendix G). Using Equation G-25 yields

$$\gamma' = \frac{2\gamma_1 h_1}{h_2} + \gamma_2 + \frac{2\gamma_1(\tan\beta_1 - \tan\beta_2)}{\tan\alpha_2 - \tan\beta_2}$$

$$\gamma' = \frac{2(0.12)(13)}{10} + 0.13 + \frac{2(0.12)(\frac{1}{3} - \frac{1}{6})}{\tan\alpha_2 - \frac{1}{6}}$$

$$= 0.442 + \frac{0.04}{\tan\alpha_2 - \frac{1}{6}}$$

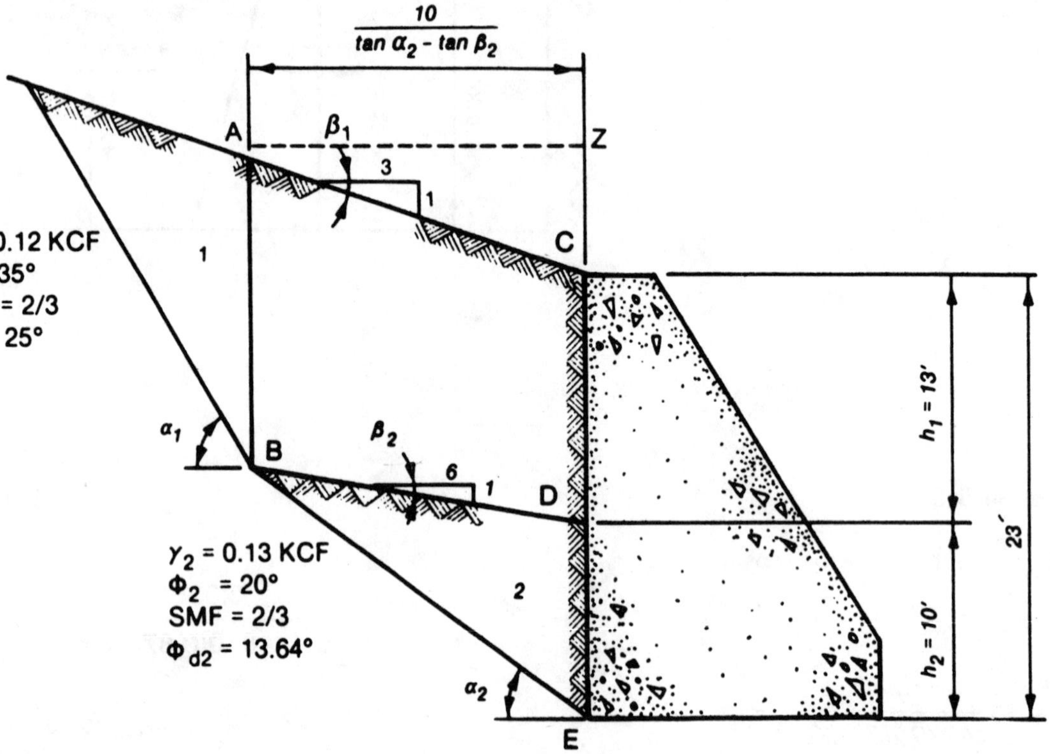

The weight of the surcharge in triangle ACZ is calculated by using Equation G-26:

$$V_\alpha = -\frac{\gamma_1 h_2^2(\tan\beta_1 - \tan\beta_2)}{2(\tan\alpha_2 - \tan\beta_2)^2} = \frac{-0.12(10)^2(\frac{1}{6})}{2(\tan\alpha_2 - \frac{1}{6})^2}$$

$$= \frac{-1}{(\tan\alpha_2 - \frac{1}{6})^2}$$

First trial: $\alpha_2 = 39°, \gamma' = 0.5042, V_\alpha = -2.4178$

From Equation G-27

$$A' = \tan \phi_{d2} - \frac{2V_\alpha(1 + \tan^2 \phi_{d2})}{\gamma' h_2^2}$$

$$A' = 0.242665 - \frac{2(-2.4178)(1.058886)}{0.5042(10)^2}$$

$$= 0.344219$$

From Equation G-28

$$c_1' = \frac{2 \tan^2 \phi_{d2} - \dfrac{4V_\alpha \tan \beta_2 (1 + \tan^2 \phi_{d2})}{\gamma' h_2^2}}{A'}$$

$$c_1' = \frac{2(0.242665)^2 - \dfrac{4(-2.4178)(\frac{1}{6})(1.058886)}{0.5042(10)^2}}{0.344219}$$

$$= 0.440486$$

From Equation G-29

$$c_2' = \frac{\tan \phi_{d2}(1 - \tan \phi_{d2} \tan \beta_2) - \tan \beta_2 + \dfrac{2V_\alpha \tan^2 \beta_2 (1 + \tan^2 \phi_{d2})}{\gamma' h_2^2}}{A'}$$

$$c_2' = \frac{0.242665(1 - 0.242665 \times \frac{1}{6}) - \frac{1}{6} + \dfrac{2(-2.4178)(\frac{1}{36})(1.058886)}{0.5042(10)^2}}{0.344219}$$

$$c_2' = 0.184077$$

From G-30

$$\alpha_2 = \tan^{-1}\left(\frac{c_1' + \sqrt{c_1'^2 + 4c_2'}}{2}\right)$$

$$\alpha_2 = 35.1^\circ \neq 39^\circ$$

Second trial: $\alpha_2 = 33^\circ, \gamma' = 0.5249, V_\alpha = -4.2911$

$$A' = 0.242665 - \frac{0.021178(-4.2911)}{0.5249}$$

$$= 0.415797$$

$$c_1' = \frac{0.117773 - \dfrac{0.007059(-4.2911)}{0.5249}}{0.415797}$$

$$= 0.422035$$

$$c_2' = \frac{0.066814 + \dfrac{0.000588(-4.2911)}{0.5249}}{0.415797}$$

$$= 0.149128$$

$\underline{\underline{\alpha_2 = 33^\circ}}$

B. CALCULATE α_1.

From Equation 3-26

$$c_1 = 2 \tan \phi_{d1} = 2(0.466308) = 0.932616$$

From Equation 3-27

$$c_2 = 1 - \tan \phi_{d1} \tan \beta - \left(\frac{\tan \beta}{\tan \phi_{d1}}\right)$$

$$c_2 = 1 - 0.466308(\tfrac{1}{3}) - \frac{(\frac{1}{3})}{0.466308} = 0.129729$$

$\alpha_1 = \underline{\underline{46.547^\circ}}$

C. CALCULATE LATERAL EARTH PRESSURE COEFFICIENTS (see Appendix H).

Wedge 1:

$$K = \frac{1 - 0.466308(0.947407)}{1 + 0.466308(1.055513)} = 0.374091$$

$$K_1 = 0.374091\left(\frac{1.055513}{0.722180}\right) = 0.5468$$

Wedge 2:

$$K = \frac{1 - 0.242665(1.539865)}{1 + 0.242665(0.649408)} = 0.541063$$

$$K_1 = 0.541063\left(\frac{0.649408}{0.482741}\right) = 0.7279$$

$$K_v = 0.541063(0.649408) = 0.3514$$

D. CALCULATE LATERAL EARTH FORCES AND PRESSURES.

Wedge 1:

$$h = h_{AB} = h_1 + \frac{h_2(\tan \beta_1 - \tan \beta_2)}{(\tan \alpha_2 - \tan \beta_2)}$$

$$h_{AB} = 13 + \frac{(10)(\frac{1}{3} - \frac{1}{6})}{0.649408 - 0.166667} = 16.452 \text{ ft}$$

$$P_{EE1} = \tfrac{1}{2}K_1\gamma_1 h^2$$

$$P_{EE1} = \tfrac{1}{2}(0.5468)(0.12)(16.452)^2 = \underline{\underline{8.88 \text{ k/ft}}}$$

$$p_{HB} = K_1\gamma_1 h_{AB} = 0.5468(0.12)(16.452) = 1.0795$$

Wedge 2:
Treat weight of ABCD as a strip surcharge on Surface BD.

$$V = \tfrac{1}{2}(h_1 + h_{AB})(\overline{AZ})(\gamma_1)$$

$$V = \tfrac{1}{2}(13 + 16.452)(20.715)(0.12)$$

$$= 36.606 \text{ k/ft}$$

$$P_\gamma = \tfrac{1}{2}K_1\gamma_2 h_2^2$$

$$P_\gamma = \tfrac{1}{2}(0.7279)(0.13)(10)^2 = 4.73 \text{ k/ft}$$

$$P_V = K_v V$$

$$P_V = 0.3514(36.606) = 12.86 \text{ k/ft}$$

$$p_{HD} = K_1 p_{VD}$$

$$p_{HD} = 0.7279 p_{VD} = 1.1355 \text{ ksf}$$

$$= \text{horizontal pressure at D}$$

$$\tfrac{1}{2}(p_{HD} + p_{HE})h_2 = P_{EE2}$$

$$p_{HE} = \frac{2P_{EE2}}{h_2} - p_{HD} = \text{horizontal pressure at E}$$

$$p_{HE} = \frac{2(17.59)}{10} - 1.1355 = 2.3825 \text{ ksf}$$

Pressure distributions on Surfaces AB and DE:

E. EARTH PRESSURE ON SURFACES CD AND DE. The earth force on AB may be transferred to Surface CD by assuming that no shear resistance is developed on Surface BD. Then to obtain moment equilibrium for Block ABCD, a vertical shear force must be developed on Surfaces

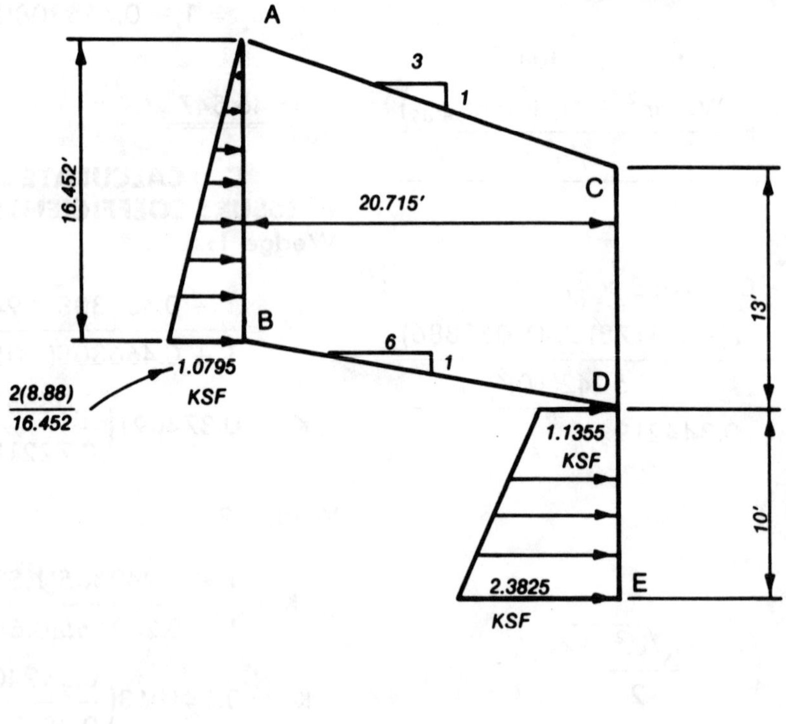

$$P_{EE2} = P_\gamma + P_V = 4.73 + 12.86 = \underline{17.59 \text{ k/ft}}$$

$$p_{VD} = \gamma_1 h_1$$

$$p_{VD} = 0.12(13) = 1.56 \text{ ksf}$$

$$= \text{vertical pressure at D}$$

AB and CD. The shear force will be

$$v_{AB} = -v_{CD} = P_{EE1}\left(\frac{\tan\beta_1 + 2\tan\beta_2}{3}\right)$$

A free body of Block ABCD is shown next.

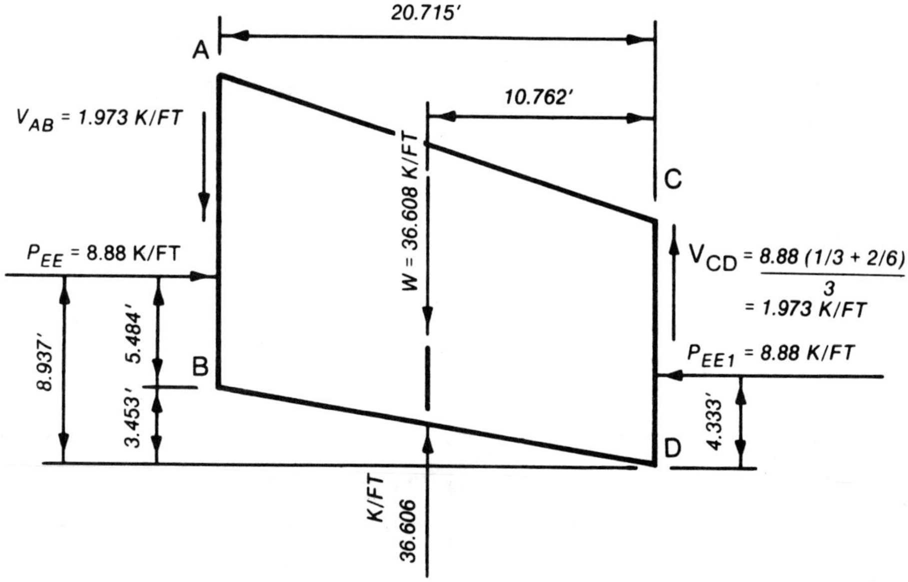

$$\sum M_D = 0$$

$$\sum M_D = 8.88(8.937 - 4.333) - 1.973(20.715)$$

$$= 0.013 \approx 0$$

The pressure distribution on Surfaces CD and DE is shown next.

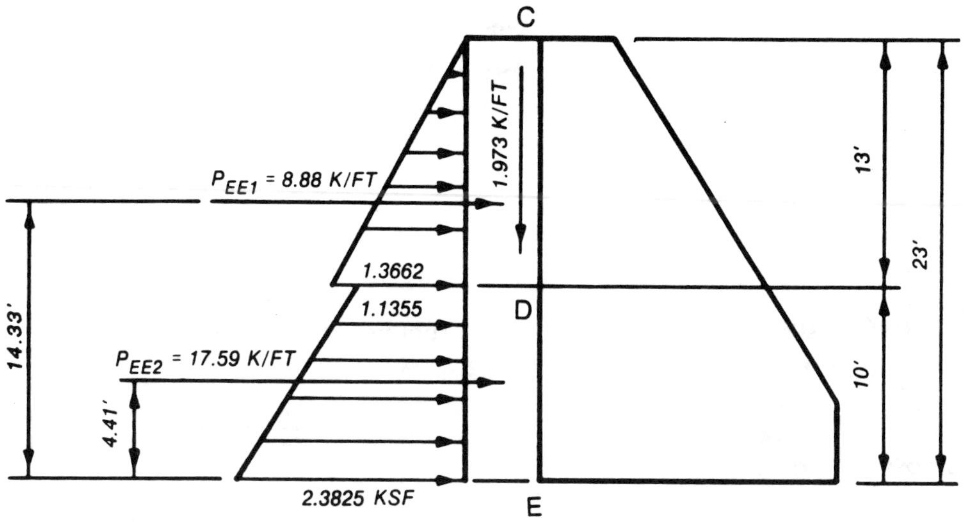

M-7. Example 7

Find the lateral earth force and pressure distribution on Surfaces FG and HI when $\phi = 35°$, $c = 0$, $\gamma = 0.12$ kcf, and SMF $= 2/3$.

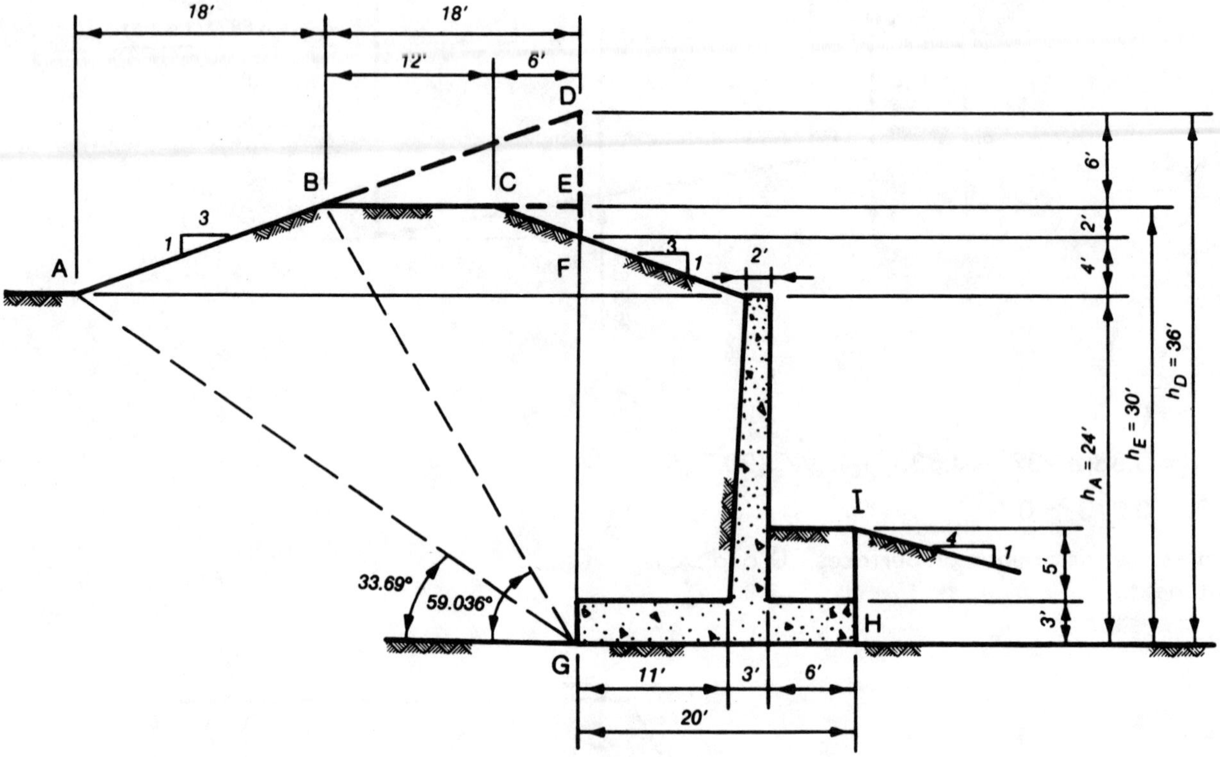

A. DRIVING SIDE

1. Assume that the critical slip plane intersects Surface BC, $\beta = 0$, $h = h_E = 30$ ft. The weight of the triangular area CEF will be taken as a negative strip surcharge and

$$V = -(\tfrac{1}{2})(0.12)(6)(2) = -0.72 \text{ k/ft.}$$

$$\phi_d = \tan^{-1}(\tfrac{2}{3}\tan 35°) = 25°, \ \tan\phi_d = 0.466308$$

From Equation 3-30

$$A = \tan\phi_d - \frac{2V(1+\tan^2\phi_d)}{\gamma h^2}$$

$$A = 0.466308 - \frac{2(-0.72)(1.217443)}{0.12(30)^2}$$

$$= 0.482541$$

From Equation 3-28

$$c_1 = \frac{2\tan^2\phi_d}{A}$$

$$c_1 = \frac{2(0.466308)^2}{0.482541} = 0.901242$$

From Equation 3-29

$$c_2 = \frac{\tan\phi_d}{A}$$

$$c_2 = \frac{0.466308}{0.482541} = 0.966359$$

From Equation 3-25

$$\alpha' = \tan^{-1}\left(\frac{c_1 + \sqrt{c_1^2 + 4c_2}}{2}\right)$$
$$= 56.866^\circ < 59.036^\circ$$

This shows that the critical slip plane does not intersect Surface BC.

2. Assume that the critical slip plane intersects Surface AB, $\tan\beta = -(1/3)$, $h = h_D = 36$ ft. The weights of areas BDE and CEF will be taken as a negative strip surcharge.

$$V = -\tfrac{1}{2}(0.12)(18)(6) - \tfrac{1}{2}(0.12)(6)(2)$$
$$= -7.2 \text{ k/ft}$$

From Equation 3-30

$$A = 0.466308 - \frac{2(-7.2)(1.217443)}{0.12(36)^2}$$
$$= 0.579034$$

From Equation 3-28

$$c_1 = \frac{2(0.466308)^2 - \dfrac{4(-7.2)(-\frac{1}{3})(1.217443)}{0.12(36)^2}}{0.579034}$$
$$= 0.621268$$

From Equation 3-29

$$c_2 = \frac{0.466308[1 - 0.466308(-\frac{1}{3})] - (-\frac{1}{3}) + \dfrac{2(-7.2)(\frac{1}{9})(1.217443)}{0.12(36)^2}}{0.579034}$$
$$c_2 = 1.484537$$

From Equation 3-25

$$\alpha = \tan^{-1}\left(\frac{c_1 + \sqrt{c_1^2 + 4c_2}}{2}\right) = \underline{\underline{57.473^\circ}}$$

$$33.69^\circ < \alpha < 59.036^\circ$$

This shows that critical slip plane does intersect Surface AB.

3. Calculate the pressure coefficients (see Appendix H).

$$K = \frac{1 - \tan\phi_d \cot\alpha}{1 + \tan\phi_d \tan\alpha}$$
$$= \frac{1 - 0.466308(0.637733)}{1 + 0.466308(1.568054)} = 0.405858$$

$$K_1 = K\left(\frac{\tan\alpha}{\tan\alpha - \tan\beta}\right)$$
$$= 0.405858\left(\frac{1.568054}{1.901387}\right) = 0.334707$$

$$K_v = K\tan\alpha = 0.405858(1.568054) = 0.636407$$

$$P_\gamma = \tfrac{1}{2}K_1\gamma h_D^2 = \tfrac{1}{2}(0.334707)(0.12)(36)^2$$
$$= 26.027 \text{ k/ft}$$

$$P_v = K_v V = 0.636407(-7.1) = -4.582 \text{ k/ft}$$

$$P_{EE} = P_\gamma + P_v = 26.027 - 4.582 = \underline{\underline{21.445 \text{ k/ft}}}$$

4. The net horizontal pressure at point F must be equal to zero, the negative pressure due to P_v cancels the positive pressure due to $P\gamma$.

$$p_{\gamma F} = 0.334707(0.12)(8) = 0.3213 \text{ ksf},$$
$$p_{vF} = -0.3213 \text{ ksf}$$

h_v (distribution length for P_v) =

$$\frac{2P_v}{p_{vF}} = \frac{2(-4.582)}{-0.3213} = 28.52 \text{ ft}$$

B. RESISTING SIDE

1. Calculate the critical slip plane angle α.

$$\tan\beta = -\tfrac{1}{4}, \tan\phi_d = 0.466308$$

From Equation 3-38

$$A = 0.466308$$

From Equation 3-36

$$c_1 = \frac{2\tan^2\phi_d}{A}$$

$$c_1 = \frac{2(0.466308)^2}{0.466308} = 0.932616$$

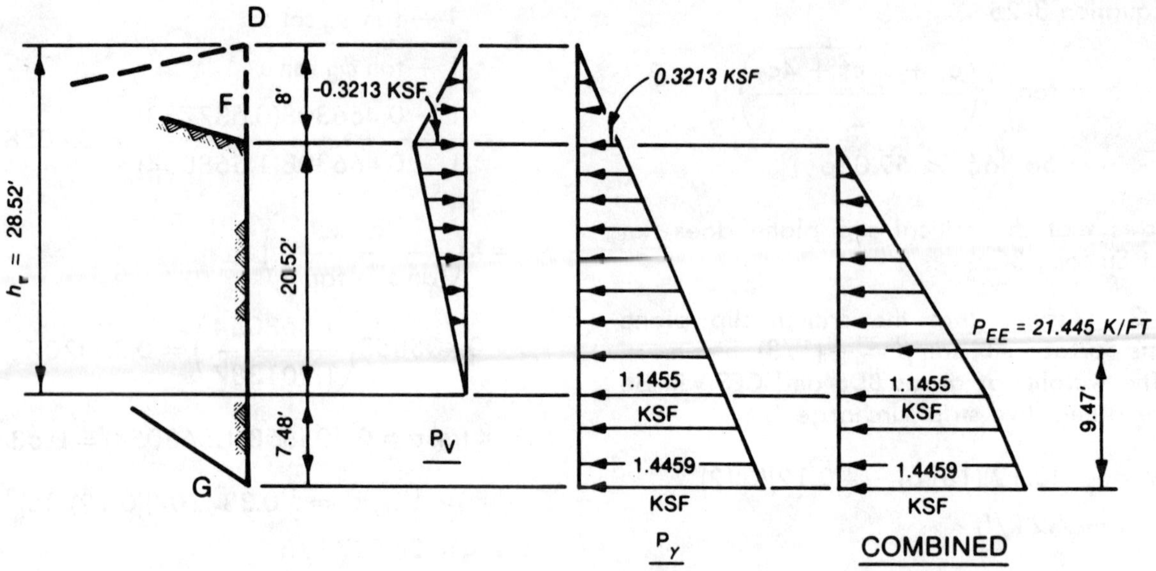

From Equation 3-37

$$c_2 = \frac{\tan \phi_d (1 + \tan \phi_d \tan \beta) + \tan \beta}{A}$$

$$c_2 = \frac{0.466308[1 + 0.466308(-\frac{1}{4})] + (-\frac{1}{4})}{0.466308}$$

$$= 0.347297$$

From Equation 3-35

$$\alpha = \tan^{-1}\left(\frac{-c_1 + \sqrt{c_1^2 + 4c_2}}{2}\right) = 15.917°$$

2. Calculate the pressure coefficients (see Appendix H) and pressure distribution.

$$K = \frac{1 + \tan \phi_d \cot \alpha}{1 - \tan \phi_d \tan \alpha}$$

$$= \frac{1 + 0.466308(3.506578)}{1 - 0.466308(0.285178)} = 3.039316$$

$$K_1 = K\left(\frac{\tan \alpha}{\tan \alpha - \tan \beta}\right)$$

$$= 3.039316\left(\frac{0.285178}{0.535178}\right) = 1.619547$$

$$P_{EE} = \tfrac{1}{2}K_1\gamma h^2 = \tfrac{1}{2}(1.619547)(0.12)(8)^2 = \underline{\underline{6.22\ \text{k/ft}}}$$

The calculation of the passive force and pressure distribution as performed above is adequate for performing a sliding analysis but should be calculated as described in paragraph 3-8 when an overturning or bearing capacity analysis is performed and for design of structural members.

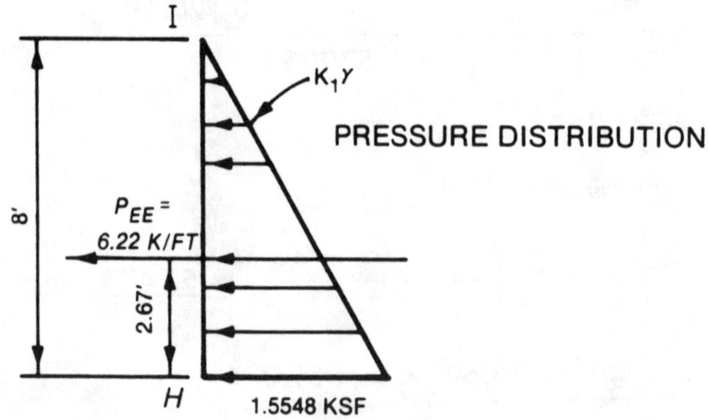

M-8. Example 8

Find the lateral earth force on the wall when

1. $\phi = 30°$, $c = 0$, $\gamma = 0.120$ kcf, SMF = 2/3
2. $\phi = 0$, $c = 0.60$ ksf, $\gamma = 0.120$ kcf, SMF = 2/3

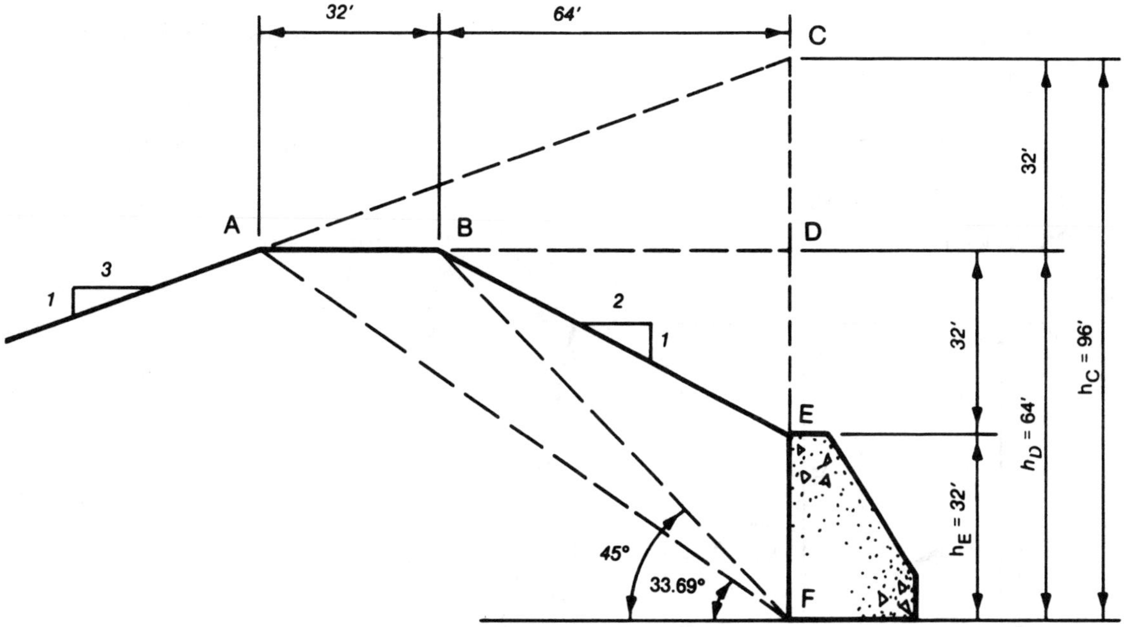

A. $\phi = 30°$, $\phi_d = \tan^{-1}$ (2/3 tan ϕ) = 21°, tan ϕ_d = 0.383864.

1. Since the tan β for Surface BE is 0.5, which is greater than tan ϕ_d, the critical slip plane will not intersect BE. Assume that the slip plane intersects Surface AB:

$$\tan\beta = 0, h = h_D = 64 \text{ ft}$$

$$V = -\tfrac{1}{2}(0.12)[64(32)] = -122.88 \text{ k/ft}$$

(negative weight of area BDE)

From Equation 3-30

$$A = \tan\phi_d - \frac{2V(1+\tan^2\phi_d)}{\gamma h_D^2}$$

$$A = 0.383864 - \frac{(2(-122.88)(1.147352)}{0.12(64)^2}$$

$$= 0.957540$$

From Equation 3-28

$$c_1 = \frac{2\tan^2\phi_d}{A}$$

$$c_1 = \frac{2(0.383864)^2}{0.957540} = 0.307771$$

From Equation 3-29

$$c_2 = \frac{\tan\phi_d}{A}$$

$$c_2 = \frac{0.383864}{0.957540} = 0.400886$$

From Equation 3-25

$$\alpha = \tan^{-1}\left(\frac{c_1 + \sqrt{c_1^2 + 4c_2}}{2}\right) = \underline{\underline{38.851°}}$$

$33.69° < \alpha < 45°$; the assumption that the slip plane intersects Surface AB is correct.

2. Calculate the pressure coefficients (see Appendix H).

$$K = \frac{1 - \tan \phi_d \cot \alpha}{1 + \tan \phi_d \tan \alpha}$$

$$= \frac{1 - 0.383864(1.241485)}{1 + 0.383864(0.805487)} = 0.399816$$

$$K_v = K \tan \alpha = 0.399816(0.805487) = 0.322047$$

$$P_\gamma = \tfrac{1}{2}K\gamma h^2 = \tfrac{1}{2}(0.399816)(0.12)(64)^2$$
$$= 98.26 \text{ k/ft}$$

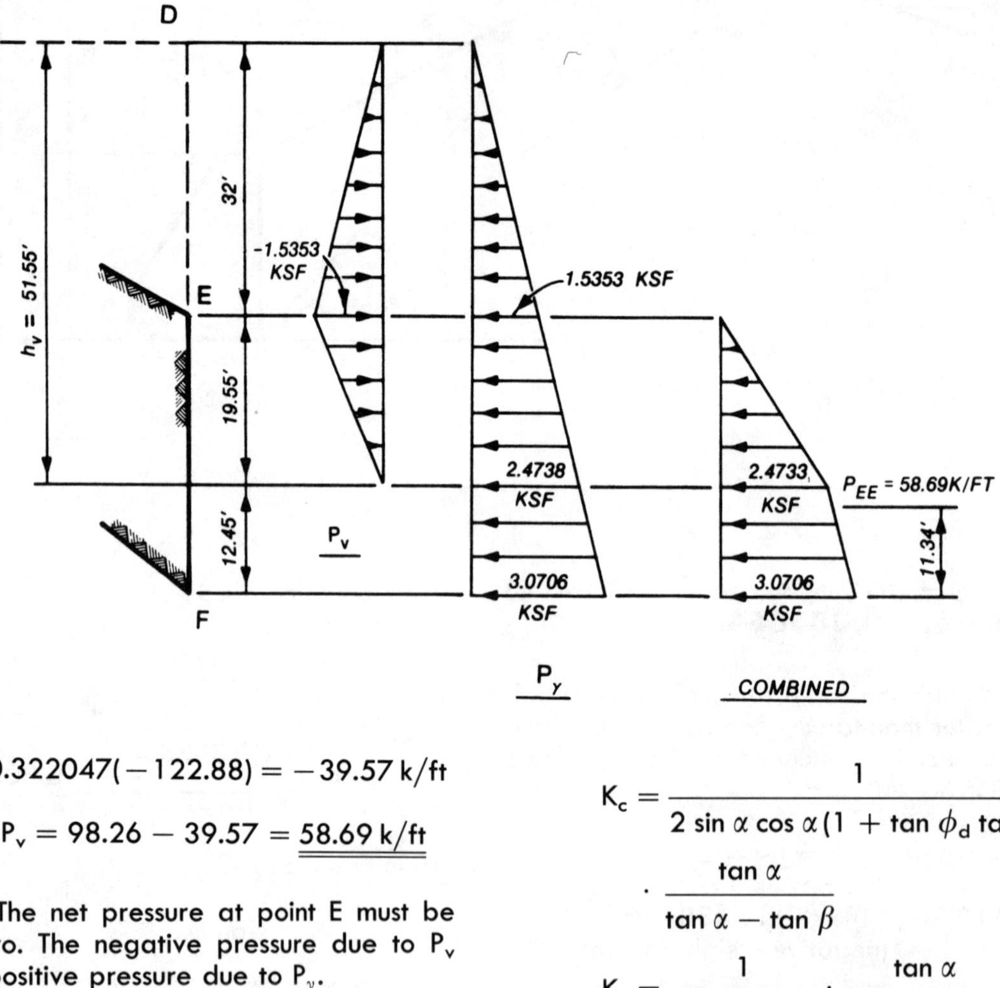

$$P_v = K_v V = 0.322047(-122.88) = -39.57 \text{ k/ft}$$

$$P_{EE} = P_\gamma + P_v = 98.26 - 39.57 = \underline{\underline{58.69 \text{ k/ft}}}$$

3. The net pressure at point E must be equal to zero. The negative pressure due to P_v cancels the positive pressure due to P_γ.

$$P_{\gamma E} = 0.399816(0.12)(32) = 1.5353 \text{ ksf}$$

$$p_{vE} = -1.5353 \text{ ksf}$$

h_v (distribution length for P_v)

$$= \frac{2P_v}{p_{vE}} = \frac{2(-39.57)}{-1.5353} = 51.55 \text{ ft}$$

B. $\phi = 0$, c = 0.6 ksf, cd = 2/3, c = 0.4 ksf.

1. Assume that the vertical crack extends upward from the critical slip plane to intersect Surface BE.

$$\tan \beta = 0.5, h = h_E = 32 \text{ ft}$$

From the equations given in Appendix H

$$K_1 = \frac{1 - \tan \phi_d \cot \alpha}{1 + \tan \phi_d \tan \alpha} \cdot \frac{\tan \alpha}{\tan \alpha - \tan \beta}$$

$$= \frac{\tan \alpha}{\tan \alpha - \tan \beta}$$

$$K_c = \frac{1}{2 \sin \alpha \cos \alpha (1 + \tan \phi_d \tan \alpha)} \cdot \frac{\tan \alpha}{\tan \alpha - \tan \beta}$$

$$K_c = \frac{1}{2 \sin \alpha \cos \alpha} \cdot \frac{\tan \alpha}{\tan \alpha - \tan \beta}$$

From Equation I-2

$$d_c = \frac{2K_c c_d}{K_1 \gamma} = \frac{c_d/\gamma}{\sin \alpha \cos \alpha}$$

$$d_c = \frac{0.4/0.12}{\sin \alpha \cos \alpha} = \frac{3.333 \text{ ft}}{\sin \alpha \cos \alpha}$$

Assume that $\alpha = 45°$, then $d_c = \dfrac{3.333}{0.5} = 6.667$ ft.

From Equation 3-30

$$A = \frac{2c_d}{\gamma(h + d_c)}$$

$$A = \frac{2(0.4)}{0.12(38.667)} = 0.172412$$

From Equation 3-28

$$c_1 = \frac{\dfrac{4c_d \tan\beta}{\gamma(h + d_c)}}{A}$$

$$c_1 = \frac{\dfrac{4(0.4)(0.5)}{0.12(38.667)}}{0.172412} = 1.00$$

From Equation 3-29

$$c_2 = \frac{-\tan\beta + \dfrac{2c_d}{\gamma(h + d_c)}}{A}$$

$$c_2 = \frac{-0.5 + \dfrac{2(0.4)}{0.12(38.667)}}{0.172412} = -1.900030$$

The term under the radical in Equation 3-25 will be negative. This makes α indeterminate, so the assumption that the crack intersects Surface BE is not correct.

2. Assume that the vertical crack intersects Surface AB.

$\tan\beta = 0, h = h_d = 64$ ft, $V = -122.88$ k/ft

(negative weight of area BDE)

Again assume that $\alpha = 45°$ and $d_c = 6.667$ ft.

From Equations 3-28, 3-29, and 3-30

$$A = \frac{2c_d}{\gamma(h + d_c)} - \frac{2V}{\gamma(h^2 - d_c^2)}$$

$$A = \frac{2(0.4)}{0.12(70.667)} - \frac{2(-1.22.88)}{0.12(4051.55)} = 0.599825$$

$$c_1 = 0, c_2 = \frac{2c_d}{\gamma(h + d_c)} \div A = \frac{2(0.4)}{0.12(70.70.667)}$$

$$\div 0.599825 = 0.157278$$

From Equation 3-25

$$\alpha = \tan^{-1}\left(\frac{c_1 + \sqrt{c_1^2 + 4c_2}}{2}\right) = 21.63° \neq 45°$$

Let $\alpha = 21.63°$ and $d_c = \dfrac{3.333}{0.342655} = 9.727$ ft:

$$A = \frac{2(0.4)}{0.12(73.727)} - \frac{2(-122.88)}{0.12(4001.39)} = 0.602246$$

$$c_1 = 0, c_2 = \frac{2(0.4)}{0.12(73.727)} \div 0.602246$$

$$= 0.150144$$

$$\alpha = \tan^{-1}\left(\frac{c_1 + \sqrt{c_1^2 + 4c_2}}{2}\right) = 21.18° \neq 21.63°$$

Let $\alpha = 21.18°$ and $d_c = \dfrac{3.333}{0.336893} = 9.893$ ft:

$$A = \frac{2(0.4)}{0.12(73.893)} - \frac{2(-122.88)}{0.12(3998.13)} = 0.602460$$

$$c_1 = 0, c_2 = \frac{2(0.4)}{0.12(73.893)} \div 0.602460$$

$$= 0.149754$$

$$\alpha = 21.15° \neq 21.18°$$

Let $\alpha = 21.15°$ and $d_c = \dfrac{3.333}{0.336506} = 9.90$ ft:

$$A = \frac{2(0.4)}{0.12(73.90)} - \frac{2(-122.88)}{0.12(3997.99)} = 0.602469$$

$$c_1 = 0, c_2 = \frac{2(0.4)}{0.12(73.90)} \div 0.602469$$

$$= 0.149737$$

$$\underline{\underline{\alpha = 21.15°}},\ \underline{\underline{d_c = 9.90 \text{ ft}}}$$

Check to see if the crack intersects Surface AB.

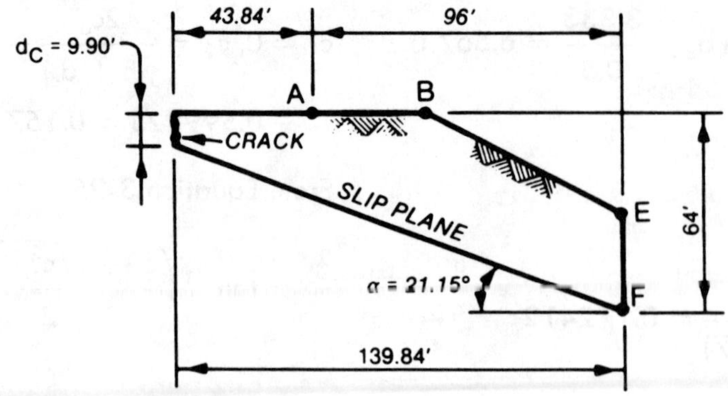

The assumption that the crack intersects Surface AB is not correct.

3. Assume that the crack intersects the sloping surface to the left of point A.

$$\tan\beta = -\tfrac{1}{3},\ h = h_c = 96 \text{ ft},$$

$$V = -\tfrac{1}{2}(0.12)[(64)(32) + (96)(32)]$$

$V = -307.20$ k/ft (negative weight of areas ACD and BDE)

$$\text{Let } \alpha = 30^\circ \text{ and } d_c = \frac{3.333}{0.433013} = 7.70 \text{ ft:}$$

From Equation 3-30

$$A = \frac{2c_d(1 - \tan\phi\tan\beta)}{\gamma(h + d_c)} - \frac{2V}{\gamma(h^2 - d_c^2)}$$

$$A = \frac{2(0.4)}{0.12(103.70)} - \frac{2(-307.20)}{0.12(9156.71)}$$

$$= 0.623441$$

From Equation 3-28

$$c_1 = \frac{\dfrac{4c_d\tan\beta}{\gamma(h + d_c)} - \dfrac{4V\tan\beta}{\gamma(h^2 - d_c^2)}}{A}$$

$$c_1 = \frac{\dfrac{4(0.4)(-\frac{1}{3})}{0.12(103.70)} - \dfrac{4(-307.20)(-\frac{1}{3})}{0.12(9156.71)}}{0.623441}$$

$$= -0.666666$$

From Equation 3-29

$$c_2 = \frac{-\tan\beta + \dfrac{2c_d}{\gamma(h + d_c)} + \dfrac{2V\tan^2\beta}{\gamma(h^2 - d_c^2)}}{A}$$

$$c_2 = \frac{-\left(-\dfrac{1}{3}\right) + \dfrac{2(0.4)}{0.12(103.7)} + \dfrac{2(-307.20)(\frac{1}{9})}{0.12(9156.71)}}{0.623441}$$

$$= 0.538132$$

From Equation 3-25

$$\alpha = \tan^{-1}\left(\frac{c_1 + \sqrt{c_1^2 + 4c_2}}{2}\right)$$

$$\alpha = 25.29^\circ \neq 30^\circ$$

$$\text{Let } \alpha = 25.29^\circ \text{ and } d_c = \frac{3.333}{0.386256} = 8.63 \text{ ft:}$$

$$A = \frac{2(0.4)}{0.12(104.63)} - \frac{2(-307.2)}{0.12(9141.52)} = 0.623799$$

$$c_1 = \frac{\dfrac{4(0.4)(-\frac{1}{3})}{0.12(104.63)} - \dfrac{4(-307.2)(-\frac{1}{3})}{0.12(9141.52)}}{0.623799}$$

$$= -0.666666$$

$$c_2 = \frac{-(-\frac{1}{3}) + \dfrac{2(0.4)}{0.12(104.63)} + \dfrac{2(-307.2)(\frac{1}{9})}{0.12(9141.52)}}{0.623799}$$

$$= 0.536741$$

$$\alpha = 25.25^\circ \neq 25.29^\circ$$

Let $\alpha = 25.25^\circ$ and $d_c = \dfrac{3.333}{0.385812} = 8.64$ ft:

$$A = \frac{2(0.4)}{0.12(104.64)} - \frac{2(-307.2)}{0.12(9141.35)} = 0.623803$$

$$c_1 = \frac{\dfrac{4(0.4)(-\frac{1}{3})}{0.12(104.64)} - \dfrac{4(-307.2)(-\frac{1}{3})}{0.12(9141.35)}}{0.623803}$$

$$= -0.666666$$

$$c_2 = \frac{-(-\frac{1}{3}) + \dfrac{2(0.4)}{0.12(104.63)} + \dfrac{2(-307.2)(\frac{1}{9})}{0.12(9141.52)}}{0.623799}$$

$$= 0.536726$$

$$\alpha = \underline{\underline{22.246^\circ \approx 25.25^\circ}},\ \underline{\underline{d_c = 8.64 \text{ ft}}}$$

4. Calculate the pressure coefficients (See Appendix H) and earth forces.

$$K = \frac{1 - \tan\phi_d \cot\alpha}{1 + \tan\phi_d \tan\alpha} = \frac{1 - 0(2.120303)}{1 + 0(0.471631)} = 1$$

$$K_1 = K\left(\frac{\tan\alpha}{\tan\alpha - \tan\beta}\right) = 1\left(\frac{0.471731}{0.804964}\right)$$

$$= 0.585903$$

$$K_v = K \tan\alpha = 1(0.471631) = 0.471631$$

$$P_\gamma = \tfrac{1}{2}K_1\gamma(h - d_c)^2$$

$$= \tfrac{1}{2}(0.585903)(0.12)(96 - 8.64)^2$$

$$= 268.29 \text{ k/ft}$$

$$P_v = K_v V = 0.471631(-307.2) = -144.89 \text{ k/ft}$$

$$P_{EE} = P_\gamma + P_v = 268.29 - 144.89 = \underline{\underline{123.40 \text{ k/ft}}}$$

5. Pressure distribution. The negative horizontal pressure, due to P_v, at point E must cancel the positive pressure, due to P_γ.

$$p_{\gamma E} = 0.585903(0.12)(64 - 8.64) = 3.8923 \text{ ksf}$$

$$p_{vE} = -3.8923 \text{ ksf}$$

$$h_v = \text{distribution length for } P_v = \frac{2(-144.89)}{-3.8923}$$

$$= 74.45 \text{ ft}$$

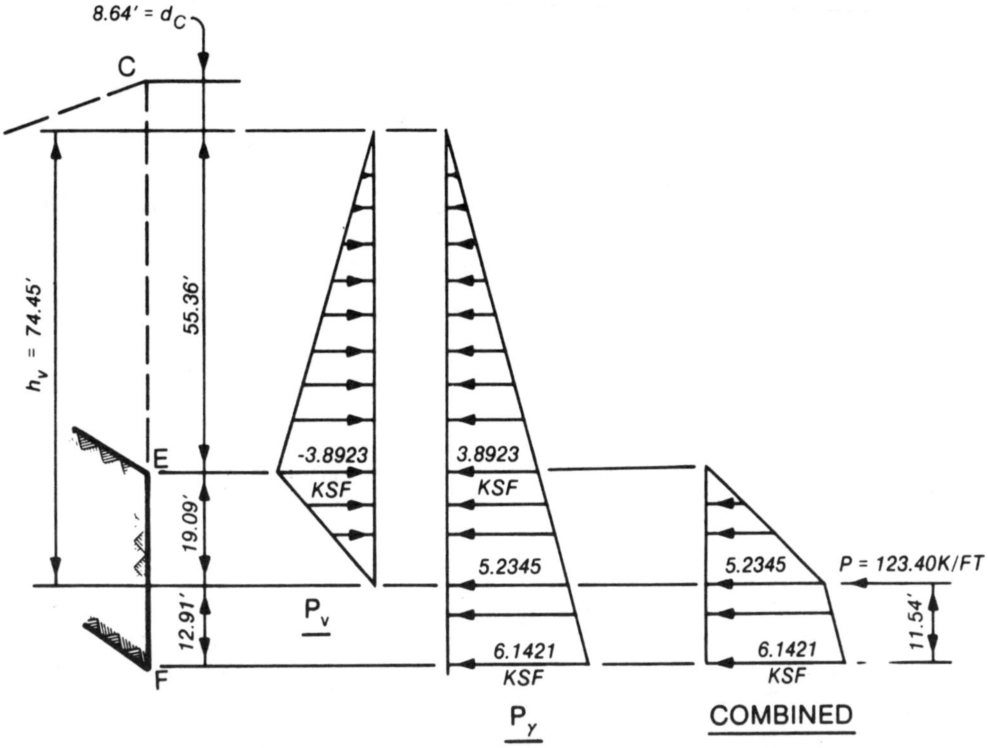

M-9. Example 9

Compute the lateral earth pressure using pressure coefficients. Check by the wedge method.

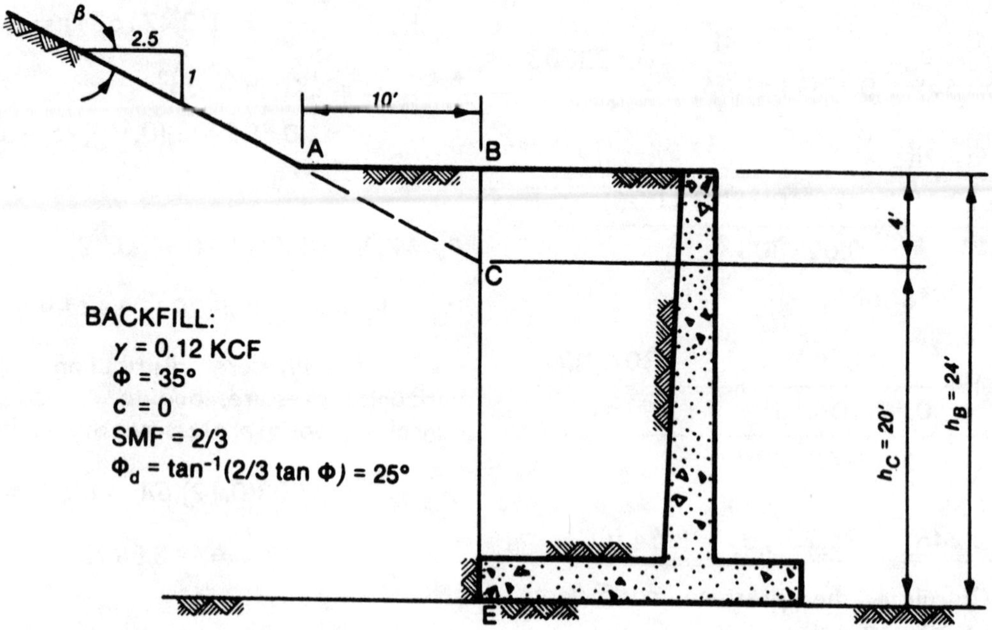

A. FIND THE CRITICAL SLIP-PLANE ANGLE. Consider the basic wedge to have a height of 20 feet (h_c), with a sloping top surface ($\tan \alpha = 0.4$). The weight of triangle ABC will be considered a finite surcharge (V).

$$V = \tfrac{1}{2}(0.12)(10)(4) = 2.4 \text{ k}$$

From Equation 3-30

$$A = \tan \phi_d - \frac{2V(1 + \tan^2)\phi_d}{\gamma h_c^2}$$

$$= 0.466308 - \frac{2(2.4)(1.217443)}{0.12(20)^2}$$

$$A = 0.344564$$

From Equation 3-28

$$c_1 = \left[2 \tan^2 \phi_d - \frac{4V \tan \beta(1 + \tan^2 \phi_d)}{\gamma h_c^2}\right] \div A$$

$$c_1 = \left[2(0.217443) - \frac{4(2.4)(0.4)(1.217443)}{0.12(20)^2}\right]$$

$$\div\, 0.344564$$

$$-c_1 = 0.979471$$

From Equation 3-29

$$c_2 = \left[\tan \phi_d (1 - \tan \phi_d \tan \beta) - \tan \beta\right.$$

$$\left. + \frac{2V \tan^2 \beta (1 + \tan^2 \phi_d)}{\gamma h_c^2}\right] \div A$$

$$c_2 = \left[0.466308(1 - 0.466308 \times 0.4) - 0.4\right.$$

$$\left. + \frac{2(2.4)(0.16)(1.217443)}{0.12(20)^2}\right] \div 0.344564$$

$$c_2 = -0.003454$$

From Equation 3.25

$$\alpha = \tan^{-1}\left(\frac{c_1 + \sqrt{c_1^2 + 4c_2}}{2}\right) = \underline{\underline{44.302°}}$$

B. CALCULATE THE EARTH PRESSURE COEFFICIENTS (Appendix H).

$$K = \frac{1 - \tan \phi_d \cot \alpha}{1 + \tan \phi_d \tan \alpha}$$

$$= \frac{1 - 0.466308 \times 1.024667}{1 + 0.466308 \times 0.975927}$$

$$K = \underline{\underline{0.3589}}$$

$$K_1 = K\left(\frac{\tan \alpha}{\tan \alpha - \tan \beta}\right) = 0.3589\left(\frac{0.975927}{0.575927}\right)$$

$$= \underline{\underline{0.6082}}$$

The earth pressure distribution is calculated in the following figures.

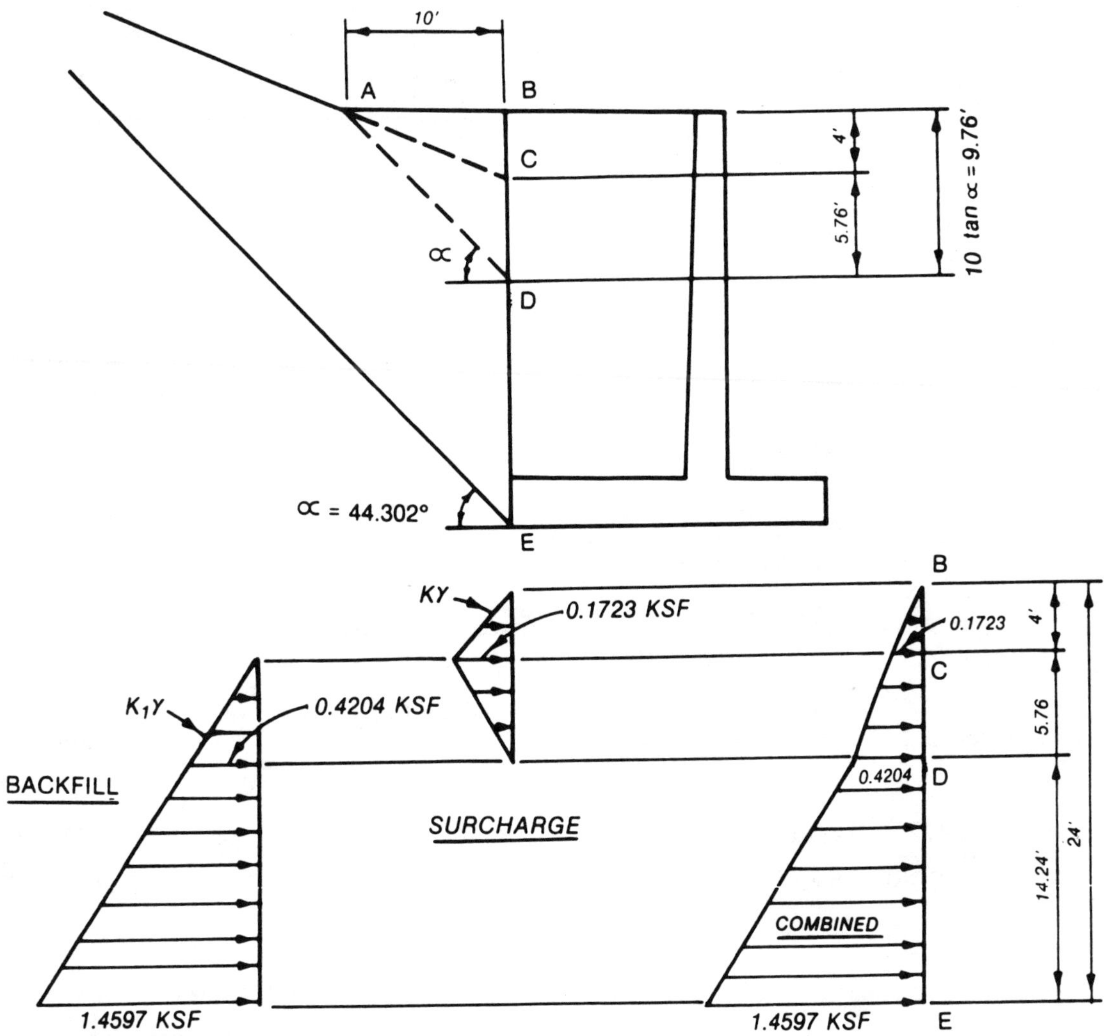

C. CHECK THE PRESSURE USING THE WEDGE METHOD.

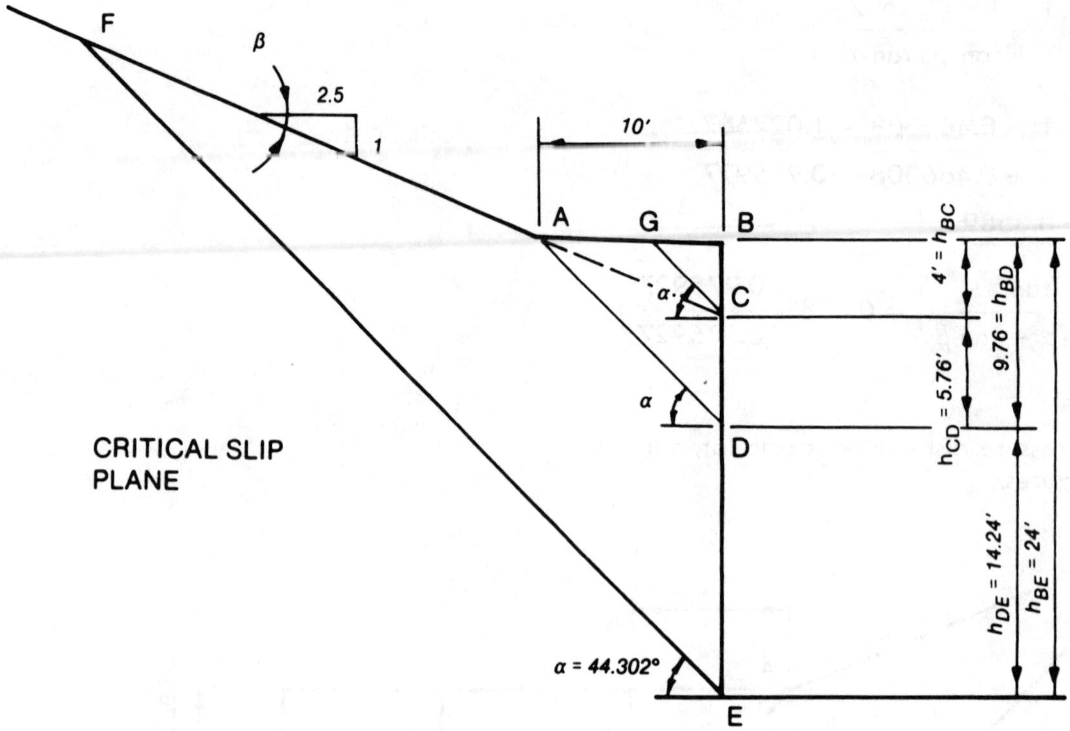

Force on Surface BC.

$$\text{Weight of wedge GBC} = W = \frac{\gamma h_{BC}^2}{2 \tan \alpha}$$

$$= \frac{0.12(4)^2}{2(0.975927)}$$

$$= 0.9837 \text{ k}$$

From Equation 3-23

$$P_{BC} = \frac{W(\tan \alpha - \tan \phi_d)}{1 + \tan \phi_d \tan \alpha}$$

$$= \frac{0.9837(0.975927 - 0.466308)}{1 + 0.466308(0.975927)}$$

$$= \underline{\underline{0.345 \text{ k}}}$$

Force on Surface BD.

$$\text{Weight of wedge ABD} = W = \tfrac{1}{2}(0.12)(10)(9.76)$$

$$= 5.8560 \text{ k}$$

$$P_{BD} = \frac{5.8560(0.975927 - 0.466308)}{1 + 0.466308(0.975927)}$$

$$= \underline{\underline{2.051 \text{ k}}}$$

Force on Surface BE:

$$\text{Weight of wedge FABE} = W = \frac{\gamma h_{CE}^2}{2(\tan \alpha - \tan \beta)} + \tfrac{1}{2}(0.12)(10)(4)$$

$$W = \frac{0.12(20)^2}{2(0.0975927 - 0.4)} + 2.4 = 44.0719 \text{ k}$$

$$P_{BE} = \frac{44.0719(0.975927 - 0.466308)}{1 + 0.466308(0.975927)}$$

$$= \underline{\underline{15.435 \text{ k}}}$$

Calculate pressures at C, D, and E.

$$p_C = \frac{2P_{BC}}{h_{BC}} = \frac{2(0.345)}{4} = 0.1725 \text{ ksf}$$

$$p_D = \frac{2(P_{BD} - P_{BC})}{h_{CD}} - p_C = \frac{2(2.051 - 0.345)}{5.76} - 0.1725 = 0.4199 \text{ ksf}$$

$$p_E = \frac{2(P_{BE} - p_{BD})}{h_{DE}} - p_D = \frac{2(15.435 - 2.051)}{14.24} - 0.4199 = 1.4599 \text{ ksf}$$

The pressure diagram-wedge method is shown next.

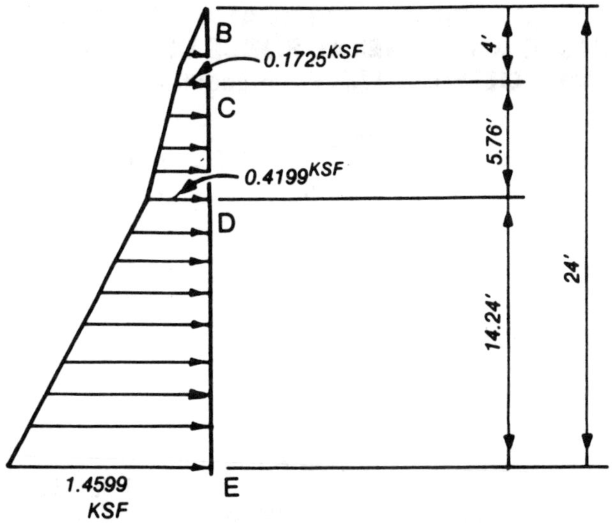

Checks combined pressure diagram obtained using pressure coefficients.

M-10. Example 10

Calculate lateral earth pressure. Use the approximate wedge method as well as the elastic method to find pressure due to the distributed finite surcharge.

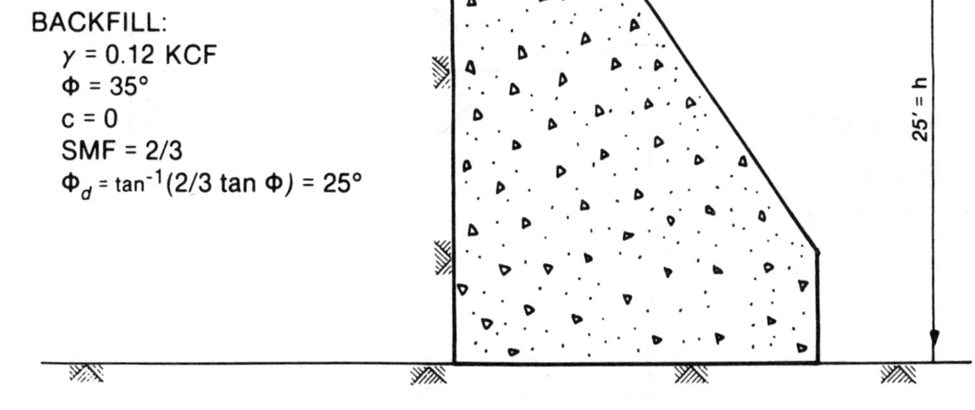

A. FIND CRITICAL SLIP-PLANE ANGLE AND LATERAL FORCE USING PRESSURE COEFFICIENTS. From Equation 3-30

$$A = \tan \phi_d - \frac{2V(1 + \tan^2 \phi_d)}{\gamma h^2} = 0.466308$$

$$- \frac{2(3)(1.217443)}{0.12(25)^2}$$

$$A = 0.368913$$

From Equation 3-28

$$c_1 = \frac{2 \tan^2 \phi_d}{A} = \frac{2(0.466308)^2}{0.368913} = 1.178832$$

From Equation 3-29

$$c_2 = \frac{\tan \phi_d}{A} = \frac{0.466308}{0.368913} = 1.264005$$

From Equation 3-25

$$\alpha = \tan^{-1}\left(\frac{c_1 + \sqrt{c_1^2 + 4c_2}}{2}\right) = \underline{\underline{61.721^\circ}}$$

From Appendix H, the pressure coefficients are

$$K = \frac{1 - \tan \phi_d \cot \alpha}{1 + \tan \phi_d \tan \alpha}$$

$$= \frac{1 - 0.466308 \times 0.537972}{1 + 0.466308 \times 1.858833}$$

$$K = \underline{\underline{0.4013}}$$

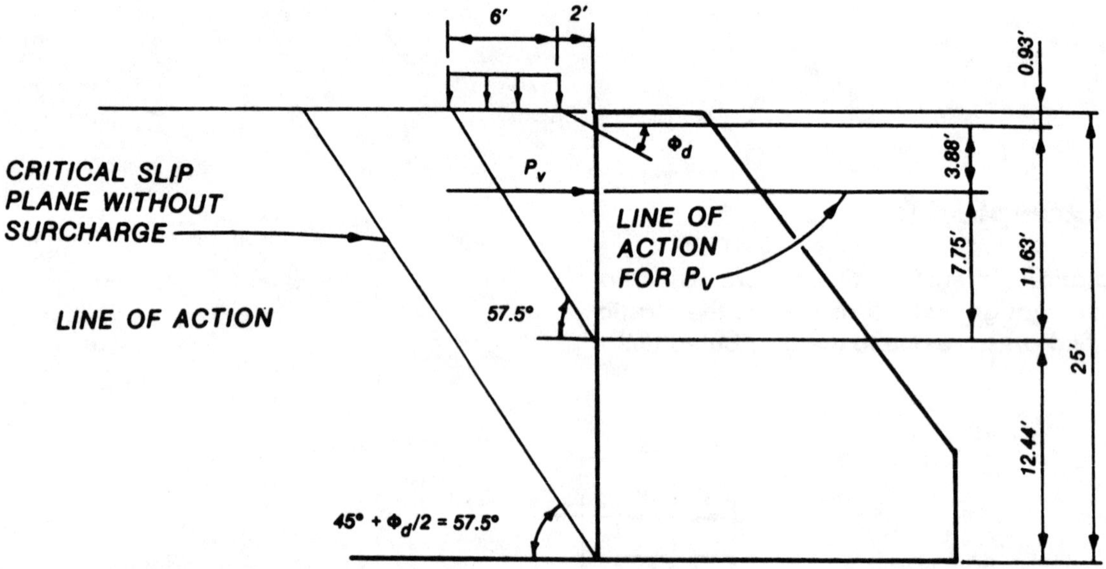

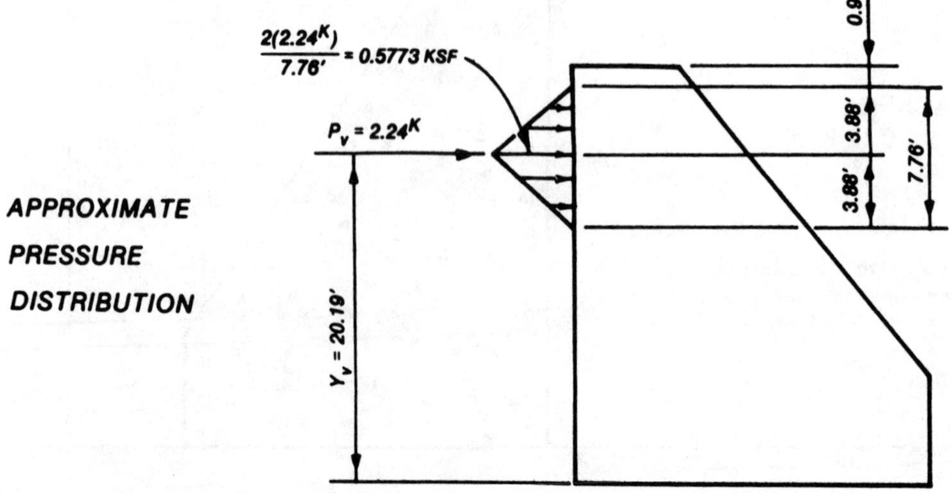

$K_v = K \tan\alpha = 0.4013(1.858833) = 0.7459$

$P_\gamma = \frac{1}{2}K\gamma h^2 = \frac{1}{2}(0.4013)(0.12)(25)^2 = 15.05\text{ k}$

$P_v = K_v V = 0.7459(3) = 2.24\text{ k}$

$P_{total} = P_\gamma + P_v = 15.05 + 2.24 = \underline{\underline{17.29\text{ k}}}$

Approximate method—pressure distribution for surcharge (see Figure 3-29).

B. FIND THE PRESSURE DISTRIBUTION DUE TO SURCHARGE BY USING THE ELASTIC METHOD (see Figure 3-27). Assume a nonyielding wall.

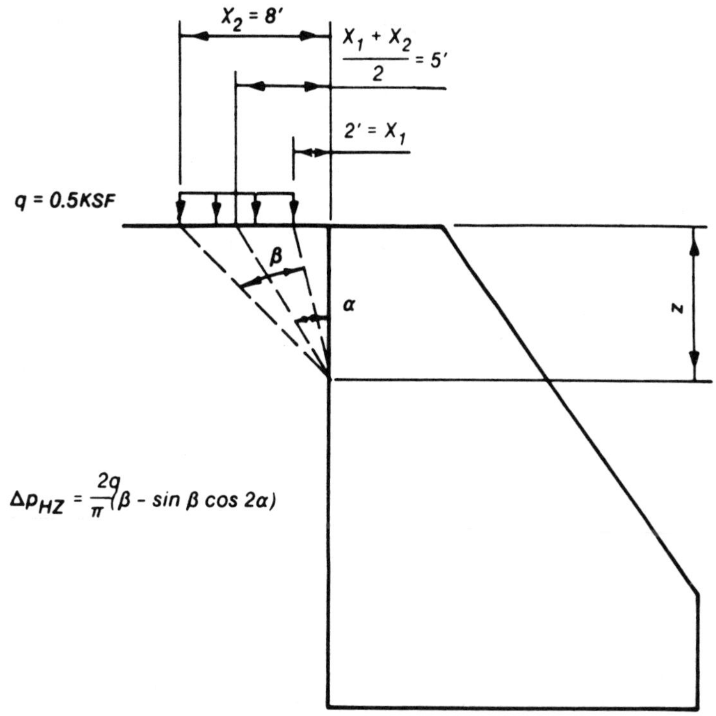

z, ft	$\frac{2q}{\pi}$	β, rad	α°	Δp_{Hz}, ksf
1.25	0.3183	0.4036	75.96	0.239
2.50	0.3183	0.5932	63.43	0.296
5.00	0.3183	0.6316	45.00	0.201
7.50	0.3183	0.5570	36.69	0.113
10.00	0.3183	0.4773	26.57	0.064
12.50	0.3183	0.4107	21.80	0.039
15.00	0.3183	0.3574	18.43	0.025
17.50	0.3183	0.3150	15.95	0.017
20.00	0.3183	0.2808	14.04	0.012
22.50	0.3183	0.2530	12.53	0.008
25.00	0.3183	0.2298	11.31	0.006

Pressure diagram-elastic method.

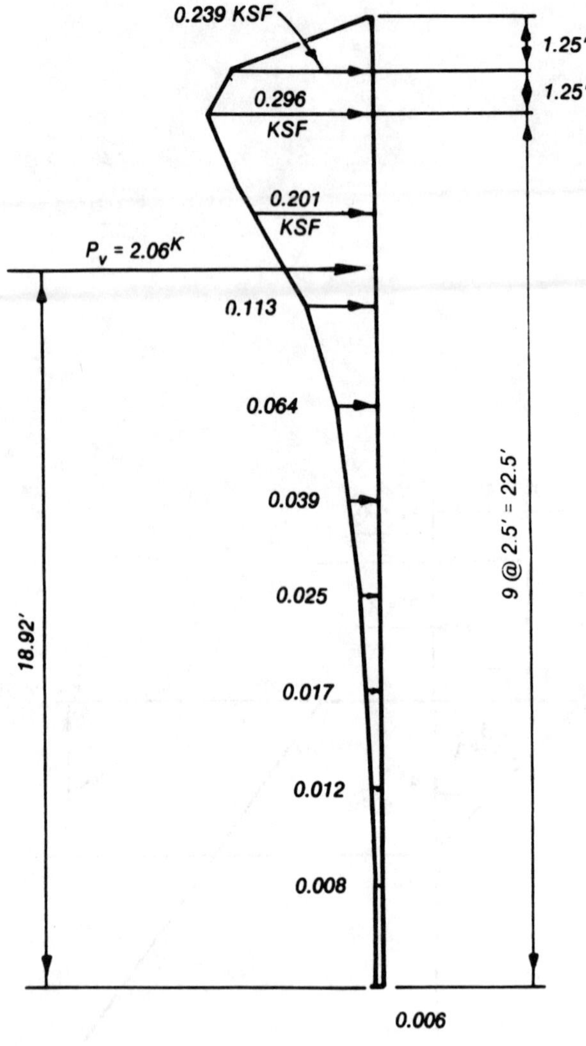

The force, due to the surcharge, determined by the approximate method is more severe. It will be combined with the backfill force to obtain the total force.

Combined pressure diagram and force:

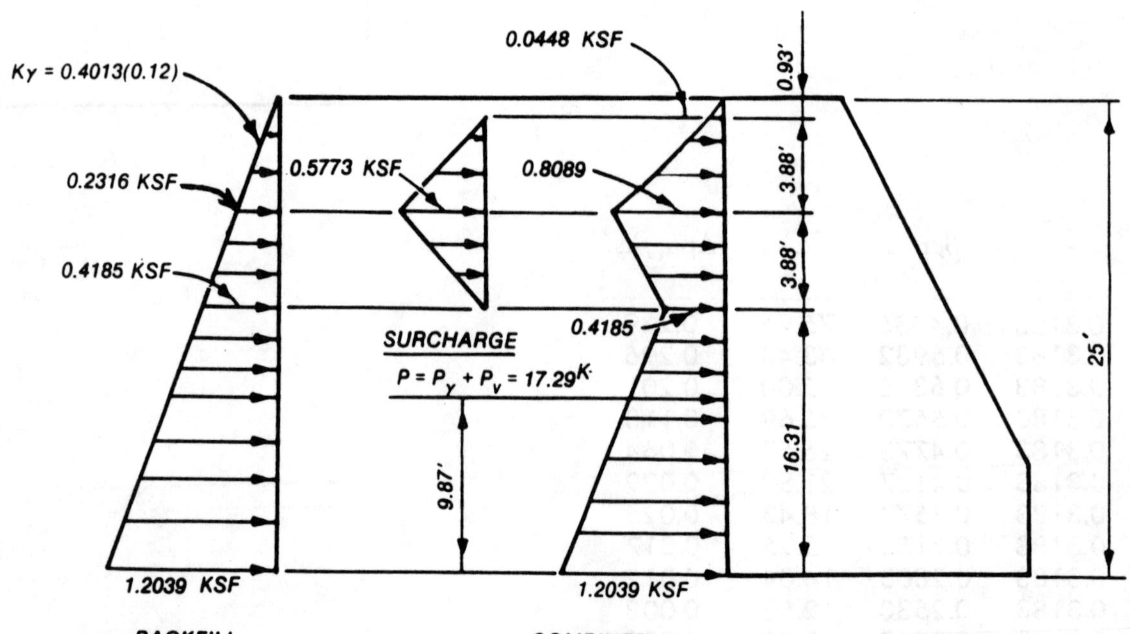

M-11. Example 11

Find the lateral forces and pressures acting on the wall for the seismic condition.
Soil properties (on both sides of wall):

$\gamma = 0.12$ k/ft^3 (moist weight)

$\gamma_v = 0.0625$ k/ft^3 (buoyancy weight)

$\gamma_s = 0.125$ k/ft^3 (saturated weight)

$\phi = 35°, c = 0$

Seismic coefficients:

$k_H = 0.20$

$k_v = 0$

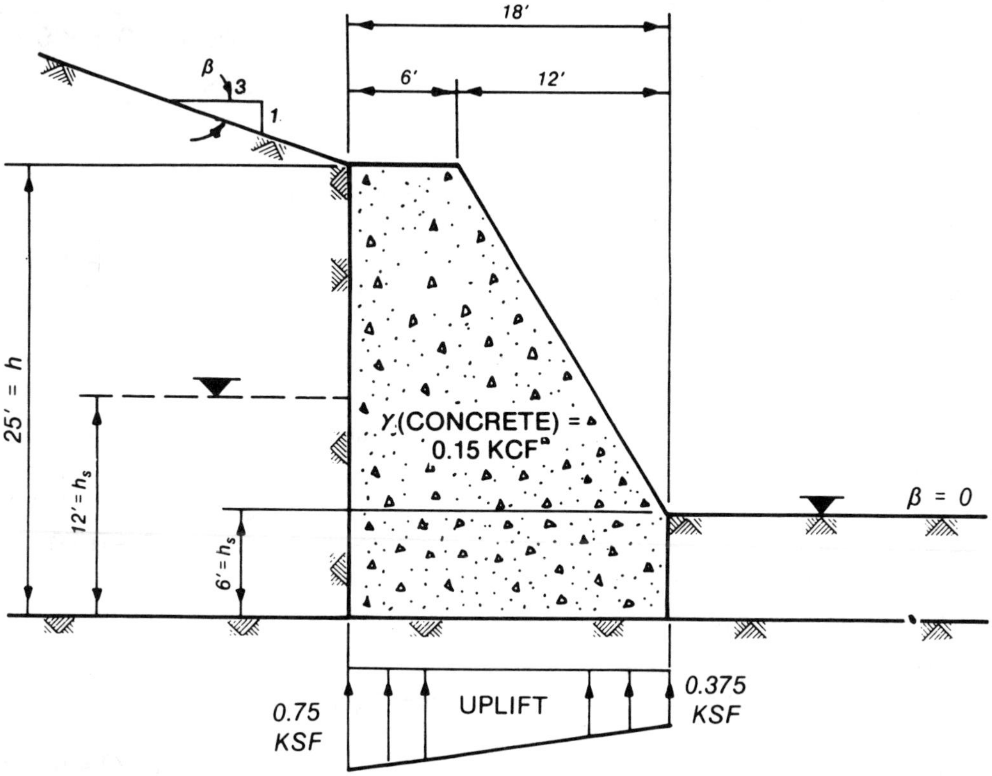

A. FIND THE FORCES ACTING ON DRIVING SIDE

From Equation 3-57

$$c_1 = \frac{2(\tan\phi - k_h)}{1 + k_h \tan\phi} = \frac{2(0.700208 - 0.2)}{1 + 0.2(0.700208)}$$

$$= 0.877526$$

From Equation 3-58

$$c_2 = \frac{\tan\phi\,(1 - \tan\phi \tan\beta) - (\tan\beta + k_h)}{\tan\phi\,(1 + k_h \tan\phi)}$$

$$c_2 = \frac{0.700208(1 - 0.700208 \times \frac{1}{3}) - (\frac{1}{3} + 0.2)}{0.700208(1 + 0.2 \times 0.700208)}$$

$$= 0.004315$$

From Equation 3-56

$$\alpha = \tan^{-1}\left(\frac{c_1 + \sqrt{c_1^2 + 4c_2}}{2}\right) = \underline{\underline{41.426°}}$$

From Appendix H

$$K = \frac{1 - \tan\phi \cot\alpha}{1 + \tan\phi \tan\alpha}$$

$$= \frac{1 - 0.700208(1.133240)}{1 + 0.700208(0.882425)}$$

$$K = 0.12763$$

From Equation 3-54

$$K_A = K\left(\frac{\tan\alpha}{\tan\alpha - \tan\beta}\right)$$

$$= 0.12763\left(\frac{0.882425}{0.882425 - \frac{1}{3}}\right) = \underline{\underline{0.2051}}$$

From Appendix H

$$K_b = K\left[1 + \left(\frac{\tan\alpha}{\tan\alpha - \tan\beta} - 1\right)\left(\frac{\gamma}{\gamma_b}\right)\right]$$

$$K_b = 0.12763\left[1 + \left(\frac{0.882425}{0.549092} - 1\right)\left(\frac{0.12}{0.0625}\right)\right]$$

$$= \underline{\underline{0.2764}}$$

From Equation 3-69

$$P_A = \tfrac{1}{2}K_A\gamma(h - h_s)^2 + \tfrac{1}{2}h_s[2K_A\gamma(h - h_s) + K_b\gamma_b h_s]$$

$$P_A = \tfrac{1}{2}(0.2051)(0.12)(13)^2 + \tfrac{1}{2}(12)[2(0.2051)(0.12)(13) + 0.2764(0.0625)(12)]$$

$$P_A = \underline{\underline{7.16\text{ k}}}$$

From Equation 3-71

$$\Delta P_{AE} = k_h\left[\frac{\gamma h^2}{2(\tan\alpha - \tan\beta)} + \frac{(\gamma_s - \gamma)h_s^2}{2\tan\alpha}\right]$$

$$\Delta P_{AE} = 0.2\left[\frac{0.12(25)^2}{2(0.549092)} + \frac{0.005(12)^2}{2(0.882425)}\right]$$

$$= \underline{\underline{13.74\text{ k}}}$$

From Equation 3-70

$$P_{ws} = \tfrac{1}{2}\gamma_w h_s^2 = \tfrac{1}{2}(0.0625)(12)^2 = \underline{\underline{4.50\text{ k}}}$$

B. FIND FORCES ACTING ON RESISTING SIDE.

From Equation 3-60

$$c_1 = \frac{2(\tan\phi - k_h)}{1 + k_h\tan\phi} = \frac{2(0.700208 - 0.2)}{1 + 0.2(0.700208)}$$

$$= 0.877526$$

From Equation 3-61

$$c_2 = \frac{\tan\phi - k_h}{\tan\phi(1 + k_h\tan\phi)}$$

$$c_2 = \frac{0.700208 - 0.2}{0.700208(1 + 0.2 \times 0.700208)}$$

$$= 0.626618$$

From Equation 3-59

$$\alpha = \tan^{-1}\left(\frac{-c_1 + \sqrt{c_1^2 + 4c_2}}{2}\right) = \underline{\underline{24.999^\circ}}$$

From Equation 3-77

$$K_P = \frac{1 + \tan\phi\cot\alpha}{1 - \tan\phi\tan\alpha} = \frac{1 + 0.700208(2.144605)}{1 - 0.700208(0.466286)}$$

$$K_P = \underline{\underline{3.7144}}$$

From Equation 3-73

$$P_P = \tfrac{1}{2}K_P\gamma_b h^2 = \tfrac{1}{2}(3.7144)(0.0625)(6)^2$$

$$= \underline{\underline{4.18\text{ k}}}$$

From Equation 3-75

$$\Delta P_{PE} = k_h = \left(\frac{\gamma_s h^2}{2\tan\alpha}\right) = 0.2\left[\frac{0.125(6)^2}{2(0.466286)}\right]$$

$$= \underline{\underline{0.97\text{ k}}}$$

From Equation 3-74

$$P_{ws} = \tfrac{1}{2}\gamma_w h_s^2 = \tfrac{1}{2}(0.0625)(6)^2 = \underline{\underline{1.13\text{ k}}}$$

C. FIND INERTIA FORCE DUE TO WEIGHT OF WALL.

$$18' \times 25' \times 0.15 = \quad 67.50 \times 12.50' = \quad 843.75$$

$$-\tfrac{1}{2} \times 12' \times 19' \times 0.15 = \underline{-17.10} \quad \times 18.67' = \underline{-319.25}$$

$$W = \quad 50.40\text{ k} \qquad\qquad 524.50$$

$$\bar{y} = \frac{524.50}{50.40} = \underline{\underline{10.41\text{ ft}}}$$

$$k_h W = 0.2(50.40\text{ k}) = \underline{\underline{10.08\text{ k}}}$$

D. SUMMARY OF FORCES AND PRESSURE DISTRIBUTIONS.

Permissible simplication for dynamic earth pressure distribution—driving side:

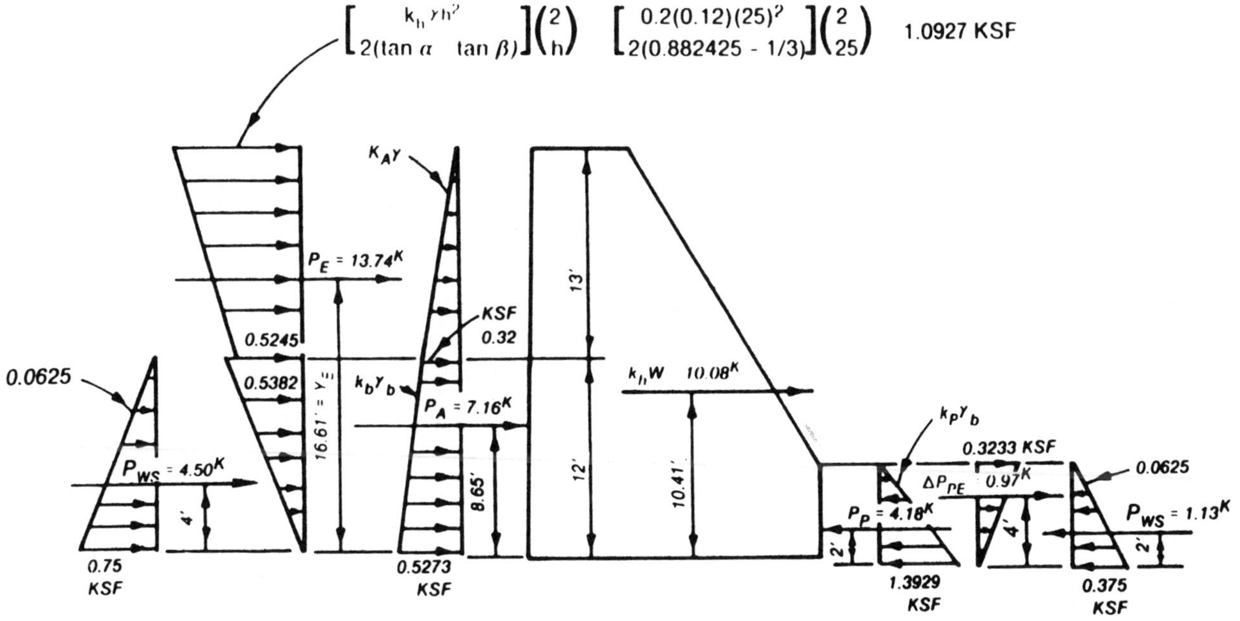

The discontinuity of this pressure diagram, at the water table, may be eliminated by considering that the soil weight above and below water is equal to the moist weight. The difference is not significant.

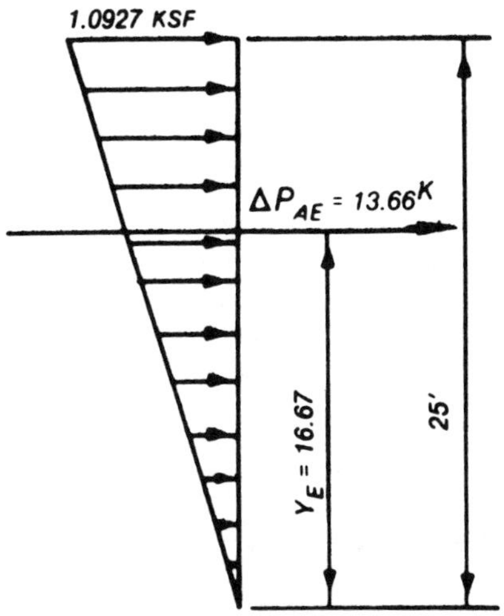

In this case, the difference in forces is -0.58% and difference in dimension, Y_E, is $+0.36\%$.

Mononobe-Okabe force and pressure distribution—resisting side.

If the last two pressure diagrams for P_P and ΔP_{PE} (on the preceding page) are combined, negative pressure will be obtained for some distance below the top of the ground. Since earth pressure cannot pull on the wall, the pressure diagram and force should be determined by setting all negative pressures to zero.

APPENDIX N

STABILITY, BEARING CAPACITY, AND REINFORCEMENT COMPUTATIONS, EXAMPLES

Summary of Examples

Example 1, page 233

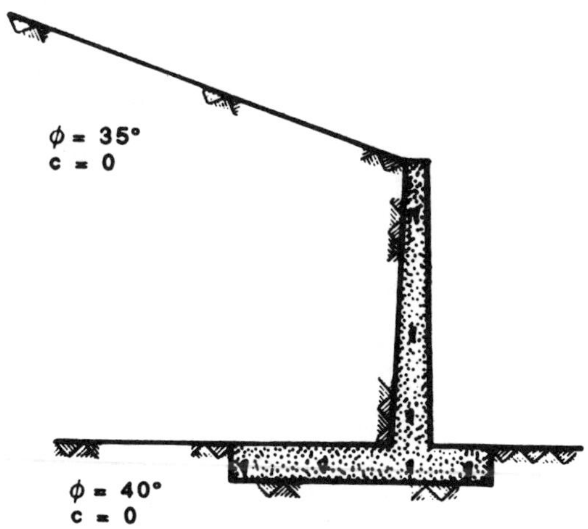

Example 2, page 241

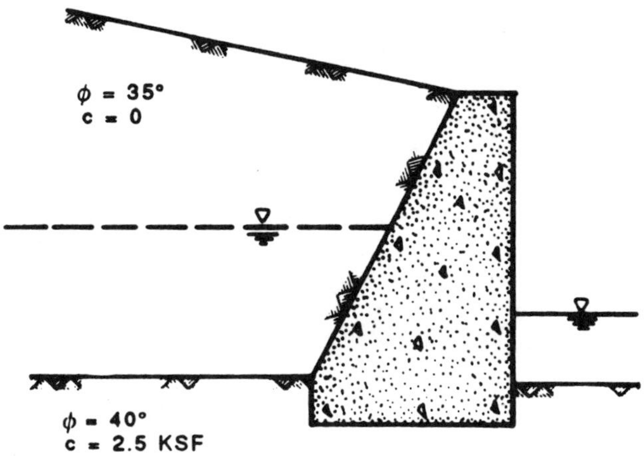

Example 3, page 244

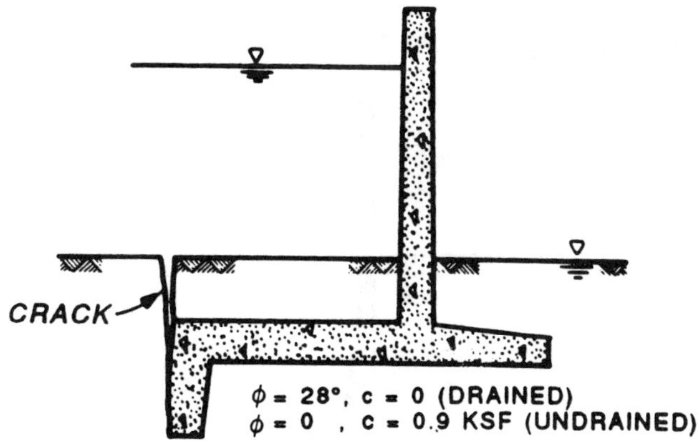

Example 4, page 252

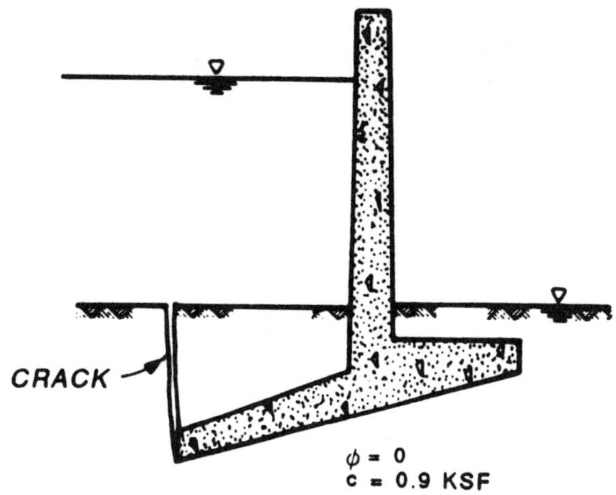

Example 5, page 257

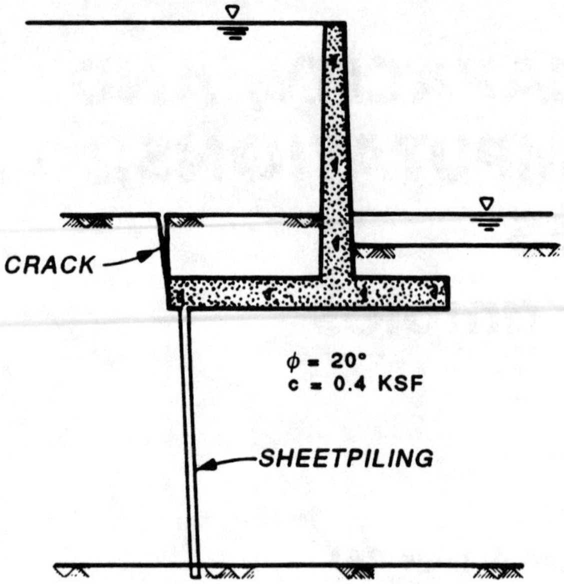

Example 7, page 264

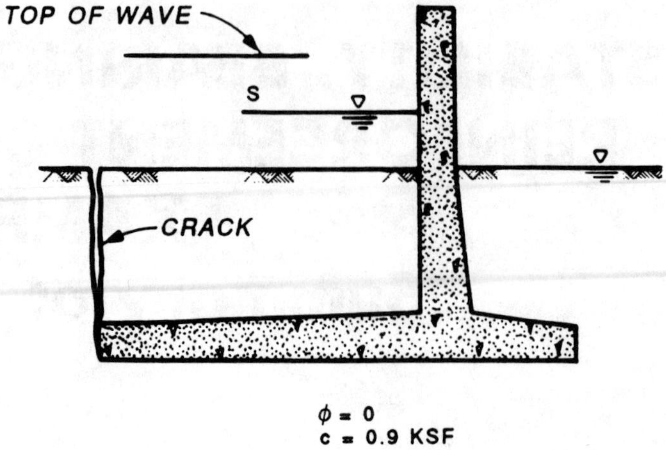

Example 6, page 261

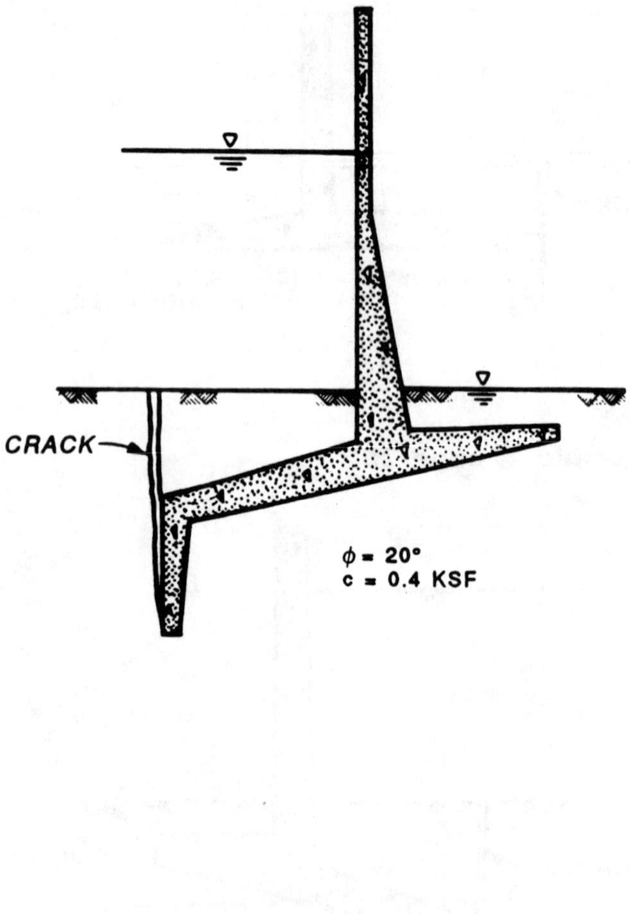

N-1. Example 1

Analyze the wall shown next for stability and bearing capacity. Find the reinforcement required at critical sections. Load Case R1.

Given:

Backfill: $\gamma = 0.120$ kcf, $\phi = 35°$, c = 0

Foundation: $\gamma = 0.135$ kcf, $\phi = 40°$, c = 0

Reinforced concrete: $\gamma = 0.150$ kcf, $f'_c = 4$ ksi, $f_y = 48$ ksi

All concrete cover = 4.5 in. (to center of gravity of steel)

Load factor = 1.9, $\phi = 0.9$

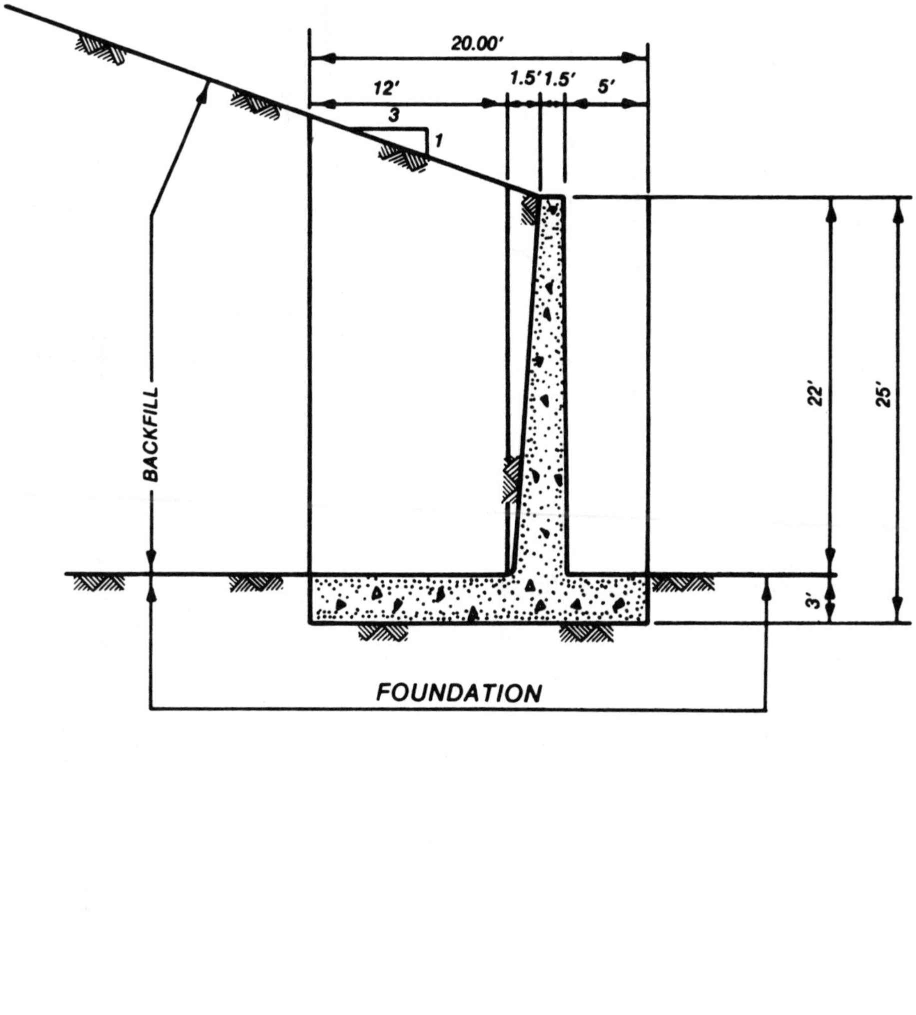

A. FIND CRITICAL SLIP-PLANE ANGLES—DRIVING SIDE

1. Foundation (see Appendix G):

$$\phi = 40°, \text{SMF} = 2/3,$$

$$\phi_d = \tan^1(2/3 \tan \phi) = 29°$$

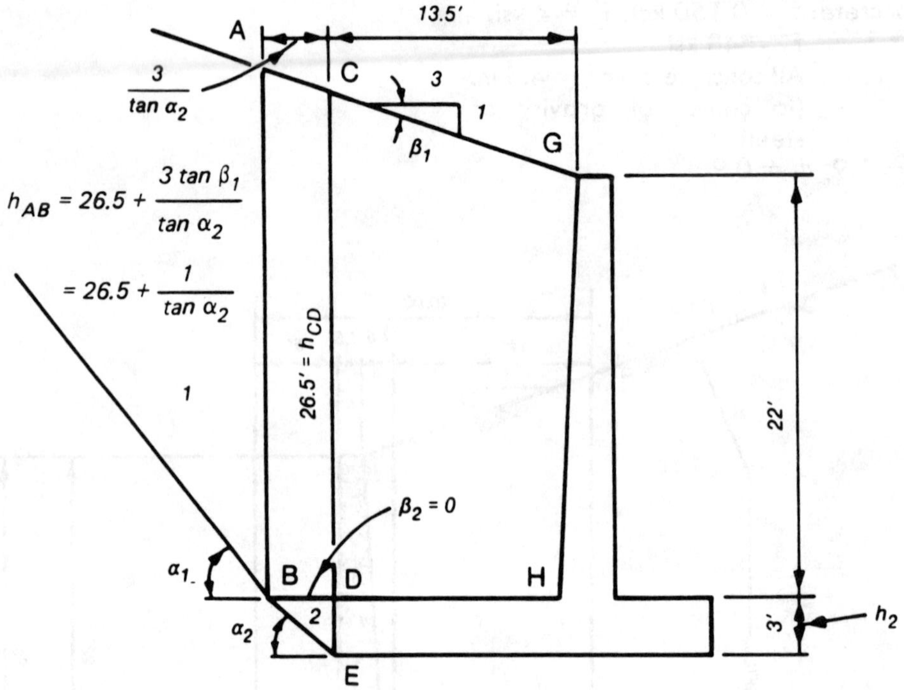

The increased unit weight in the bottom layer (layer 2) is calculated as

From Equation G-25

$$\gamma' = \frac{2\gamma_{h1}}{h_2} + \gamma_2 + \frac{2\gamma_1(\tan\beta_1 - \tan\beta_2)}{\tan\alpha_2 - \tan\beta_2}$$

$$\gamma' = \frac{2(0.12)(26.5)}{3} + 0.135 + \frac{2(0.12)(\frac{1}{3})}{\tan\alpha_2}$$

$$= 2.255 + \frac{0.08}{\tan\alpha_2}$$

From Equation G-26

$$V_\alpha = \frac{\gamma_{h_2}^2(\tan\beta_1 - \tan\beta_2)}{2(\tan\alpha_2 - \tan\beta_2)^2}$$

$$V_\alpha = \frac{0.12(3)^2(\frac{1}{3})}{2\tan^2\alpha_2} = \frac{0.18}{\tan^2\alpha_2}$$

Let $\alpha_2 = 59.5° = 45 + \frac{\phi_d}{2}$:

$$\gamma' = 2.255 + \frac{0.08}{1.697663} = 2.302$$

$$V_\alpha = \frac{0.18}{(1.697633)^2} = 0.0625$$

From Equation G-27 (omitting the c term)

$$A' = \tan\phi_d - \frac{2V_\alpha(1 + \tan^2\phi_d)}{\gamma' h_2^2}$$

$$A' = 0.554309 - \frac{2(0.0625)(1.307259)}{2.302(3)^2}$$

$$= 0.546422$$

From Equation G-28 (omitting the c term)

$$c_1' = \left[2\tan^2\phi_d - \frac{4V_\alpha \tan\beta_2(1+\tan^2\phi_d)}{\gamma' h_2^2}\right] \div A'$$

$$c_1' = 2(0.554309)^2 \div 0.546422 = 1.2462$$

From Equation G-29 (omitting the c term)

$$c_2' = \frac{\tan\phi_d(1-\tan\phi_d\tan\beta_2) - \tan\beta_2 + \dfrac{2V_\alpha\tan^2\beta_2(1+\tan^2\phi_d)}{\gamma' h_2^2}}{A}$$

$$c_2' = 0.554309 \div 0.546422 = 1.01443$$

From Equation G-30

$$\alpha_2 = \tan^{-1}\left(\frac{c_1' + \sqrt{c_1'^2 + 4c_2'}}{2}\right) = 59.77^\circ$$

$$\neq 59.5^\circ$$

Let $\alpha_2 = 59.73^\circ$, $\gamma' = 2.302$, and $V_\alpha = 0.061$:

$$A' = 0.554309 - \frac{2(0.061)(1.307259)}{2.302(3)^2}$$

$$= 0.546611$$

$$c_1' = 2(0.554309)^2 \div 0.546611 = 1.124231$$

$$c_2' = 0.554309 \div 0.546611 = 1.014083$$

$$\alpha_2 = 59.76^\circ \approx 59.73^\circ \text{ (close enough)}$$

$$\alpha_2 = \underline{\underline{59.76^\circ}}$$

2. Backfill:

$$h_{AB} = 26.5 + \frac{3\tan\beta_1}{\tan\alpha_2} = \underline{\underline{27.083 \text{ ft}}}$$

$$\phi = 35^\circ$$

$$\tan\phi_d = \tan^{-1}(2/3\tan\varphi) = 25^\circ$$

From Equation 3-26

$$c_1 = 2\tan\phi_d = 2(0.466308) = 0.932616$$

From Equation 3-27

$$c_2 = 1 - \tan\phi_d\tan\beta_1 - \frac{\tan\beta_1}{\tan\phi_d}$$

$$c_2 = 1 - 0.466308(\tfrac{1}{3}) - \frac{\frac{1}{3}}{0.466308}$$

$$= 0.129729$$

From Equation 3-25

$$\alpha_1 = \underline{\underline{46.55^\circ}}$$

B. EARTH PRESSURE COEFFICIENTS (see Appendix H).

1. Backfill:

$$K_1 = \frac{1-\tan\phi_d\cot\alpha_1}{1+\tan\phi_d\tan\alpha_1}\cdot\frac{\tan\alpha_1}{\tan\alpha_1-\tan\beta_1}$$

$$K_1 = \frac{1-0.466308\times0.947307}{1+0.466308\times1.055624}\cdot\frac{1.055624}{1.055624-\frac{1}{3}}$$

$$= \underline{\underline{0.5468}}$$

2. Foundation:

$$K = \frac{1-\tan\phi_d\cot\alpha_2}{1+\tan\phi_d\tan\alpha_2}$$

$$= \frac{1-0.554309\times0.582949}{1+0.554309\times1.715416} = \underline{\underline{0.3470}}$$

$$K_v = K\tan\alpha_2 = \underline{\underline{0.5952}}$$

C. LATERAL PRESSURE AND FORCES ON SURFACES AB AND DE.

$$P_{AB} = \tfrac{1}{2}K_1\gamma_1 h_{AB}^2 = \tfrac{1}{2}(0.5468)(0.12)(27.083)^2$$

$$= \underline{\underline{24.06 \text{ kips}}}$$

Surface DE:

$$P_{DE} = \tfrac{1}{2}K\gamma_2 h_2^2 + K_v V$$

$$V = \tfrac{1}{2}(h_{AB} + h_{CD})(\gamma_1)\left(\frac{3}{\tan\alpha_2}\right)$$

$$V = \tfrac{1}{2}(27.083 + 26.5)(0.12)\left(\frac{3}{1.715416}\right)$$

$$= 5.623 \text{ kips}$$

$$P_{DE} = \tfrac{1}{2}(0.347)(0.135)(3)^2 + 0.5952(5.623)$$

$$= 3.5576 \text{ kips}$$

$$p_B = K_1\gamma_1 hAB = 0.5468(0.12)(27.083)$$

$$= 1.7771 \text{ ksf}$$

$$p_D = K\gamma_1 h_{CD} = 0.347(0.12)(26.5) = 1.1035 \text{ ksf}$$

$$P_{DE} = \left(\frac{p_D + p_E}{2}\right)h_2$$

$$p_E = \frac{2p_{DE}}{h_2} - p_D$$

$$p_E = \frac{2(3.5576)}{3} - 1.1035$$

$$p_E = 1.2683 \text{ ksf}$$

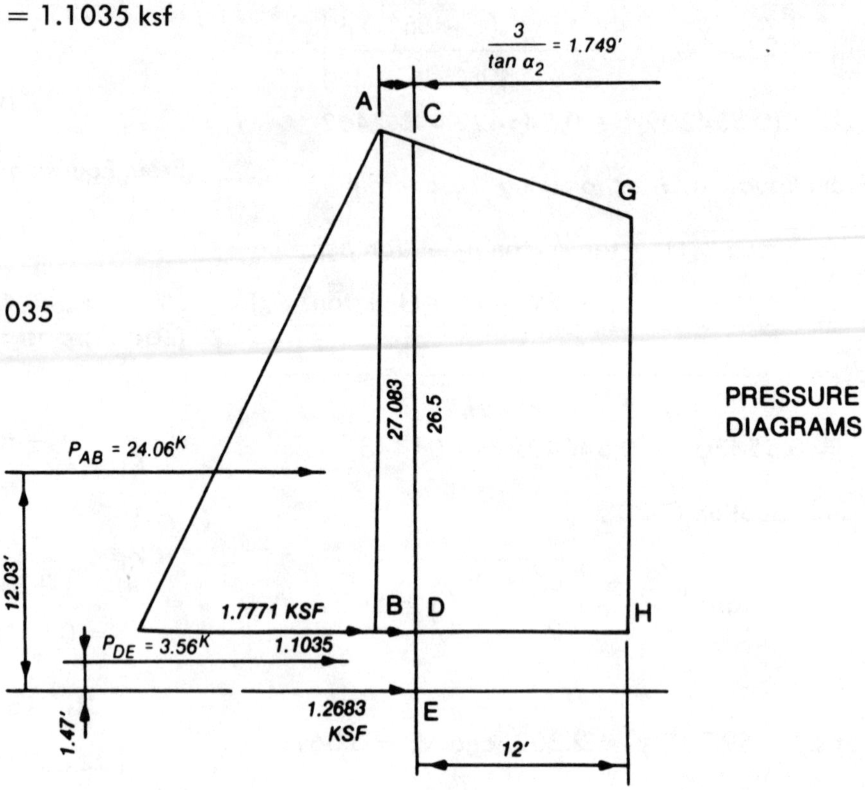

D. SHEAR FORCE ON STRUCTURAL WEDGE (paragraphs 4-8c, 9-7a).

1. The horizontal force P_{AB} will be considered to act on any vertical plane in the soil that lies on Surface AB, or on any surface to the right of Surface AB, such as surfaces CD or GH.

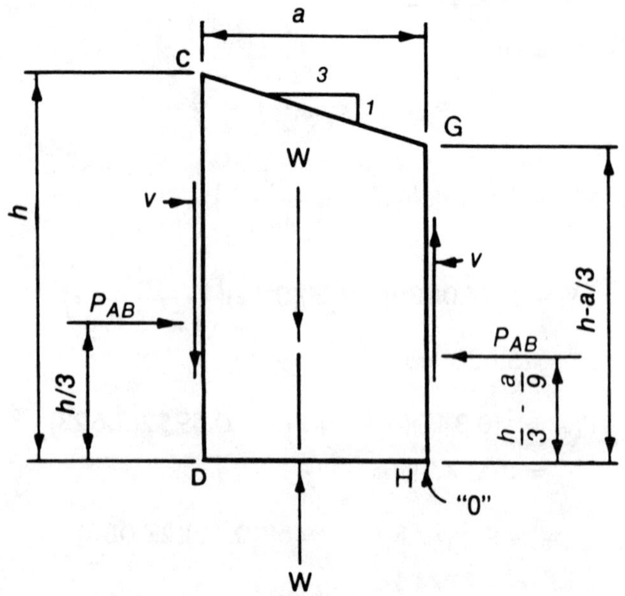

2. In order for the body of soil lying between CD and GH to be in equilibrium, a small vertical shear will be assumed to act on vertical surfaces along with the horizontal force P_{AB}.

3. The value of this shear force is found from a free body of any block of soil as shown next.

$$\sum_{o} = 0 = P_{AB}\left[\frac{h}{3} - \left(\frac{h}{3} - \frac{a}{9}\right)\right] - va$$

$$va = P_{AB} = \left(\frac{a}{9}\right),$$

$$v = \frac{P_{AB}\tan\beta}{3} = \frac{P_{AB}}{9} = \frac{24.06}{9} = 2.673 \text{ kips}$$

E. WEIGHT AND CENTER OF GRAVITY OF STRUCTURAL WEDGE.

$$
\begin{aligned}
1.5\text{ ft} \times 22\text{ ft} \times 0.15\text{ kcf} &= 4.95 \times 5.75\text{ ft} = 28.46\text{ kips} \\
\tfrac{1}{2} \times 1.5\text{ ft} \times 22\text{ ft} \times 0.15\text{ kcf} &= 2.48 \times 7.00\text{ ft} = 17.36\text{ kips} \\
20\text{ ft} \times 3\text{ ft} \times 0.15\text{ kcf} &= \underline{9.00} \times 10.00\text{ ft} = \underline{90.00}\text{ kips} \\
\text{Concrete} &= 16.43\text{ k/ft} \qquad 135.83 \div 16.43 \\
&= 8.27\text{ ft} = \bar{x}
\end{aligned}
$$

$$
\begin{aligned}
\tfrac{1}{2} \times 13.5\text{ ft} \times 4.5\text{ ft} \times 0.12\text{ kcf} &= 3.65 \times 15.50\text{ ft} = 56.58 \\
12\text{ ft} \times 22 \times 0.12\text{ kcf} &= 31.68 \times 14.00\text{ ft} = 443.52 \\
\tfrac{1}{2} \times 1.5\text{ ft} \times 22\text{ ft} \times 0.12\text{ kcf} &= \underline{1.98} \times 7.50\text{ ft} = \underline{14.85} \\
\text{Soil} &= 37.31\text{ k/ft} \qquad 514.95 \div 37.31 \\
&= 13.80\text{ ft} \\
&= \bar{x}
\end{aligned}
$$

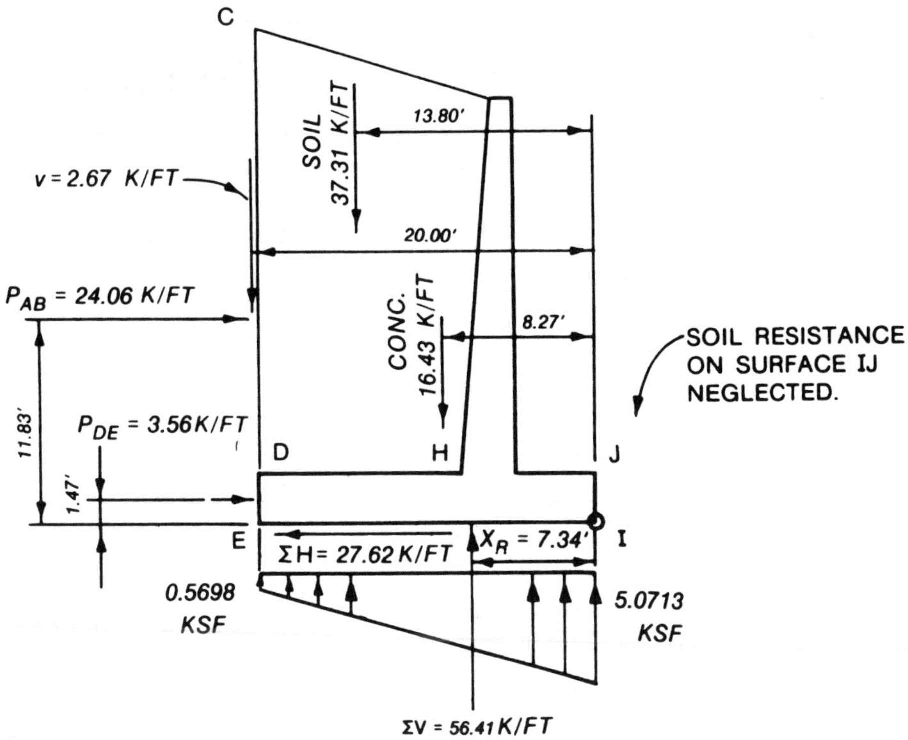

F. OVERTURNING STABILITY (paragraph 4-8).

$$
\begin{aligned}
2.67\text{ kips} \times 20.00\text{ ft} &= 53.400\text{ ft-kips} \\
37.31\text{ kips} \times 13.80\text{ ft} &= 514.878\text{ ft-kips} \\
\underline{16.43}\text{ kips} \times 8.27\text{ ft} &= 135.876\text{ ft-kips} \\
56.41 &= \sum V
\end{aligned}
$$

$$x_R = \frac{414.291}{56.41} = 7.34\text{ ft}$$

$$
\begin{aligned}
-24.06 \times 11.83 &= -284.63 \\
\underline{-\ 3.56} \times 1.47 &= \underline{-5.233} \\
-27.62 = \sum H \qquad & 414.291 = \sum M_I
\end{aligned}
$$

One hundred percent of the base is in compression; overturning stability requirement is satisfied.

G. SLIDING STABILITY ANALYSIS (paragraph 4-15).

$$N' = \sum V = 56.41\text{ kips},\ T = \sum H = 27.62\text{ kips}$$

From Equation 4-12

$$T \leq \frac{N' \tan \phi + cL}{FS}$$

Using the minimum FS of 1.5 from Table 4-1 yields

$$27.62 \le \frac{56.41 \tan 40° + 0(20)}{1.5}$$

$$27.62 \le 31.56$$

The sliding stability requirement is satisfied.

H. CHECK BEARING CAPACITY (Chapter 5).
Refer to Figure 5-1a.

$$\delta = \tan^{-1}\left(\frac{27.62}{56.41}\right) = 26° = 0.4538 \text{ rad}$$

$$e = \frac{B}{2} - x_R = \frac{20}{2} - 7.34 = 2.66 \text{ ft}$$

$$\bar{B} = B - 2e = 20 - 2(2.66) = 14.68 \text{ ft}$$

From Equation 5-8a

$$q_o = \gamma_d = 0.135(3) = 0.405 \text{ ksf}$$

From Equation 5-5a

$$\xi_{qi} = \left(1 - \frac{\delta}{90}\right)^2 = 0.5057$$

From Equation 5-56

$$\xi_{\gamma i} = \left(1 - \frac{\delta}{\phi}\right) = \left(1 - \frac{26}{40}\right)^2 = 0.1225$$

From Table 5-1

$$N_q = 64.20, N_\gamma = 93.69$$

Using Equation 5-2 yields

$$Q = \bar{B}\left(\xi_{qi} q_o N_q + \frac{\xi_{\gamma i}\bar{B}\gamma N_\gamma}{2}\right)$$

$$Q = 14.68\left[0.5057(0.405)(64.20) + \frac{0.1225(14.68)(0.135)(93.69)}{2}\right]$$

$$Q = 359.97 \text{ kips}$$

For this problem $\sum V = N'$. From Equation 5-1 the factor of safety is calculated as

$$FS = \frac{Q}{N'} = \frac{359.97}{56.41} = 6.38 > 3.00$$

The bearing capacity requirement is satisfied (Table 4-1).

I. REINFORCEMENT AT BASE OF STEM (Chapter 9). Neglect vertical shear component v.

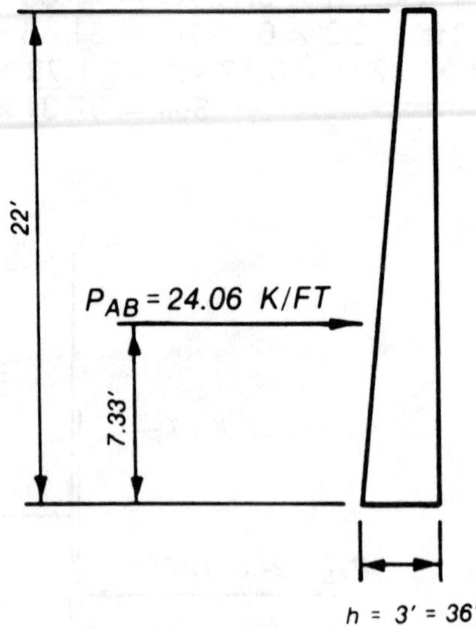

$$M = 24.06(7.33 \text{ ft}) = 176.36 \text{ k-ft}$$

From Figure 9-2

$$k_u = 1 - \sqrt{1 - \frac{\frac{M_u}{\phi}}{0.425 f'_c b d^2}}$$

$$\frac{M_u}{\phi} = \frac{1.9(176.36)(12)}{0.90} = 4468 \text{ k-in.}$$

$$b = 12 \text{ in.}, d = h - 4.5 = 31.5 \text{ in.}$$

$$0.425 f'_c b d^2 = 0.425(4)(12)(31.5)^2 = 20{,}241.9$$

$$k_u = 1 - \sqrt{1 - \frac{4468}{20{,}241.9}} = 0.1174$$

$$C_u = T_u = 0.85 f'_c k_u b d$$

$$C_u = T_u = 0.85(4)(0.1174)(12)(31.5)$$
$$= 150.88 \text{ kips}$$

$$A_s = \frac{T_u}{f_y} = \frac{150.88}{48} = \underline{\underline{3.14 \text{ in.}^2/\text{ft}}}$$

Check if the ductility requirement of $\rho_{min} \le \rho \le$

p_{max} is satisfied:

From paragraph 9-8b(3)

$$\rho_{max} = \lambda \rho_b, \lambda = 0.25$$

From paragraph 9-8b(4) from ACI 318

$$\rho_{min} = 200/f_y = 200/48{,}000 = 0.00417$$

The reinforcement ratio ρ_b in the balanced condition may be obtained by applying equilibrium and compatibility conditions. From the linear strain condition shown below

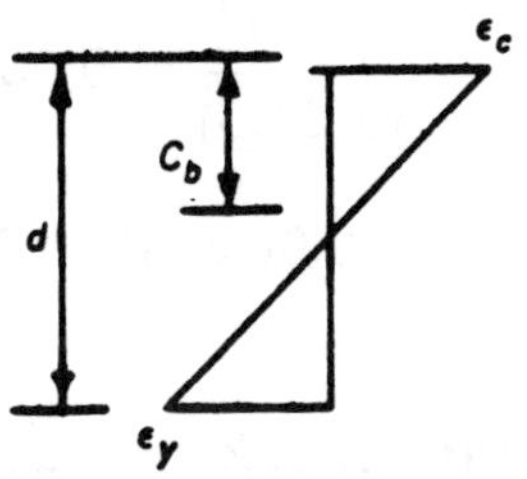

$$\frac{c_b}{d} = \frac{\varepsilon_c}{\varepsilon_c + \varepsilon_y} = \frac{0.0003}{0.003 + \dfrac{f_y}{29{,}000 \text{ ksi}}}$$

The compressive force C is

$$C = 0.85'_c \beta_1 b c_b$$

The tensile force is

$$T = f_y A_{sb} = \rho_b b d f_y$$

Equating C to T yields

$$0.85 f'_c b \beta_1 c_b = \rho_b b d f_y$$

$$\rho_b = \left(\frac{0.85 f'_c}{f_y}\right) \beta_1 \left(\frac{c_b}{d}\right)$$

$$= \left(\frac{0.85 f'_c}{f_y}\right) \beta_1 \left(\frac{\varepsilon_c}{\varepsilon_c + \dfrac{f_y}{29{,}000 \text{ ksi}}}\right)$$

$\beta_1 = 0.85$, $\varepsilon_c = 0.003$
(for nonhydraulic structure, paragraph 9-8e)

$$\rho_b = \frac{0.85(4)}{48}(0.85)\left(\frac{0.003}{0.003 + \dfrac{48}{29{,}000}}\right)$$

$$= 0.03880$$

$$\rho_{max} = 0.25(0.03880) = 0.009700$$

$$\rho = \frac{A_s}{bd} = \frac{3.14}{12(31.5)} = 0.008307$$

$$0.00417 < \rho < 0.008307 < 0.009700$$

Ductility is adequate.

J. REINFORCEMENT IN HEEL AT FACE OF STEM.

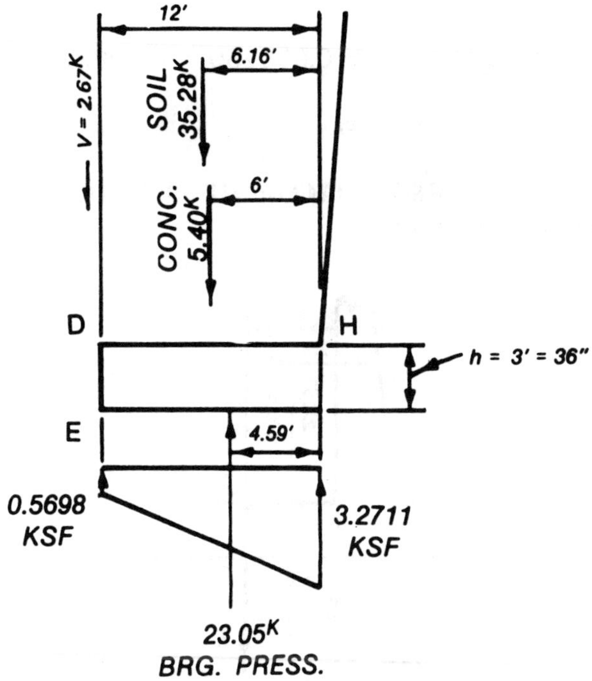

2.67	×	12.00	=	32.04
35.28	×	6.16	=	217.32
5.40	×	6.00	=	32.40
−23.05	×	4.59	=	−105.80
		M	=	175.96 k-ft

From Figure 9-2

$$\frac{M_u}{\phi} = \frac{1.9(175.96)(12)}{0.9} = 4458 \text{ k-in.}$$

$$b = 12 \text{ in.}, d = 36 - 4.5 = 31.5 \text{ in.}$$

$$0.425 f'_c b d^2 = 20{,}241.9$$

$$k_u = 1 - \sqrt{1 - \frac{4458}{20{,}241.9}} = 0.1170$$

$$C_u = T_u = 0.85(4)(0.1170)(12)(31.5)$$
$$= 150.37 \text{ kips}$$

$$A_s = \frac{150.37}{48} = \underline{\underline{3.13 \text{ in.}^2/\text{ft}}}$$

Check ductility

$$\rho_{max} = 0.009700, \rho_{min} = 0.00417$$

$$\rho = \frac{A_s}{bd} = \frac{3.13}{12(31.5)} = 0.008280$$

$$0.00417 < \rho = 0.008280 < 0.009700$$

Ductility is adequate.

K. REINFORCEMENT IN TOE AT FACE OF STEM.

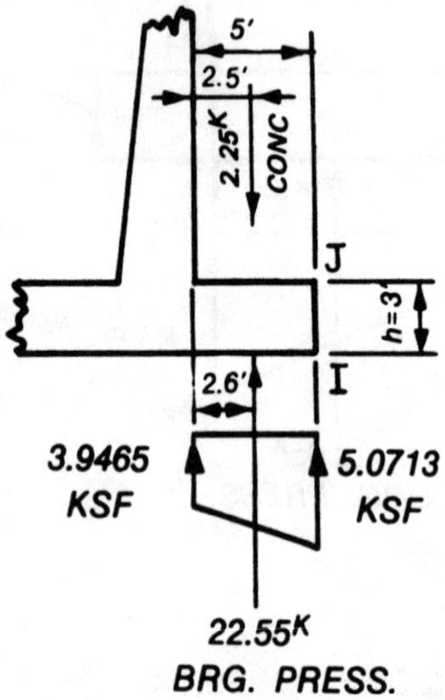

$$22.55 \times 2.60 = 58.63$$
$$-2.25 \times 2.50 = \underline{-5.625}$$
$$M = 53.01 \text{ k-ft}$$

From Figure 9-2

$$\frac{M_u}{\phi} = \frac{1.9(53.01)(12)}{0.9} = 1342.8 \text{ k-in.}$$

$$b = 12 \text{ in.}, d = 36 \text{ in.} - 4.5 = 31.5 \text{ in.}$$

$$0.425f'_c bd^2 = 20{,}241.9$$

$$k_u = 1 - \sqrt{1 - \frac{1342.8}{20{,}241.9}} = 0.0337$$

$$C_u = T_u = 0.85(4)(0.337)(12)(31.5)$$
$$= 43.31 \text{ kips}$$

$$A_s = \frac{43.31}{48} = \underline{\underline{0.90 \text{ in}^2/\text{ft}}}$$

Check ductility:

$$\rho_{max} = 0.02910, \rho_{min} = 0.00417$$

$$\rho = \frac{A_s}{bd} = \frac{0.90}{12(31.5)} = 0.002381$$

$$0.00417 < \rho = 0.002381 < 0.009700$$

Ductility is adequate.

1. Shear check (paragraph 9-8f). The shear capacity of the concrete will be checked at a distance d from the base of the stem according to ACI 318.

$$d_v = 31.5 - \frac{1.5}{22} d_v, \; 1.06818 d_v = 31.5$$

$d_v = 29.5$ in. $= 2.46$ ft (member depth for shear at distance d_v above base)

Since the shear has a quadratic variation, the shear at distance d_v can be calculated as shown below.

$$V = \frac{(22 - 2.46)^2}{(22)^2}(24.06) = 18.98 \text{ kips}$$

$$V_u = 1.9(18.98) = 36.06 \text{ kips}$$

From ACI 318

$$\phi V_c = (0.85)(2)\sqrt{f'_c}bd_v$$
$$\phi V_c = (0.85)(2)\sqrt{4000}(12)(29.5)$$
$$= 38{,}061 \text{ lb} > V_u$$

Check the shear capacity of the heel at the base of the stem:

$$V = 2.67 + 5.40 + 32.28 - 23.05 = 20.3 \text{ kips}$$

$$V_u = 1.9(20.3) = 38.57 \text{ kips}$$

$$\phi V_c = 0.85)(2)\sqrt{f'_c}bd_v$$
$$\phi V_c = (0.85)(2)\sqrt{4000}(12)(31.5)$$
$$= 40{,}642 \text{ lb} > V_u$$

N-2. Example 2

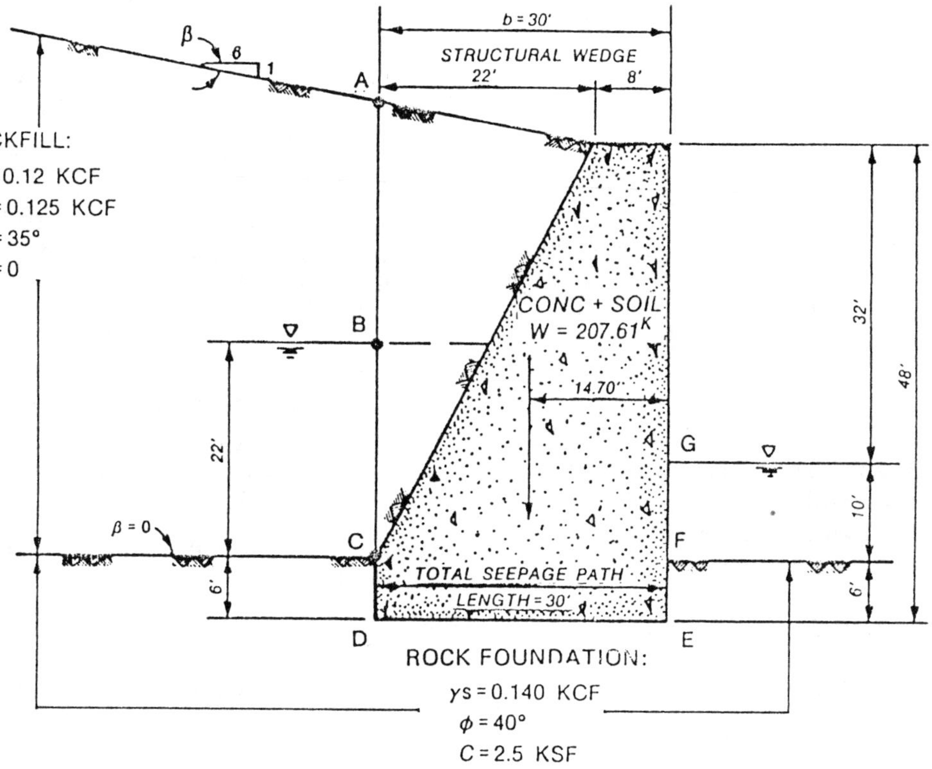

Analyze the wall above for overturning, sliding, and bearing capacity. Case R2, unusual loading condition.

A. DRIVING-SIDE LATERAL PRESSURE COEFFICIENTS.

1. Foundation:

$$\text{SMF} = \tfrac{2}{3},\ \phi_d = \tan^{-1}(\tfrac{2}{3}\tan\phi) = 29.22^\circ$$

From Equation 3-15

$$K_A = \tan^2\left(45^\circ - \frac{\phi_d}{2}\right) = 0.3439$$

$$K_c = \sqrt{K} = 0.5865$$

$$p_{VD} = 23.67(0.12) + 22(0.125) - 28(0.0625) + 6(0.14) = 4.68 \text{ ksf}$$

$$p_{HD} = Kp_v - 2K_c c_d,\ c_d = \tfrac{2}{3}c = 1.67 \text{ ksf}$$

$$p_{HD} = 0.3439(4.68) - 2(0.5865)(1.67) = -0.3495 \text{ ksf}$$

Negative horizontal earth pressure at D shows that there will be no earth pressure on Surface CD. Also, no at-rest earth pressure can act on Surface EF.

2. Backfill:

$$\text{SMF} = \tfrac{2}{3}\ ,\ \phi_d = \tan^{-1}(\tfrac{2}{3}\tan\phi) = 25^\circ$$

From Equation 3-26

$$c_1 = 2\tan\phi_d = 2(0.466308) = 0.932616$$

From Equation 3-27

$$c_2 = 1 - \tan\phi_d\,\beta - \left(\frac{\tan\beta}{\tan\phi_d}\right)$$

$$c_2 = 1 - 0.466308 \times \tfrac{1}{6} - \left(\frac{\tfrac{1}{6}}{0.466308}\right)$$

$$= 0.564864$$

From Equation 3-25

$$\alpha = \tan^{-1}\left(\frac{c_1 + \sqrt{c_1^2 + 4c_2}}{2}\right) = \underline{\underline{53.49^\circ}}$$

From Appendix H

$$K_1 = \frac{1 - \tan\phi_d \cot\alpha}{1 + \tan\phi_d \tan\alpha} \cdot \frac{\tan\alpha}{\tan\alpha - \tan\beta}$$

$$K_1 = \frac{1 - 0.466308 \times 0.740231}{1 + 0.466308 \times 1.350929} \cdot \frac{1.350929}{1.350929 - \tfrac{1}{6}}$$

$$= \underline{\underline{0.4583}} \text{ (use above water table)}$$

B. UPLIFT AND WATER PRESSURES (paragraph 3-22, rock foundation assumed). Since the total seepage path length is the base width, full hydrostatic pressures will exist at points D and E (see paragraph 3-22).

$$u_D = 28 \text{ ft } (0.0625 = \underline{\underline{1.75 \text{ ksf}}}$$

$$u_E = 16 \text{ ft } (0.0625) = \underline{\underline{1.00 \text{ ksf}}}$$

For the same reason, the buoyant weights for backfill and foundation materials are

$$\gamma_b(\text{Backfill}) = 0.125 - 0.0625 = 0.0625 \text{ kcf}$$

$$\gamma_b(\text{Foundation}) = 0.140 - 0.0625 = 0.0775 \text{ kcf}$$

C. BUOYANT EARTH PRESSURE COEFFICIENT (Appendix H):

$$K = \frac{1 - \tan\phi_d \cot\alpha}{1 + \tan\phi_d \tan\alpha} = 0.40176$$

$$K_b = K\left[1 + \left(\frac{\tan\alpha}{\tan\alpha - \tan\beta} - 1\right)\left(\frac{\gamma_m}{\gamma_b}\right)\right]$$

$$K_b = 0.40176\left[1 + \left(\frac{1.350929}{1.350929 - \tfrac{1}{6}} - 1\right)\right.$$

$$\left.\times \left(\frac{0.12}{0.0625}\right)\right] = \underline{\underline{0.5103}}$$

D. OVERTURNING STABILITY ANALYSIS (paragraph 4-8). For the first trial solution, all of the base is assumed to be in compression (full hydrostatic pressure is applied at points D and E). For the second trial, full hydrostatic pressure is assumed to act over the portion of the base which is not in compression. This procedure is continued until the assumed amount of the base in compression equals the calculated value. The final solution is shown next.

$$\begin{aligned}
207.61 \text{ kips} \times 14.70 &= \quad 3051.87 \text{ ft-kips} \\
\underline{-45.72} \text{ kips} \times 16.33 &= \quad -746.61 \text{ ft-kips} \\
161.89 \text{ kips} &= \textstyle\sum V \\
-51.77 \text{ kips} \times 22.08 &= -1143.081 \text{ ft-kips} \\
-24.50 \text{ kips} \times 9.33 &= \quad -228.59 \text{ ft-kips} \\
\underline{8.00} \text{ kips} \times 5.33 &= \quad \underline{42.64} \text{ ft-kips} \\
\underline{\underline{-68.27 \text{ kips}}} = \textstyle\sum H & \quad\quad 976.23 \div 161.89 \\
&= \underline{\underline{6.030 \text{ ft}}} = x_R
\end{aligned}$$

The width of the base in compression $= 3x_R =$ 18.09 ft. The percent in compression

$$= \left(\frac{18.09}{30}\right)(100) = 60.3\% > 50\%$$

required by Table 4-1. The overturning criterion is satisfied.

E. SLIDING STABILITY ANALYSIS (paragraph 4-15).

$$N' = \textstyle\sum V = 161.89 \text{ kips}, \; T = \sum H = 68.27 \text{ kips}$$

From Equation 4-12

$$T \leq \frac{N' \tan\phi + cL}{FS}$$

Using the minimum required FS of 1.33 as given in Table 4-1 yields

$$68.27 \leq \frac{161.89 \tan 40^\circ + 2.5(18.09)}{1.33}$$

$$68.27 \leq 136.14$$

The sliding criterion is satisfied.

F. BEARING CAPACITY ANALYSIS (Chapter 5). From Figure 5-1a

$$\delta = \tan^{-1}\left(\frac{\sum H}{\sum V}\right) = \tan^{-1}\left(\frac{68.27}{161.89}\right) = 22.87^\circ$$

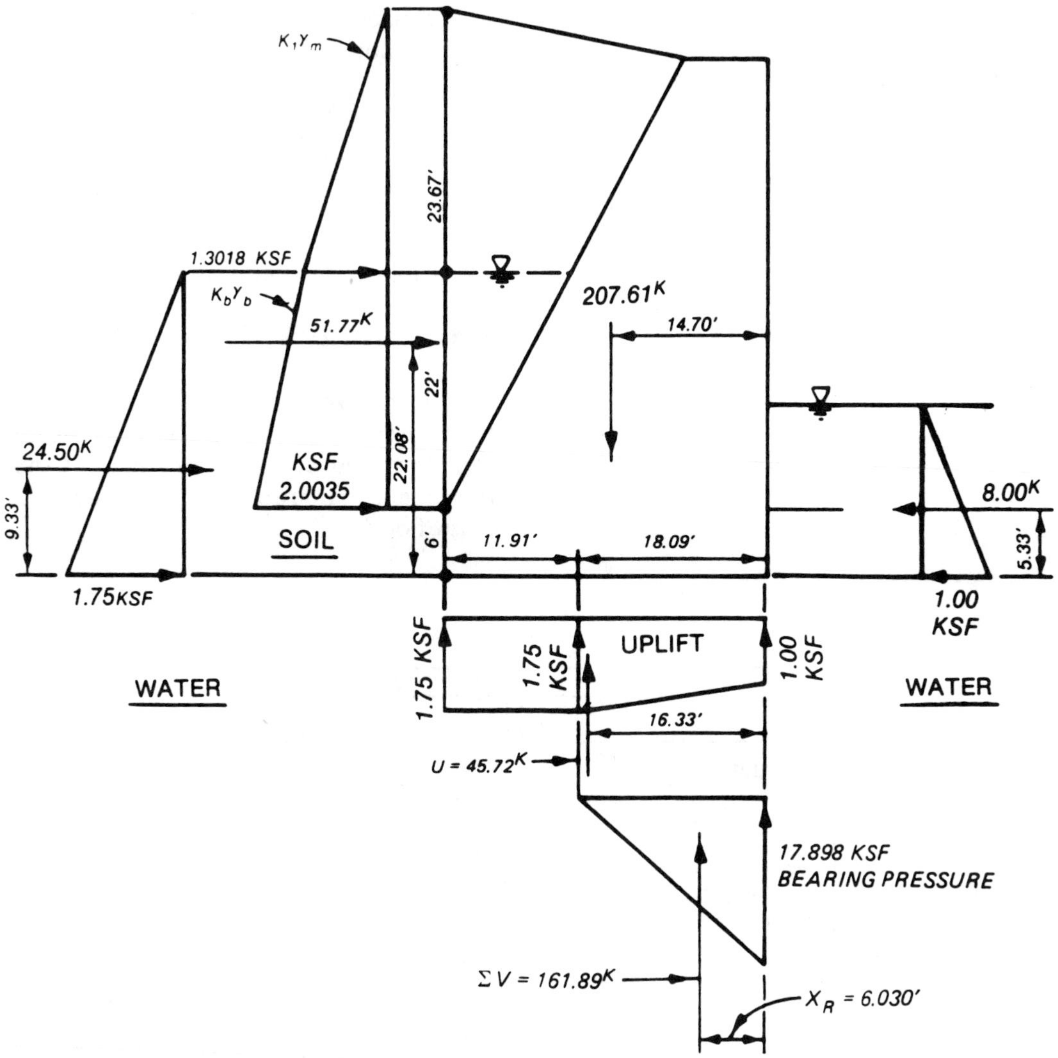

$$e = \frac{B}{2} - x_R = 15 - 6.030 = 8.970 \text{ ft}$$

$$\bar{B} = B - 2e = 30 - 2(8.970) = 12.060 \text{ ft}$$

$$\gamma' = 0.0775 \text{ kcf}, D = 6 \text{ ft}$$

From Equation 5-8a

$$q_o = \gamma' D = 0.0775(6 \text{ ft}) = 0.465 \text{ ksf}$$

From Equation 5-4a

$$\xi_{cd} = 1 + 0.2(D/\bar{B}) \tan\left(45^\circ + \frac{\phi}{2}\right)$$

$$\xi_{cd} = 1 + 0.2(0.4975)(2.1445) = 1.213$$

From Equation 5-4c

$$\xi_{qd} = \xi_{\gamma d} = 1 + 0.1(D/\bar{B}) \tan\left(45^\circ + \frac{\phi}{2}\right)$$
$$= 1.107$$

From Equation 5-5a

$$\xi_{qi} = \xi_{ci} = \left(1 - \frac{\delta^\circ}{90^\circ}\right)^2 = \left(1 - \frac{22.87}{90}\right)^2$$
$$= 0.5564$$

From Equation 5-56

$$\xi_{\gamma i} = \left(1 - \frac{\delta^\circ}{\phi}\right)^2 = \left(1 - \frac{22.87}{40}\right)^2 = 0.1834$$

From Table 5-1

$$N_c = 75.31, N_q = 64.20, N_\gamma = 93.69$$

From equation 5-2

$$Q = \bar{B}\left(\xi_{cd}\xi_{ci}cN_c + \xi_{qd}\xi_{qi}q_oN_q + \frac{\xi_{\gamma d}\xi_{\gamma i}\bar{B}\gamma N_\gamma}{2}\right)$$

$$Q = 12.060\Big[1.213(0.5564)(2.5)(75.31) + 1.107(0.5564)(0.465)(64.2) + \frac{1.107(0.1834)(12.060)(0.0775)(93.69)}{2}\Big]$$

$$Q = 1861.41 \text{ kips}$$

For this problem $\sum V = N'$. The FS is calculated with Equation 5-1.

$$FS = \frac{Q}{N'} = \frac{1861.41}{161.89} = 11.50 > 2.0$$

(required by Table 4-1)

The bearing criterion is satisfied.

N-3. Example 3

Analyze the wall for overturning stability, sliding, and bearing capacity. Determine water pressure using the line-of-creep method. Calculate reinforcement required at critical sections. Case I1, design flood.

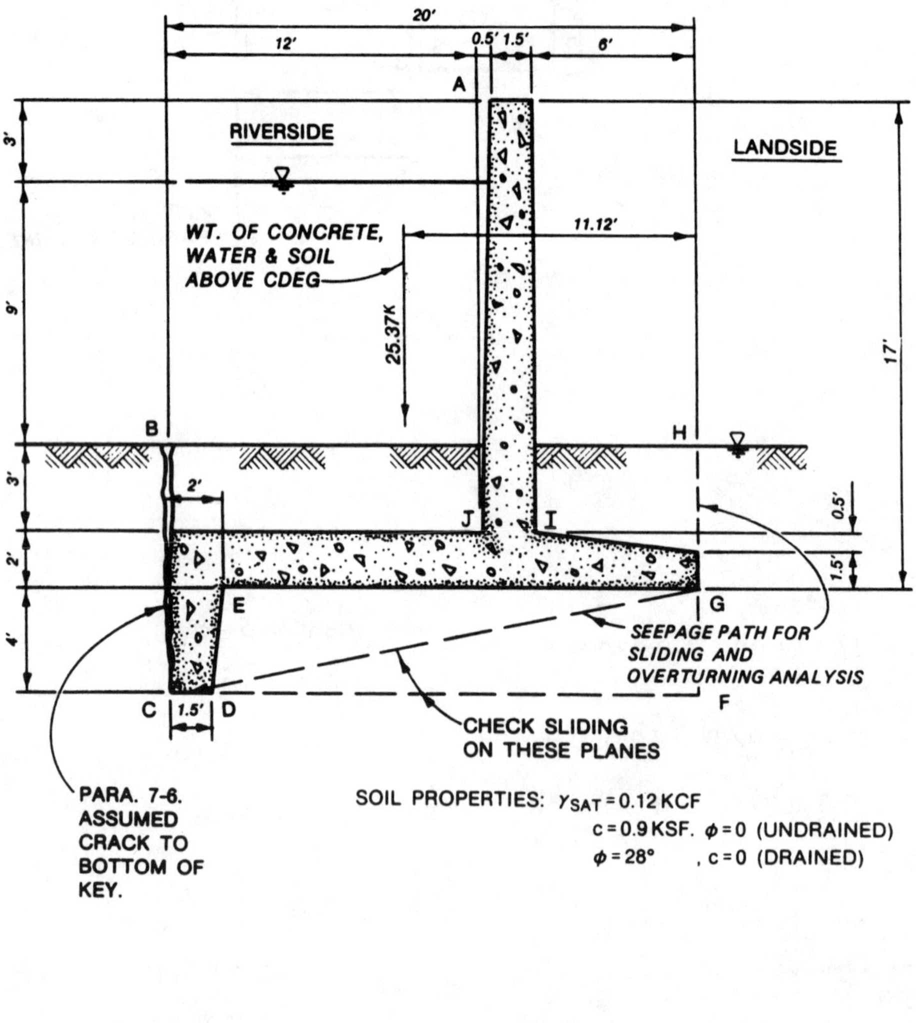

A. WATER PRESSURES (line of creep, paragraph 3-19).

$$L_{CG} = \sqrt{4^2 + 20^2} = 20.40 \text{ ft},\ L_{BC} = 0,\ \Delta_h = 9 \text{ ft}$$

$$L_{GH} = 5 \text{ ft},\ L_S = 20.4 \text{ ft} + 5 \text{ ft} = 25.4 \text{ ft}$$

$$u_C = \left(\begin{matrix}\text{Dist. below}\\ \text{Headwater}\end{matrix} - \frac{\Delta_h(L_{BC})}{L_S}\right)$$

$$\gamma_w = \left(18 \text{ ft} - \frac{9 \times 0}{25.4}\right)0.0625 = 1.125 \text{ ksf}$$

$$u_G = \left(\begin{matrix}\text{Dist. below}\\ \text{Headwater}\end{matrix} - \frac{\Delta_h(L_{CG})}{L_S}\right)$$

$$\gamma_w = \left(14 \text{ ft} - \frac{9 \times 20.4}{25.4}\right)0.0625 = 0.4232 \text{ ksf}$$

Compute the water pressure at D and E. Prorate the head loss along path CDEG. Total the head loss along the path.

$$CG = \frac{9(20.4)}{25.4} = 7.228 \text{ ft}$$

$$L_S(\text{concrete surface}) = L_{CD} + L_{DE} + L_{EG}$$

$$L_S = 1.5 + 4.03 + 18$$

$$= 23.53 \text{ ft}$$

$$\text{Head loss at D} = \left(\frac{1.5}{23.53}\right)(7.228) = 0.4608 \text{ ft}$$

$$u_D = (18 - 0.4608)0.0625 = 1.0962 \text{ ksf}$$

$$\text{Head loss at E} = \left(\frac{(1.5 + 4.03)}{23.53}\right)(7.228)$$

$$= 1.699 \text{ ft}$$

$$u_E = (14 - 1.699)0.0625 = 0.7688 \text{ ksf}$$

B. OVERTURNING STABILITY (paragraph 4-8).

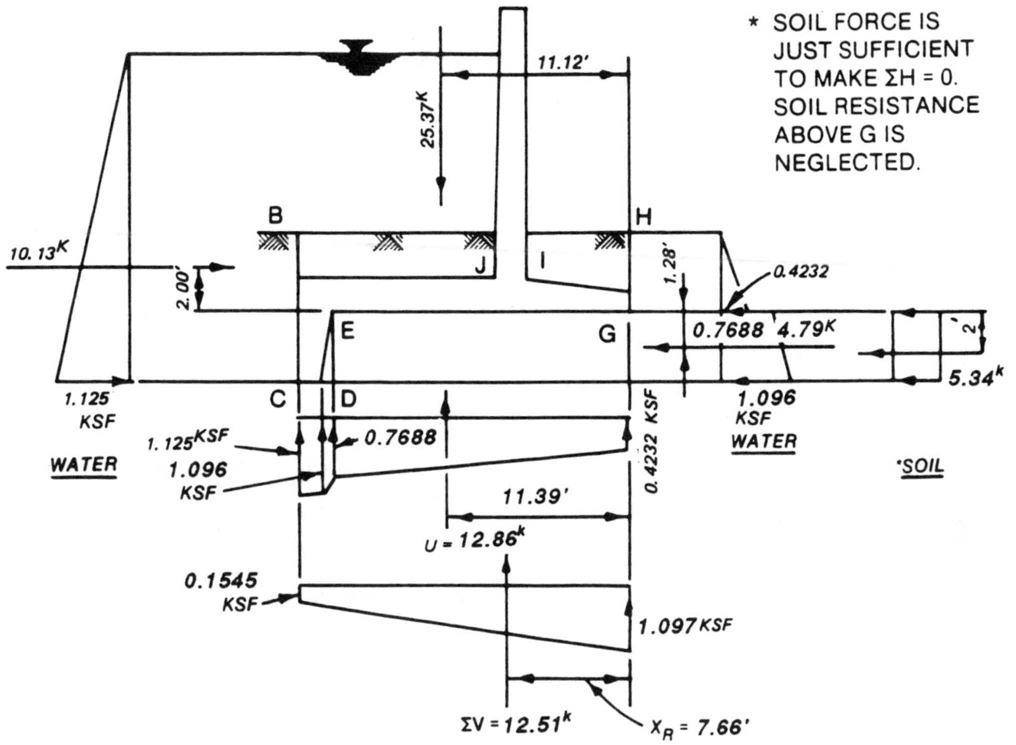

Overturning calculations:

$$
\begin{array}{rrr}
25.37 \text{ kips} \times & 11.12 \text{ ft} = & 282.11 \text{ ft-k} \\
-12.86 \text{ kips} \times & 11.39 \text{ ft} = & -146.48 \text{ ft-k} \\
\sum V = 12.51 \text{ kips} & & \\
-10.13 \text{ kips} \times & 2.00 \text{ ft} = & -\ 20.26 \text{ ft-k} \\
-\ 4.79 \text{ kips} \times & -1.28 \text{ ft} = & -6.13 \text{ ft-k} \\
5.34 \text{ kips} \times & -2.00 \text{ ft} = & -10.68 \text{ ft-k} \\
\sum H = 0.00 \text{ kip} & & 98.56 \div 12.51 = 7.87 \text{ ft} = x_R
\end{array}
$$

One hundred percent of the base is in compression, which satisfies the requirements of Table 4-2.

C. SLIDING STABILITY ANALYSIS (paragraph 4-10).

1. On Plane CG:

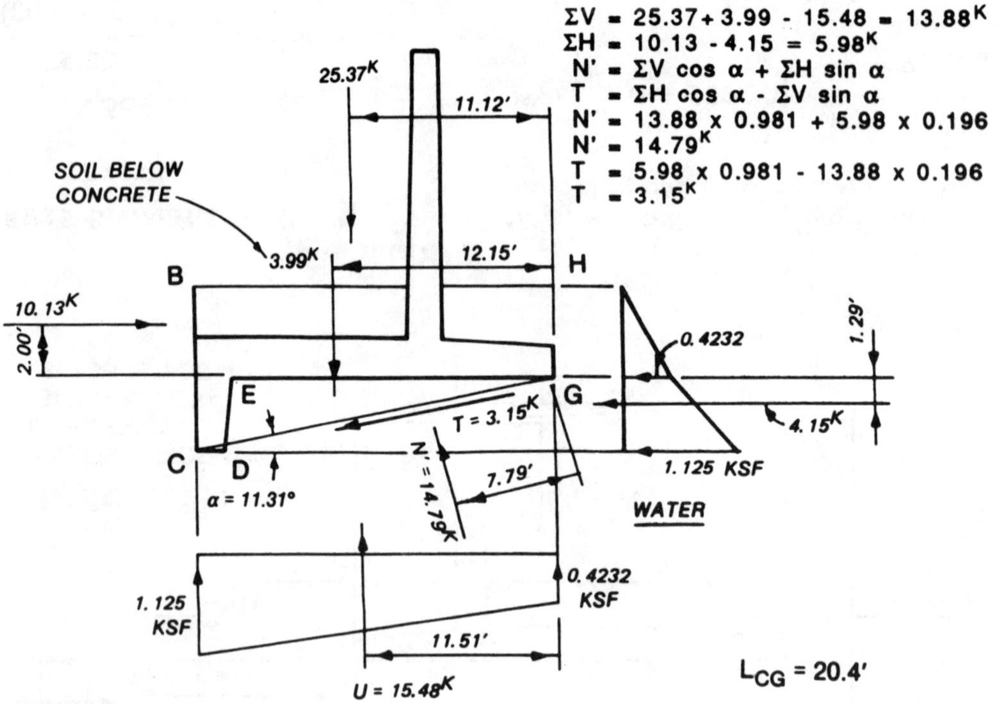

For undrained shear strength

$$cL_{CG} = 0.9 \times 20.4 = 18.36 \text{ kips}$$

From Equation 4-12

$$T \leq \frac{N' \tan \phi + cL}{FS}$$

From Table 4-2, the minimum FS is 1.5.

$$3.15 \leq \frac{18.36}{1.5}$$

$$3.15 \leq 12.24$$

For drained shear strength

$$N' \tan \phi = 14.79 \times 0.532 = 7.86 \text{ kips}$$

From Equation 4-12

$$T \leq \frac{N' \tan \phi + cL}{FS}$$

Using the minimum FS of 1.5 from Table 4-2 yields

$$3.15 \leq \frac{7.86}{1.5}$$

$$3.15 \leq 5.24$$

The sliding resistance is adequate without considering soil resistance on the landside.

2. On Plane CF:

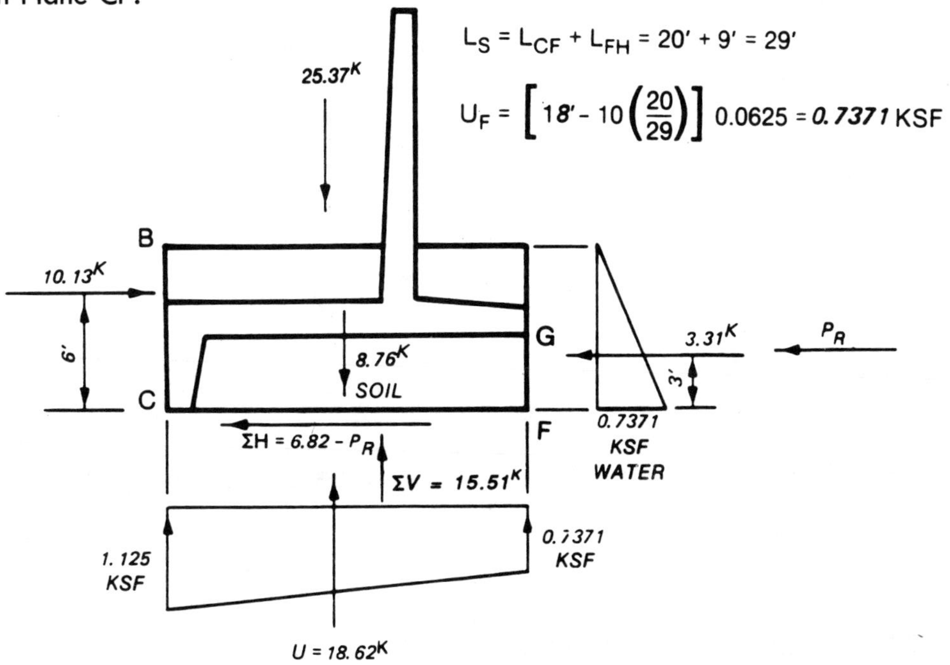

γ' = effective weight on land side

$= [9 \text{ ft}(0.12) - 0.7371] \div 9$

$= 0.0381$ kcf

Undrained condition

$$c = 0.9 \text{ ksf}$$

From Equation 3-10

$$c_d = \text{SMF}(c)$$

$$c_d = \tfrac{2}{3}(0.9) = 0.6 \text{ ksf}$$

From Equation 3-4

$$K_o = 1.0$$

The at-rest earth force may be calculated as

$$P_R = \tfrac{1}{2}\gamma_b h^2 = \tfrac{1}{2}(0.0381)(9)^2 = 1.54 \text{ kips}$$

$$c_d L_{CF} = 0.6(20) = 12 \text{ kips}$$

From Equation 4-12

$$T \le \frac{N' \tan\phi + cL}{FS}$$

Using the minimum FS of 1.5 from Table 4-2 yields

$$T = \sum H = 6.82 - 1.54 = 5.28 \text{ kips}$$

$$5.28 \le \frac{12}{1.5}$$

$$5.28 \le 8$$

The sliding resistance is adequate.

Drained condition

$$\phi = 28^\circ$$

From equation 3-10

$$\text{SMF} = \frac{\tan\phi_d}{\tan\phi} = \frac{2}{3}$$

$$\phi_d = \tan^{-1}(\tfrac{2}{3}\tan\phi) = 19.52^\circ$$

$$K_P = \tan^2\left(45^\circ + \frac{\phi}{2}\right) = 2.77$$

Let $K = \frac{1}{2}K_p = 1.385$ (paragraph 3-8b):

$$P_R = \frac{1}{2}K\gamma_b h^2 = \frac{1}{2}(1.385)(0.0381)(9)^2 = 2.14 \text{ kips}$$

$$N' = \sum V = 25.37 + 8.76 - 18.62 = 15.51$$

$$T = 10.13 - 3.31 - 2.14 = 4.68 \text{ kips}$$

From equation 4-12

$$T \leq \frac{N' \tan \phi + cL}{FS}$$

Using the minimum FS of 1.5 from Table 4-2 yields

$$4.68 \leq \frac{15.51 \tan 28^\circ}{1.5}$$

$$4.68 \leq 5.50$$

The sliding resistance is adequate.

D. BEARING CAPACITY ANALYSIS (ON PLANE CG).

$$N' = 14.79 \text{ kips}$$

$$T = 3.15 \text{ kips}$$

$$\alpha = 11\text{-}31^\circ = 0.197 \text{ rad}$$

$$e = \frac{L_{CG}}{2} - a = \frac{20.4}{2} - 7.79 = 2.41$$

$$\bar{B} = L_{CG} - 2e = 15.58 \text{ ft}$$

From Figure 5-1a

$$\delta = \tan^{-1}\left(\frac{T}{N'}\right) = \tan^{-1}\left(\frac{3.15}{14.79}\right) = 12.02^\circ$$

$$\gamma' = 0.120 - 0.0625 = 0.0575 \text{ kcf}$$

$$D = 5 \text{ ft}$$

From Equation 5-8a

$$q_o = \gamma' D = 0.0575(5) = 0.2875 \text{ ksf}$$

From Equation 5-4a

$$\xi_{cd}(\phi = 0) = 1 + 0.2\left(\frac{D}{\bar{R}}\right) = 1 + 0.2\left(\frac{5}{15.58}\right)$$
$$= 1.064$$

From Equation 5-46

$$\xi_{qd} = \xi_{\gamma d} = 1 \text{ (for } \phi = 0^\circ)$$

From Equation 5-4c

$$\xi_{qd} = \xi_{\gamma d} = 1 + 0.1\left(\frac{D}{\bar{R}}\right)\tan\left(45^\circ + \frac{\phi}{2}\right) \quad (\text{for } \phi = 28^\circ)$$

$$\xi_{qd} = \xi_{\gamma d} = 1 + 0.1\left(\frac{5}{15.58}\right)(1.6643) = 1.053 \quad (\phi = 28^\circ)$$

From Equation 5-5a

$$\xi_{qi} = \xi_{ci} = \left(1 - \frac{\delta^\circ}{90^\circ}\right)^2 = \left(1 - \frac{12.02}{90}\right)^2$$
$$= 0.7507$$

From Equation 5-5b

$$\xi_{\gamma i} = \left(1 - \frac{12.02}{28}\right)^2 = 0.3257$$

From Equation 5-6a

$$\xi_{qt} = \xi_{\gamma t} = (1 - \alpha \tan \phi)^2$$
$$= (1 - 0.197 \times 0.532)^2 = 0.8015$$

From Equation 5-6b

$$\xi_{ct} = 1 - \left(\frac{2\alpha}{\pi + 2}\right) = 1 - \left(\frac{0.394}{\pi + 2}\right)$$
$$= 0.9232 \text{ (for } \phi = 0^\circ)$$

Undrained condition

$$c = 0.9 \text{ ksf}$$

From Equation 5-2

$$Q = \bar{B}(\xi_{cd}\xi_{ci}\xi_{ct}cN_c + \xi_{qd}\xi_{qi}\xi_{qt}q_oN_q)$$

From Table 5-1

$$N_c = 5.14, N_q = 1.00$$

$$Q = 15.58[1.064(0.7507)(0.9232)(5.14)(0.9)$$
$$+ 1(0.7507)(0.8015)(0.2875)(1)]$$

$$Q = 55.84 \text{ kips, FS} = \frac{Q}{N'} = \frac{55.84}{14.79} = 3.78 > 3.00$$

(required by Table 4-2)

Drained condition

$$\phi = 28^\circ$$

$$Q = \bar{B}\left(\xi_{qd}\xi_{qi}\xi_{qt}q_oN_q + \frac{\xi_{\gamma d}\xi_{\gamma i}\xi_{\gamma t}\bar{B}_\gamma N_\gamma}{2}\right)$$

From Table 5-1

$$N_q = 14.72, N_\gamma = 11.19$$

$$Q = 15.58[1.053(0.7507)(0.8015)(0.2875) \times (14.72) + \frac{1.053(0.3257)(0.8015)(15.58)(0.0575)(11.19)}{2}]$$

$$= 63.24 \text{ kips}$$

$$FS = \frac{Q}{N'} = \frac{63.24}{14.79} = 4.28 > 3.00$$

(required by Table 4-1)

Bearing capacity is adequate.

E. REINFORCEMENT (Chapter 9).

$f'_c = 3$ ksi, $f_y = 40$ ksi, cover = 4.5 in.

1. At base of stem

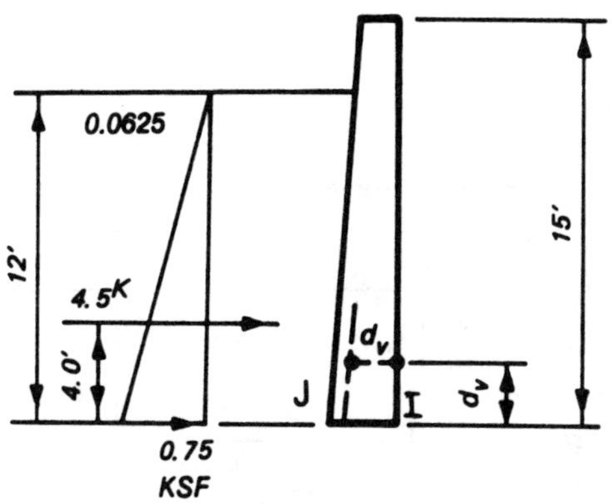

$$M = 4.5 \times 4 = 18.00 \text{ k-ft}$$

$$\frac{M_u}{\phi} = \frac{1.9(18.00)12}{0.9} = 456.0 \text{ k-in.}$$

$$d = 24 \text{ in.} - 4.5 \text{ in.} = 19.5 \text{ in.}$$

$$b = 12 \text{ in.}$$

From Figure 9-2

$$k_u = 1 - \sqrt{1 - \frac{M_u/\phi}{0.425f'_cbd^2}}$$

$$k_u = 1 - \sqrt{1 - \frac{456.0}{5817.825}} = 0.0400$$

$$C_u = T_u = 0.85f'_ck_ubd = 0.85(3)(0.040)(12)(19.5) = 23.87 \text{ kips}$$

$$A_s = \frac{T_u}{f_y} = \frac{23.87}{40} = \underline{\underline{0.60 \text{ in.}^2/\text{ft}}}$$

Check ductility requirements:

$$\beta_1 = 0.85, \varepsilon_c = 0.003$$

See Example 1, paragraph i, this appendix:

$$\rho_b = \left(\frac{0.85f'_c}{f_y}\right)\beta_1\left(\frac{\varepsilon_c}{\varepsilon_m + \frac{f_y}{29{,}000 \text{ ksi}}}\right)$$

$$\rho_b = \left(\frac{0.85(3)}{40}\right)(0.85)\left(\frac{0.003}{0.003 + \frac{40}{29{,}000}}\right)$$

$$= 0.03712$$

From paragraph 9-8c(2)(a)

$$\rho_{max} = \lambda\rho_b, \lambda = 0.25$$

$$\rho_{max} = 0.25(0.03712) = 0.000928$$

From paragraph 9-86(4), from ACI 318

$$\rho_{min} = 200/f_y = 200/40{,}000 \text{ ksi} = 0.005$$

$$\rho = \frac{A_s}{bd} = \frac{0.60}{12(19.5)} = 0.002564$$

$$\rho < \rho_{min}$$

Instead of using ρ_{min}, ACI 318 allows the minimum area of reinforcement to be one-third greater than that required by analysis. Therefore

$$A_s = 4/3(0.60) = 0.80 \text{ in.}^2/\text{ft}$$

$$\rho = \frac{A_s}{bd} = \frac{0.80}{12(19.5)} = 0.00342$$

$$\rho = \rho_{max}$$

Ductility is adequate.

2. Check the shear at distance d_v above base (paragraph 9-8f).

$$d_v = d - \left(\frac{0.5}{15}\right)d_v = 19.5 - 0.0333\ d_v$$

$$d_v = \frac{19.5}{1.03333} = 18.87 \text{ in.} = 1.573 \text{ ft}$$

$$V = \tfrac{1}{2}(0.0625)(12 - 1.573)^2 = 3.40 \text{ kips}$$

$$V_u = 1.9\ V = 1.9(3.40) = 6.46 \text{ kips} = 6460 \text{ lb}$$

From ACI 318

$$\phi V_c = \phi 2\sqrt{f'_c}\ bd_v$$
$$\phi V_c = 2(0.85)(\sqrt{3000})(12)(18.87)$$
$$= 21{,}084 \text{ lb} > 6460$$

The shear capacity is adequate. Shear is most critical for the stem.

3. Toe at face of stem (Figure 9-2):

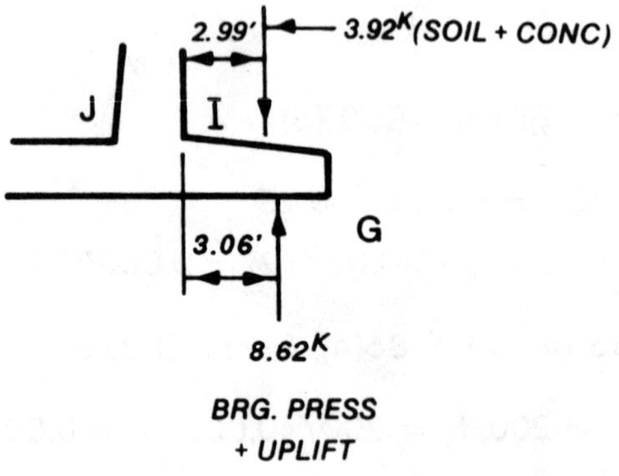

$$\begin{aligned} 8.62 \times 3.06 &= 26.38 \\ -3.92 \times 2.99 &= -11.72 \\ V_I = 4.70 \text{ kips} \quad M &= 14.66 \text{ k-ft} \end{aligned}$$

b = 12 in., d = 19.5 in.

$$0.425 f'_c bd^2 = 5817.825$$

$$\frac{M_u}{\phi} = \frac{1.9(14.66)(12)}{0.9} = 371.39 \text{ k-in.}$$

$$k_u = 1 - \sqrt{1 - \frac{371.39}{5817.825}} = 0.03244$$

$$C_u = T_u = 0.85(3)(0.03244)(12)(19.5)$$
$$= 19.36 \text{ kips}$$

$$A_s = \frac{T_u}{f_y} = \frac{19.36}{40} = \underline{\underline{0.48 \text{ in.}^2/\text{ft}}}$$

Check ductility:

$$\rho_{max} = 0.00928,\ \rho_{min} = 0.005$$

$$\rho = \frac{A_s}{bd} = \frac{0.48}{12(19.5)} = 0.00205$$

Again, since $\rho < \rho_{min}$, use a one-third increase in the area of reinforcement required by analysis. Therefore,

$$A_s = 4/3(0.48) = 0.64 \text{ in.}^2/\text{ft}$$

$$\rho = \frac{A_s}{bd} = \frac{0.64}{12(19.5)} = 0.002735$$

$$\rho < \rho_{max}$$

Ductility is adequate.

4. Heel and key reinforcement:

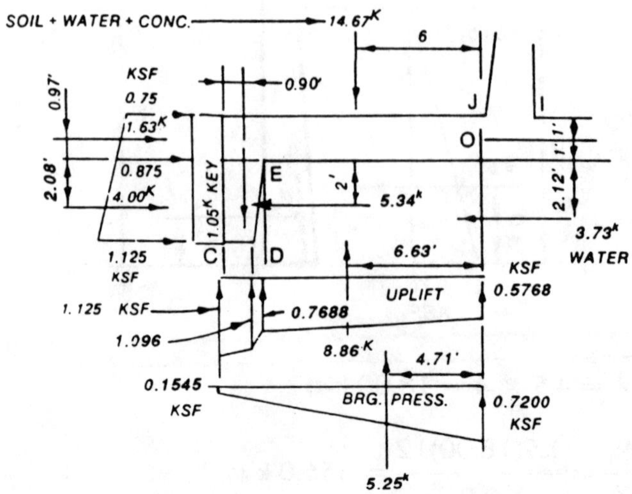

$$
\begin{aligned}
14.67 \times 6.00 &= 88.02 \\
1.05 \times 11.10 &= 11.66 \\
-8.86 \times 6.63 &= -58.74 \\
-5.25 \times 4.71 &= -24.73
\end{aligned}
$$

V = 1.61 kips

$$
\begin{aligned}
-3.73 \times 3.12 &= -11.64 \\
-5.34 \times 3.00 &= -16.02 \\
1.63 \times 0.02 &= 0.03 \\
4.00 \times 3.08 &= 12.32
\end{aligned}
$$

P = −3.44 kips

$$M_o = -0.90 \text{ k-ft}$$

b = 12 in., h = 24 in., d = 19.5 in.

$$0.425f'_cbd^2 = 5817.825$$

Reinforcement at O with 5.34 kips force on key (Figure 9-4):

$$M_u = 1.9M_o = 1.9(0.90)(12) = 20.52 \text{ in.-k}$$

$$P_u = 1.9P = 1.9(3.44) = 6.54 \text{ kips (tension)}$$

$$\frac{M_u}{P_u} = \frac{20.52}{6.54} = 3.14$$

$$d - \frac{h}{2} = 19.5 - \frac{24}{2} = 7.5$$

$$\frac{M_u}{P_u} < d - \frac{h}{2}$$

Therefore, the conditions of Figure 9-5 apply.

$$M_{ue} = -M_u + P_u\left(d - \frac{h}{2}\right)$$

$$= -20.52 + 6.54(7.5)$$

$$= 28.53 \text{ in.-k}$$

$$A'_s = \frac{M_{ue}}{\phi f_y(d - d')} = \frac{28.53}{0.9(40)(19.5 - 4.5)}$$

$$= 0.053 \text{ in.}^2/\text{ft}$$

$$A_s = \frac{\dfrac{P_u}{\phi} - f_yA'_s}{f_y} = \frac{\dfrac{6.54}{0.9} - 40(0.053)}{40}$$

$$= 0.13 \text{ in.}^2/\text{ft}$$

Reinforcement at O neglecting 5.34 kips force on key (Figure 9-3):

$$M_o = 0.90 + 5.34(3.00)$$

$$= 16.92 \text{ ft-k} = 203.04 \text{ in.-k}$$

$$M_u = 1.9M_o = 385.78 \text{ in.-k}$$

$$P = 1.9 \text{ kips (compression)}$$

$$P_u = 1.9P = 1.9(1.9) = 3.61 \text{ kips}$$

$$\phi = 0.9 - \left(\frac{P_u}{0.1f'_cbh}\right)(0.2)$$

$$= 0.9 - \left(\frac{3.61}{0.1(3)(12)(19.5)}\right)(0.2)$$

$$= 0.89$$

$$M_{ue} = M_u + P_u\left(d - \frac{h}{2}\right)$$

$$= 385.78 + 3.61\left(19.5 - \frac{24}{2}\right)$$

$$= 412.86 \text{ in-k}$$

$$k_u = 1 - \sqrt{1 - \frac{M_{ue}}{\phi 0.425f'_cbd^2}}$$

$$= 1 - \sqrt{1 - \frac{412.86}{0.89(5817.825)}}$$

$$= 0.04070$$

$$A_s = \frac{0.85f'k_ubd - \dfrac{P_u}{\phi}}{fy}$$

$$= \frac{0.85(3)(0.04070)(12)(19.5) - \dfrac{3.61}{0.89}}{40}$$

$$= 0.51 \text{ in.}^2/\text{ft}$$

Check ductility:

$$\rho_{max} = 0.00928, \rho_{min} = 0.005$$

$$\rho = \frac{A_s}{bd} = \frac{0.51}{12(19.5)} = 0.00218$$

$$\rho < \rho_{min}$$

Use $A_s = 4/3(0.51) = 0.68$ in.^{2}/ft:

$$\rho = \frac{A_s}{bd} = \frac{0.68}{12(19.5)} = 0.00291$$

$$\rho < \rho_{max}$$

Ductility is adequate.

Reinforcement in key at E (Figure 9-2):

$$M = 5.34(2) + 3.73(2.12) - 4.00(2.08)$$
$$= 10.27 \text{ ft-k}$$

$$\frac{M_u}{\phi} = \frac{1.9(10.27)(12)}{0.9} = 260.11 \text{ k-in.}$$

$$k_u = 1 - \sqrt{1 - \frac{260.11}{5817.825}} = 0.02261$$

$$C_u = T_u = 0.85(3)(0.02261)(12)(19.5)$$
$$= 13.49 \text{ kips}$$

$$A_s = \frac{T_u}{f_y} = \frac{13.49}{40} = \underline{\underline{0.34 \text{ in.}^2/\text{ft}}}$$

Check ductility:

$$\rho_{max} = 0.00928, \rho_{min} = 0.005$$

$$\rho = \frac{A_s}{bd} = \frac{0.34}{12(19.5)} = 0.00145$$

$$\rho < \rho_{min}$$

Use $A_s = 4/3(0.34) = 0.45$ in.^{2}/ft:

$$\rho = \frac{A_s}{bd} = \frac{0.45}{12(19.5)} = 0.00194$$

$$\rho < \rho_{max}$$

Ductility is adequate.

N-4. Example 4

Analyze the wall for overturning, sliding, and bearing capacity. Use the line-of-creep method to determine water pressures. Load Case 11, design flood. Calculate the reinforcement at critical sections.

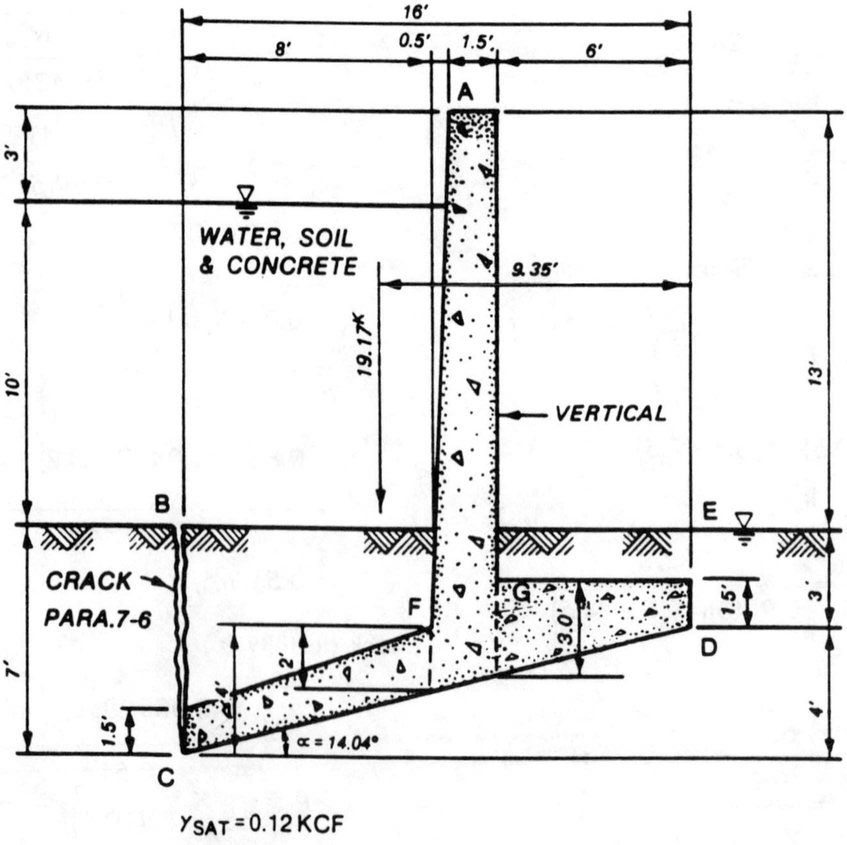

$\gamma_{SAT} = 0.12$ KCF
$\phi = 0$
c = 0.9 KSF

A. WATER PRESSURES (paragraph 3-19).

$$u_C = 17(0.0625) = 1.0625 \text{ ksf}$$

$$\Delta_h = 10 \text{ ft}$$

$$L_{CD} = 16.49 \text{ ft}$$

$$X_R = \frac{M_D}{N'} = \frac{55.42}{9.58} = 5.78 \text{ ft} > \frac{L_{CD}}{3} = \frac{16.49}{3}$$
$$= 5.50 \text{ ft}$$

One hundred percent of Surface CD is in compression (Table 4-2). The overturning criterion is satisfied.

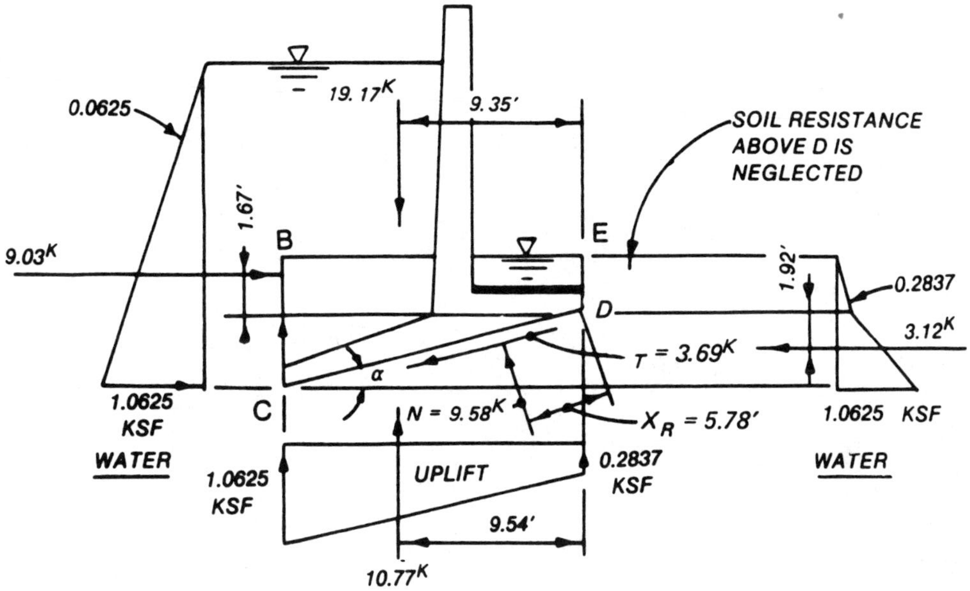

$$L_{DE} = 3 \text{ ft}$$

$$L_S = L_{CD} + L_{DE} = 19.49 \text{ ft}$$

$$U_D = \left(13 - \frac{10 \times 16.49}{19.49}\right)0.0625 = 0.2837 \text{ ksf}$$

B. OVERTURNING STABILITY.

19.17 ×	9.35 =	179.24
−10.77 ×	9.54 =	−102.75
8.40 kips = $\sum V$		
−9.03 ×	1.67 =	−15.08
3.12 ×	−1.92 =	−5.99
−5.91 kips		55.42 k-ft = M_D

$$N' = \sum V \cos\alpha + \sum H \sin\alpha$$
$$T = \sum H \cos\alpha - \sum V \sin\alpha$$
$$N' = 8.40(0.970) + 5.91(0.243)$$
$$N' = 9.58 \text{ kips}$$
$$T = 5.91(0.97) - 8.40(0.243)$$
$$T = 3.69 \text{ kips}$$

C. SLIDING STABILITY (paragraph 4-15). On Surface CD

$$N' = 9.58 \text{ kips}, T = 3.69 \text{ kips}, \alpha = 14.04°,$$

$$\phi = 0, c = 0.9 \text{ ksf}$$

$$L_{CD} = 16.49 \text{ ft}$$

From Equation 4-12

$$T \le \frac{N' \tan\phi + cL}{FS}$$

Using the minimum required FS of 1.5, as given in Table 4-2, yields

$$3.69 \le \frac{9.58 \tan 0° + 0.9(16.49)}{1.5}$$

$$3.69 \le 9.89$$

The sliding criterion is satisfied.

D. BEARING CAPACITY ANALYSIS.

$$\alpha = 14.04^\circ$$

$$\alpha = 0.245 \text{ rad}$$

$$\delta = \tan^{-1}\left(\frac{T}{N'}\right) = \tan^{-1}\left(\frac{3.69}{9.58}\right) = 21.07^\circ$$

$$e = \frac{L_{CD}}{2} - x_R = \frac{16.49}{2} - 5.78 = 2.465 \text{ ft}$$

See Figure 5-1b

$$\bar{B} = L_{CD} - 2e = 16.49 - 2(2.465) = 11.56 \text{ ft}$$

$$\gamma' = \gamma_{sat} - \gamma_w = 0.120 - 0.0625 = 0.0575 \text{ kcf}$$

From Equation 5-8a

$$q_o = \gamma' D = 0.0573(3) = 0.1725 \text{ ksf}$$

From Equation 5-4a

$$\xi_{cd} = 1 + 0.2\left(\frac{D}{\bar{B}}\right) = 1 + 0.2\left(\frac{3}{11.56}\right) = 1.05$$

From Equation 5-46

$$\xi_{\gamma d} = \xi_{qd} = 1.00$$

From Equation 5-5a

$$\xi_{ci} = \xi_{qi} = \left(1 - \frac{\delta^\circ}{90^\circ}\right)^2 = \left(1 - \frac{21.07}{90}\right)^2 = 0.587$$

From Equation 5-5b

$$\xi_{\gamma i} = 0$$

From Equation 5-6a

$$\xi_{qt} = \xi_{\gamma t} = 1$$

From Equation 5-6b

$$\xi_{ct} = 1 - \left(\frac{2\alpha}{\pi + 2}\right) = 1 - \left(\frac{2 \times 0.245}{5.14}\right) = 0.905$$

From Table 5-1

$$N_c = 5.14, N_q = 1, N_\gamma = 0$$

From Equation 5-7a

$$\xi_{\gamma g} = \xi_{qg} = [1 - \tan(\beta)]^2 = [1 - \tan(0)]^2 = 1$$

From Equation 5-7b

$$\xi_{cg} = 1 - \frac{2(\beta)}{(\pi + 2)} = 1 - \frac{2(0)}{(\pi + 2)} = 1$$

From Equation 5-2

$$Q = \bar{B}\left[(\xi_{cd}\xi_{ci}\xi_{ct}\xi_{cg} c N_c) + (\xi_{qd}\xi_{qi}\xi_{qt}\xi_{qg} q_o N_q) + \frac{(\xi_{\gamma d}\xi_{\gamma i}\xi_{\gamma t}\xi_{\gamma g}\bar{B}\gamma N_\gamma)}{2}\right]$$

$$= 11.56\left[(1.05)(0.587)(0.905)(1)(0.9)(5.14) + (1)(0.587)(1)(1)(0.1725)(1) + \frac{(1)(0)(1)(1)(11.56)(0.0575)(0)}{2}\right]$$

$$= 31.00 \text{ kips}$$

From Equation 5-1

$$FS = \frac{Q}{N'} = \frac{31.00}{9.58} = 3.24$$

3.24 > 3.0, so the bearing criterion is satisfied (Table 4-2)

E. REINFORCEMENT (Chapter 9).

$f'_c = 3$ ksi, $f_y = 40$ ksi, load factor = 1.9, $\phi = 0.9$, cover = 4.5 in.

1. At base of stem (Figure 9-2): b = 12 in., h = 23.44 in., d = 23.44 − 4.5 = 18.94 in.

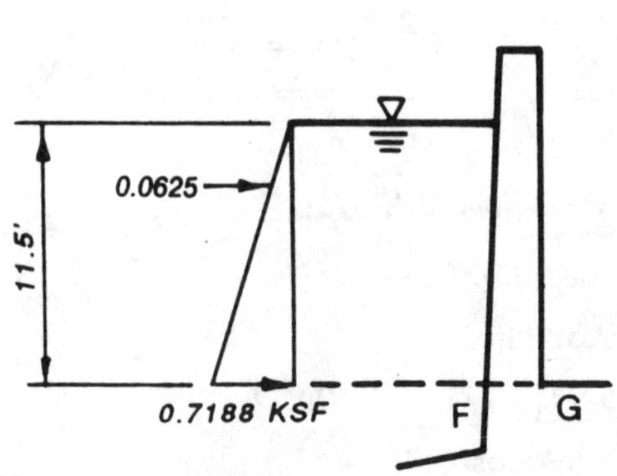

$$M = 0.0625(11.5)\left(\frac{11.5}{2}\right)\left(\frac{11.5}{3}\right)$$

$$M = 0.0625(11.5)^3(\tfrac{1}{6}) = 15.84 \text{ ft-k}$$

$$\frac{M_u}{\phi} = \frac{1.9(15.84)(12)}{0.9} = 401.28 \text{ k-in.}$$

$$0.425f'_c bd^2 = 5488.471$$

$$k_u = 1 - \sqrt{1 - \frac{401.28}{5488.471}} = 0.03725$$

$$C_u = T_u = 0.85f'_c k_u bd$$

$$C_u = T_u = 21.59 \text{ kips}$$

$$A_s = \frac{T_u}{f_y} = \frac{21.59}{40} = \underline{\underline{0.54 \text{ in.}^2/\text{ft}}}$$

For shear check see Examples 1 and 3, this appendix.

Check ductility and concrete section strength: calculate ρ_b for $\varepsilon_c = 0.003$, $f_y = 40$ ksi, $f'_c = 3$ ksi, $\beta_1 = 0.85$.
See Example 1, this appendix, for the derivation of ρ_b.

$$\rho_b = \frac{0.85f'_c\beta_1}{f_y}\left(\frac{\varepsilon_c}{\dfrac{\varepsilon_1 + f_y}{29{,}000 \text{ ksi}}}\right)$$

$$= \frac{0.85(3)(0.85)}{40}\left(\frac{0.003}{0.003 + \dfrac{40}{29{,}000}}\right)$$

$$= 0.03712$$

$$\rho_{max} = 0.25\rho_b = 0.00928$$

See paragraph 9-8b(4), from ACI 318

$$\rho_{min} = 200/f_y = 200/40{,}000 = 0.005$$

$$\rho_{actual} = \frac{A_s}{bd} = \frac{0.54}{12(18.94)} = 0.00238$$

Since $\rho < \rho_{min}$, use a one-third increase in the required area of reinforcement (from ACI 318).

$$A_s = \tfrac{4}{3}(0.54) = 0.72 \text{ in.}^2/\text{ft}$$

$$\rho = \frac{A_s}{bd} = \frac{0.72}{12(18.94)} = 0.00317$$

$$\rho < \rho_{max}$$

Ductility is adequate.

2. In toe at stem (Chapter 9, Figure 9-3): $b = 12$ in., $d = 36 - 4.5 = 31.5$ in.

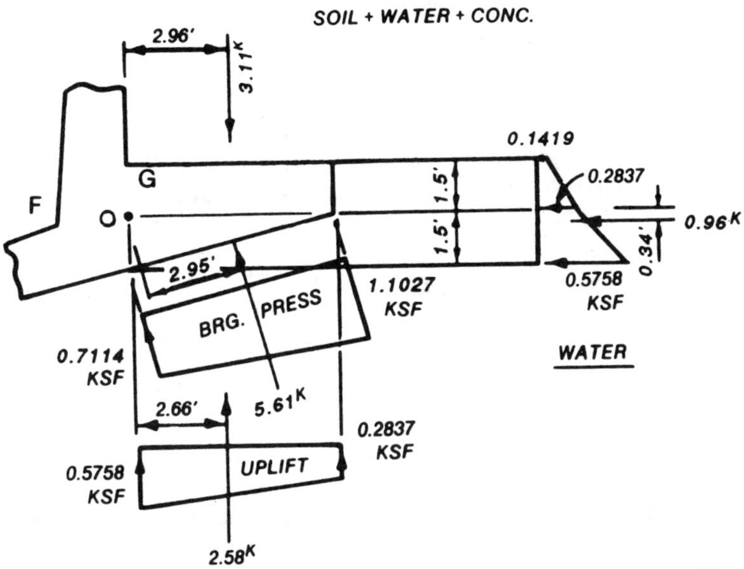

$$2.58(2.66) = 6.863$$
$$5.61(2.95) = 16.550$$
$$-3.11(2.96) = -9.206$$
$$0.96(-0.34) = \underline{-0.326}$$
$$M_o = 13.881 \text{ k-ft}$$

$$P_o = 0.96 + 5.61 \sin 14.04° = 2.32 \text{ kips}$$

$$M_u = 1.9(13.881) = 26.37 \text{ ft-k} = 316.44 \text{ in.-k}$$

$$P_u = 1.9(2.32) = 4.41 \text{ kips}$$

$$M_{ue} = M_u + P_u\left(d - \frac{h}{2}\right)$$
$$= 316.44 + 4.41\left(31.5 - \frac{36}{2}\right)$$
$$= 375.98 \text{ in.-k}$$

$$\frac{M_{ue}}{\phi} = \frac{375.98}{0.9} = 417.75 \text{ in.-k}$$

$$0.425f'_c bd^2 = 0.423(3)(12)(31.5)^2 = 15{,}181.425$$

$$k_u = 1 - \sqrt{1 - \left(\frac{M_{ue}}{\phi}\right)\left(\frac{1}{0.425f'_c bd^2}\right)}$$
$$= 1 - \sqrt{1 - \frac{417.75}{15{,}181.425}}$$
$$= 0.01385$$

$$C_u = 0.85f'_c k_u bd = 0.85(3)(0.01385(12)(31.5)$$
$$C_u = 13.35 \text{ kips} = T_u$$

$$A_s = \frac{T_u}{f_y} = \frac{13.35}{40} = 0.33 \text{ in.}^2/\text{ft}$$

$$\rho = \frac{0.33}{12(31.5)} = 0.000873 < \rho_{max} = 0.00928$$

Since $\rho < \rho_{min}$, use a one-third increase in the area of reinforcement (from ACI 318).

$$A_s = 4/3(0.33) = 0.44 \text{ in.}^2/\text{ft}$$

3. In heel at stem (Figure 9-3):

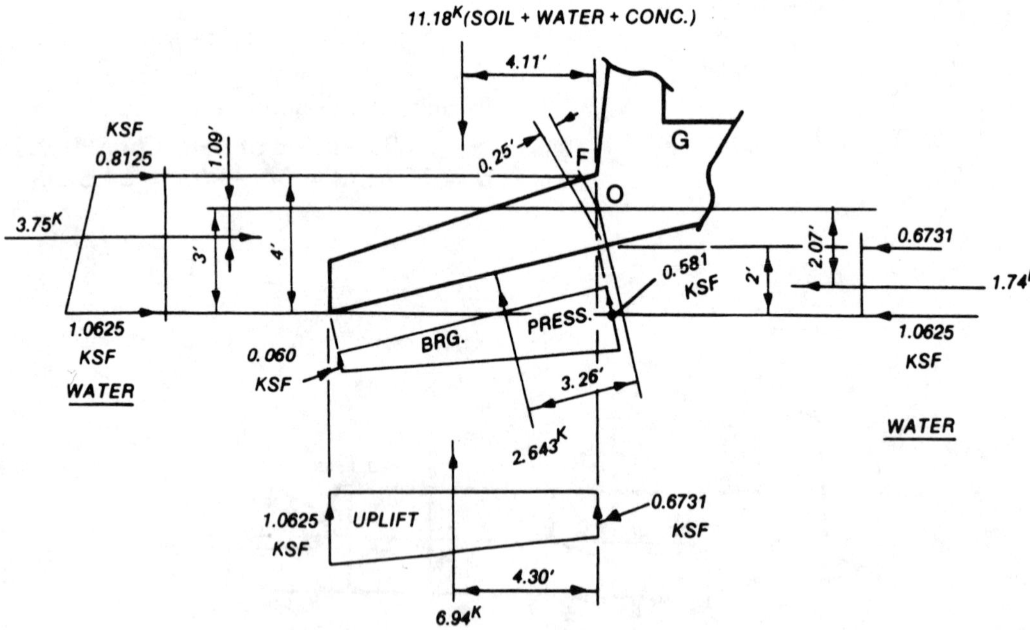

−1.74	× 2.07 =	−3.60	
11.18	× 4.11 =	45.95	
3.75	× 1.09 =	4.09	
−2.643	× 3.26 =	−8.62	
−6.94	× 4.30 =	−29.84	
	M_o =	7.98	k-ft

$b = 12$ in., $d = 19.5$ in.

$0.425f'_c bd^2 = 5817.825$

$$M_u = 1.9(7.98) = 15.16 \text{ ft-k} = 181.92 \text{ in.-k}$$

$$P = 3.75 - 1.74 - 2.643(\sin 14.04°)$$
$$= 1.37 \text{ kips (compression)}$$

$$P_u = 1.9(1.37) = 2.60 \text{ kips}$$

$$M_{ue} = M_u + P_u\left(d - \frac{h}{2}\right)$$

$$= 181.92 + 2.60\left(19.5 - \frac{24}{2}\right)$$

$$= 201.42 \text{ in.-k}$$

$$\phi = 0.9 - \left(\frac{P_u}{0.1f'_c bh}\right)(0.2)$$

$$= 0.9 - \left(\frac{2.6}{0.1(3)(12)(24)}\right)(0.2)$$

$$= 0.89$$

$$\frac{M_{ue}}{\phi} = \frac{201.42}{0.89} = 226.31 \text{ in.-k}$$

Check d_{min} from Equation 9-13:

$$\rho_{max} = 0.25\rho_b = 0.00928$$

$$k_m = \frac{f_y \rho_{max}}{0.85f'_c} = \frac{40{,}000(0.00928)}{0.85(3000)} = 0.14557$$

$$d_{min} = \sqrt{\frac{M_u/\phi}{0.85f'_c k_m b\left(1 - \frac{k'_m}{2}\right)}}$$

$$= \sqrt{\frac{226.31}{0.85(3)(0.14557)(12)\left(1 - \frac{0.14557}{2}\right)}}$$

$$= 7.40 \text{ in.}$$

This is an approximate value of d_{min} because Equation 9-13 is valid only for flexure.

$$h_{min} = 7.4 + 4.5 = 11.9 \text{ in.}$$

The heel could have a constant thickness of 12 inches if desired. The steel will be sized for a d = 19.5 inches.

$$k_u = 1 - \sqrt{1 - \frac{M_{ue}}{\phi 0.425f'_c bd^2}}$$

$$= 1 - \sqrt{1 - \frac{226.31}{5817.825}}$$

$$= 0.01964$$

$$A_s = \frac{0.85f'_c k_u bd - \frac{P_u}{\phi}}{f_y}$$

$$= \frac{0.85(3)(0.01964)(12)(19.5) - \frac{2.60}{0.89}}{40}$$

$$= 0.22 \text{ in.}^2/\text{ft}$$

$$\rho = \frac{A_s}{bd} = \frac{0.22}{12(19.5)} = 0.0009402 < \rho_{min} = 0.005$$

Since $\rho < \rho_{min}$, use a one-third increase in the area of reinforcement.

$$A_s = 4/3(0.22) = 0.29 \text{ in.}^2/\text{ft}$$

N-5. Example 5

Analyze the wall for overturning, sliding, and bearing capacity. Determine water pressure using the line-of-creep method. Load Case 12. Assume that sheet pile cutoff wall is 50 percent effective.

A. WATER PRESSURE WITH STEEL SHEET PILE CUTOFF WALL. Refer to paragraph 7-4a. Full-head pressure on headwater side of sheet pile at point C = (18 ft)(0.0625 pcf) = 1.125 ksf.

Pressure at point F, based on seepage path from cutoff wall (point D) to point G

$$= \left(\begin{array}{l}\text{pressure head at F}\\ \text{due to position}\end{array} + \text{total head differential} \times \frac{FG}{DFG}\right)(0.0625)$$

$$= \left(6.0 + 12.0 \times \frac{4}{21.25}\right)(0.0625)$$

$$= 0.516 \text{ ksf at point F}$$

Pressure on tailwater side of sheet pile at point D

= full-head pressure at D − 50%
(full-head pressure at D − pressure at point F)

= 1.125 − 0.5(1.125 − 0.516) = 0.821 ksf

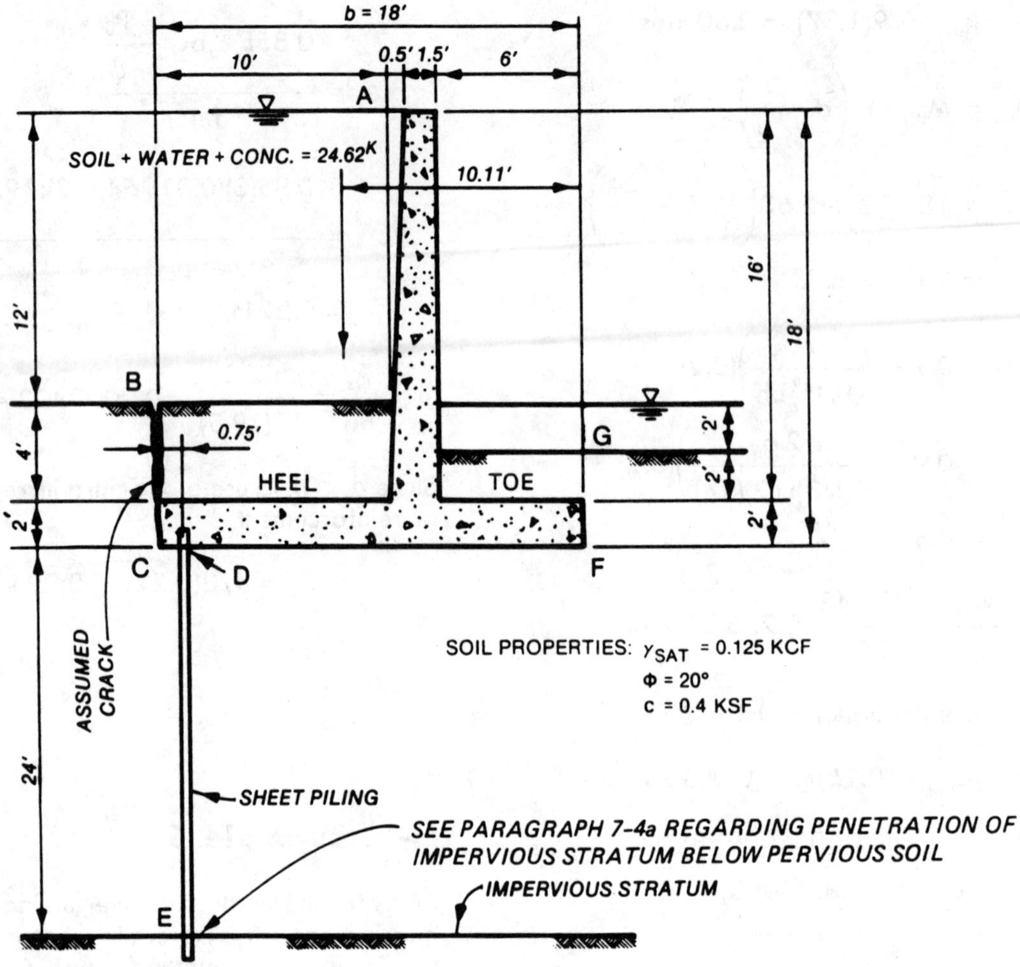

B. LATERAL SOIL FORCE. Resisting side:

$\gamma' = $ [2 ft(weight of water)
+ 4(weight of saturated earth)
− seepage at F] ÷ 4

$= [2(0.0625) + 4(0.125) - 0.516] \div 4$

$= 0.0273$ kcf average, point G to point F

From Equation 3-4

$$K_o = 1 - \sin\phi = 1 - 0.342 = 0.658$$

$$P_o = \tfrac{1}{2}K_o\gamma_b h^2 = \tfrac{1}{2}(0.658)(0.0273)(4)^2$$
$$= 0.1437 \text{ kips}$$

C. OVERTURNING (paragraph 4-8).

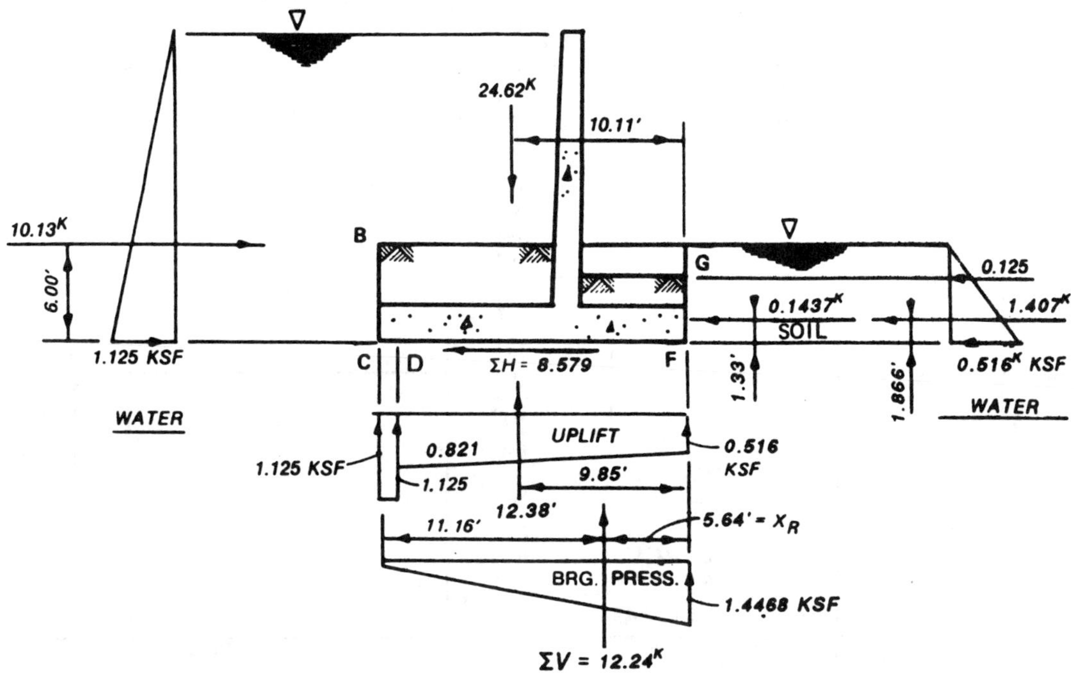

$$\begin{aligned} 24.62 \times 10.11 &= 248.91 \\ -12.38 \times 9.85 &= -121.94 \\ \textstyle\sum V = 12.24 \text{ kips} & \\[6pt] -10.13 \times 6.00 &= -60.78 \\ 0.14 \times 1.33 &= 0.19 \\ 1.41 \times 1.87 &= 2.64 \\ \textstyle\sum H = -8.57 \text{ kips} \qquad & 69.02 \div 12.24 = 5.64 \text{ ft} = x_R \end{aligned}$$

$$a = 3x_R = 3(5.64) = 16.92 \text{ ft} < 18.00 \text{ ft} = b$$

$$\text{Percent of base in compression} = \left(\frac{a}{b}\right)(100) = \left(\frac{16.92}{18}\right)(100) = 94\% > 75\%$$

(required by Table 4-2)

The criterion is satisfied.

D. SLIDING STABILITY (FS = 1.33 required by Table 4-2, paragraph 4-15).

$$N' = \sum V = 12.24 \text{ kips}, \ T = \sum H = 8.58 \text{ kips}$$

From Equation 4-12

$$T \leq \frac{N' \tan \phi + cL}{FS}$$

Using the minimum required FS of 1.33 from Table 4-2 yields

$$8.58 \leq \frac{12.24 \tan 20^\circ + 0.4(16.92)}{1.33}$$

$$8.58 = 11.22$$

The sliding criterion is satisfied.

E. BEARING CAPACITY (Chapter 5).

From Figure 5-1

$$\delta = \tan^{-1}\left(\frac{\sum H}{\sum V}\right) = \tan^{-1}\left(\frac{8.58}{12.24}\right) = 35.03^\circ$$

$$e = \frac{B}{2} - x_R = \frac{18}{2} - 5.64 = 3.36 \text{ ft}$$

$$\bar{B} = B - 2e = 18 - 2(3.36) \doteq 11.28 \text{ ft}$$

$$\gamma' = \gamma_b = 0.125 - 0.0625 = 0.0625, \ D = 4 \text{ ft}$$

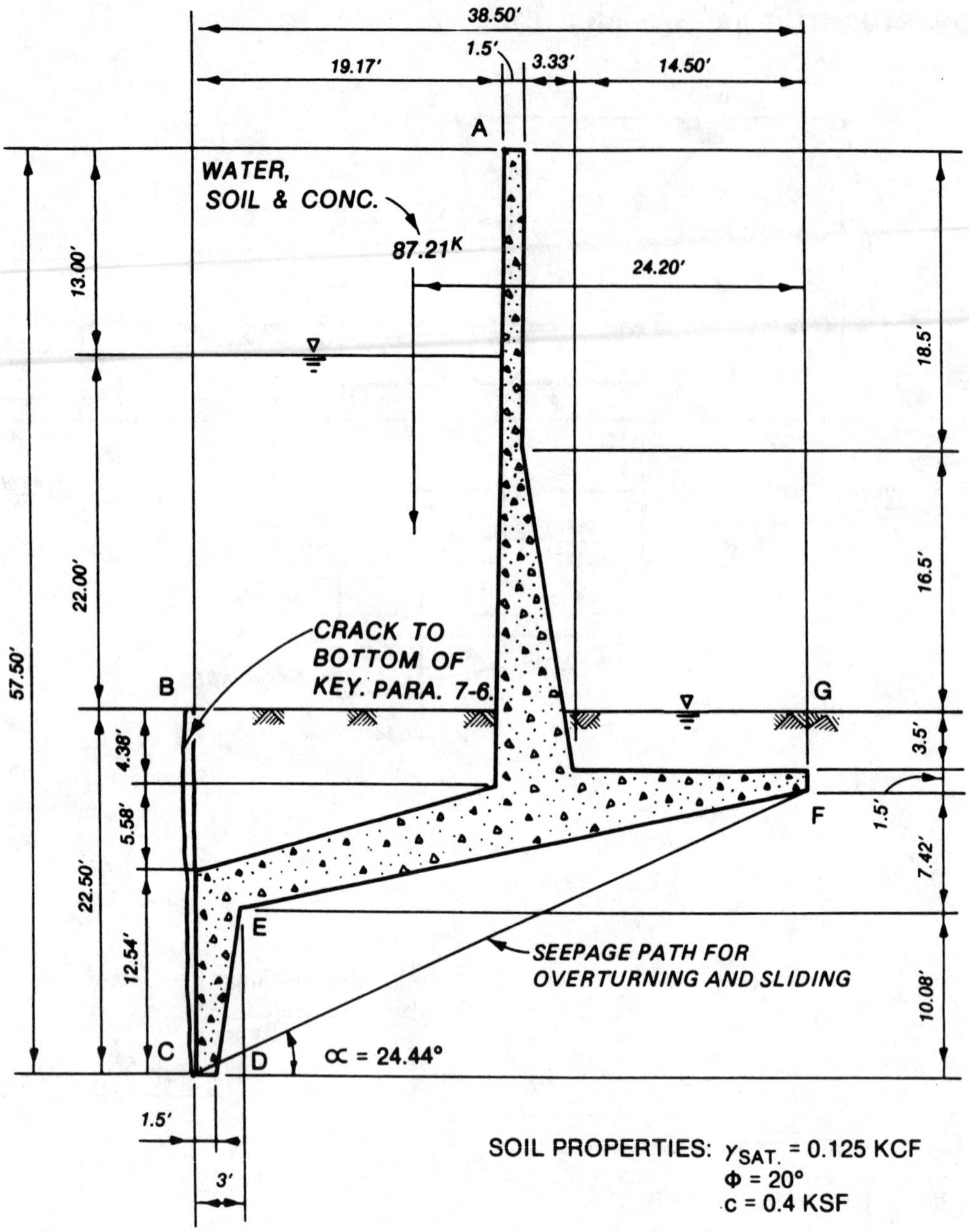

From Equation 5-8a

$$q_o = \gamma' D = 0.0625(4) = 0.25 \text{ ksf}$$

From Equation 5-4a

$$\xi_{cd} = 1 + 0.2\left(\frac{D}{B}\right)\tan\left(45^\circ + \frac{\phi}{2}\right)$$

$$= 1 + 0.2\left(\frac{4}{11.28}\right)(1.428) = 1.101$$

From Equation 5-4c

$$\xi_{qd} = \xi_{\gamma d} = 1 + 0.1\left(\frac{D}{B}\right)\tan\left(45^\circ + \frac{\phi}{2}\right)$$

$$= 1 + 0.1\left(\frac{4}{11.28}\right)(1.428) = 1.051$$

From Equation 5-5a

$$\xi_{qi} = \xi_{ci} = \left(1 - \frac{\delta^\circ}{90^\circ}\right)^2 = \left(1 - \frac{35.03}{90}\right)^2 = 0.373$$

From Equation 5-5b

$$\xi_{\gamma i} = 0, \delta > \phi$$

From Equation 5-7a

$$\xi_{\gamma g} = \xi_{qg}[1 - \tan(\beta)]^2 = [1 - \tan(0)]^2 = 1$$

From Equation 5-7d

$$\xi_{cg} = \xi_{qg} - \frac{1 - \xi_{qg}}{N_c \tan\phi} = 1 - \frac{1 - 1}{14.83\,(\tan 20^\circ)} = 1$$

From Equation 5-6a

$$\xi_{qt} = \xi_{\gamma t} = (1 - \alpha \tan\phi)^2 = (1 - 0)^2 = 1$$

From Equation 5-6c

$$\xi_{ct} = \xi_{qt} - \frac{1 - \xi_{qt}}{N_c \tan\phi} = 1 - \frac{1 - 1}{14.83\,(\tan 20^\circ)} = 1$$

From Table 5-1

$$N_c = 14.83, N_q = 6.40, N_\gamma = 2.87$$

From Equation 5-2

$$Q\bar{B}\left[(\xi_{cd}\xi_{ci}\xi_{ct}\xi_{cg}cN_c) + \approx \xi_{qcd}\xi_{qi}\xi_{qt}\xi_{qg}q_oN_q) + \frac{(\xi_{\gamma d}\xi_{\gamma i}\xi_{\gamma t}\xi_{\gamma g}\bar{B}\gamma N_\gamma)}{2}\right]$$

$$= 11.28[(1.101)(0.373)(1)(1)(0.4)(14.83) + 1.051(0.373)(1)(1)(0.25)(6.4) + 1.051(0)(1)(1)(11.28)(0.0625)(2.87)(\tfrac{1}{2})]$$

$$= 34.55 \text{ kips}$$

$$N' = 12.24 \;(\textstyle\sum V \text{ for this problem})$$

From Equation 5-1

$$FS = \frac{Q}{N'} = \frac{34.55}{12.24} = 2.82 > 2$$

The bearing criterion is satisfied (Table 4-2).

N-6. Example 6

Analyze the wall for overturning, sliding, and bearing capacity. Use the line-of-creep method to determine water pressures. Load Case I1, design flood.

A. WATER PRESSURES (paragraph 3-19).

$$L_{CF} = \sqrt{(38.5)^2 + (17.5)^2} = 42.29 \text{ ft},$$

$$L_{FG} = 5 \text{ ft}, \Delta_h = 22 \text{ ft}$$

$$L_S = L_{CF} + L_{FG} = 42.29 + 5 = 47.29 \text{ ft}$$

$$u_C = 44.5\,(0.0625) = 2.7813 \text{ ksf}$$

$$u_F = \left(27 \text{ ft} - \frac{22 \times 42.29}{47.29}\right)0.0625 = 0.4579 \text{ ksf}$$

Compute the water pressure at D and E. Prorate the head loss along the path CDEF.

$$\text{Head loss along path CF} = \left(\frac{42.29}{47.29}\right)(22) = 19.674 \text{ ft.}$$

$$L_S(\text{concrete surface}) = L_{CD} + L_{DE} + L_{EF}$$

$$L_S = 1.5 + 10.19 + 36.27 = 47.96 \text{ ft}$$

$$\text{Head loss at D} = \left(\frac{1.5}{47.96}\right)(19.674) = 0.6153 \text{ ft}$$

$$u_D = (44.5 - 0.6153)0.0625 = 2.743 \text{ ksf}$$

$$\text{Head loss at E} = \left(\frac{(1.5 + 10.19)}{47.96}\right)(19.674) = 4.795 \text{ ft}$$

$$u_E = (34.42 - 4.795)0.0625 = 1.852 \text{ ksf}$$

B. LATERAL SOIL FORCE (RESISTING SIDE, ABOVE POINT F).

From Equation 3-4

$$K_o = 1 - \sin\phi = 1 - 0.342 = 0.658$$

$$\gamma' = [5(0.125) - 0.4579] \div 5 = 0.0334 \text{ kcf}$$

$$P_o = \tfrac{1}{2}K_o\gamma_b h^2 = \tfrac{1}{2}(0.658)(0.0334)(5)^2 = \underline{\underline{0.275 \text{ kips}}}$$

C. LATERAL WATER FORCE (DRIVING SIDE).

$h = 22.00 + 22.50 = 44.5$ ft, $\gamma_w = 0.0625$ ksf

$P_w = \frac{1}{2}\gamma_w h^2 = \frac{1}{2}(0.0625)(44.5)^2 = \underline{\underline{61.88 \text{ kips}}}$

D. OVERTURNING STABILITY (paragraph 4-8, Figure 4-6).

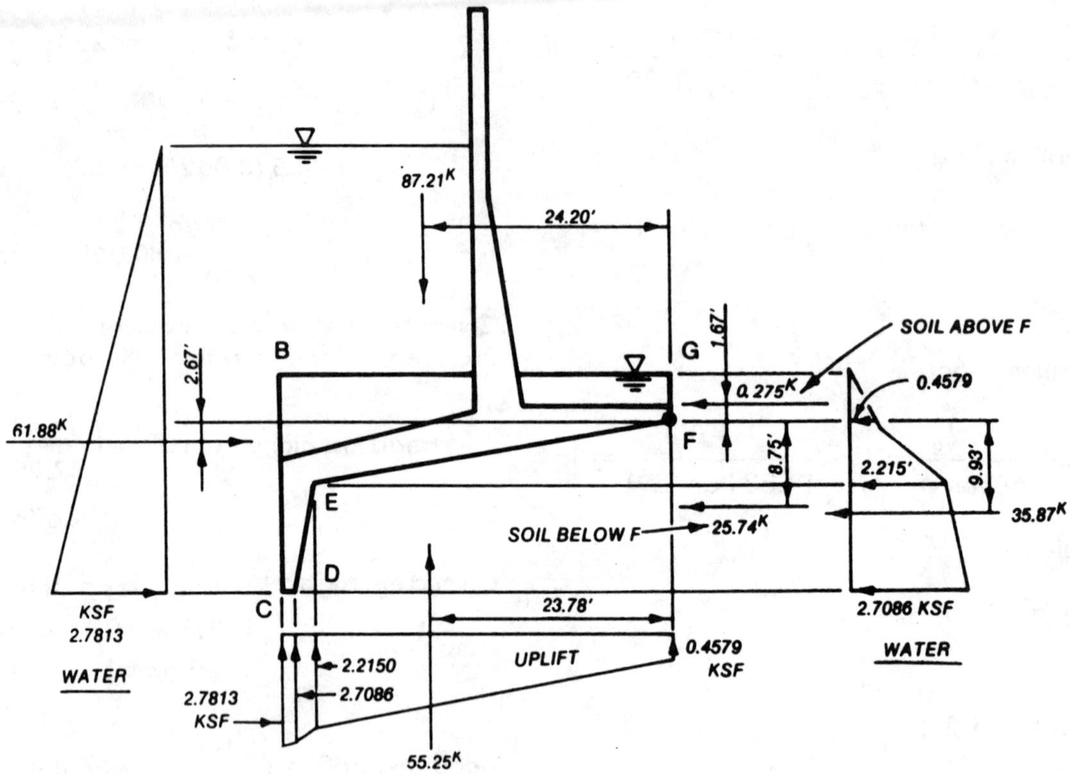

$$
\begin{array}{rcrcr}
87.21 & \times & 24.20 = & & 2110.48 \\
-48.59 & \times & 23.78 = & & -1155.47 \\
\underline{\underline{38.62}} & = \sum V & & & \\
-61.88 & \times & -2.67 = & & 165.22 \\
0.275 & \times & 1.67 = & & 0.45 \\
28.74 & \times & -8.75 = & & -251.48 \\
32.87 & \times & -10.11 = & & -332.32 \\
\underline{\underline{0}} & = \sum H & M_F = & & 536.88 \text{ k-ft}
\end{array}
$$

$$x_R = \frac{536.88}{38.62} = \underline{\underline{13.90 \text{ ft}}} > \frac{b}{3}$$

One hundred percent of the base is in compression; the overturning criteria are satisified (Table 4-2).

E. SLIDING STABILITY (on Plane CF, paragraph 4-15).

$$N' = \sum V \cos\alpha + \sum H \sin\alpha$$
$$T = \sum H \cos\alpha - \sum V \sin\alpha$$
$$N' = 44.63(0.9104) + 32.13(0.4137) = 53.92 \text{ kips}$$
$$T = 32.13(0.9104) - 44.63(0.4137) = 10.79 \text{ kips}$$
$$L_{CF} = 42.29 \text{ ft}$$

For the sliding stability to be satisfied

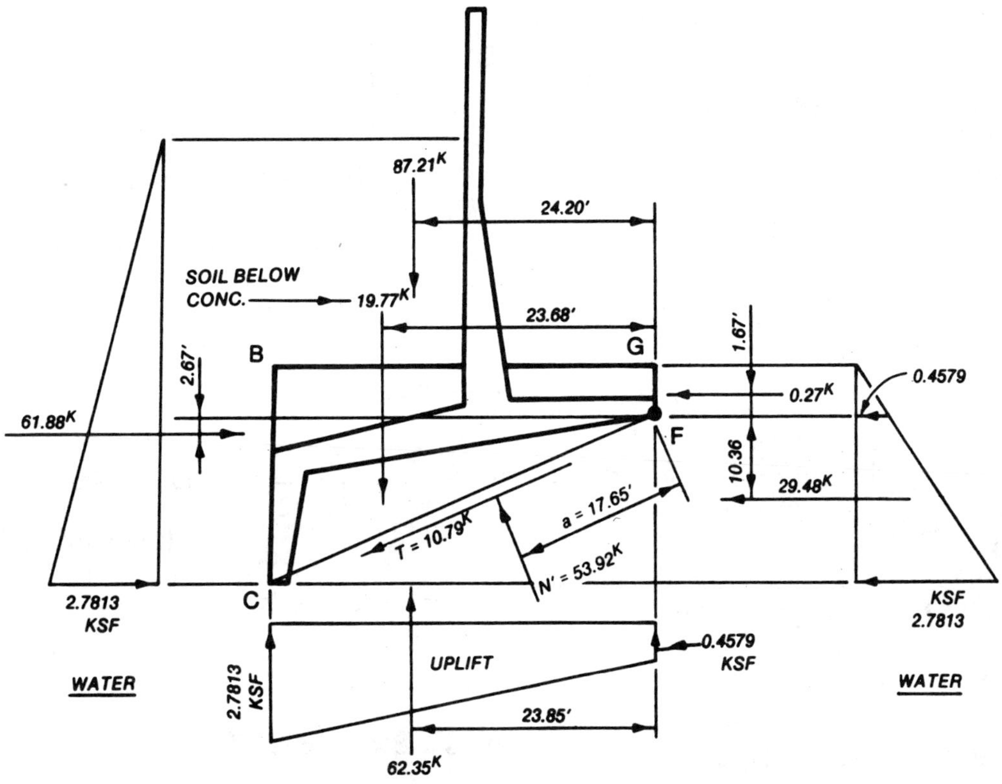

From Equation 4-12

$$T \leq \frac{N' \tan \phi + cL}{FS}$$

$$L = L_{CF}$$

$$N' \tan \phi + cL_{CF} = 53.92(0.364) + 0.4(42.29)$$

$$= 36.54 \text{ kips}$$

From Table 4-2, the minimum FS required is 1.5.

$$T \leq \frac{36.54}{1.5}$$

$$10.79 \leq 24.36$$

The criterion are satisfied.

F. BEARING CAPACITY (Chapter 5).

From Figure 5-1

$$\alpha = 24.44^\circ, \delta = \tan^{-1}\left(\frac{T}{N'}\right) = \tan^{-1}\left(\frac{10.79}{53.92}\right)$$

$$= 11.32^\circ$$

$$\alpha = 0.427 \text{ rad}$$

$$e = \frac{L_{CF}}{2} - a = \frac{42.29}{2} - 17.65 = 3.495$$

$$\bar{B} = L_{CF} - 2e = 42.29 - 2(3.495) = 35.30 \text{ ft}$$

From Equation 5-8a

$$D = 5 \text{ ft}, \gamma_b = 0.0625, q_o = \gamma_b D = 0.3125 \text{ ksf}$$

From Equation 5-4a

$$\xi_{cd} = 1 + 0.2\left(\frac{D}{\bar{B}}\right) \tan\left(45^\circ + \frac{\phi}{2}\right)$$

$$= 1 + 0.2\left(\frac{5}{35.3}\right)(1.428) = 1.04$$

From Equation 5-4c

$$\xi_{qtd} = \xi_{\gamma d} = 1 + 0.1\left(\frac{D}{\bar{B}}\right) \tan\left(45^\circ + \frac{\phi}{2}\right) = 1.02$$

From Equation 5-6a

$$\xi_{qt} = \xi_{\gamma t} = (1 - \alpha \tan \phi)^2$$
$$= (1 - 0.427 \times 0.364)^2 = 0.7133$$

From Equation 5-6c

$$\xi_{ct} = \xi_{qt} - \frac{1 - \xi_{qt}}{N_c \tan \phi} = 0.7133 - \frac{1 - 0.7133}{14.83 \times 0.364}$$

$$= 0.66$$

From Equation 5-5a

$$\xi_{qi} = \xi_{ci} = \left(1 - \frac{\delta^\circ}{90^\circ}\right)^2 = \left(1 - \frac{11.32}{90}\right)^2 = 0.764$$

From Equation 5-5b

$$\xi_{\gamma i} = \left(1 - \frac{\delta^\circ}{\phi}\right)^2 = \left(1 - \frac{11.32}{20}\right)^2 = 0.188$$

From Table 5-1

$$N_c = 14.83, N_q = 6.40, N_\gamma = 2.87$$

From Equation 5-2

$$Q = \bar{B}\left(\xi_{cd}\xi_{ci}\xi_{ct}cN_c + \xi_{qd}\xi_{qi}\xi_{qt}q_oN_q + \frac{\xi_{\gamma d}\xi_{\gamma i}\xi_{\gamma t}\bar{B}\gamma N_\gamma}{2}\right)$$

$$Q = 35.3\left[1.04(0.764)(0.66)(0.4)(14.83) + 1.02(0.764)(0.7133)(0.3125)(6.4) + \frac{1.02(0.188)(0.7133)(35.3)(0.0625)(2.87)}{2}\right]$$

$$= 164.34 \text{ kips}$$

From Equation 5-1

$$FS = \frac{Q}{N'} = \frac{164.34}{53.92} = 3.05 > 3.00$$

(required by Table 4-2)

The bearing criterion are satisfied.

N-7. Example 7

Analyze the wall for overturning, sliding, and bearing capacity. Load Case C2b, breaking wave condition.

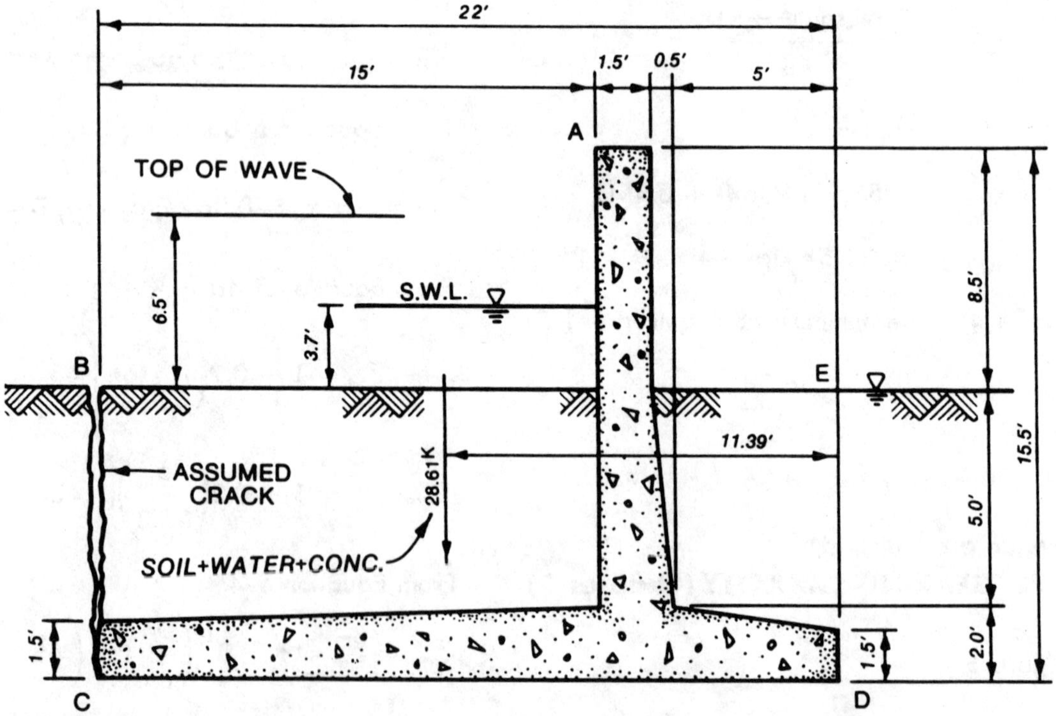

SOIL PROPERTIES: $\gamma_{SAT} = 0.125$ KCF

$\phi = 0°$

c = 0.90 KSF

WATER: $\gamma_W = 0.064$ KCF

A. WATER PRESSURES (paragraphs 3-19 and 3-24d,f). Seepage from still water level (swl) (paragraph 3-24f):

$$L_{CD} = 22 \text{ ft}, L_{DE} = 7.0 \text{ ft},$$

$$L_S = 22 + 7 = 29 \text{ ft}$$

$$u_B = 3.7(0.064) = 0.237 \text{ ksf}$$

$$u_C = 10.7(0.064) = 0.685 \text{ ksf}$$

$$u_D = \left[10.7 - \frac{3.7(22)}{29}\right]0.064 = 0.505 \text{ ksf}$$

(on line of creep)

$$u_E = 0.0 \text{ (end creep, tail water)}$$

Wave force (paragraph 3-24): Increase in static force, due to wave crest, above point C:

$$\tfrac{1}{2}(6.5 + 7.0)^2(0.064) - \tfrac{1}{2}(3.7 + 7.0)^2(0.064)$$
$$= 5.832 - 3.664 = 2.168 \text{ kips}$$

$$5.832(13.5)\tfrac{1}{3} - 3.664(10.7)\tfrac{1}{3} - 13.176 \text{ ft-k}$$

$$2.168 \text{ k/LF acting at } \frac{13.176}{2.168} = 6.077 \text{ ft above point C}$$

The dynamic force of the breaking wave is calculated from procedures found in Chapter 7 of the *Shore Protection Manual.*

Dynamic force of breaking wave = 5.176 kips at swl

B. SOIL FORCE (RESISTING SIDE).

$$K_o = 1.00 \text{ (at-rest)}$$

$$\gamma' = \gamma_{sat} - \text{seepage uplift}$$

$$\gamma' = [7(0.125) - 0.505] \div 7 = 0.0529 \text{ kcf}$$

$$P_o = \tfrac{1}{2}K_o\gamma_b h^2 = \tfrac{1}{2}(1.00)(0.0529)(7)^2 = \underline{\underline{1.296 \text{ kips}}}$$

C. OVERTURNING (paragraph 4-8). Initially, all of the base is assumed to be in compression. Full hydrostatic uplift is applied to any part of the base not in compression. The percent of the base in compression is recomputed until the assumed value equals the computed one. The final solution is shown next.

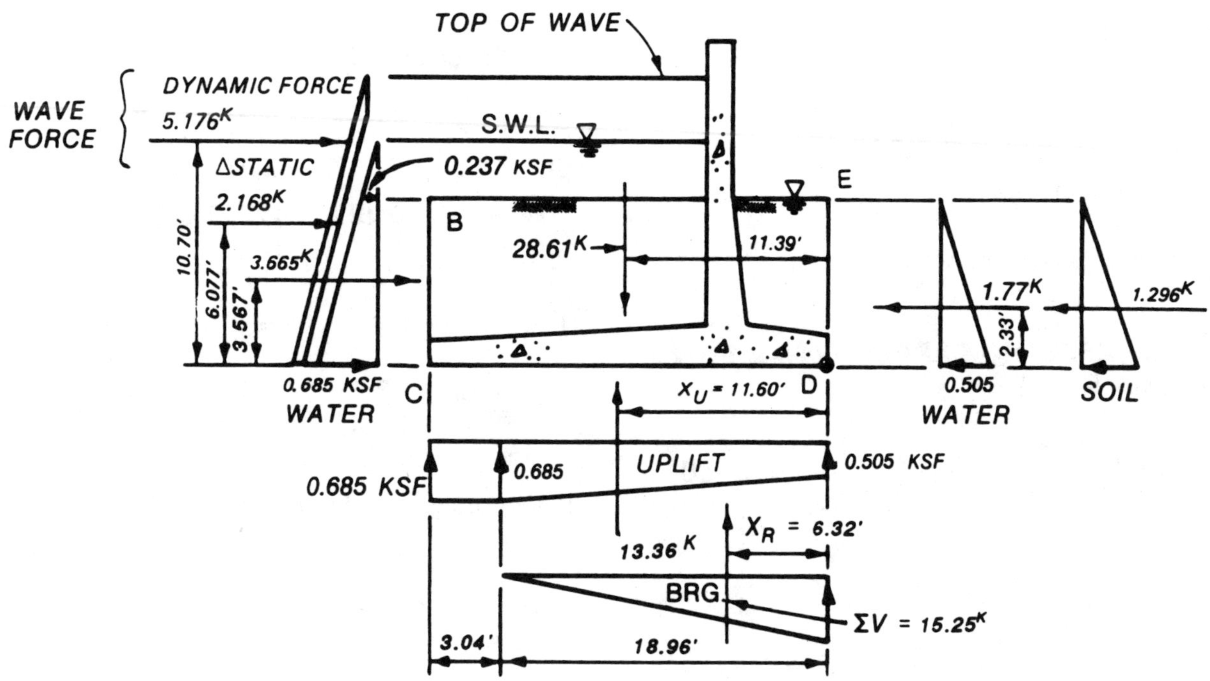

$$
\begin{aligned}
& 28.61 \times 11.39 = 325.87 \\
& \underline{-13.36} \times 11.60 = -154.98 \\
\sum V = \; & 15.25 \text{ kips}
\end{aligned}
$$

$$
\begin{aligned}
& -5.176 \times 10.700 = -55.38 \\
& -2.168 \times 6.077 = -13.18 \\
& -3.665 \times 3.567 = -13.07 \\
& 1.77 \times 2.333 = 4.12 \\
& \underline{1.296} \times 2.333 = \underline{3.02} \\
\sum H = \; & -7.943 \text{ kips} \qquad 96.40 \div 15.25 = 6.32 \text{ ft} = x_R
\end{aligned}
$$

$$a = 3x_R = 3(6.32) = 18.96 \text{ ft}$$

$$\text{Percent of base in compression} = \left(\frac{a}{b}\right)100 = \left(\frac{18.96}{22}\right)100$$

$$= 86.18\% > 60\% \text{ (Table 4-3)}$$

overturning criterion are satisfied.

D. SLIDING STABILITY (paragraph 4-15).

From Equation 4-12

$$T \leq \frac{N' \tan \phi + cL}{FS}$$

$$N' = \sum V = 15.25 \text{ kips}, \; T = \sum H = 7.94 \text{ kips}$$

Using the minimum required FS of 1.25 given in Table 4-3 yields

$$7.40 \leq \frac{15.25 \tan 0^\circ + 0.9(18.96)}{1.25}$$

$$7.40 \leq 13.65$$

The sliding criterion are satisfied.

E. BEARING CAPACITY (Chapter 5).

From Figure 5-1

$$\delta = \tan^{-1}\left(\frac{\sum H}{\sum V}\right) = \tan^{-1}\left(\frac{7.94}{15.25}\right) = 27.50^\circ$$

$$e = \frac{B}{2} - x_R = \frac{22}{2} - 6.32 = 4.68 \text{ ft}$$

$$\bar{B} = B - 2e = 22 - 2(4.68) = 12.64 \text{ ft}$$

From Table 5-1

$$N_c = 5.14, N_q = 1.00$$

$$D = 7 \text{ ft}, \; \gamma_b = \gamma' = 0.125 - 0.064 = 0.061 \text{ kcf}$$

From Equation 5-8a

$$q_o = \gamma' D = 0.061(7) = 0.427 \text{ ksf}$$

From Equation 5-4a

$$\xi_{cd} = 1 + 0.2\left(\frac{D}{\bar{B}}\right) = 1 + 0.2\left(\frac{7}{12.64}\right) = 1.111$$

From Equation 5-4b

$$\xi_{qd} = 1.00$$

From Equation 5-5a

$$\xi_{ci} = \xi_{qi} = \left(1 - \frac{\delta^\circ}{90^\circ}\right)^2 = \left(1 - \frac{27.50}{90}\right)^2 = 0.482$$

From Equation 5-5

$$
\begin{aligned}
Q &= \bar{B}(\xi_{cd}\xi_{ci} c N_c + \xi_{qd}\xi_{qi} q_o N_q) \\
Q &= 12.64[1.111(0.482)(0.9)(5.14) \\
&\quad + 1.00(0.482)(0.427)(1.00)] \\
Q &= 33.91 \text{ kips}
\end{aligned}
$$

Using Equation 5-1 yields

$$FS = \frac{Q}{N'} = \frac{33.91}{15.25} = 2.22 > 1.50 \text{ (Table 4-3)}$$

The bearing criterion are satisfied.

APPENDIX O

COMPUTER PROGRAM ABSTRACTS FOR REFERENCED PROGRAMS

O-1. Computer Program Abstracts

The following computer program abstracts are provided: Bearing Capacity Analysis of Shallow Foundations (CBEAR); CFRAG—Seepage Analysis of Confined Flow Problems by the Method of Fragments; CSLIDE—Sliding Stability Analysis of Concrete Structures; Seepage Package; UTEXAS2—University of Texas Analysis of Slopes—Version 2; An Interactive Graphics Three-Dimensional Geometry Program; CSTR—Concrete Strength Investigation and Design of Hydraulic Structures; User's Guide for Concrete Strength Investigation and Design (CASTR) in Accordance with ACI 318-83.

CATEGORY A

CBEAR	ELECTRONIC COMPUTER PROGRAM ABSTRACT		
TITLE OF PROGRAM	Bearing Capacity Analysis of Shallow Foundations (CBEAR) (I0017)	PROGRAM NO. 741-F3-R0107	
PREPARING AGENCY U.S. Army Engineer Waterways Experiment Station, Automatic Data Processing Center, P.O. Box 631, Vicksburg, MS 39180			
AUTHOR(S)	DATE PROGRAM COMPLETED	STATUS OF PROGRAM	
		PHASE	STAGE
Reed L. Mosher and Michael E. Pace	June 1982	FINAL	OPER

A. PURPOSE OF PROGRAM

This program can be used for the analysis of the bearing capacity of shallow strip, rectangular, square, or circular foundations on one- or two-layer soil systems. The bearing capacity can be computed considering the effects of embedment of the foundation, inclination of the foundation base, inclined loads, a sloping soil surface, eccentric loads in three dimensions, submerged soil, or surcharge.

B. PROGRAM SPECIFICATIONS

Time-sharing FORTRAN Program.

C. METHODS

The bearing capacity of an infinite strip footing is derived based on the classical theory of plasticity, using limit equilibrium analysis. The soil behavior is assumed to be as follows: (a) Mohr-Coulomb failure criteria govern; (b) shear strength at any point is independent of strain; (c) elastic deformations are negligible with respect to plastic deformation; and (d) volume change due to stress is negligible.

D. EQUIPMENT DETAIL

E. INPUT–OUTPUT

Data are input from a prepared data file in free-field format or from the user's terminal during execution. If the data are input from a terminal the user may enter data by using key command words or by following a prompting sequence. Output from the program may be directed to a file or printed at the user's terminal.

F. ADDITIONAL REMARKS

Program is available through the *CORPS* on WES DPS/1, CSC H6000 at Macon, GA.

A copy of the program and documentation may be obtained from the Engineering Computer Programs Library (ECPL), WES; telephone number: commercial (601) 634-2581 or FTS 542-2581.

WES FORM 1 JUL 80 2205 REPLACES ENG FORM 2883 WHICH IS OBSOLETE.

CATEGORY A

CFRAG	ELECTRONIC COMPUTER PROGRAM ABSTRACT		
TITLE OF PROGRAM	CFRAG—Seepage Analysis of Confined Flow Problems by the Method of Fragments (I0018)	PROGRAM NO. 741-F3-R0108	
PREPARING AGENCY U.S. Army Engineer Waterways Experiment Station, Automatic Data Processing Center, P.O. Box 631, Vicksburg, MS 39180			
AUTHOR(S)	DATE PROGRAM COMPLETED	STATUS OF PROGRAM	
		PHASE	STAGE
Michael E. Pace, Dennis R. Williams, Thomas F. Wolff, and Reed Mosher	October 1983		OP

A. PURPOSE OF PROGRAM

CFRAG is a seepage program designed to analyze groundwater flow using the method of fragments. The program can be used to compute: (1) seepage through soil mediums that can be modeled using fragments; (2) head losses; (3) exit gradients; and (4) resultant uplift and lateral forces.

B. PROGRAM SPECIFICATIONS

The CFRAG program is written in FORTRAN 66. The Corps time-sharing library file name is I0018.

C. METHODS

The method of fragments is an approximate analytical procedure for computing groundwater flow. The principal assumptions used in the derivation of this method are: (1) the flow is confined and of finite depth; (2) Darcy's law is valid, therefore, laminar flow exists; (3) steady-state flow exists; (4) the soil medium is homogeneous and isotropic; and (5) equipotential lines at certain locations of the flow region can be approximated by vertical lines.

D. EQUIPMENT DETAILS

Time-sharing computer (Honeywell level 66, CDC Cyber, or Harris 500).

E. INPUT–OUTPUT

Input—Data may be supplied from a prepared data file or from the user's terminal during execution. If the data are input from the terminal, the user may enter data by using key command words or by following a prompting sequence.

Output—Provides an echoprint of the input data and results of the analysis.

F. ADDITIONAL REMARKS

A copy of the program and documentation may be obtained from the Engineering Computer Programs Library, WES; telephone: (601) 634-2581 or FTS 542-2581.

CSLIDE	**ELECTRONIC COMPUTER PROGRAM ABSTRACT**		
TITLE OF PROGRAM	CSLIDE—Sliding Stability Analysis of Concrete Structures (X0075)	PROGRAM NO. 713-F3-R0075	
PREPARING AGENCY U.S. Army Engineer Waterways Experiment Station, Information Technology Laboratory, P.O. Box 631, Vicksburg, MS 39180			
AUTHOR(S)	DATE PROGRAM COMPLETED	STATUS OF PROGRAM	
Michael E. Pace and Virginia R. Noddin	July 1986	PHASE	STAGE OP

A. PURPOSE OF PROGRAM

This program was developed to assess the sliding stability of concrete structures using the limit equilibrium method described in the Engineering Technical Letter (ETL) 1110-2-256.

B. PROGRAM SPECIFICATIONS

CSLIDE is written in FORTRAN 77.

C. METHODS

The program utilizes the limit equilibrium wedge method described in ETL 1110-2-256. The program performs an iterative search to find the failure surface with the minimum factor of safety.

(Continued)

D. EQUIPMENT DETAILS

The program is operational on the Honeywell DPS-8 CDC Cyber and Harris 500 computers in the time-sharing mode. Any ASCII time-sharing terminal may be used, but if graphics are desired a Tektronix 4014 terminal must be used.

E. INPUT–OUTPUT

Input—Data may be supplied from a prepared data file or from the user's terminal during execution. If the data are input from the terminal, the user may enter data by using key command words or by following a prompting sequence. All data are read in free-field format.

Output—An echoprint of the input data and the results of the analysis in tabular and graphical form may be obtained.

F. ADDITIONAL REMARKS

A copy of the program and documentation may be obtained from the Engineering Computer Programs Library (ECPL), WES; telephone: (601) 634-2581 or FTS 542-2581.

This program is designated X0075 in the *CORPS* Library.

C. METHODS (Continued)

CSLIDE can compute the factor of safety against sliding considering the effects of

a. Multiple soil layers with irregular surfaces.

b. Water and seepage effects.

c. Applied vertical surcharge loads, which include line, strip, triangular, uniform, and ramp loads.

d. Applied horizontal point loads.

e. Irregular shaped structural geometry with either a horizontal or sloped base.

f. A percentage of the base of the structure in compression due to overturning effects.

g. Single or multiple failure planes.

h. Horizontal and vertical induced loads due to earthquake accelerations.

i. Factors which require the user to predetermine the failure surface.

CATEGORY B

FEMSEEP	**ELECTRONIC COMPUTER PROGRAM ABSTRACT**		
TITLE OF PROGRAM Seepage Package (X8202)		PROGRAM NO. 704-F3-R0009	
PREPARING AGENCY U.S. Army Engineer Waterways Experiment Station, Information Technology Laboratory, P.O. Box 631, Vicksburg, MS 39180			
AUTHOR(S)	DATE PROGRAM COMPLETED	STATUS OF PROGRAM	
		PHASE	STAGE
Fred T. Tracy	January 1983		OP

A. PURPOSE OF PROGRAM

(1) To interactively generate a finite element (FE) grid, (2) to perform a FEM seepage analysis, and (3) to plot the results.

B. PROGRAM SPECIFICATIONS

The driver runs in time-sharing and calls X8200 (704-F3-R0006) for grid generation, 704-F3-R0011 for seepage analysis, and X8201 (704-F3-R0005) for postprocessing.

C. METHODS

Four-sided isoparametric elements are used primarily in grid generation, the finite element method is used for the seepage analysis, and flow nets, contouring, vector, and number plots are used in the postprocessing.

D. EQUIPMENT DETAILS

Output is displayed on a storage tube terminal such as the Tektronix 4012 or 4014. The program is operational on Honeywell, CDC, and Harris computers.

E. INPUT–OUTPUT

Input–output is performed using a graphics terminal in the time-sharing environment.

F. ADDITIONAL REMARKS

The seepage analysis consists of steady-state solutions to plane or axisymmetric, confined or unconfined, and homogeneous or inhomogenous problems.

Documentation may be obtained from the Engineering Computer Programs Library (ECPL), WES; telephone number: commercial (601) 634-2581 or FTS 542-2581.

ELECTRONIC COMPUTER PROGRAM ABSTRACT

TITLE OF PROGRAM	UTEXAS2-University of Texas Analysis of Slopes-Version 2 (I0029)	PROGRAM NO.

PREPARING AGENCY U.S. Army Engineer Waterways Experiment Station, Information Technology Lab and Geotechnical La, P.O. Box 631, Vicksburg, MS 39180-0631

AUTHOR(S)	DATE PROGRAM COMPLETED	STATUS OF PROGRAM	
		PHASE	STAGE
(See reverse)			OP

A. PURPOSE OF PROGRAM

UTEXAS2 is a slope stability program designed to analyze slopes by any of four methods. The program will calculate the safety factor either for a prescribed shear surface or for a search of the critical shear surface. Both circular and noncircular shear surfaces can be evaluated.

B. PROGRAM SPECIFICATIONS

The UTEXAS2 program is written in FORTRAN 77. The CORPS time-sharing library file name is I0029.

C. METHODS

The four analysis procedures are Spencer's method, Simplified Bishop's procedure, Modified Swedish procedure with the Corps' side force assumption, and Modified Swedish procedure with Lowe and Karafiath's side force assumption. There are five options for type of shear strength data utilized and six options for specifying pore pressures. All analysis procedures and major features can be run in a single data file.

D. EQUIPMENT DETAILS

Microcomputer with at least 512K memory, a hard disk, and a math coprocessor. Time-sharing computer (CDC Cyber, or Harris 500) with Tektronix 4014 terminal or emulator for graphics.

E. INPUT–OUTPUT

Input—Data supplied from a prepared data file that allows for free-field input and requires command words.

Output—Provides an echoprint of the input data and results of the analysis in an output file divided into a series of tables.

F. ADDITIONAL REMARKS

Graphics capability for displaying the input data and the final shear surface is available. A copy of the program and documentation may be obtained from the Engineering Computer Programs Library, WES; telephone: (601) 634-2581 or FTS 542-2581.

WES FORM 2205 1 JUL 80 REPLACES ENG FORM 2883 WHICH IS OBSOLETE.

AUTHOR(S) (Continued).
Dr. Stephen G. Wright—Univ. of Texas, Austin; POC-Earl Edris, Geotech Lab and Reed Mosher, Information Tech Lab

CATEGORY B

3DSAD	**ELECTRONIC COMPUTER PROGRAM ABSTRACT**		
TITLE OF PROGRAM	An Interactive Graphics Three-Dimensional Geometry Program (X8100)	PROGRAM NO. 713-F3-R0008	
PREPARING AGENCY U.S. Army Engineer Waterways Experiment Station, Automatic Data processing Center, P.O. Box 631, Vicksburg, MS 39180			
AUTHOR(S)	DATE PROGRAM COMPLETED	STATUS OF PROGRAM	
		PHASE	STAGE
Fred T. Tracy	October 1978	OP	

A. PURPOSE OF PROGRAM

Program allows user to describe the geometry of a three-dimensional structure, interactively plot the described structure, and compute weight and centroid information for individual pieces or the sum total for the structure. Other modules of the program apply generalized loading forces and pressures and perform overturning and sliding analyses.

B. PROGRAM SPECIFICATIONS

FORTRAN, Time-sharing program.

C. METHODS

Use a right-handed coordinate system. Solid pieces of the structures may be described as (1) block: a two-dimensional cross-section extended in the third direction; (2) an eight node brick element; (3) a cluster of surfaces to form a solid.

D. EQUIPMENT DETAILS

Low-speed graphics terminal, central processor.

E. INPUT—OUTPUT

Input must be by a basic data file with addition and display commands entered. Output will be displayed directly on the terminal.

F. ADDITIONAL REMARKS

Program is available through the *CORPS* on WES G-635, CSC H600 at Macon, GA, and Boeing Computer Services.

Reports include:
1. General Geometry Module
2. General Loads Module
3. General Analyses Module

WES FORM 2205 1 JUL 80 REPLACES ENG FORM 2883 WHICH IS OBSOLETE.

CATEGORY B

ELECTRONIC COMPUTER PROGRAM ABSTRACT

TITLE OF PROGRAM	CSTR—Concrete Strength Investigation and Design of Hydraulic Structures (X0066)	PROGRAM NO. 713-F3-R0066

PREPARING AGENCY U.S. Army Engineer Waterways Experiment Station, Automation Technology Center, P.O. Box 631, Vicksburg, MS 39180-0631

AUTHOR(S)	DATE PROGRAM COMPLETED	STATUS OF PROGRAM	
		PHASE	STAGE
C. C. Hamby (LMKED-DS) and W. A. Price III (WESKA-E)	September 1984 September 1984	Operational	

A. PURPOSE OF PROGRAM

To perform investigation or design of concrete beams or columns in accordance with ETL 1110-2-265 for hydraulic structures.

B. PROGRAM SPECIFICATIONS

Written in FORTRAN IV using the Graphics Compatibility System (GCS). The CORPS time-sharing library file name is X0066.

C. METHODS

Strength analysis for investigation or design of rectangular cross sections of hydraulic structures subjected to axial load plus uniaxial flexure. Analysis is based on the rectangular stress block described in ETL 1110-2-265.

D. EQUIPMENT DETAILS

Tektronix 4014 terminal, if graphics output is desired, otherwise, any ASCII time-sharing terminal.

E. INPUT-OUTPUT

Input is from a data file; output is to a Tektronix 4014 graphics terminal or regular printing terminal.

F. ADDITIONAL REMARKS

A copy of the program and documentation may be obtained from the Engineering Computer Programs Library, WES; telephone: (601) 634-2581 or FTS 542-2581.

The User's Guide for this program is WES IR K-84-9.

ELECTRONIC COMPUTER PROGRAM ABSTRACT

TITLE OF PROGRAM	User's Guide for Concrete Strength Investigation and Design (CASTR) in Accordance with ACI 318-83	PROGRAM NO. 713-F3-R0067

PREPARING AGENCY U.S. Army Engineer Waterways Experiment Station, Information Technology Laboratory, P.O. Box 631, Vicksburg, MS 39180-0631

AUTHOR(S)	DATE PROGRAM COMPLETED	STATUS OF PROGRAM	
		PHASE	STAGE
C. C. Hamby (LMKED-DS) and W. A Price III (WESKA-E)	October 1985 October 1985	Operational	

A. PURPOSE OF PROGRAM

To perform investigation or design of concrete beams or columns in accordance with ACI Code 318 for nonhydraulic structures.

B. PROGRAM SPECIFICATIONS

Written in FORTRAN IV using the Graphics Compatibility System (GCS). The CORPS time-sharing library file name is X0067.

C. METHODS

Strength analysis for investigation or design of rectangular cross sections of nonhydraulic structures subjected to axial load plus uniaxial flexure. Analysis is based on the rectangular stress block approximation described in Section 10.2.7 of ACI 318-83.

D. EQUIPMENT DETAILS

Tektronix 4014 terminal, if graphics output is desired, otherwise, any ASCIII time-sharing terminal.

E. INPUT–OUTPUT

Input is from a data file; output is to a Tektronix 4014 graphics terminal or regular printing terminal.

F. ADDITIONAL REMARKS

Input data are prepared the same as for program 713-F3-R0066, "CSTR-Concrete Strength Investigation and Design of Hydraulic Structures (X0066)." Differences between the two programs lie only in the stress block depth and other parameters. Call WES, (601) 634-2300 or FTS 542-2300 for more information.

WES FORM 2205 1 JUL 80 REPLACES ENG FORM 2883 WHICH IS OBSOLETE.

GLOSSARY

Active earth pressure Minimum horizontal pressure condition which develops when a wall rotates about its base and away from the backfill 0.001 to 0.003 radian (see paragraphs 3-4a and 3-5 and Figure 3-1).

At-rest earth pressure Lateral pressure condition when no wall movement occurs (see paragraphs 3-4c, 3-5, and 3-10 and Figure 3-1).

Bond breaker A coating or sheet of a substance that does not adhere to concrete, placed on one monolith in the area where concrete will be placed later, to maintain separate movement of the two monoliths.

Breaking wave See **wave parameter definitions.**

Broken wave See **wave parameter definitions.**

Buttress A vertical fin, turned perpendicular to the stem, placed monolithic with the stem and heel so as to brace the stem into a series of wall slabs supported at the base and the buttresses.

c Symbol for cohesion value (see paragraph 3-5 and Figure 3-8).

Closed-face structure (Chapter 10) A prefabricated facing system for an alternative type of retaining wall, where the facing has no openings. The joints between precast panels will form a pattern but vegetation cannot grow through them. See open-face structure.

Coastal flood wall A flood wall that is resisting the effects of the surge tide and waves accompanying a storm. See paragraphs 4-1, 4-5, and 7-1.

Counterfort Like a buttress, except located over the toe instead of over the heel. The fin is in tension instead of the compression of the buttress.

Diffraction (of waves) The effect of bending of the direction of propagation (travel) of a wave crest as it passes by the end of a jetty or through an opening. Thus the wave pattern spreads outward into the sheltered region within the barrier's geometric shadow.

Drained phi value Phi (ϕ) value of a soil, as determined by an S test.

Drained soil tests Tests in which the confining and shear stresses are applied so slowly that porewater pressure does not build up. An increase in applied stress produces an equal increase in effective stress.

Driving force A force acting to move the wall, usually caused by earth and water beyond the end of the heel. See paragraph 3-7.

Driving wedge A wedge of soil which produces a horizontal force (a driving force) tending to cause instability of a wall. See paragraph 3-7.

Fetch The area of water in which waves are generated by a wind having a fairly constant direction and speed. Sometimes used synonymously with *fetch length*, the horizontal distance over which a wind generates waves.

Frequency of a wave See **wave parameter definitions, wave frequency.**

Heel The heel of a wall is the base slab projection pointing toward the net driving force.

Kern The portion of the area under the base such that when the resultant normal force (N or V in the figures in this manual) is inside the kern, the entire base is in compression contact with the subgrade. When the resultant normal force is located outside the kern, some portion of the base will be tending to lift off the subgrade instead of being in compression contact. For a rectangular base, the edges of the kern will lie at the one-third points of the base width, as shown in Figure 4-4.

Nonbreaking wave See **wave parameter definitions.**

Open-faced structure (Chapter 10) A prefabricated facing system for an alternative type of retaining wall, where the precast facing units have openings between or through them. Vegetation, usually evergreen, is then planted so as to grow through the openings and present a more natural appearance of the wall's exposed face.

Passive earth pressure Pressure condition that develops when a wall is moved toward backfill, causing horizontal stresses to increase and shear stresses to reverse direction (see paragraphs 3-4b and 3-5).

Phi Angle of internal friction of earth (see Figure 3-1).

Protected side (of a flood wall or seawall) The side facing the area protected from flooding.

Q test (of soil) Also known as an unconsolidated undrained or UUtest. Both the confining and shearing stresses are applied so rapidly that the neutral stress supports all the added load, and there is no change in water content.

R test (of soil) Also known as a consolidated undrained or CU test. The confining stresses are applied to the soil so slowly that the neutral stress is not changed and the soil consolidates fully. The shear stress, however, is applied so quickly that the neutral stress carries all this change, and there is no further consolidation or change in water content.

Refraction (of a wave) The process by which the direction of movement is changed for a wave moving in shallow water at an angle to the underwater contours: the part of the wave advancing in shallower water moves more slowly than that part still advancing in deeper water, causing the wave crest to bend toward alignment with the underwater contours.

Resisting force A force tending to resist movement of a wall. See paragraph 3-8.

Resisting wedge A wedge of soil which produces a lateral reaction (a resisting force) which tends to resist instability of a wall. See paragraph 3-8.

S test (of soil) Also known as a consolidated drained or CD test. The confining stresses are applied so slowly that the neutral stress does not change. The soil consolidates with no change in neutral stress.

Seawall See paragraph 1-5a, especially the comparison with a coastal flood wall. A seawall is usually a gravity structure for the purpose of protecting the area behind it from the action of tide and waves in front of it, sometimes with a face shaped to dissipate wave energy.

Shoaling Point at which the water depth becomes more shallow as the wave crest approaches the shoreline or a wall.

Stem The vertical wall portion of a cantilever inverted T-wall.

Structural wedge The structural wedge consists of the structure itself with any soil and water contained within the boundaries of the structure or over the base.

Surge stillwater level A rise above normal water level on the open coast due to the action of wind stress on the water surface. Surges resulting from hurricanes also include the rise in level due to atmospheric pressure reduction.

Toe The toe of a wall is the portion of the base slab pointing toward the net resisting force.

Train, wave See **wave parameter definitions.**

Undrained soil tests Tests where the pore-water pressure is not allowed to dissipate and the water carries all of the applied stresses.

Unprotected side of a flood wall or seawall The side of the wall facing the storm tide and waves or the rising water.

Wall friction The angle of friction between a soil mass tending to move parallel along the interface between the soil mass and a wall. See paragraph 3-14.

Water stop A strip of material, cast into each of two adjacent concrete monoliths and spanning the space between them, for the purpose of preventing the flow of water through the space. Usually made of elastomeric material for civil works structures, but also made of sheet metal. See paragraph 7-13 and Figure 7-9.

Wave parameter definitions These definitions (see paragraph 3-24) are best obtained from Chapter 7 of the *Shore Protection Manual* (US Army Engineer Waterways Experiment Station 1984).* Abbreviated definitions follow.

Breaking wave While there are several types of breaking waves, the type associated with the high-impulse effects used in this manual is the plunging breaker, where the wave crest becomes more and more steep, until the crest curls forward and falls over a pocket of trapped air. This is characterized by a very large dynamic impulse of short (0.2 second, approximately) duration and a rise in hydrostatic pressure from still water (no wave) up to the crest one-half of the wave height above still water. See paragraph 3-24d.

Broken wave A somewhat confused mass of water surging forward after the wave has broken some distance away from the wall. See paragraph 3-24e.

Nonbreaking wave A wave impacting on a wall under conditions where there is no breaking tendency but the mass of water is stopped from forward motion by the wall and imparts its energy to the vertical face. See paragraph 3-24c.

Wave frequency The number of wave crests passing a given point in one second, the reciprocal of the wave period.

Wave height The vertical distance from the crest of a wave to the adjacent trough (lowest elevation). A continuously changing value in a typical train of waves, frequently assumed to have a Gaussian distribution. See paragraph 3-24b for the assumed values as fractions of the significant wave height (an average of the highest one-third of all of the waves in the wave train is assumed for design of the structure).

Wave length The horizontal distance from one wave crest to the next crest in the group (train) of waves. A continuously changing value in a typical train of waves, frequently assumed to have a Gaussian distribution.

* The reference cited in this appendix may be found in Appendix A, "References."

Wave period The time in seconds required for a wave crest to travel a distance of one wave length, the reciprocal of the frequency.

Wave steepness The ratio of wave height to wave length.

Wave train A group of waves coming from the same direction over the same fetch.

DEPARTMENT OF THE ARMY ETL 1110-2-322

CECW-ED
CECW-EG

U.S. Army Corps of Engineers

Washington, D.C. 20314-1000

Engineer Technical
Letter No. 1110-2-322

15 October 1990

Engineering and Design

RETAINING AND FLOOD WALLS

1. Purpose

This engineer technical letter (ETL) provides supplemental material to engineer manual EM 1110-2-2502, "Retaining and Flood Walls."

2. Applicability

This ETL applies to HQUSACE/OCE elements, major subordinate commands, districts, laboratories, and field operating activities (FOA) having responsibilities for the design of civil works projects.

3. References

A. EM 1110-2-2103, "Details of Reinforcement—Hydraulic Structures."

B. EM 1110-2-2502, "Retaining and Floodwalls."

4. Summary

A. SIMPLIFIED PRESSURE COEFFICIENT METHOD. A simplified method of computing an earth pressure distribution using pressure coefficients which were derived in Appendix H of EM 1110-2-2502 is presented. With this method forces may be computed which are equal to the forces from a wedge analysis.

B. INCLUSION OF WALL FRICTION. The general wedge equation as defined in EM 1110-2-2502 (Equation 3-23) is expanded to include the effects of wall friction. The corresponding equations for computing the critical slip plane angle α and pressure coefficients, including the effects of wall friction, are also discussed.

C. LIMITATION OF 2/3 SMF. EM 1110-2-2502 requires the use of a strength mobilization factor (SMF) of 2/3 to compute forces and pressures, using Coulomb's equation (EM 1110-2-2502, Equation 3-11) or the general wedge equation. The value of the SMF equal to 2/3 was developed by comparing the earth pressure coefficient computed using Coulomb's equation for a horizontal backfill with the earth pressure coefficient calculated using Jaky's empirical relationship for predicting at-rest earth pressure coefficients. Using an SMF of 2/3 in Coulomb's equation results in an earth pressure coefficient approximately equal to the value obtained from Jaky's equation. When the slope of the backfill (β) is large compared to the angle of internal friction (ϕ) of the soil (about 60 percent of ϕ), the earth pressures and forces computed using an SMF of 2/3 will exceed the at-rest pressures as computed by the Danish Code equation (EM 1110-2-2502, Equation 3-5). Guidelines are presented to allow the computation of earth pressures for backfills having $\tan \beta / \tan \phi > 0.56$ which agree more closely with the Danish Code equation.

D. DETAILS OF REINFORCEMENT. Criteria for steel detailing pertaining to the main flexural reinforcement and the temperature and shrinkage reinforcement for concrete cantilever walls are presented.

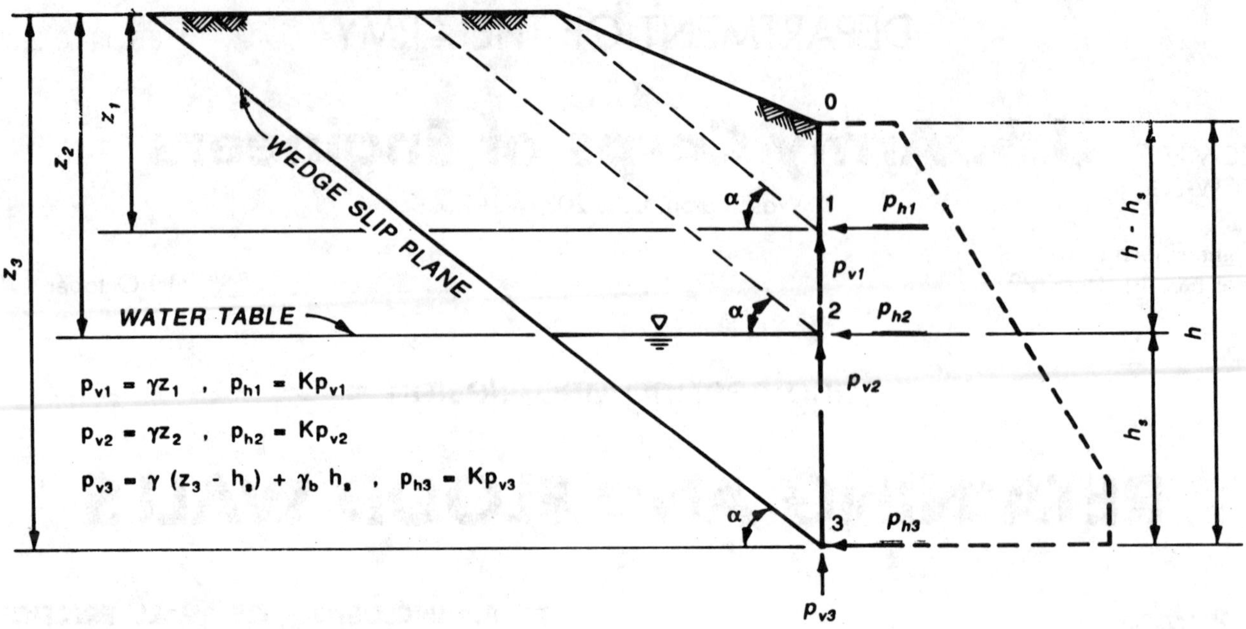

Figure 1. Simplified pressure coefficient method

5. Simplified Pressure Coefficient Method for Computing Earth Pressures and Forces

A. BACKGROUND. In EM 1110-2-2502, pressure coefficients were derived for the calculation of earth pressures. Several coefficients were required to calculate the lateral earth pressures and are defined below:

K = the basic coefficient

K_1 = the coefficient for a sloping ground surface above the water table

K_b = the coefficient to use below the water table

K_v = the coefficient to multiply by V to obtain the horizontal force due to a vertical surcharge

K_c = the coefficient to multiply by 2c to obtain the negative pressure due to cohesion

In this ETL, one additional pressure coefficient will be derived called K_H. This coefficient is defined as:

K_H = the coefficient to multiply by $(H_L - H_R)$ to obtain the horizontal force due to external horizontal forces acting on a wedge

A simpler method using only the coefficients K and K_c can be used to compute earth pressures due to the soil geometry. If external forces are present on the wedge, such as a strip surcharge or a horizontal line load, then the pressure coefficients K_v and K_H must be used to determine the lateral force due to these external loads.

B. APPLICABILITY. This method can be applied to any soil geometry possessing an angle of internal friction and a cohesion value. The method may be applied to irregular backfills to compute lateral earth pressures in an easy manner. All results from this method compare exactly with values calculated using Coulomb's or the general wedge equations.

C. PROCEDURE. The procedure to calculate the lateral earth pressure distribution due to the geometry of the backfill is shown in Figure 1. The equation used to compute a horizontal stress at a certain depth is given as

$$p_{hz} = Kp_{vz} - 2K_c c_d \tag{1}$$

where

p_{hz} = the effective horizontal stress at depth z

p_{vz} = the vertical effective stress at depth z

c_d = the developed cohesion value of the soil, equal to $c \cdot SMF$

z = depth below surface of backfill as shown in Figure 1

This procedure uses a combination of pressure coefficients and backfill geometry to calculate effective horizontal pressures at various depths along the face of the backfill. The vertical pressures shown in Figure 1 are based on the geometry of the backfill. Enclosure 1 demonstrates the relationship between the geometry of the backfill, the vertical effective stresses, and the pressure coefficients.

D. EXAMPLES. Examples demonstrating the above method are provided in Enclosure 2.

6. Inclusion of Wall Friction in General Wedge Equation

A. BACKGROUND. Friction along the wall-to-soil interface and/or along a soil-to-soil interface is a result of the differential movement of the backfill relative to the movement of the wall and/or differential movement within the backfill. The magnitude and distribution of this shear force are dependent upon several factors, including: the soil type used in the backfill, the relative movement of the backfill to the movement of the wall, the settlement of the soil relative to the wall resulting from compaction, the distance between the plane on which shear forces are to be computed and the back of the wall, the slope of the ground surface of the backfill, and the presence of any permanent surcharge loadings. EM 1110-2-2502 states in paragraph 3-14 that the friction between the backfill and wall, or on a plane within the backfill, may have a value *between* zero and one-half of the internal friction angle (unfactored) of the backfill material for use in the design. Due to the influence of the previously mentioned factors, which affect the development of this friction force, guidelines for the selection of this friction force used for various categories of walls and backfills will be addressed in subsequent engineering guidance.

B. GENERAL WEDGE EQUATION. The derivation for the general wedge equation including wall friction for a backfill with a sloping surface possessing both a ϕ and a c is given in Enclosure 3. The resulting equation for the horizontal effective earth force is defined as:

$$P_{EE} = \frac{(W + V)(\tan\alpha - \tan\phi_d) + (H_L - H_R - P_w)(1 + \tan\alpha \tan\phi_d) + \dfrac{U \tan\phi_d}{\cos\alpha} - \dfrac{c_d L}{\cos\alpha}}{1 + \tan\alpha \tan\phi_d + \tan\delta\,(\tan\alpha - \tan\phi_d)} \tag{2}$$

where

P_{EE} = horizontal effective earth force contributed by wedge or wedge segment

W = total wedge weight, including water

V = any vertical force applied to wedge

α = angle between slip plane and horizontal

ϕ_d = developed angle of internal friction of the soil, equal to $\tan^{-1}$ (SMF $\tan\phi$)

H_L = any external horizontal force applied to the wedge from the left, acting to the right

H_R = any external horizontal force applied to the wedge from the right, acting to the left

P_w = internal water force acting on the side of the wedge free body (P_w is equal to the net difference of the water force for wedge segments with water on two vertical sides)

U = uplift or buoyant force acting on and normal to the wedge slip plane

c_d = developed cohesion of the soil, equal to $c \cdot$ SMF

L = length along the slip plane of the wedge

δ = angle of friction on vertical plane of wedge

The force computed using Equation 2 is the horizontal component of the total effective earth force. The total effective earth force P acts at an angle δ from the horizontal and is given by the following equation

$$P = \frac{P_{EE}}{\cos\delta} \tag{3}$$

C. EQUATIONS FOR DETERMINING CRITICAL SLIP PLANE ANGLE. Based on Equation 2, the equations for the calculation of the critical slip plane angle may be derived by the procedure shown in Appendix G of EM 1110-2-2502. The resulting equation for the critical slip plane angle is defined as:

$$\alpha_{critical} = \tan^{-1}\left(\frac{c_1 + \sqrt{c_1 + 4c_2}}{2}\right) \tag{4}$$

where

$$c_1 = \left\{2 \tan\phi_d (\tan\phi_d + \tan\delta) - \frac{4V \tan\beta\,(1 + \tan^2\phi_d)}{\gamma(h^2 - d_c^2)} - \frac{4(H_R - H_L)\tan\beta\,(1 - \tan^2\phi_d)\tan\delta}{\gamma(h^2 - d_c^2)} + \frac{4c_d[\tan\phi_d + \tan\beta + (1 - \tan\phi_d \tan\beta)\tan\delta]}{\gamma(h - d_c)}\right\} \div A \tag{5}$$

$$c_2 = \Bigg\{ \tan\phi_d\,[1 - \tan\phi_d\,(\tan\beta + \tan\delta)] - \tan\beta$$

$$+ \frac{2V\tan^2\beta\,(1 + \tan^2\phi_d)}{\gamma(h^2 - d_c^2)}$$

$$+ \frac{2(H_R - H_L)\,[\tan^2\beta\,(1 - \tan^2\phi_d)\tan\delta]}{\gamma(h^2 - d_c^2)}$$

$$+ \frac{2c_d[1 - \tan\phi_d\tan\beta - (\tan\phi_d + \tan\beta)\tan\delta]}{\gamma(h - d_c)} \Bigg\} \div A \quad (6)$$

Terms used in Equation 5 and 6 are defined as

γ = effective unit weight of the soil (use moist or saturated weight above the water table and buoyant weight below the water table)

d_c = the depth of cracking in a cohesive soil

$$A = \tan\phi_d + \tan\delta - \frac{2V(1 + \tan^2\phi_d)}{\gamma(h^2 - d_c^2)}$$

$$- \frac{2(H_R - H_L)(1 + \tan^2\phi_d)\tan\delta}{\gamma(h^2 - d_c^2)}$$

$$+ \frac{2c_d[1 - \tan\phi_d\tan\beta - (\tan\phi_d + \tan\beta)\tan\delta]}{\gamma(h + d_c)} \quad (7)$$

The derived equations are valid for a backfill with a planar (flat or inclined) top surface, a vertical strip surcharge V, and a horizontal surcharge equal to $(H_L - H_R)$. The equations for α assume that the backfill is completely above or completely below the water table, but can be used when the water table is anywhere within the backfill with sufficient accuracy for design. The surcharge loads can have any arbitrary shape but must be contained entirely within the driving wedge. By using positive and negative surcharges, Equations 4 through 7 may be used to calculate the critical slip plane angle for irregular-shaped backfills. The effect of water may be taken into account by using the average unit weight of soil for the γ term in Equations 5 through 7. The effect of seepage should be considered in determining the buoyant weight. An example showing the calculation of the buoyant weight is shown in Example 6 of Enclosure 2.

D. EQUATIONS FOR DETERMINING EARTH PRESSURE COEFFICIENTS. The following equations for the pressure coefficients were derived from the general wedge equation following the same procedure found in Appendix H of EM 1110-2-2502. The equation for the effective horizontal earth force written in terms of the derived pressure coefficients is given as:

$$P_{EE} = \frac{K_1\gamma(h^2 - d_c^2)}{2} + K_vV + K_H(H_L - H_R)$$

$$- 2K_cc_d(h - d_c) \quad (8)$$

where

$$K = \frac{1 - \tan\phi_d\cot\alpha}{1 + \tan\phi_d\tan\alpha + \tan\delta\,(\tan\alpha - \tan\phi_d)} \quad (9)$$

$$K_1 = K\frac{\tan\alpha}{\tan\alpha - \tan\beta} \quad (10)$$

$$K_v = K\tan\alpha \quad (11)$$

$$K_H = \frac{1 + \tan\phi_d\tan\alpha}{1 + \tan\phi_d\tan\alpha + \tan\delta\,(\tan\alpha - \tan\phi_d)} \quad (12)$$

$$K_c = \frac{1}{2\sin\alpha\cos\alpha\,[1 + \tan\phi_d\tan\alpha + \tan\delta\,(\tan\alpha - \tan\phi_d)]}$$

$$\cdot \frac{\tan\alpha}{\tan\alpha - \tan\beta} \quad (13)$$

The effects of the water table may be considered by using a pressure coefficient K_b below the water table. K_b is defined as

$$K_b = K\left[1 + \left(\frac{\tan\alpha}{\tan\alpha - \tan\beta} - 1\right)\frac{\gamma}{\gamma_b}\right] \quad (14)$$

The forces and pressures obtained by using the above coefficients are the horizontal components of the total forces and pressures. The total force and pressure acts at an angle δ with respect to the horizontal.

E. EXAMPLES. Examples demonstrating the use of the above equations are contained in Enclosure 4.

7. Limitation of Use of 2/3 SMF for Sloping Backfills

A. BACKGROUND. EM 1110-2-2502 states in paragraph 3-13b that an SMF of 2/3 should be used with the general wedge equation or Coulomb's equation to predict the at-rest earth forces acting on the driving side of a wall. If an SMF of 1.0 is used with the general wedge equation or Coulomb's equation, then the force computed would be for the active condition. As the SMF is reduced below 1.0, the magnitude of the assigned material strength properties (ϕ and c) decreases, which causes an increase in the lateral earth force computed. The use of an SMF of 2/3 produces a force from the general wedge equation or Coulomb's equation which approximates the force for the at-rest condition. As stated in paragraph 4c, the concept of an SMF of 2/3 was developed by comparing Coulomb's equation for level backfills with Jaky's empirical relationship for predicting at-rest earth pressure coefficients. The actual magnitude of the force computed depends on several factors, which include the soil geometry (e.g., surface slope, irregular), soil properties, and surcharge loading.

B. APPLICABILITY. When the slope of the backfill β is large compared with the developed angle of internal friction of the soil ϕ_d , the horizontal earth pressure coefficient computed from the general wedge equation or Coulomb's equation exceeds the value computed using the Danish Code equation. This condition occurs approximately where the ratio $\tan \beta/\tan \phi$ exceeds 0.56. When the ratio $\tan \beta/\tan \phi$ exceeds 0.56, the value of the SMF should increase from 2/3 to cause the coefficient computed from the Danish Code equation and Coulomb's equation to be approximately equal. Also, since the factor of safety is the reciprocal of the SMF, as the SMF increases from 2/3, the factor of safety for the soil wedge will decrease from 1.5. The above discussion pertains to positive values of β. For negative values of β, the SMF should remain at 2/3. For some combinations of β and ϕ, where β is negative, the horizontal earth pressure coefficient computed using the Danish Code equation is less than the coefficient computed using Coulomb's equation for the active condition (SMF = 1.0). The earth pressure coefficient should never be less than the coefficient obtained for the active condition. Therefore, for negative values of β, the more conservative earth pressure coefficient computed using an SMF of 2/3 should be used.

C. CRITERIA. Tables of values of SMF's to use for various combinations of ϕ's and β's are given in Enclosure 5. EM 1110-2-2502 in paragraph 4-15 describes a procedure for using the soil wedge forces computed by the SMF procedure to determine the factor of safety against sliding. In the table of SMF values, the corresponding factor of safety for the soil wedge will be less than 1.5, as discussed previously. The reciprocal of the SMF for the soil wedge is the limiting value for the factor of safety against sliding when using the single wedge equation procedure as described in EM 1110-2-2502 in paragraph 4-15. For cases where the required factor of safety for sliding is greater than this limiting value, the single wedge procedure cannot be used as a check for sliding stability and a multiple wedge analysis will be required.

D. EXAMPLES. An example demonstrating the variance of the horizontal earth pressure coefficient with ϕ and β is shown in Enclosure 6.

8. Details of Reinforcement

A. MAIN FLEXURAL REINFORCEMENT. The amount of reinforcement in each part of a reinforced concrete wall should be determined by the currently approved engineering procedure. The minimum clearance and minimum spacing of bars, and the minimum concrete cover requirements should be in accordance with EM 1110-2-2103. Maximum spacing of bars is given in EM 1110-2-2502. More than two rows or layers of reinforcing bars should not be used except where unusual conditions make this procedure necessary, in which case, a special supporting arrangement for holding the bars in their designed locations should be provided in the design and shown on the drawings. Bending and splicing of reinforcement should be in accordance with EM 1110-2-2103.

B. TEMPERATURE AND SHRINKAGE REINFORCEMENT. Temperature and shrinkage reinforcement should be in accordance with EM 1110-2-2103. The amount of longitudinal reinforcement for volume changes resulting from shrinkage and temperature changes is intended to control cracking between properly spaced contraction joints in reinforced concrete walls.

C. TYPICAL STEEL DETAILS. Typical steel details for a retaining wall and a floodwall are shown in Enclosure 7.

FOR THE COMMANDER:

7 Encl JOHN A. MCPHERSON
Acting Chief, Engineering Division
Directorate of Civil Works

WHY THE SIMPLIFIED PRESSURE COEFFICIENT METHOD WORKS

1. The following discussion will show how the simplified *K* method is able to predict horizontal earth pressures that equal the pressures computed when using the *K*'s discussed in paragraph 6d of this ETL. Essentially, the basic coefficient K and the coefficient K_1 for a sloping backfill are related by the geometry of the backfill.

2. A basic wedge with a sloping cohesionless backfill as shown below will be examined.

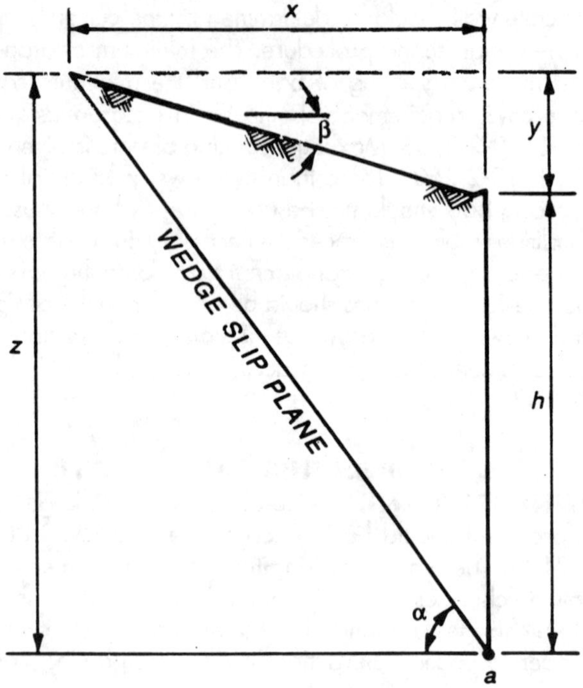

From the figure,

$$\frac{y}{x} = \tan \beta \tag{1}$$

$$\frac{z}{x} = \tan \alpha \tag{2}$$

Solving for *x* in both Equations 1 and 2 and setting the results equal to each other gives Equation 3:

$$\frac{z}{\tan \alpha} = \frac{y}{\tan \beta} \tag{3}$$

Noting that $z = y + h$ and by substituting in Equation 3 for *z* and simplifying the following equation results

$$\frac{y - y \tan \beta}{\tan \alpha} = \frac{h \tan \beta}{\tan \alpha} \tag{4}$$

Substituting $y = x \tan \beta$ into Equation 4 and solving for *x* yields

$$x = \frac{h}{\tan \alpha - \tan \beta} \tag{5}$$

From Equation 2, substituting in for *x*, *z* equals

$$z = h\left(\frac{\tan \alpha}{\tan \alpha - \tan \beta}\right) \tag{6}$$

If the pressure at the bottom of the wedge (point *a*) is calculated using both *K* and K_1, the following pressures result

$$p_{ah} = K_1 \gamma h = K\left(\frac{\tan \alpha}{\tan \alpha - \tan \beta}\right)\gamma h \tag{7}$$

$$p_{ah} = K \gamma z = K\left(\frac{\tan \alpha}{\tan \alpha - \tan \beta}\right)\gamma h \tag{8}$$

The pressures at point *a* calculated in Equations 7 and 8 are identical.

3. A basic wedge with a sloping cohesive backfill as shown below will be examined.

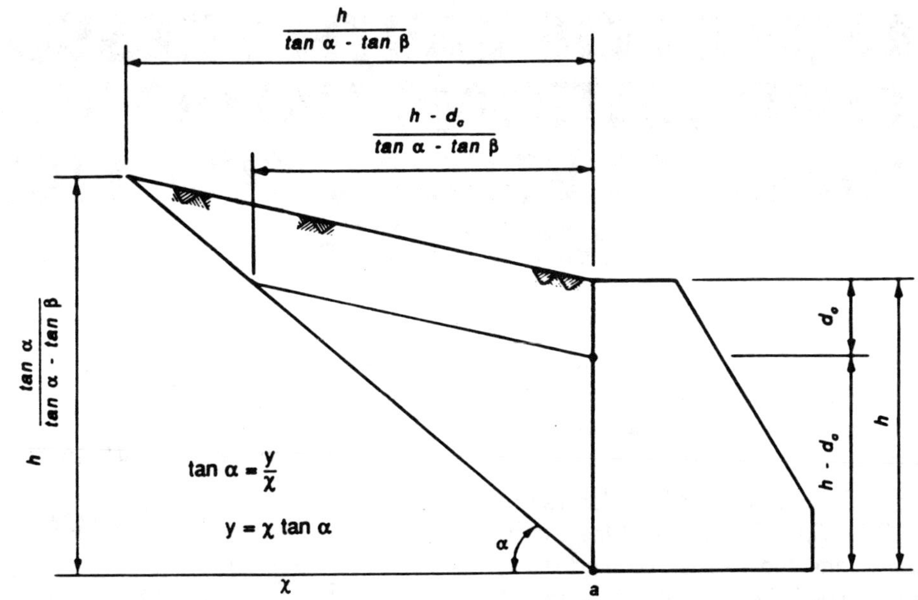

If the pressure at the bottom of the wedge (point a) is calculated using both K and K_1, the following pressures result

$$p_{ah} = K_1\gamma(h - d_c) = K\left(\frac{\tan\alpha}{\tan\alpha - \tan\beta}\right)\gamma(h - d_c) \quad (9)$$

where

$$d_c = \frac{2K_c c_d}{K_1\gamma} \quad (10)$$

Thus

$$p_{ah} = K\left(\frac{\tan\alpha}{\tan\alpha - \tan\beta}\right)\gamma h - 2K_c c_d \quad (11)$$

Using the basic coefficient K and the distance z defined previously

$$p_{ah} = K\gamma z - 2K_c c_d$$

$$= K\left(\frac{\tan\alpha}{\tan\alpha - \tan\beta}\right)\gamma h - 2K_c c_d \quad (12)$$

The pressures at point a calculated in Equations 11 and 12 are identical.

EXAMPLES OF EARTH PRESSURE COMPUTATIONS USING THE SIMPLIFIED PRESSURE COEFFICIENT METHOD

1. The following examples demonstrate the method of computing earth pressures using the simplified pressure coefficient method described in paragraph 5c of this ETL.

2. Examples 1 through 3 of this enclosure are taken from EM 1110-2-2502. The results of the examples are compared against the results obtained using the simplified pressure coefficient method. Example 4 demonstrates additional geometry configurations.

EXAMPLE 1

3. This example is taken from EM 1110-2-2502, Example 3, page M-7. The lateral earth pressures will be computed by the simplified pressure coefficient method and compared to the results from EM 1110-2-2502.

4. The soil geometry and properties are shown in the figure below.

The following values were determined for the problem from EM 1110-2-2502.

$$\alpha_{critical} = 45.466°$$

$$K = 0.44767$$

5. The pressures are calculated as follows for the points shown in the figure on the following page

$$p_{v1} = \gamma z_1 = 0.12(13.594) = 1.6313 \text{ ksf}$$

$$p_{v2} = \gamma(z_2 - h_s) + \gamma_b h_s$$

$$= 0.12(16.856) + 0.0625(10) = 2.6477 \text{ ksf}$$

$$p_{h1} = Kp_{v1} = 0.44767(1.6313) = 0.7303 \text{ ksf}$$

$$p_{h2} = Kp_{v2} = 0.44767(2.6477) = 1.1853 \text{ ksf}$$

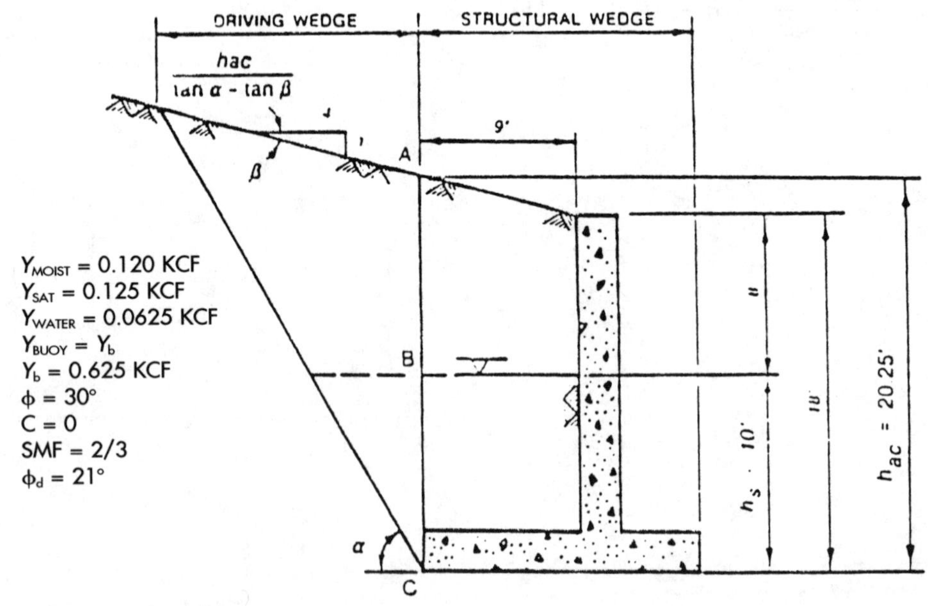

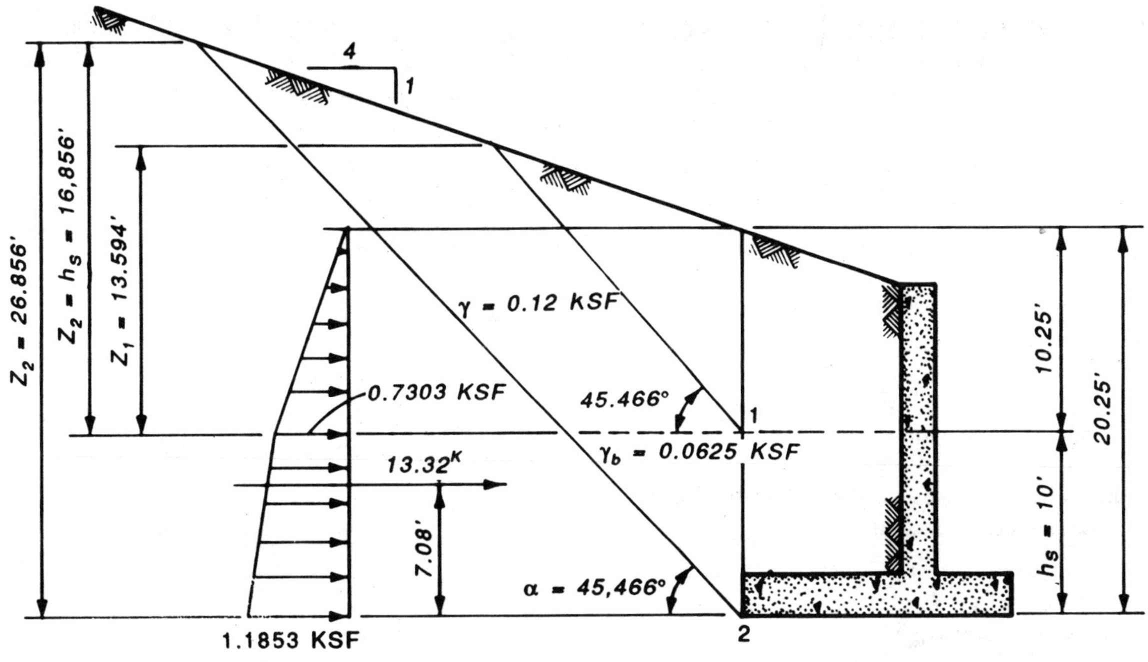

EARTH PRESSURE

The solution agrees with the solution found in EM 1110-2-2502.

EXAMPLE 2

6. This example is taken from EM 1110-2-2502, Example 5, page M-21. The lateral earth pressures will be computed by the simplified pressure coefficient method and compared to the results from EM 1110-2-2502.

7. The soil geometry and properties are shown in the figure below.

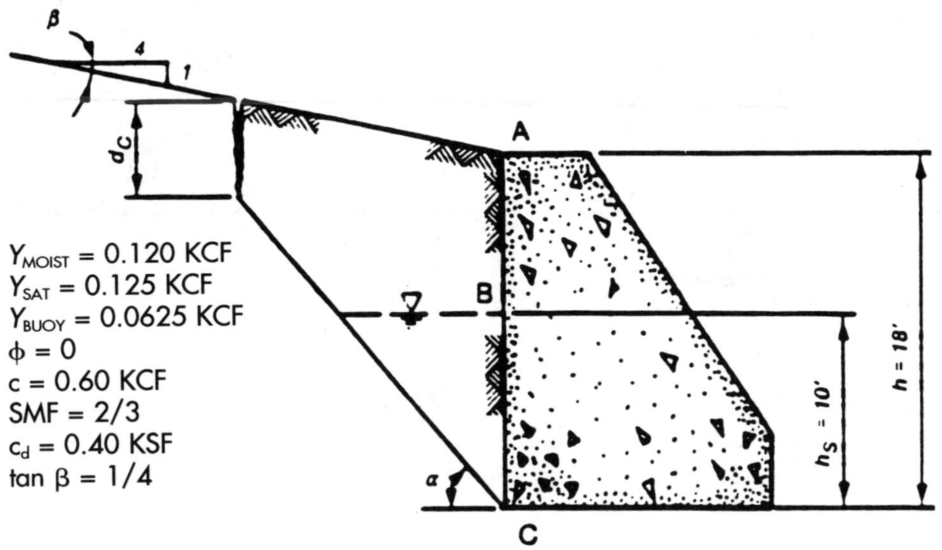

The following values were determined for the problem from EM 1110-2-2502

$$\alpha_{critical} = 29°$$

$$d_c = 7.86 \text{ ft}$$

$$K = 1.00$$

Using the equation for K_c in Appendix H from EM 1110-2-2502 yields

$$K_c = \frac{1}{2 \sin \alpha \cos \alpha} \cdot \frac{\tan \alpha}{\tan \alpha - \tan \beta} = 2.14791$$

8. The pressures for the points shown in the figure below are calculated as follows:

$$p_{v1} = \gamma z_1 = 0.12(14.572) = 1.7486 \text{ ksf}$$

$$p_{v2} = \gamma(z_2 - h_s) + \gamma_b h_s = 0.12(22.788)$$

$$+ 0.0625(10) = 3.3596 \text{ ksf}$$

$$p_{ho} = -2K_c c_d = -2(2.1479)(0.4) = -1.7183 \text{ ksf}$$

$$p_{h1} = Kp_{v1} - 2K_c c_d = 1.0(1.7486)$$

$$- 1.7183 = 0.0303 \text{ ksf}$$

$$p_{h2} = Kp_{v2} - 2K_c c_d = 3.33596$$

$$- 1.7183 = 1.6413 \text{ ksf}$$

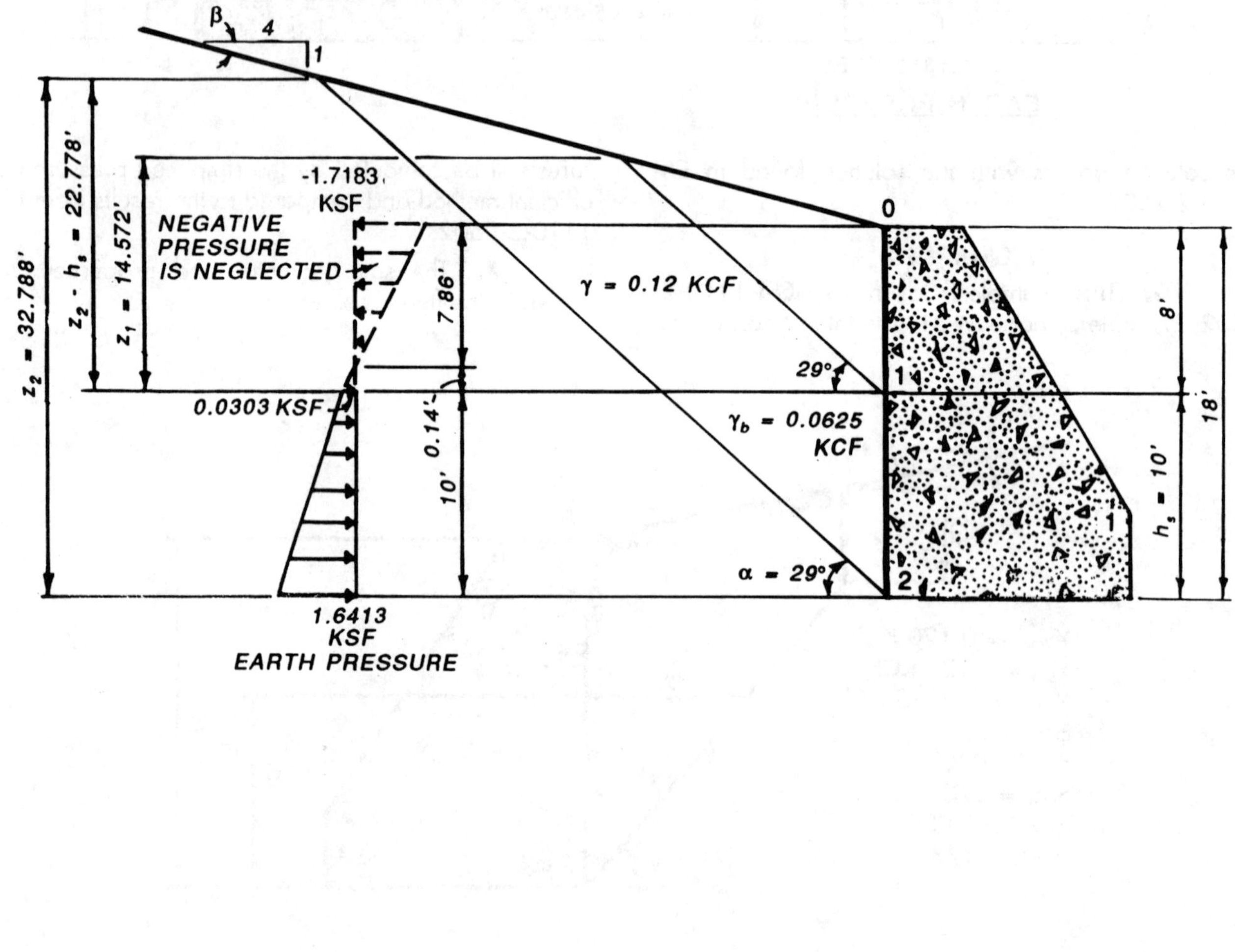

The negative pressures down to depth d_c are neglected. The positive pressures agree with the EM 1110-2-2502 solution.

EXAMPLE 3

9. This example is taken from EM 1110-2-2502, Example 9, page M-52. The lateral earth pressures will be computed by the simplified pressure coefficient method and compared to the results from EM 1110-2-2502.

10. The soil geometry and properties are shown in the figure below.

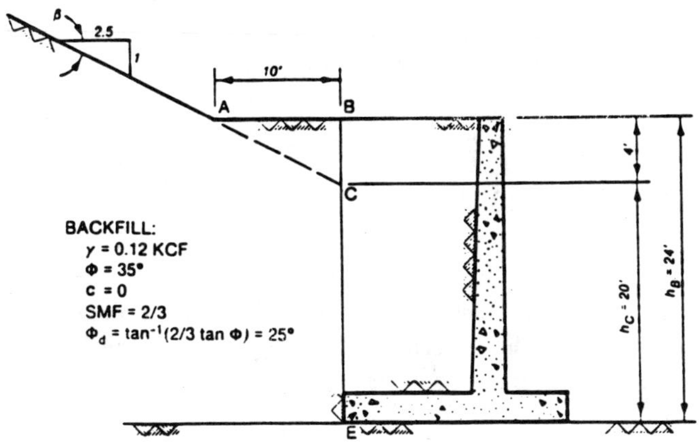

The following values were calculated in EM 1110-2-2502

$$\alpha_{critical} = 44.302°$$

$$K = 0.3589$$

11. The pressures for the points shown on the figure on the following page are calculated as follows:

$$p_{v1} = \gamma z_1 = 0.12(4.0) = 0.4800 \text{ ksf}$$

$$p_{v2} = \gamma z_2 = 0.12(9.76) = 1.1712 \text{ ksf}$$

$$p_{v3} = \gamma z_3 = 0.12(33.891) = 4.0669 \text{ ksf}$$

$$p_{h1} = Kp_{v1} = 0.3589(0.4800) = 0.1723 \text{ ksf}$$

$$p_{h2} = Kp_{v2} = 0.3589(1.1712) = 0.4203 \text{ ksf}$$

$$p_{h3} = Kp_{v3} = 0.3589(4.0669) = 1.4596 \text{ ksf}$$

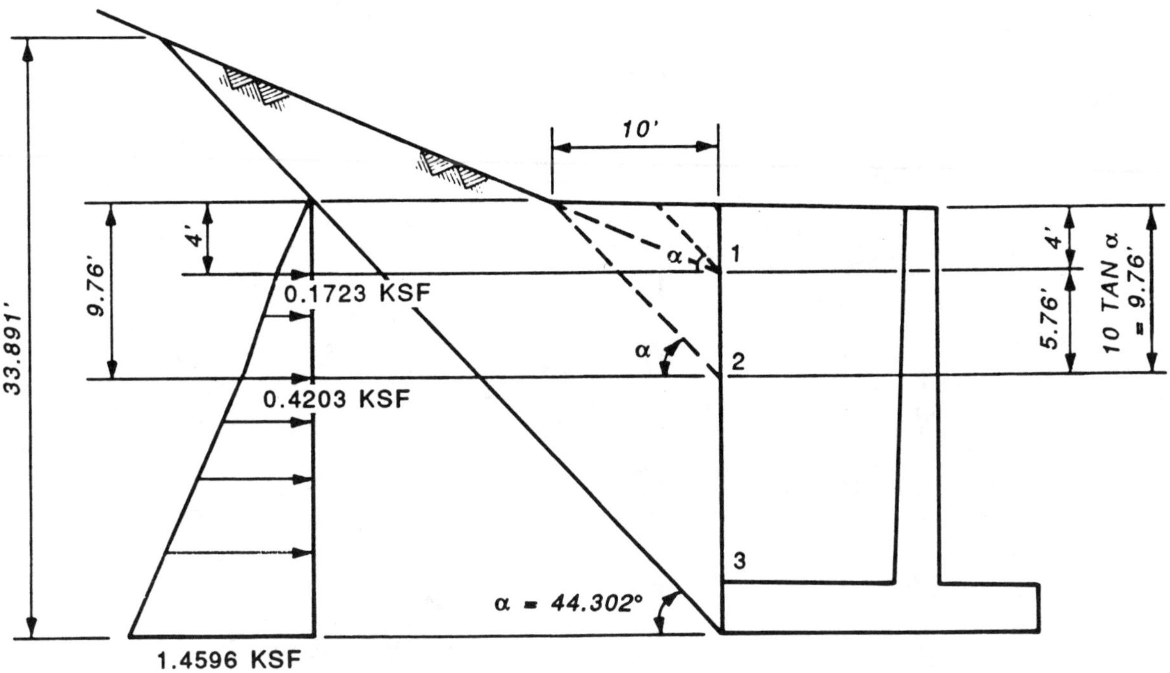

The pressures calculated agree with the solution found in EM 1110-2-2502.

EXAMPLE 4

12. The lateral earth pressures for the driving side will be computed by the simplified pressure coefficient method for the problem shown in the figure below.

13. The water pressures are calculated using the line of creep method.

Differential head = 16 ft − 6 ft = 10 ft

L_s = Total length of the seepage path = $L_{AB} + L_{BC} + L_{CD}$

$= 16 + 22 + 6 = 44$ ft

$$\text{Seepage gradient } i = \frac{10 \text{ ft}}{44 \text{ ft}} = 0.2273$$

$$u_b = [16 - 0.2273(16)]0.0625$$

$$= 0.773 \text{ ksf}$$

$$u_c = [16 - 0.2273(16 + 22)]0.0625$$

$$= 0.460 \text{ ksf}$$

14. The buoyant unit weights of the soil may now be calculated.

For the left side:

Effective unit weight of water γ_{we}

$$= \gamma_w(1 - i) = 0.0483 \text{ kcf}$$

$$\gamma_{bL} = \gamma_s - \gamma_{we} = 0.125 - 0.0483$$

$$= 0.0767 \text{ kcf}$$

For the right side:

Effective unit weight of water γ_{we}

$$= \gamma_w(1 + i) = 0.07671 \text{ kcf}$$

$$\gamma_{bR} = \gamma_s - \gamma_{we} = 0.125 - 0.0767$$

$$= 0.0483 \text{ kcf}$$

15. The critical slip plane angle α and K may now be calculated. To calculate the critical slip plane angle, the average unit weight of the soil in the wedge must be used. This is calculated as

$$\gamma_{avg} = \frac{\gamma h^2 - (\gamma - \gamma_{bL})h_s^2}{h^2}$$

$$= \frac{0.12(38)^2 - (0.12 - 0.0767)(16)^2}{38^2}$$

$$= 0.1123 \text{ kcf}$$

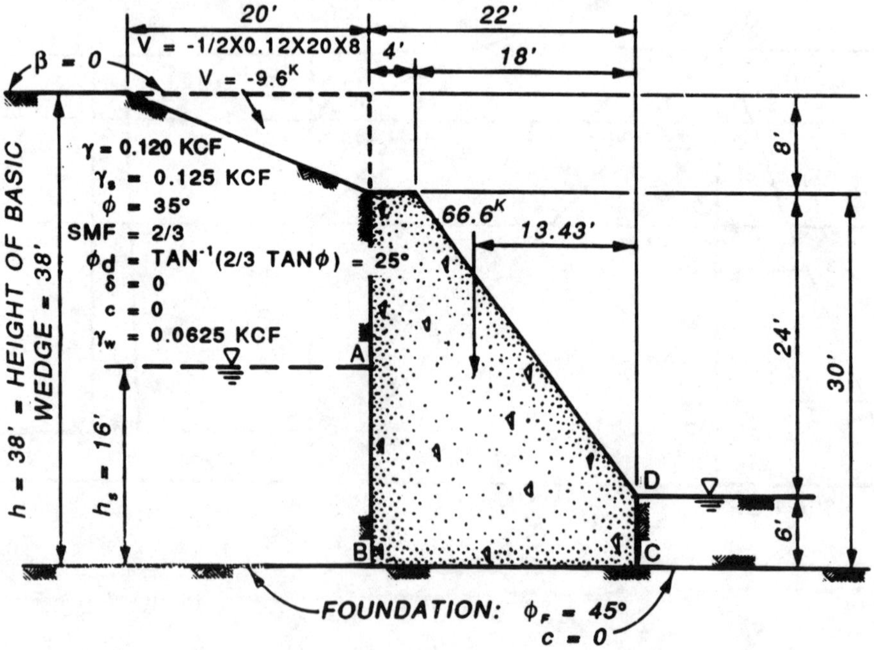

Also the soil wedge EFG is modeled as a negative surcharge equal to

$$V = -\tfrac{1}{2}(0.12)(20)(8) = -9.6 \text{ kips}$$

The average unit weight is now used in Equation 3-30 from EM 1110-2-2502

$$A = \tan \phi_d - \frac{2V(1 + \tan^2 \phi_d)}{\gamma h^2}$$

$$= \tan 25° - \frac{2(-9.6)(1 + \tan^2 25°)}{0.1123(38)^2}$$

$$= 0.610454$$

From Equations 3-28 and 3-29 from EM 1110-2-2502

$$c_1 = \frac{2 \tan^2 \phi_d}{A}$$

$$= \frac{2 \tan^2 25°}{0.610454}$$

$$= 0.712398$$

$$c_2 = \frac{\tan \phi_d}{A}$$

$$= \frac{\tan 25°}{0.610454}$$

$$= 0.763871$$

The critical slip plane angle α may now be calculated from Equation 3-25 from EM 1110-2-2502.

$$\alpha_{\text{critical}} = \tan^{-1}\left(\frac{c_1 + \sqrt{c_1 + 4c_2}}{2}\right)$$

$$= 52.4313°$$

The basic pressure coefficient K may now be calculated using the following equation from Appendix H of EM 1110-2-2502.

$$K = \frac{1 - \tan \phi_d \cot \alpha}{1 + \tan \phi_d \tan \alpha}$$

$$= \frac{1 - \tan 25° \cot 52.4313°}{1 + \tan 25° \tan 52.4313°}$$

$$= 0.39927$$

16. The earth pressures may now be calculated for the points shown in the figure below.

$$p_{v1} = 0.12(20.222) = 2.4266 \text{ ksf}$$

$$p_{v2} = 0.12(22) + 0.0767(4) = 2.9468 \text{ ksf}$$

$$p_{v3} = 0.12(22) + 0.0767(16) = 3.8672 \text{ ksf}$$

$$p_{h1} = Kp_{v1} = 0.33927(2.4266) = 0.9689 \text{ ksf}$$

$$p_{h2} = Kp_{v2} = 0.33927(2.9468) = 1.1766 \text{ ksf}$$

$$p_{h3} = Kp_{v3} = 0.33927(3.8672) = 1.5441 \text{ ksf}$$

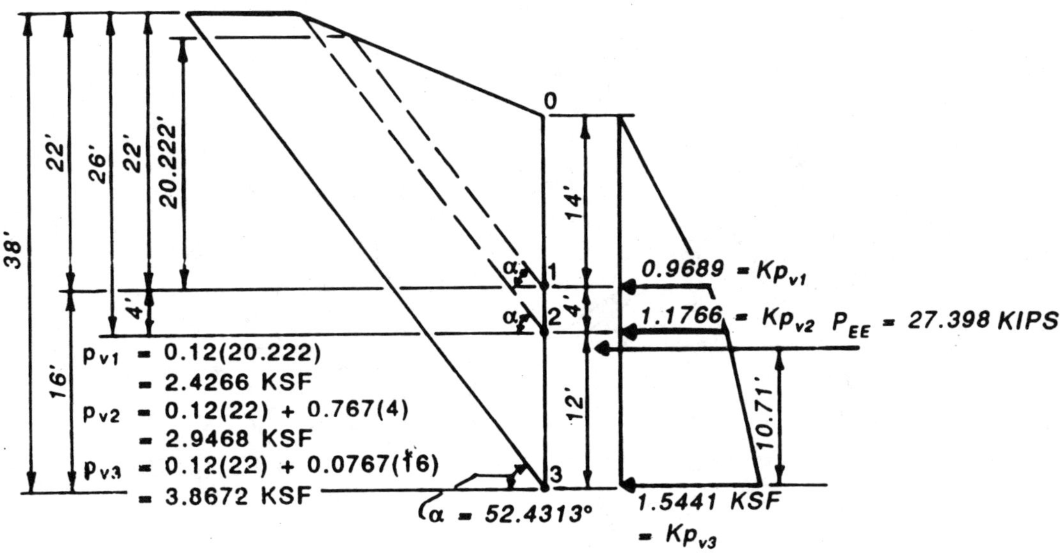

DERIVATION OF GENERAL WEDGE EQUATION FOR SINGLE WEDGE ANALYSIS INCLUDING THE EFFECT OF WALL FRICTION (δ)

1. Effective Horizontal Earth Force.

Given the following driving wedge which possesses both a c and a ϕ, an equation for P_{EE}, the horizontal effective earth force, will be derived.

Summing forces normal to the slip plane yields,

$$H_L \sin\alpha + U + N' - V\cos\alpha - H_R \sin\alpha + P_{EE}\tan\delta\cos\alpha - P_{EE}\sin\alpha - W\cos\alpha - P_w \sin\alpha = 0$$

Solving for N' yields,

$$N' = (H_R + P_{EE} + P_w - H_L)\sin\alpha + (V + W - P_{EE}\tan\delta)\cos\alpha - U = 0 \quad (1)$$

Summing forces tangent to the slip plane yields,

$$H_L\cos\alpha - T + V\sin\alpha + W\sin\alpha - H_R\cos\alpha - P_{EE}\tan\delta\sin\alpha - P_{EE}\cos\alpha - P_w\cos\alpha = 0$$

Solving for T yields,

$$T = (H_L - H_R - P_{EE} - P_w)\cos\alpha + (V + W - P_{EE}\tan\delta)\sin\alpha \quad (2)$$

According to the Mohr-Coulomb failure criterion,

$$T = N'\tan\phi + cL \quad (3)$$

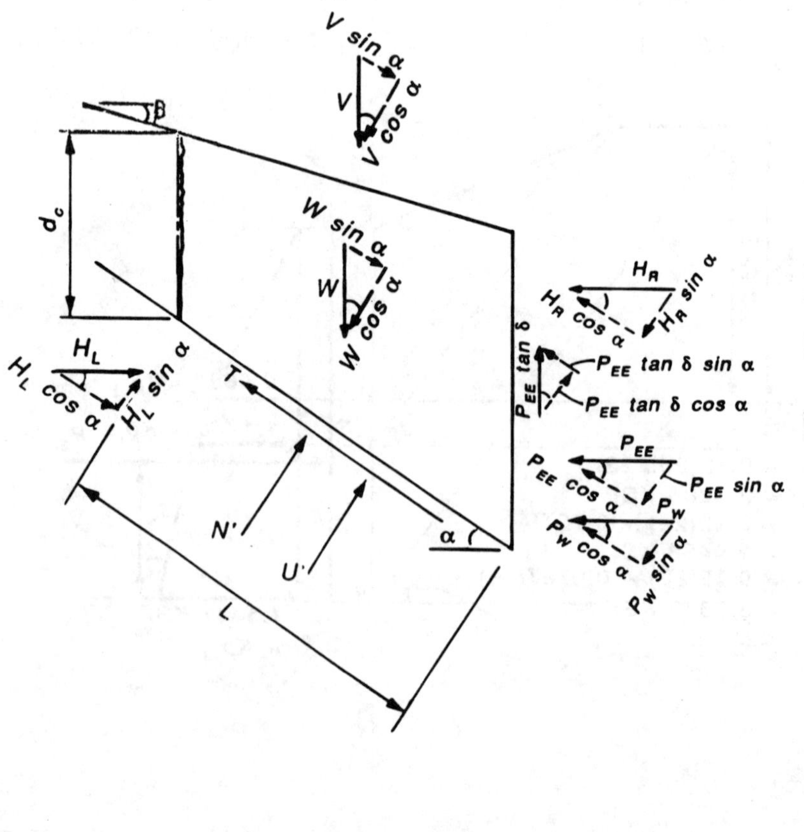

Inserting Equations 1 and 2 into Equation 3 yields,

$$(H_L - H_R - P_{EE} - P_w) \cos \alpha$$

$$+ (V + W - P_{EE} \tan \delta) \sin \alpha$$

$$= [(H_R + P_{EE} + P_w - H_L) \sin \alpha$$

$$+ (V + W - P_{EE} \tan \delta) \cos \alpha - U] \tan \phi + cL$$

Simplifying and solving for P_{EE} yields,

$$P_{EE} = \frac{(W + V)(\sin \alpha - \cos \alpha \tan \phi) + (H_L - H_R - P_w)(\cos \alpha + \sin \alpha \tan \phi) + U \tan \phi - cL}{\sin \alpha \tan \phi + \sin \alpha \tan \delta + \cos \alpha - \cos \alpha \tan \delta \tan \phi}$$

$$P_{EE} = \frac{(W + V)(\tan \alpha - \tan \phi) + (H_L - H_R - P_w)(1 + \tan \alpha \tan \phi) + \dfrac{U \tan \phi}{\cos \alpha} - \dfrac{cL}{\cos \alpha}}{1 + \tan \alpha \tan \phi + \tan \delta (\tan \alpha - \tan \phi)} \quad (4)$$

The total effective earth force acts at an angle δ with respect to the horizontal. The total effective earth force is defined as

$$P = \frac{P_{EE}}{\cos \delta}$$

2. Soil Parameters. The horizontal effective earth force for a particular SMF can be calculated by inserting the factored soil parameters ϕ_d and c_d into Equation 5. This yields,

$$P_{EE} = \frac{(W + V)(\tan \alpha - \tan \phi_d) + (H_L - H_R - P_w)(1 + \tan \alpha \tan \phi_d) + \dfrac{U \tan \phi_d}{\cos \alpha} - \dfrac{cL}{\cos \alpha}}{1 + \tan \alpha \tan \phi_d + \tan \delta (\tan \alpha - \tan \phi_d)} \quad (5)$$

EXAMPLES OF EARTH PRESSURE COMPUTATIONS USING SIMPLIFIED PRESSURE COEFFICIENT METHOD INCLUDING THE EFFECTS OF WALL FRICTION

1. The following examples demonstrate the method of computing earth pressures using the simplified pressure coefficient method described in paragraph 5c of this ETL. The effect of wall friction is taken into account. The equations for the critical slip plane angle and the pressure coefficients are defined in paragraphs 6c and 6d.

2. Example 1 is taken from EM 1110-2-2502, Example 2, page M-5. Example 2 in this enclosure is the same as Example 4 in Enclosure 2 with the addition of wall friction.

EXAMPLE 1

3. The earth pressures will be calculated for the soil geometry and properties given in the figure below. The earth pressure coefficients will be computed using pressure coefficients and compared against Coulomb's equation. The effects of wall friction will be taken into account in this example. Since the value of wall friction is affected by a number of factors, as described in paragraph 6a of this ETL, the value assigned to the wall friction was selected to demonstrate the mechanics of the procedure. Recommended values of wall friction will be addressed in subsequent engineering guidance.

4. Coulomb's equation (Equation 3-12 from EM 1110-2-2502) with $\theta = 90°$ reduces to

$$K_o = \frac{\cos^2 \phi_d \cos \delta}{\cos \delta \left[1 + \sqrt{\dfrac{\sin \phi_d \sin (\phi_d - \beta)}{\cos \delta \cos \beta}}\right]^2}$$

$$= \frac{\cos^2 21° \cos 10.5°}{\cos 10.5° \left[1 + \sqrt{\dfrac{\sin 21° \sin (21° - 14°)}{\cos 10.5° \cos 14°}}\right]^2}$$

$$= 0.5510$$

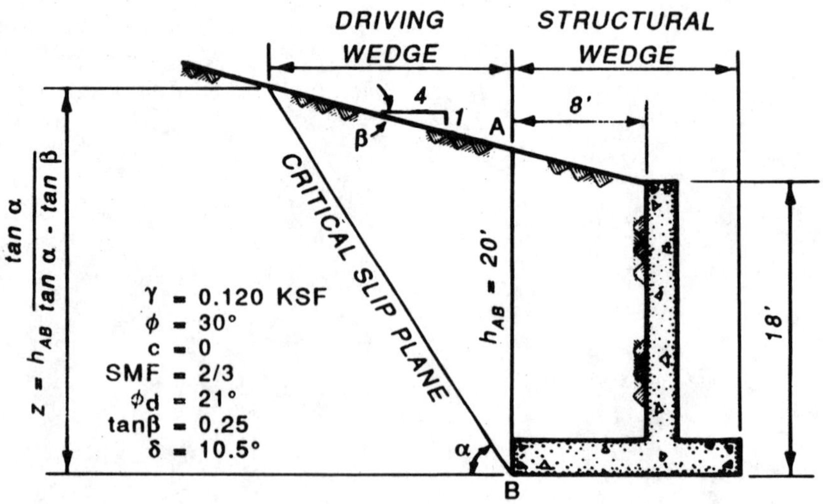

The horizontal effective earth force is computed as

$$P_{EE} = \tfrac{1}{2} K_o \gamma h^2 = \tfrac{1}{2} (0.5510)(0.12)(20)^2$$

$$= 13.224 \text{ kips/ft}$$

The vertical shear force is equal to

$$P_V = P_{EE} \tan \delta = 13.224 \tan 10.5^\circ = 2.451 \text{ kips/ft}$$

5. The pressures and forces will now be computed using the simplified coefficient method. The critical slip plane angle is calculated using Equations 4 through 7 of this ETL as shown below.

$$A = \tan \phi_d + \tan \delta = \tan 21^\circ + \tan 10.5^\circ$$

$$= 0.569203$$

$$c_1 = \frac{2 \tan \phi_d (\tan \phi_d + \tan \delta)}{A}$$

$$= \frac{2 \tan 21^\circ (\tan 21^\circ + \tan 10.5^\circ)}{0.569203}$$

$$= 0.767728$$

$$c_2 = \frac{\tan \phi_d [1 - \tan \phi_d (\tan \beta + \tan \delta)] - \tan \beta}{A}$$

$$= \frac{\tan 21^\circ [1 - \tan 21^\circ (\tan 14^\circ + \tan 10.5^\circ)] - \tan 14^\circ}{0.569203}$$

$$= 0.122480$$

$$\alpha_{critical} = \tan^{-1}\left(\frac{c_1 + \sqrt{c_1 + 4c_2}}{2}\right) = 42.092^\circ$$

6. The pressure coefficients K and K_1 may now be calculated using Equations 9 and 10 of this ETL.

$$K = \frac{1 - \tan \phi_d \cot \alpha}{1 + \tan \phi_d \tan \alpha + \tan \delta (\tan \alpha - \tan \phi_d)}$$

$$= \frac{1 - \tan 21^\circ \cot 42.092^\circ}{1 + \tan 21^\circ \tan 42.092^\circ + \tan 10.5^\circ (\tan 42.092^\circ - \tan 21^\circ)}$$

$$= 0.3985$$

$$K_1 = K \cdot \frac{\tan \alpha}{\tan \alpha - \tan \beta} = \frac{0.3985 \tan 42.092^\circ}{\tan 42.092^\circ - \tan 14^\circ} = 0.5510$$

The horizontal effective earth force P_{EE} is calculated as

$$P_{EE} = \tfrac{1}{2} K_1 \gamma h^2 = \tfrac{1}{2}(0.5510)(0.12)(20)^2 = 13.224 \text{ kips/ft}$$

This agrees with the results obtained in paragraph 2.

7. The results will now be compared against the general wedge equation given in Equation 2 of this ETL.

$$W = \frac{\gamma h^2}{2(\tan \alpha - \tan \beta)} = \frac{0.12}{2(\tan 42.092^\circ - \tan 14^\circ)}$$

$$= 36.7357 \text{ kips/ft}$$

$$P_{EE} = \frac{W(\tan \alpha - \tan \phi_d)}{1 + \tan \alpha \tan \phi_d + \tan \delta (\tan \alpha - \tan \phi_d)}$$

$$= \frac{36.7357(\tan 42.092^\circ - \tan 21^\circ)}{1 + \tan 42.092^\circ \tan 21^\circ + \tan 10.5^\circ (\tan 42.092^\circ - \tan 21^\circ)}$$

$$= 13.224 \text{ kips/ft}$$

This agrees with the previous solutions.

8. The lateral earth pressure distribution on the backfill may now be calculated. The pressures are calculated for two of the points shown on the following figure. The pressures can be calculated using the simplified method as follows:

$$p_{vC} = \gamma z_C = 0.12(21.65) = 3.318 \text{ ksf}$$

$$p_{hC} = K p_{vC} = 0.3985(3.318) = 1.322 \text{ ksf}$$

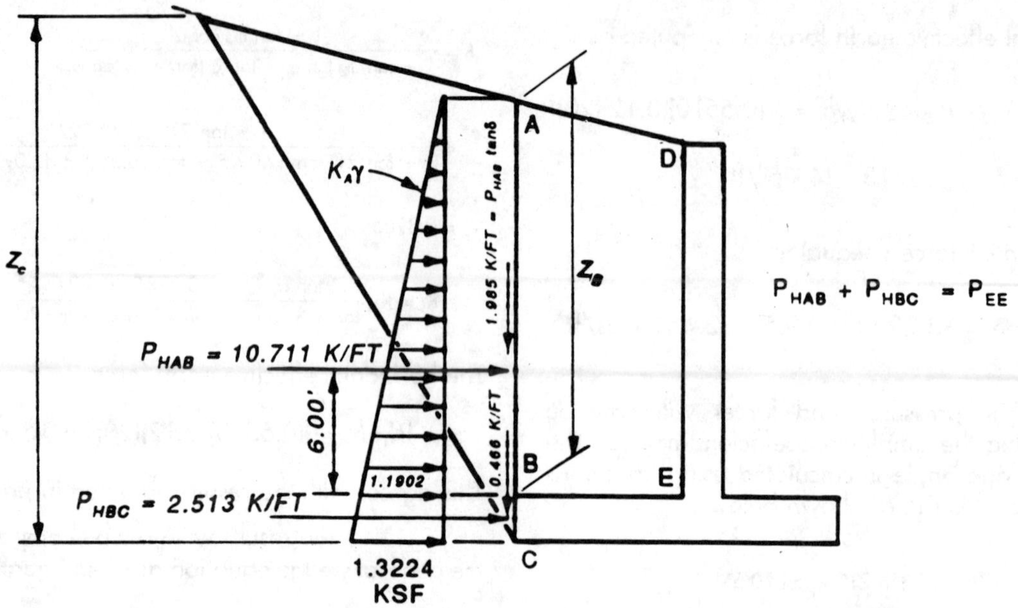

Since the pressure diagram is linear down to point *C*, the value for the pressure at point *B* may be calculated by proportions as follows:

$$p_{hB} = p_{hC}\frac{z_B}{z_C} = 1.322(18/20) = 1.190 \text{ ksf}$$

EXAMPLE 2

9. The earth pressures will be calculated for the soil geometry and properties given in the figure below. The earth pressures will be computed using the simplified coefficient method. The effects of wall friction will be taken into account in this example. Since the value of wall friction is affected by a number of factors, as described in paragraph 6a of this ETL, the value assigned to the wall friction was selected to demonstrate the mechanics of the procedure. Recommended values of wall friction will be addressed in subsequent engineering guidance.

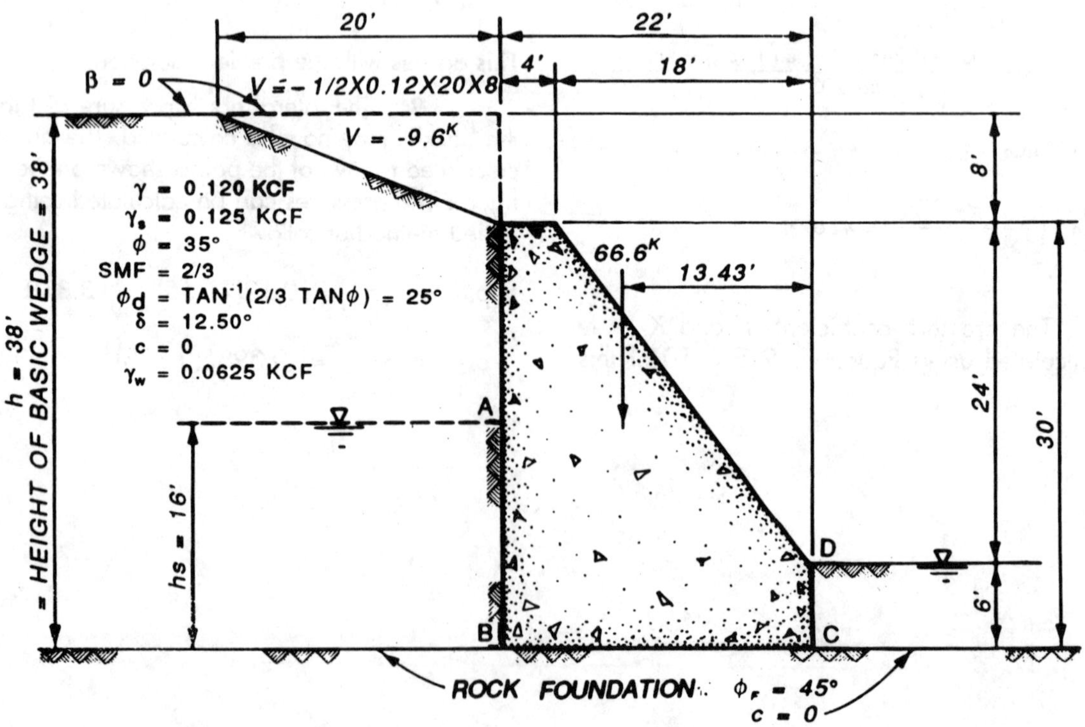

10. The critical slip plane angle is calculated from Equations 4 through 7 of this ETL as shown below. Values for γ and V are taken from Example 4 in Enclosure 2.

$$A = \tan \phi_d + \tan \delta - \frac{2V(1 + \tan^2 \phi_d)}{\gamma h^2}$$

$$= \tan 25° + \tan 12.5° - \frac{2(-9.6)(1 + \tan^2 25°)}{0.1123(38)^2}$$

$$= 0.832149$$

$$c_1 = \frac{2 \tan \phi_d (\tan \phi_d + \tan \delta)}{A}$$

$$= \frac{2 \tan 25° (\tan 25° + \tan 12.5°)}{0.832149}$$

$$= 0.771067$$

$$c_2 = \frac{\tan \phi_d (1 - \tan \phi_d \tan \delta)}{A}$$

$$= \frac{\tan 25° (1 - \tan 21° \tan 12.5°)}{0.832149}$$

$$= 0.502436$$

$$\alpha_{critical} = \tan^{-1}\left(\frac{c_1 + \sqrt{c_1 + 4c_2}}{2}\right) = 50.016°$$

11. The pressure coefficient K may now be calculated using Equation 9 of this ETL.

$$K = \frac{1 - \tan \phi_d \cot \alpha}{1 + \tan \phi_d \tan \alpha + \tan \delta (\tan \alpha - \tan \phi_d)}$$

$$= \frac{1 - \tan 25° \cot 50.016°}{1 + \tan 25° \tan 50.016° + \tan 12.5° (\tan 50.016° - \tan 25°)}$$

$$= 0.35465$$

12. The lateral earth pressure distribution may now be calculated. The pressures are calculated for the points shown on the figure on the following page. The pressures are calculated as follows:

$$p_{v1} = 0.12(21.067) = 2.5280 \text{ ksf}$$

$$p_{v2} = 0.12(22) + 0.0767(1.849) = 2.7818 \text{ ksf}$$

$$p_{v3} = 0.12(22) + 0.0767(16) = 3.8672 \text{ ksf}$$

$$p_{h1} = Kp_{v1} = 0.35465(2.5280) = 0.8966 \text{ ksf}$$

$$p_{h2} = Kp_{v2} = 0.35465(2.7818) = 0.9866 \text{ ksf}$$

$$p_{h3} = Kp_{v3} = 0.35465(3.8672) = 1.3715 \text{ ksf}$$

The vertical shear force is calculated as

$$P_V = P_{EE} \tan \delta = 24.702 \tan 12.5° = 5.476 \text{ kips}$$

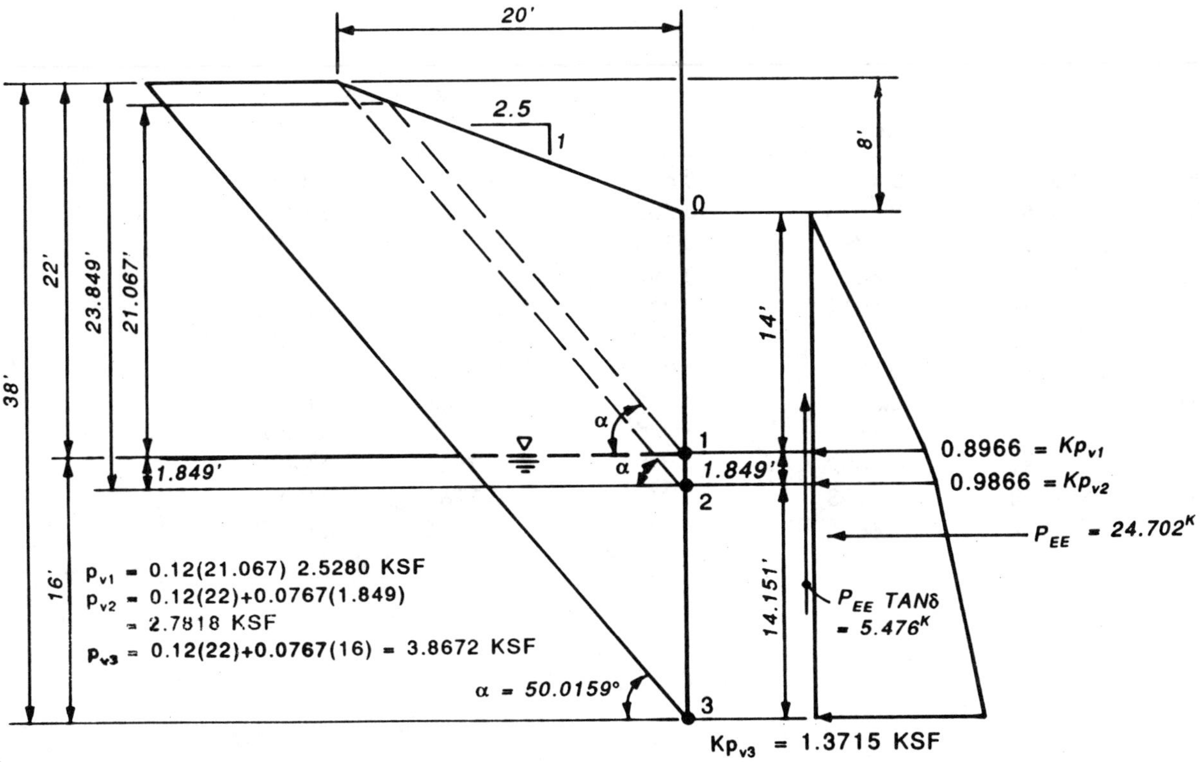

SMF VALUES WHEN THE RATIO $\tan\beta/\tan\phi$ EXCEEDS 0.56

The following tables indicate the appropriate SMF value to use in Coulomb's equation (or the general wedge equation) to compute a K_o that is close to the value computed from the Danish Code Equation. The last two columns contain the recommended SMF and the corresponding K_o value calculated using Coulomb's equation. Note that a row of asterisks denotes the condition $\beta > \phi_d$; Coulomb's equation is not valid for this condition.

PHI = 26.00°

β	$K_{o\beta}$	K_o	K_o	Recommended	
(Deg)	(Danish)	(SMF = 1)	(SMF = 2/3)	K_o	SMF
15.00	0.7070	0.4822	0.7086	0.7086	0.6667
16.00	0.7164	0.4920	0.7390	0.7164	0.6865
17.00	0.7258	0.5026	0.7818	0.7257	0.7070
18.00	0.7352	0.5143	0.8895	0.7351	0.7296
19.00	0.7445	0.5273	******	0.7444	0.7546
20.00	0.7537	0.5420	******	0.7536	0.7821
21.00	0.7629	0.5589	******	0.7628	0.8120
22.00	0.7720	0.5786	******	0.7719	0.8444
23.00	0.7811	0.6026	******	0.7810	0.8793
24.00	0.7901	0.6333	******	0.7900	0.9169
25.00	0.7990	0.6776	******	0.7994	0.9570

PHI = 27.00°

β	$K_{o\beta}$	K_o	K_o	Recommended	
(Deg)	(Danish)	(SMF = 1)	(SMF = 2/3)	K_o	SMF
15.00	0.6873	0.4608	0.6805	0.6805	0.6667
16.00	0.6965	0.4696	0.7059	0.6965	0.6756
17.00	0.7056	0.4792	0.7387	0.7056	0.6937
18.00	0.7147	0.4897	0.7874	0.7147	0.7139
19.00	0.7238	0.5012	******	0.7237	0.7362
20.00	0.7328	0.5141	******	0.7327	0.7608
21.00	0.7417	0.5286	******	0.7416	0.7876
22.00	0.7505	0.5453	******	0.7505	0.8168
23.00	0.7594	0.5649	******	0.7593	0.8484
24.00	0.7681	0.5887	******	0.7681	0.8825
25.00	0.7768	0.6193	******	0.7769	0.9190
26.00	0.7854	0.6635	******	0.7855	0.9582

PHI = 28.00°

β (Deg)	$K_{o\beta}$ (Danish)	K_o (SMF = 1)	K_o (SMF = 2/3)	Recommended K_o	Recommended SMF
16.00	0.6768	0.4483	0.6764	0.6764	0.6667
17.00	0.6856	0.4570	0.7033	0.6856	0.6824
18.00	0.6945	0.4665	0.7389	0.6944	0.7004
19.00	0.7033	0.4768	0.7959	0.7032	0.7203
20.00	0.7120	0.4882	******	0.7119	0.7423
21.00	0.7207	0.5009	******	0.7206	0.7664
22.00	0.7293	0.5153	******	0.7293	0.7927
23.00	0.7378	0.5317	******	0.7378	0.8213
24.00	0.7463	0.5511	******	0.7462	0.8522
25.00	0.7547	0.5747	******	0.7547	0.8854
26.00	0.7631	0.6052	******	0.7632	0.9210
27.00	0.7714	0.6491	******	0.7718	0.9592

PHI = 29.00°

β (Deg)	$K_{o\beta}$ (Danish)	K_o (SMF = 1)	K_o (SMF = 2/3)	Recommended K_o	Recommended SMF
17.00	0.6658	0.4359	0.6722	0.6722	0.6667
18.00	0.6744	0.4445	0.7008	0.6743	0.6888
19.00	0.6829	0.4538	0.7398	0.6828	0.7066
20.00	0.6914	0.4639	0.8095	0.6913	0.7263
21.00	0.6998	0.4751	******	0.6998	0.7480
22.00	0.7082	0.4877	******	0.7081	0.7717
23.00	0.7165	0.5018	******	0.7164	0.7975
24.00	0.7247	0.5181	******	0.7247	0.8254
25.00	0.7329	0.5373	******	0.7328	0.8556
26.00	0.7410	0.5607	******	0.7410	0.8880
27.00	0.7491	0.5909	******	0.7492	0.9228
28.00	0.7571	0.6346	******	0.7572	0.9601

PHI = 30.00°

β (Deg)	$K_{o\beta}$ (Danish)	K_o (SMF = 1)	K_o (SMF = 2/3)	Recommended K_o	Recommended SMF
17.00	0.6462	0.4159	0.6440	0.6440	0.6667
18.00	0.6545	0.4236	0.6681	0.6544	0.6788
19.00	0.6628	0.4320	0.6985	0.6627	0.6947
20.00	0.6710	0.4411	0.7415	0.6710	0.7124
21.00	0.6792	0.4511	0.8394	0.6792	0.7319
22.00	0.6873	0.4621	******	0.6872	0.7532
23.00	0.6954	0.4745	******	0.6953	0.7765
24.00	0.7034	0.4884	******	0.7033	0.8018
25.00	0.7113	0.5045	******	0.7113	0.8291
26.00	0.7192	0.5235	******	0.7191	0.8587
27.00	0.7270	0.5466	******	0.7270	0.8904
28.00	0.7347	0.5765	******	0.7348	0.9245
29.00	0.7424	0.6200	******	0.7424	0.9610

PHI = 31.00°

β (Deg)	$K_{o\beta}$ (Danish)	K_o (SMF = 1)	K_o (SMF = 2/3)	Recommended K_o	Recommended SMF
18.00	0.6348	0.4037	0.6386	0.6386	0.6667
19.00	0.6429	0.4113	0.6638	0.6428	0.6845
20.00	0.6508	0.4195	0.6964	0.6508	0.7004
21.00	0.6588	0.4285	0.7444	0.6587	0.7179
22.00	0.6666	0.4383	******	0.6666	0.7371
23.00	0.6745	0.4491	******	0.6744	0.7581
24.00	0.6822	0.4612	******	0.6821	0.7810
25.00	0.6899	0.4750	******	0.6898	0.8058
26.00	0.6976	0.4908	******	0.6976	0.8326
27.00	0.7051	0.5096	******	0.7051	0.8615
28.00	0.7126	0.5324	******	0.7126	0.8926
29.00	0.7201	0.5620	******	0.7202	0.9260
30.00	0.7274	0.6052	******	0.7276	0.9617

PHI = 32.00°

β (Deg)	$K_{o\beta}$ (Danish)	K_o (SMF = 1)	K_o (SMF = 2/3)	Recommended K_o	Recommended SMF
19.00	0.6231	0.3917	0.6331	0.6331	0.6667
20.00	0.6309	0.3991	0.6595	0.6308	0.6899
21.00	0.6385	0.4071	0.6944	0.6385	0.7056
22.00	0.6462	0.4159	0.7488	0.6461	0.7229
23.00	0.6538	0.4255	******	0.6537	0.7419
24.00	0.6613	0.4361	******	0.6612	0.7626
25.00	0.6687	0.4481	******	0.6687	0.7851
26.00	0.6762	0.4616	******	0.6761	0.8094
27.00	0.6835	0.4772	******	0.6835	0.8357
28.00	0.6908	0.4956	******	0.6907	0.8641
29.00	0.6980	0.5182	******	0.6979	0.8946
30.00	0.7051	0.5475	******	0.7053	0.9273
31.00	0.7122	0.5902	******	0.7122	0.9624

PHI = 33.00°

β (Deg)	$K_{o\beta}$ (Danish)	K_o (SMF = 1)	K_o (SMF = 2/3)	Recommended K_o	Recommended SMF
19.00	0.6036	0.3730	0.6051	0.6051	0.6667
20.00	0.6111	0.3797	0.6274	0.6110	0.6808
21.00	0.6185	0.3869	0.6552	0.6185	0.6949
22.00	0.6259	0.3948	0.6926	0.6259	0.7105
23.00	0.6333	0.4034	0.7558	0.6332	0.7276
24.00	0.6406	0.4128	******	0.6405	0.7463
25.00	0.6478	0.4232	******	0.6477	0.7667
26.00	0.6550	0.4349	******	0.6549	0.7888
27.00	0.6621	0.4482	******	0.6620	0.8128
28.00	0.6691	0.4635	******	0.6691	0.8386
29.00	0.6761	0.4817	******	0.6761	0.8665
30.00	0.6830	0.5039	******	0.6830	0.8964
31.00	0.6899	0.5329	******	0.6899	0.9286
32.00	0.6967	0.5751	******	0.6970	0.9630

PHI = 34.00°

β (Deg)	$K_{o\beta}$ (Danish)	K_o (SMF = 1)	K_o (SMF = 2/3)	Recommended K_o	Recommended SMF
20.00	0.5916	0.3612	0.5984	0.5984	0.6667
21.00	0.5988	0.3678	0.6215	0.5987	0.6856
22.00	0.6059	0.3748	0.6507	0.6058	0.6997
23.00	0.6130	0.3825	0.6911	0.6130	0.7151
24.00	0.6201	0.3909	0.7679	0.6201	0.7320
25.00	0.6271	0.4001	******	0.6271	0.7504
26.00	0.6340	0.4103	******	0.6340	0.7705
27.00	0.6409	0.4218	******	0.6409	0.7923
28.00	0.6478	0.4348	******	0.6478	0.8158
29.00	0.6545	0.4499	******	0.6545	0.8413
30.00	0.6612	0.4677	******	0.6612	0.8686
31.00	0.6678	0.4896	******	0.6678	0.8981
32.00	0.6744	0.5182	******	0.6745	0.9297
33.00	0.6809	0.5600	******	0.6814	0.9636

PHI = 35.00°

β (Deg)	$K_{o\beta}$ (Danish)	K_o (SMF = 1)	K_o (SMF = 2/3)	Recommended K_o	Recommended SMF
21.00	0.5792	0.3495	0.5914	0.5914	0.6667
22.00	0.5862	0.3559	0.6154	0.5861	0.6901
23.00	0.5930	0.3628	0.6460	0.5930	0.7040
24.00	0.5999	0.3703	0.6899	0.5998	0.7193
25.00	0.6066	0.3785	0.7989	0.6066	0.7360
26.00	0.6134	0.3875	******	0.6133	0.7542
27.00	0.6200	0.3975	******	0.6200	0.7740
28.00	0.6266	0.4087	******	0.6266	0.7955
29.00	0.6332	0.4214	******	0.6331	0.8187
30.00	0.6396	0.4362	******	0.6396	0.8437
31.00	0.6460	0.4538	******	0.6461	0.8706
32.00	0.6524	0.4753	******	0.6523	0.8996
33.00	0.6587	0.5034	******	0.6588	0.9307
34.00	0.6649	0.5447	******	0.6654	0.9641

PHI = 36.00°

β (Deg)	$K_{o\beta}$ (Danish)	K_o (SMF = 1)	K_o (SMF = 2/3)	Recommended K_o	Recommended SMF
22.00	0.5666	0.3380	0.5842	0.5842	0.6667
23.00	0.5733	0.3442	0.6090	0.5732	0.6942
24.00	0.5799	0.3509	0.6412	0.5798	0.7081
25.00	0.5864	0.3582	0.6891	0.5864	0.7232
26.00	0.5929	0.3662	******	0.5928	0.7398
27.00	0.5994	0.3749	******	0.5993	0.7577
28.00	0.6057	0.3847	******	0.6057	0.7773
29.00	0.6121	0.3956	******	0.6120	0.7984
30.00	0.6183	0.4081	******	0.6183	0.8212
31.00	0.6245	0.4226	******	0.6245	0.8458
32.00	0.6307	0.4398	******	0.6306	0.8724
33.00	0.6367	0.4610	******	0.6366	0.9009
34.00	0.6427	0.4887	******	0.6428	0.9316
35.00	0.6487	0.5294	******	0.6491	0.9645

PHI = 37.00°

β (Deg)	$K_{o\beta}$ (Danish)	K_o (SMF = 1)	K_o (SMF = 2/3)	Recommended K_o	Recommended SMF
22.00	0.5473	0.3209	0.5558	0.5558	0.6667
23.00	0.5538	0.3265	0.5766	0.5537	0.6857
24.00	0.5601	0.3325	0.6023	0.5601	0.6981
25.00	0.5665	0.3391	0.6362	0.5664	0.7119
26.00	0.5727	0.3461	0.6889	0.5727	0.7269
27.00	0.5790	0.3539	******	0.5789	0.7432
28.00	0.5851	0.3625	******	0.5850	0.7610
29.00	0.5912	0.3720	******	0.5911	0.7802
30.00	0.5973	0.3827	******	0.5973	0.8010
31.00	0.6033	0.3949	******	0.6033	0.8235
32.00	0.6092	0.4091	******	0.6091	0.8478
33.00	0.6151	0.4259	******	0.6150	0.8740
34.00	0.6208	0.4467	******	0.6208	0.9021
35.00	0.6266	0.4739	******	0.6267	0.9324
36.00	0.6322	0.5140	******	0.6327	0.9649

PHI = 38.00°

β (Deg)	$K_{o\beta}$ (Danish)	K_o (SMF = 1)	K_o (SMF = 2/3)	Recommended K_o	Recommended SMF
23.00	0.5345	0.3097	0.5474	0.5474	0.6667
24.00	0.5407	0.3151	0.5687	0.5406	0.6894
25.00	0.5468	0.3210	0.5953	0.5467	0.7018
26.00	0.5528	0.3273	0.6310	0.5528	0.7154
27.00	0.5588	0.3342	0.6894	0.5588	0.7302
28.00	0.5648	0.3417	******	0.5647	0.7464
29.00	0.5707	0.3501	******	0.5707	0.7639
30.00	0.5765	0.3593	******	0.5765	0.7829
31.00	0.5823	0.3698	******	0.5822	0.8035
32.00	0.5880	0.3817	******	0.5880	0.8256
33.00	0.5937	0.3956	******	0.5937	0.8496
34.00	0.5993	0.4120	******	0.5993	0.8754
35.00	0.6048	0.4324	******	0.6048	0.9032
36.00	0.6102	0.4591	******	0.6104	0.9331
37.00	0.6156	0.4986	******	0.6160	0.9653

PHI = 39.00°

β (Deg)	$K_{o\beta}$ (Danish)	K_o (SMF = 1)	K_o (SMF = 2/3)	Recommended K_o	Recommended SMF
24.00	0.5214	0.2986	0.5387	0.5387	0.6667
25.00	0.5273	0.3038	0.5605	0.5273	0.6928
26.00	0.5332	0.3095	0.5879	0.5331	0.7052
27.00	0.5390	0.3157	0.6255	0.5389	0.7186
28.00	0.5447	0.3224	0.6913	0.5447	0.7333
29.00	0.5504	0.3297	******	0.5503	0.7493
30.00	0.5560	0.3378	******	0.5560	0.7666
31.00	0.5616	0.3468	******	0.5616	0.7853
32.00	0.5671	0.3570	******	0.5671	0.8056
33.00	0.5726	0.3686	******	0.5726	0.8275
34.00	0.5780	0.3821	******	0.5779	0.8512
35.00	0.5833	0.3982	******	0.5832	0.8767
36.00	0.5886	0.4182	******	0.5886	0.9041
37.00	0.5938	0.4444	******	0.5937	0.9337
38.00	0.5989	0.4831	******	0.5993	0.9656

PHI = 40.00°

β (Deg)	$K_{o\beta}$ (Danish)	K_o (SMF = 1)	K_o (SMF = 2/3)	Recommended K_o	Recommended SMF
25.00	0.5082	0.2876	0.5297	0.5297	0.6667
26.00	0.5138	0.2927	0.5519	0.5137	0.6960
27.00	0.5194	0.2982	0.5802	0.5193	0.7082
28.00	0.5249	0.3041	0.6197	0.5249	0.7216
29.00	0.5304	0.3106	0.6954	0.5303	0.7361
30.00	0.5358	0.3177	******	0.5357	0.7519
31.00	0.5412	0.3256	******	0.5411	0.7690
32.00	0.5465	0.3344	******	0.5464	0.7876
33.00	0.5518	0.3442	******	0.5518	0.8076
34.00	0.5570	0.3556	******	0.5569	0.8292
35.00	0.5621	0.3687	******	0.5621	0.8526
36.00	0.5672	0.3845	******	0.5672	0.8778
37.00	0.5722	0.4040	******	0.5723	0.9049
38.00	0.5771	0.4296	******	0.5773	0.9342
39.00	0.5820	0.4677	******	0.5825	0.9658

EXAMPLE OF LIMITATION OF 2/3 SMF

1. The following example will demonstrate that if β is large compared to ϕ (the ratio $\tan\beta/\tan\phi > 0.56$), the value of the horizontal earth pressure coefficient computed from Coulomb's equation using an SMF of 2/3 will exceed the value of the horizontal pressure coefficient computed from the Danish Code equation. In some cases, this increase will be overly conservative.

2. For a smooth (no wall friction), vertical wall with a soil having $\phi = 30°$ and $\beta = 21°$, the horizontal earth pressure coefficients will be computed by the Danish Code Equation and Coulomb's equation.

Danish Code Equation (Equation 3-5, EM 1110-2-2502)

$$K_o = (1 - \sin\phi)(1 + \sin\beta)$$

$$= (1 - \sin 30°)(1 + \sin 21°)$$

$$= 0.6792$$

Coulomb's Equation with SMF = 2/3 (Equation 3-14, EM 1110-2-2502)

$$\phi_d = \tan^{-1}(\text{SMF} \tan\phi)$$

$$= \tan^{-1}(2/3 \tan 30°)$$

$$= 21.0517°$$

$$K_o = \frac{\cos^2\phi_d}{\left[1 + \sqrt{\dfrac{\sin\phi_d \sin(\phi_d - \beta)}{\cos\beta}}\right]^2}$$

$$= \frac{\cos^2 21.0517°}{\left[1 + \sqrt{\dfrac{\sin 21.0517° \sin(21.0517° - 21°)}{\cos 21°}}\right]^2}$$

$$= 0.8394$$

Coulomb's Equation with SMF = 0.7319 (from Table in Enclosure 5 for ϕ = 30°)

$$\phi_d = \tan^{-1}(\text{SMF} \tan\phi)$$

$$= \tan^{-1}(0.7319 \tan 30°)$$

$$= 22.9071$$

$$K_o = \frac{\cos^2 22.9071°}{\left[1 + \sqrt{\dfrac{\sin 21.0517° \sin(22.9071° - 21°)}{\cos 21°}}\right]^2}$$

$$= 0.6791$$

3. From the summary of the results below, Coulomb's equation with an SMF of 2/3 results in an at-rest earth pressure coefficient which exceeds the Danish Code equation.

At-rest, Danish Code Equation: $K_o = 0.6792$

Coulomb's equation with SMF = 2/3: $K_o = 0.8394$

Coulomb's equation with SMF = 0.7319: $K_o = 0.6791$

TYPICAL STEEL DETAILS

1. A typical reinforcement for a cantilever retaining wall is shown below.

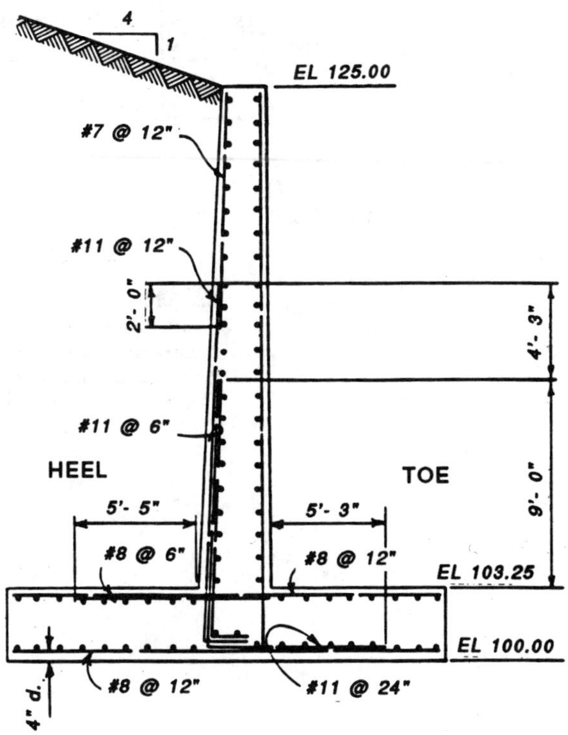

NOTES:
1. ALL REINF. TO BE #6 @ 12", EXCEPT AS SHOWN.
2. ALL HOOKS SHOWN ARE ACI STANDARD HOOKS.
3. MINIMUM CLEAR CONCRETE COVER TO BE 3" EXCEPT AS SHOWN.

2. A typical reinforcement for a floodwall is shown below.

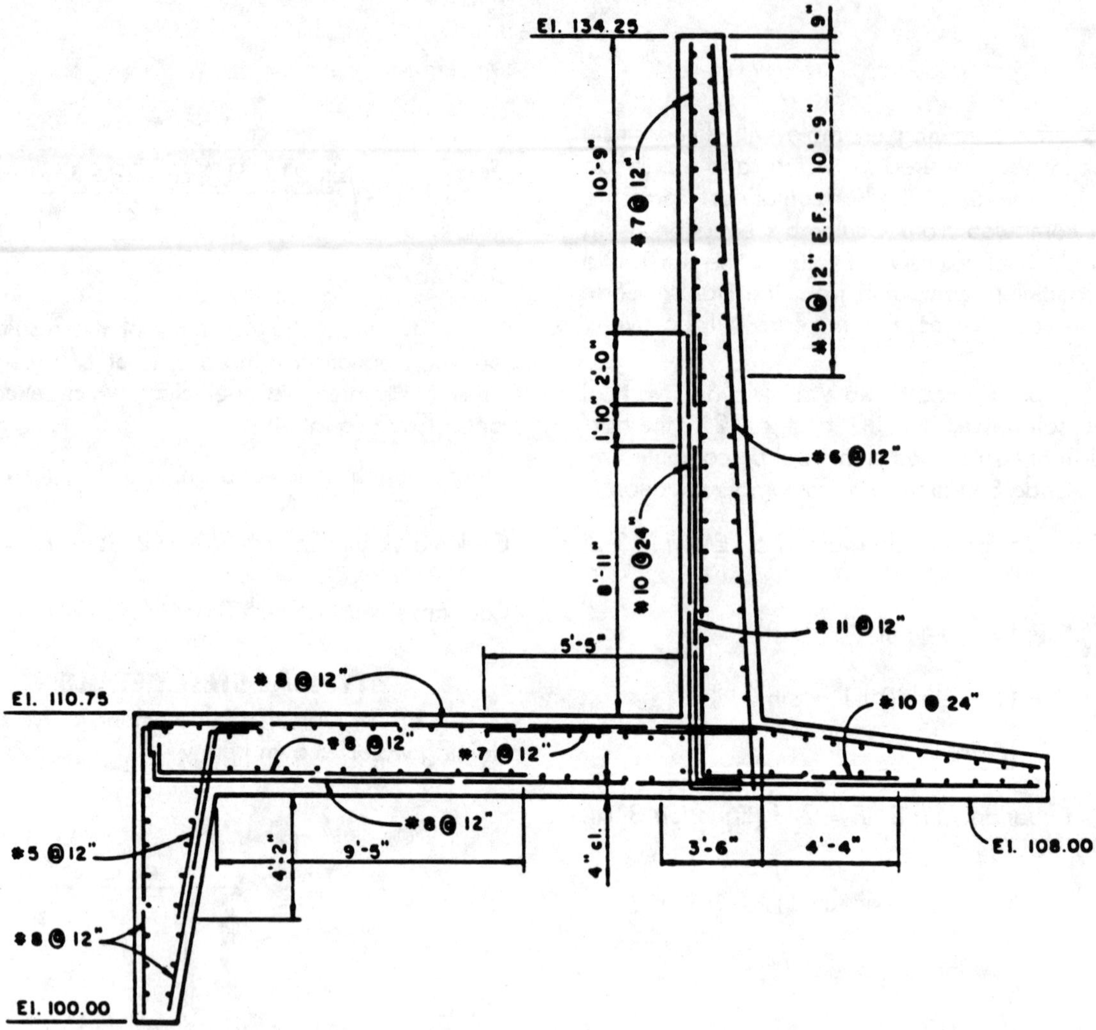

NOTES:

1. All longitudinal reinforcement to be #6 @ 12" E.F. except as noted.
2. Concrete cover for reinforcement to be 3 inches except as noted.
3. All hooks shown to be ACI standard hooks.

INDEX